Autonomic and Enteric Ganglia

Transmission and Its Pharmacology

Autonomic and Enteric Ganglia

Transmission and Its Pharmacology

Edited by

Alexander G. Karczmar

Loyola University Stritch School of Medicine
Maywood, Illinois

Kyozo Koketsu

and

Syogoro Nishi

Kurume University School of Medicine
Kurume, Japan

PLENUM PRESS • NEW YORK AND LONDON

Library of Congress Cataloging in Publication Data

Main entry under title:

Autonomic and enteric ganglia.

Includes bibliographies and index.

1. Ganglia, Autonomic. 2. Neural transmission. 3. Ganglionic blocking agents. 4. Ganglionic stimulating agents. I. Karczmar, A. G. (Alexander George), 1918– . II. Koketsu, Kyozo, 1922– . III. Nishi, Syogoro. [DNLM: 1. Ganglia, Autonomic–physiology. 2. Ganglionic Blockaders–pharmacodynamics. 3. Gastrointestinal System–innervation. 4. Neural Transmission. 5. Neuroregulators–pharmacodynamics. WL 600 A9388]

QP368.8.A96 1985 612′.89 85-19398

ISBN-13: 978-1-4615-9438-3

Softcover reprint of the hardcover 1st edition 1973
A Division of Plenum Publishing Corporation
233 Spring Street, New York, N.Y. 10013

ISBN-13: 978-1-4615-9438-3 e-ISBN-13: 978-1-4615-9436-9
DOI:10.1007/978-1-4615-9436-9

Foreword

In the early 1960s, Dr. Alexander G. Karczmar, Professor of Pharmacology and Experimental Therapeutics at the Stritch School of Medicine of the Medical Center at Loyola University of Chicago, was confronted with a certain technical problem concerning his studies of synaptic transmission by means of microelectrode methods. He thought that the problem might be resolved if he could interest a microelectrode expert such as Dr. Kyozo Koketsu in his studies. Dr. Koketsu was a past member of the Faculty of the Kurume University School of Medicine who as a Research Fellow at the Australian National University had helped Sir John Eccles, subsequently a Nobel Prize winner, in developing microelectrode procedures. After further considering the matter, Dr. Karczmar was pleasantly surprised to discover that by coincidence Dr. Koketsu was his neighbor, serving at that time as a Research Professor at the Neuropsychiatry Institute of the University of Illinois, College of Medicine of Chicago. This was the beginning of a long relationship, as Dr. Koketsu joined Dr. Karczmar at Loyola as Professor of Pharmacology and Therapeutics and Director of the Neurophysiology Laboratory at the Stritch School of Medicine. It was not long before Dr. Syogoro Nishi—Dr. Koketsu's former colleague on the Faculty of Medicine at Kurume University, and at that time a Research Fellow in Neurophysiology at the Rockefeller Institute in New York—joined Drs. Koketsu and Karczmar at Loyola. Although in due time Drs. Nishi and Koketsu returned to their alma mater (where Dr. Nishi is today a Professor of Physiology and Dr. Koketsu is Dean and President of the Kurume University School of Medicine), the friendship and teamwork of these three men lasted beyond their time together at Loyola and continues today.

Working together, Drs. Karczmar, Koketsu, and Nishi established at the Stritch School of Medicine a nucleus of synaptic (including ganglionic) researchers, and their laboratories became the center of activities for their students, research associates, postdoctoral fellows, and visiting investigators. Among the members of their team were scientists from Europe (including Poland, France, Russia, England, and Scotland), Australia, Japan, and China. And the research in these truly international laboratories led to many discoveries that helped clarify processes of synaptic transmission that constitute the basis of CNS function and behavior. After the return of Drs. Koketsu and Nishi to Japan, the research and training activities of what then became a Loyola-Kurume group were continued, and there was a continuous exchange of personnel between Kurume and Chicago.

This book is a result of these many years of friendship and teamwork, as the editors of this book–which I trust is both unique and remarkable–and the authors are men and women who have worked together, whether at Loyola or Kurume.

As the Provost of the Medical Center of Loyola University of Chicago, I am pleased to recall this history of research and training development involving both this Center and Kurume University Medical School. I am happy that Loyola University Medical Center contributed to the support of the investigations of Drs. Karczmar, Koketsu and Nishi, and of their associates. Similarly, we were happy to provide all the help we could during the planning session for this book and during its final preparation.

I know that this book constitutes only one stage in the research and training activities of Kurume University School of Medicine-Loyola University Medical Center, and that more papers and texts will result in the future from this notable international cooperation.

Richard W. Matre, Ph.D.
Provost
Loyola University Medical Center
Maywood, Illinois

Preface

This book was planned by the editors with a specific aim and a defined philosophy. The aim was to describe the transmission processes and related phenomena in vertebrate autonomic and enteric ganglia; the philosophy was to present the pertinent information as a unified whole, with as focused an approach as possible, avoiding any degree of tangentiality, even at the risk of occasional incompleteness.

There was an additional aspect to this book: It was to be authored by a team of friends of long standing who were in a good position to prepare a cohesive, homogeneous text, since they had been co-workers for many years in ganglionic research and had published many papers and reviews together.

Once the book's philosophy and aims had been decided on by the editors, its plan could be laid down almost automatically and its chapters assigned as appropriate for the interests and research experience of the members of the team. First of all, it appeared self-evident that to render ganglionic transmission and its processes intelligible, an understanding of the anatomy, electrophysiology, and neuropharmacology of autonomic and enteric ganglia was needed. Accordingly, this book emphasizes the anatomy, histology, and cytology as well as the cytochemistry of the sympathetic and parasympathetic ganglia and of the enteric neurons. With this as the substrate for the dynamic events of autonomic ganglionic and enteric transmission, the characteristics of the membranes involved in this transmission are described. The responses, both pre- and postsynaptic, of the ganglionic and enteric membranes are described in detail, as are the electrophysiological characteristics of these responses and their ionic generation. The transmitters involved in the various responses and their presynaptic or intraneuronal origin, the release mechanisms, and the postsynaptic sites of their action are also discussed. Furthermore, the molecular bases of these responses—the receptors and the channels—are described. Thus, the concepts of ganglionic and/or enteric nicotinic, muscarinic, and peptidergic transmission are explored in terms of their generation, electrophysiology, and types of responses. To this end, attention is focused on bioactive substances, whether active as endogenous transmitters and modulators or as exogenous substances that affect the ganglia pre- or postsynaptically when applied to the ganglia. Thus, this topic merges with a consideration of the pharmacology of ganglionic transmission. Furthermore—since it is clear that endogenous bioactive substances as well as pharmacological agents ultimately affect the ganglia via certain energy processes that they generate—the role of second messengers in the generation of ganglionic potentials is considered. The numerous references in this text to the many substances known today as transmitters and putative transmitters—such as histamine, catecholamines, serotonin, GABA and other amino acids, bioactive peptides including substance P and enkephalins, and, of course, acetylcholine as well as second messengers (cyclic nucleotides and the components of the phosphatidylinositol cycle)—reflect the concept, first enunciated in the eighteenth century, that ganglia constitutes "little brains." In fact, one chapter in this book concerns the ganglia as a convenient model for the study of the CNS.

Finally, it was necessary to include a discussion of the presynaptic spinal neurons and reflexes, including the afferent cells, since the ganglia and the enteric system are activated preganglionically at the afferent end of the reflex. In this same context, it was important to cover physiological aspects of transmission, particularly in organs endowed with as much autonomic peripheral control as the neurons of the enteric system, including ascending and decending excitations and inhibitions.

On the other hand, to achieve the desired focus, the editors decided that such subjects as embryonic development of the ganglia and the enteric system, ganglionic metabolism, invertebrate ganglia, and the physiology and pharmacology of effector organs did not belong in this book. Nor does this book include descriptions of specific research topics.

Yet, within this entity, there is complexity, as behooves the "little brains." The full significance of this complexity and of the "little brain" concept gradually became apparent in the course of ganglionic and enteric studies beginning during the nineteenth century. The historical development and emergence of this concept, presented in the first chapter of this book, not only provide an exciting insight into the heuristic thinking and experimental elegance of the work of our predecessors—such as Langley and Dale; Feldberg, Kibjakow, and MacIntosh; and Eccles, Bülbring, and Zamis—but also serve to clarify the components of this complexity. In addition, Chapter 3 seeks to define and classify the various processes and their interactions that bring about point-to-point control of ganglionic and enteric transmission, and to provide a bird's-eye view of the current status of concepts of ganglionic transmission and its multifactorial regulation.

Who may benefit from reading this book? Since mathematical, biophysical, and transmitter aspects of membrane function and ganglionic potentials and their generation are described in some detail, synaptologists and other neuroscientists should be included among its readers. Since the ganglia clearly exemplify the processes of synaptic transmission and of their modulation—processes that cannot be as easily analyzed in the CNS—neurologists should also profit from reading this volume. Furthermore, since this book concerns the classic subject of autonomic and enteric transmission, pharmacologists of any ilk should find some interest and pleasure in perusing these chapters. By the same token, clinicians who are concerned with several aspects of autonomic disease and do not as yet forego the hope that hypertension may one day be treated pharmacologically might be interested in paging through this book. Finally, medical students should look to this book for basic descriptions of synaptic mechanisms, transmitter function, peristaltic motion, and gastrointestinal processes.

We would be remiss if we did not at this time express our thanks to the many people who contributed to the creation of this volume. The editors wish to include in this list, of course, our colleagues, who with us are the authors. We should also thank our staffs, which include Mary Ann Jurgus in Chicago and Blockaden Machen in Kurume. Also, our thanks are due to the administrations of Loyola University of Chicago and its Stritch School of Medicine and of Kurume University Medical College, who made our past research as well as this endeavor possible. Special thanks are due to Dr. Charles Visokay of Karger Publishers, who helped us in organizing the planning sessions, and to Mr. Kirk Jensen, Senior Editor of Plenum Publishing Corporation, and his firm for their help and for their patience and support during the long process of creating and producing this book.

Alexander G. Karczmar, K. Koketsu, and S. Nishi
Maywood, Illinois, and Kurume, Japan

Contents

I. HISTORY AND ANATOMICAL BASES OF GANGLIONIC AND ENTERIC TRANSMISSION

II. TRANSMISSION AND MODULATION IN SYMPATHETIC GANGLIA AND THEIR NEUROPHARMACOLOGY

V. CLINICAL AND CNS-RELATED ASPECTS OF GANGLIONIC TRANSMISSION

I

History and Anatomical Bases of Ganglionic and Enteric Transmission

1

Historical Development of Concepts of Ganglionic Transmission

ALEXANDER G. KARCZMAR

This historical and conceptual introduction to this book on ganglionic transmission in the vertebrates is organized under the following three headings: (1) anatomical and morphological features of the ganglia and their functional aspects, (2) neurotransmitter and neurochemical characteristics of ganglionic transmission, and (3) synaptic responses of the ganglia.

I. ANATOMICAL AND MORPHOLOGICAL ASPECTS OF PAST GANGLIONIC RESEARCH

It is customary to begin the history of research concerning anatomical features of ganglionic transmission with the work of Gaskell (1886) and Langley (1878, 1921). The history of such research actually begins much earlier, however, because in the 2nd century, A.D., Galen (129–200)(see Galen, 1528) described seven pairs of "ganglionic swellings" of the cranial nerves that corresponded to superior cervical and nodose ganglia, and to stellate and celiac ganglia; he also described the vagosympathetic trunk

ALEXANDER G. KARCZMAR • Department of Pharmacology and Experimental Therapeutics, Stritch School of Medicine, Loyola University Medical Center, Maywood, Illinois 60153.

and its rami communicantes. His description is presumably (O'Malley, 1964) based on dissections of the Barbary ape (cf. also Pick, 1970). But even Galen's studies were preceded by those of Herophilus (ca. 300 B.C.) and Erasistratus (ca. 290 B.C.), since they described peripheral nerves, some of which may have been autonomic; they also described the activity of these nerves as depending on their conveying, tubelike, vital, natural, and animal spirits (Garrison, 1929; A. Castiglioni, 1947; Pick, 1970; Holmstedt and Liljenstrand, 1963).

Galen's descriptions of cranial nerves and of the vagosympathetic trunk were used or plagiarized by Vesalius (1543)(see Saunders and O'Malley, 1947); the first clear separation of the vagus and sympathetic nerves was due to C. Estienne (1545), B. Eustachio (1564), and C. Reid (1634), who referred to the latter as "intercostal nerves" (cf. Pick, 1970). Willis (1664) described the sympathetic trunk in more detail, although he considered it as a branch of abducens or as a cranial nerve, splanchnic nerves, prevertebral ganglia [which he referred to, like Reid (1634), as intercostal nerve], and visceral plexus; like his predecessors, he ascribed their function to a flow of spirits originating in the brain, and he was the first to distinguish between voluntary (volitional) and "involuntary" movement. That the "intercostal" chain is of subcranial rather than cranial origin was determined by Petit (1727), who also showed that the "intercostals" innervate (and carry the "animal spirits" to) the eye (cf. Brooks and Seller, 1981; Pick, 1970). The first finding as to the nature of vagal influence on the heart had to wait for Weber and Weber (1845), who applied a Faraday stimulator to the vagus nerve to demonstrate vagal inhibitory cardiac action (cf. Brazier, 1959).

The work of the Webers constitutes an important step in the development of investigations that link morphology with function; however, theirs was not the first study of this linkage. Winslow (1732), Whytt (1751, 1765), Meckel (1751), Johnstone (1764), and Monro (1783)(cf. Pick, 1970) contributed to morphological studies of mammalian including human ganglia, as they first used the term "ganglia" (Meckel, 1751), described plexi and possible sensory contribution to the ganglia as well as efferent ganglionic nerves as different from preganglionic, and renamed "intercostals" as "nervi sympathici maximi" (Winslow, 1732). They also emphasized the "animal" generation of the effects via these nerves on the heart, intestine, bladder, and pupil; sometimes, they referred to this function as due to "sympathies" that the pertinent nerves carry (Winslow, 1732)(cf. Pick, 1970). Furthermore, they assigned "sensibility" to these nerves and recognized what would today be called the reflex character of the involuntary function (Whytt, 1765)(cf. Pick, 1970). Finally, stressing the complex morphology of the ganglia and their sympathetic outflow and their importance with respect to involuntary and visceral (Bichat, 1802) activity, they referred to the ganglia as "little brains"* (Winslow, 1732; Johnstone, 1764; see Pick, 1970). Altogether, an amazing degree of morphological and functional insight into the ganglia was developed by the 18th century, since the investigators in question had a good understanding of the homeostatic nature of the activity of ganglia (Bichat, 1802), their control of visceral function, and their relay role with respect to linking emotions to their effects on function. Thus, Bichat wrote of "le système des ganglions . . . controlant la vie organique," and his student Reil (1807) described this system as "vegetative."

The 19th century brought additional descriptions of unmyelinated postganglionic nerves (Ehrenberg, 1833; Remak, 1838), rami communicantes, and pathways relaying ganglia to the spinal cord and to the effectors (Remak, 1854), as well as of special ganglia such as otic and of plexi (Meissner, 1857; Auerbach, 1864).

The functional meaning of these structures became clearer with the discovery of the smooth muscle in the iris, uterus, bladder, arterial wall, and gastrointestinal tract (for references, cf. Pick, 1970). However, the major impetus to the understanding of the functional role of what we call today the autonomic nervous system was due to the discovery of C. Bernard (1852) that in dogs, unilateral sectioning of "the cervical sympathetic nerve . . ." led to ipsilateral "heating of the head" and "more active circulation of blood; the arteries seem to be fuller."† This finding was considered by Bernard in connection with the effect on body circulation of CNS lesions described by P. Flourens (1824) and F. Magendie (1823). The full meaning of these experiments became apparent when Brown-Sequard (1852) showed that in animals, galvanic stimulation of sympathetic nerves cooled the skin and attenuated cutaneous blood flow (see also Pick, 1970).

Subsequent work by Ludwig (1851), Pflüger (1875)(see Brooks, 1979; Brooks and Sellers, 1981), and particularly Bernard (1858, 1878) led to establishing the role of what would be referred to today as the parasympathetic and sympathetic systems in the control of the iris, salivation, vasodilation, gastrointestinal tract, visceral and genital organs, and other functions [the sudomotor, pilomotor, and cardioaccelerator action of the sympathetic nerves was described by Luchsinger (1876), Goltz *et al.* (1875), and Schiff (cf. Brooks and Seller, 1981)]. Furthermore, Bernard demonstrated the relationship between these phenomena and the medulla and, foreshadowing Cannon, emphasized the homeostatic role of the systems

*More than ten years ago, this author and his associates innocently used the same term to denote the same concept (Karczmar *et al.*, 1972); we thought we had coined a new term! Our apologies, Professors Winslow and Johnstone.

†These most important studies by C. Bernard of autonomic ganglia are generally unrecognized compared to his other studies.

involved. In this context, it is important to stress Ludwig's discovery in 1866, of the depressor nerve and reflex (cf. Brooks and Seller, 1981); in fact, Ludwig may be credited with developing the feedback concept (Brooks and Seller, 1981).

Altogether, the German physiologists of the 19th century, particularly the students of Muller and Ludwig (see Rothschuh, 1973), shared with the French investigators such as Bichat and Bernard the credit for setting the stage for the work, familiar to students of the autonomic nervous system, of Gaskell (1847–1914) and Langley (1852–1925). Again, Gaskell and Langley combined a thorough study of the morphology of the peripheral nervous system with pertinent functional interpretation.

The studies of Gaskell provide an ideal example of such an interdisciplinary approach. Gaskell (1886, 1916) traced the anterior and posterior roots of cranial nerves of the "vertebral or lateral ganglionic chain" and the rami communicantes as he described the three separate outflows of the medullated (which today are called preganglionic) fibers—bulbar, thoraco-lumbar, and sacral (in the rat, these fibers are nonmyelinated); he referred to this system as the involuntary nervous system. Furthermore, he described "prevertebral or collateral ganglia"; finally, he discussed the relationship between posterior (i.e., sensory) roots and the ganglia and their "splanchnic" roots. Since he also studied the functional role of the "vagosympathetic" (Gaskell, 1886) and cranial nerves in the control of cardiac, pupillary, and glandular (salivary and lachrymal) activities, he ultimately suggested the existence of a "reflex arc" analogous to the "voluntary reflex arc," an afferent or receptor nerve terminating in the lateral horn or the vagal nuclei, and a connector nerve uniting the lateral horn or the vagal nuclei with the sympathetic or vagal ganglion cells. It is also of particular importance that Gaskell (1886, 1916) emphasized his findings and those of others as suggesting that "every . . . involuntary . . . tissue is innervated by two sets of nerve fibers of opposite characters"; he also stressed that his "explanation of inhibition is applicable to that occurring in the central nervous system" and that the concept of "excitor and inhibitory . . . activity" concerns the latter.

These findings served as a solid basis for the work of Langley (1852–1925), who was the first investigator to make systemic use of chemical or pharmacological tools (others including Gaskell used these tools more sporadically, although not less heuristically) to investigate nerve or synaptic functions and their sites. He first used pilocarpine in 1876 (cf. Langley, 1879, 1921) in this fashion; even more important was his subsequent work with Dickinson (Langley and Dickinson, 1889), in which they utilized the finding of Hirschmann (1863) that "after a moderate dose of nicotine, stimulation of the sympathetic nerve in the neck causes no dilation of the pupil," to map the cells of the peripheral nervous system that innervate the "involuntary" effectors. These and other studies led

Langley to the definition of "preganglionic and postganglionic neurons" and of the "autonomic nervous system" that embraced the "parasympathetic" system, described earlier by Gaskell as consisting of sacral and cranial outflows, as well as the prevertebral and vertebral ganglionic chains. Langley also demonstrated, via experiments in which nicotine was applied directly to the ganglia, that the site of the blocking action of nicotine is the ganglion itself rather than, as believed by Hirschmann (1863), the nerve endings of the sympathetic outflow. Finally, Langley (1921) recognized that the plexi of Auerbach and Meissner constitute the independent "enteric system" with "no significant connections with the autonomic outflow from the spinal cord" (North, 1982)(see also Chapters 2 and 16); an early contribution of Ramón y Cajal (1892) to this subject is of importance in this context.

The description of the autonomic nervous system generated in the late 1880s by Gaskell and Langley and clarified during the next 25 years by the investigations of the pertinent neurotransmitter mechanisms (see Section II) had finalized the essential understanding of the morphological and functional aspects of the autonomic system, yet much was left to be completed. Among important way stations in the pertinent development must be cited the demonstration that not only skin afferents (Gaskell, 1886) but also and particularly the visceral afferents (Ranson and Billingsley, 1916; Hering and Breuer, 1868; see Pick, 1970) contribute to Gaskell's reflex arc; Langley (1921) himself contributed to these studies, and the early studies of "visceral pain" (Lennander, 1901) are important in this context. In fact, as early as 1900, E. A. Schafer (1900) wrote at length of axon and ganglionic reflexes, ganglionic block, and other phenomena (see also Brooks, 1982). Subsequent studies of sympathetic and depressor responses, including his own work on the Bezold-Jarisch reflex (Schaefer and Jarisch, 1940), are interestingly described by Schaefer (1981) in his letter to Seller (see Brooks and Seller, 1981).

Other important findings concerned ganglionic interconnections and their histology and cytology. Thus, the axonal and dendritic processes of the autonomic cells identified as early as 1881 by Ramón y Cajal (1881) served subsequently for the anatomical classification of the autonomic cells (Dogiel, 1895; Carpenter and Conel, 1914; Ramón y Cajal, 1911; see also Russian investigators such as Lawrentjew, 1929; see also Botar, 1966). Thus, many investigators (cf. Pick, 1970) evaluated the fiber content of the rami, the ratio of pre- to postganglionic fibers and to the cell counts, and the presence of aberrant ganglia, synapses, or fibers (Kuntz, 1938). Furthermore, these studies revealed that ganglia may constitute complex relay stations that may subserve subtle interactions among several anatomical inputs, subsequently revealed as multitransmitter in nature. Such complexity had already been revealed, in the case of the pelvic ganglia, by Langley (Langley and Anderson, 1895, 1896); subsequently, it was demonstrated that their cells receive both parasympathetic and sympathetic input and contain both adrenergic and cholinergic neurons (Kuntz and Moseley, 1936). Still later, the presence of the small intensely fluorescent (SIF) cells (Dail *et al.*, 1975)(see below) and purinergic neurons (Burnstock, 1972) was demonstrated, this complex morphology accounting for complex transmission interplay exhibited by the pelvic ganglia (e.g., DeGroat *et al.*, 1979). Finally, the fine cytological structure of the ganglion cells that includes: mitochondria, Golgi apparatus, neuroglia, subsynaptic and presynaptic specialization, and presynaptic vesicles, was described by Palay and Palade (1955) and DeRobertis and Bennett (1954, 1955); their identification of synaptic vesicles was challenged initially by Rosenbluth (1963) and Taxi (1961, 1965, 1967)(see also Pick, 1970).

A particular morphological advance, the consequences of which are still with us, as illustrated by several chapters of this book (e.g., Chapters 3, 4, 9, and 13), relates to SIF cells. The story begins with the chromaffin cells, i.e., cells stainable with dichromate; that such cells are present in the ganglia and that they may be innervated preganglionically was summarized already by Smirnow (1890). Later, the chromaffin stain was related to epinephrine (Watzka, 1943), the latter being the identifiable catecholamine at that time (see below). When fluorescence microscopy was developed and made specific for catecholamines, it could be shown that chromaffin cells correspond to catecholamine-containing SIF cells, although the latter may not show chromaffin stain, since their catecholamines may be washed out during the chromaffin test (Norberg *et al.*, 1966; Eranko and Harkonen, 1965).

An important aspect of ganglionic organization concerns ganglionic spinal and supraspinal connections. While Gaskell described the intrasegmental "connector" pool of the preganglionic neurons already in 1886, their exact location and distribution, while studied by many investigators (Biedl, 1895; Poljak, 1922; cf. Pick, 1970), were not fully described until recent times, and Gabella (1976) stressed the complexity of the matter; this matter is taken up by Lees in Chapter 2. The task remains even more incomplete with respect to the supraspinal control of autonomic ganglia. Yet the existence of such control has been known since the investigations and heuristic insights, already mentioned, of Magendie, Flourens, and Bernard (see above)(see also Marey, 1863, 1881). Central control of blood pressure and other autonomic reactions was subsequently demonstrated by Dittmar (1873), Owsjannikov (1869), and Karplus and Kreidl (1909); these investigators were concerned particularly with the hypothalamic control of the autonomic system, and their work was expanded brilliantly by Bard (1929) and by the Nobel Prize Winner Hess (1925, 1954). The role of other central supraspinal areas present in the cortex (Korner, 1971) and the limbic system, medulla, and bulbospinal axis (Lim *et al.*, 1938) and of the baro- and chemoreceptors (Heymans and Neil, 1958) was investigated subsequently. It should be added in this context that while the spinal ascending and descending autonomic pathways have been studied continually beginning with the work of Bruce (1906), Greving (1926), and Bok (1928)(cf. Pick, 1970), their exact location and central origin or terminations are not exactly known; thus, central origin of direct and indirect hypothalamic or limbic control of preganglionic nuclei is not precisely understood (Saper *et al.*, 1976; cf. Brodal, 1981). This constitutes the subject matter of Chapters 17 and 18.

Autonomic effects exerted by these and related CNS areas combine with behavioral influences in preserving autonomic homeostasis as reflected in patterns of blood pressure, temperature, and other body functions in a manner that is only partially understood. Of course, this central control has to be considered within the framework of the concept of the classic "involuntary" function of the autonomic nervous system and of the autonomic reflex (see Chapters 17 and 18).

II. NEUROTRANSMITTERS

The history of ganglionic transmission is linked with the development of the concept of chemical transmission as a rival of the concept of electrical transmission. The concept of chemical transmission has developed from that of the "animal spirits" that according to Galen (1528) were fluids which traveled in hollow nerves to act on the body (see above)(see also Brazier, 1959; Brooks and Seller, 1981). This speculation was shared for the next two hundred years by Descartes (1649), Croone (1664), and Haller (1755), who spoke of "spiritous liquids" that descend from tubular nerves to swell the muscle like a balloon during its contraction (cf. Brazier, 1959). The concept of animal spirits was challenged by that of irritability of the effectors (Glisson, 1677) and of "animal electricity" (Galvani, 1791; cf. Brazier, 1959). Finally, in the 19th century, the choice was stated as that between "nervous" and "electric" (or galvanic) force (Mateucci, 1840; DuBois-Reymond, 1843; cf. Brazier, 1959; G. Castiglioni, 1932). While DuBois-Reymond preferred electricity as the active principle, since he related the electrical phenomena to the latency—which he discovered—between the nerve action current and its invasion of the skeletal muscle, he also envisaged the possibility that the effect (the muscle contraction) may depend on the elaboration of substances such as ammonia or lactic acid. The next step was made when the latency was related to histologically demonstrated discontinuity between nerve and muscle, noticed by Kuhne (1862) and Ramón y Cajal (1881); of course, this discontinuity is analogous to that subsequently demonstrated for the central as well as ganglionic neurons by Ramón y Cajal (1911).

The controversy between the two views continued into the next century. While Loewi suggested as early as 1914 (Loewi and Gettwert, 1914) the presence of chemical cholinergic transmission at the parasympathetic vagocardiac junction (he referred at that time to "Cholinwirkung" at the vagus) and proved it in 1921 (cf. Lembeck and Giere, 1968), Eccles preferred, contrary to the views of Feldberg, Brown, Bacq, and others (see Bacq, 1975), the electrical to the chemical theory of transmission with respect to the neuromyal junction. This is when the autonomic ganglion entered the controversy, since Eccles emphasized that anticholinesterases (anti-ChEs) did not cause the changes in ganglionic responses that could be expected on the basis of the chemical hypothesis (J. C. Eccles, 1934, 1936; cf. Koelle, 1966). Also, he doubted the capacity of ganglionic ChEs and the diffusion process to terminate the effects of released acetylcholine (ACh) and to prevent occurrence of ACh block; he pointed out that if ACh were to serve as transmitter, a mechanism for the limitation of its action should be available. The argument concerning disposition of ACh was subsequently resolved by Emmelin and MacIntosh (1956). At any rate, in the 1940s, Eccles yielded to the arguments of others; his friend Karl Popper's emphasis on flexibility (Bacq, 1975) as a correlate of creativity was a causative factor in Eccles's "conversion."

This newly found belief in chemical synaptic transmission was subsequently expanded by Eccles to the central nervous system as he demonstrated the existence of cholinergic transmission at the synapse between the spinal Renshaw cell and the motoneuron; this proof (J. C. Eccles *et al.*, 1954) was based on electrophysiological and pharmacological analysis, and it was consistent with the corollary neuroanatomical and biochemical evidence. The extension of this proof to other spinal and supraspinal sites and to transmitters other than ACh is technically difficult, and the pertinent research is continuing (cf. McGeer *et al.*, 1978; Cole and Nicoll, 1983).

Notwithstanding this controversy between the proponents of the chemical and the electrical hypotheses of transmission, data concerning chemical transmission to sympathetic and parasympathetic effector sites continued to accumulate through the 19th and the first decades of the 20th century. In the case of parasympathetic transmission, the proof of chemical transmission is linked with the demonstration by E. and E. H. Weber (see above) of vagal cardial slowing; it is interesting that in the 1850s, Purkinje, among whose many discoveries and descriptions were those of the histological characteristics of human sweat glands (cf. Brooks and Seller, 1981; these glands were not studied again with respect to the relationship between function and structure until the investigations of Sperling and Koppanyi, 1949) also described the effects of belladonna on the activity of these and other structures. Subsequently, Schmiedeberg and Koppe (1869), Kobert and Sobot (1887), Gaskell (1887), and Dixon

(1907) pointed out that muscarine and pilocarpine produced effects similar to those of vagal stimulation, and Gaskell stated that "muscarin does not act as an excitant to inhibitory mechanisms but as a depressant to motor activity," while Dixon and Hamill (1909) tried to provide, but could not, a Loewi-like proof of this idea. At that time (perhaps with the work of Elliott and speculations of DuBois-Reymond in mind), Langley (1905)(cf. Langley, 1921) presented clearly the idea of chemical neurotransmitters and also, in fact, of receptors. The next links in the chain of evidence were forged by Hunt and Taveau (1906), who stressed the extraordinarily potent vasodepressor ["3/100 mg (per kg) animal"] action of acetylated choline (ACh), as compared with that of choline, and by Dale (1914), who pointed out the evanescence of cardiovascular action of intravenously administered ACh; to explain it, he postulated the presence of an esterase (ChE) in the blood. At that time, Dale (1914) also described the pressor effects of choline derivatives, and it is interesting that more than 30 years later, Kosterlitz and his associates demonstrated nicotinic actions of choline at the ganglia. Dale (1914) recognized that earlier Boehm (1885)(cf. Dale, 1914, 1953) demonstrated that nitrous ester of choline evokes a depressor response that is converted into a pressor response by atropine; guided by this finding and his own results, Dale coined the terms "muscarinic" and "nicotinic" for the excitatory actions of ACh at the parasympathetic sites and at the autonomic ganglia, respectively. This then led to the demonstration by Loewi (1921) of the cholinergic nature, in the frog, of the vagal effect. Subsequently, Loewi and Navratil (1924) used physostigmine in Loewi's vagocardiac preparation and confirmed Dale's concept of evanescent action of ACh. This then explained the effects of physostigmine at many sites described toward the turn of the century by Hamer, Lenz, Bezold and Gotz, and Arnstein and Sustschinsky (see Karczmar, 1970; Holmstedt, 1972) and led to the concept of anti-ChE mechanism of the action of physostigmine. It is interesting that it is not easy to duplicate the original experiment of Loewi, and Friedman (1971) pointed out that Loewi was lucky in carrying out his demonstration during the seasonal peak of the sensitivity of the frog heart to ACh or vagal stimulation.

In parallel, the work of Elliott (1905) initiated studies of catecholamines as neurotransmitters by demonstrating at such sites as the urethra a parallelism between the effect of sympathetic nerve stimulation and that of epinephrine (adrenaline), which had shortly before been crystallized from an adrenal gland extract. The demonstration in the 1920s by Loewi, Barger and Dale, Cannon and Frederick, and many others (cf. Bacq, 1975) of the release of an epinephrine like substance at several sites indicated that a "sympathin" or "sympathins" (the term coined by Cannon) (cf. Cannon and Rosenblueth, 1937) exert in the autonomic nervous system a function that is analogous to, but opposite from, that of ACh. The problem that soon emerged was that the effects of epinephrine differed in several respects from the effects of sympathins; for example, sympathins that were supposedly released following the stimulation of sympathetic nerves, such as hepatic or cardiac accelerans, exhibited only excitatory (such as cardiac acceleration) rather than also inhibitory (as on the intestinal smooth muscle) effect. Cannon and Rosenblueth (1937) explained these phenomena by differences between receptors that, coupled with sympathin ["adrenaline or adrenaline-like substance" (Rosenblueth, 1950)], transformed it into sympathin E (excitatory) or sympathin I (inhibitory) depending on the receptor; on the other hand, Bacq early advanced the view that sympathin E effect is due to norepinephrine that is formed from epinephrine by demethylation (cf. Bacq, 1975); this hypothesis was ultimately proven by von Euler (1946) (cf. von Euler, 1956).

The main subject of this book, ganglionic transmission, is a relative latecomer in the development of the concept of the chemical nature of transmission. The crucial experiment was performed in 1933 by Kibjakow (1933), who reported that the perfusate of preganglionically stimulated superior cervical ganglion of the cat contained an element capable of causing, when reintroduced into the vasculature of this ganglion, the contraction of the nictitating membrane, just as would the stimulation of the ganglion itself. It is of interest that Kibjakow did not employ eserine (physostigmine) in his experiments; actually, it is not easy to duplicate his results without improving the perfusate setup, as was done by Emmelin and MacIntosh (1956), or using physostigmine (Feldberg and Gaddum, 1934), and Kibjakow was repeatedly unsuccessful in trying to demonstrate the effect of the perfusate at scientific meetings (F. MacIntosh, personal communication). In fact, there is still controversy as to whether pharmacologically effective concentrations of ACh can be recovered from the brain or peripheral sites without employment of an anti-ChE. Subsequently, by using a bioassay and certain chemical tests (acid stability vs. instability in alkaline solutions), Feldberg and Gaddum (1934) and Feldberg and Vartiainen (1934) identified Kibjakow's substance as ACh; in fact, already in 1925, Witanowski (1925) described the presence of ACh in the ganglia. The release of ACh on preganglionic stimulation was subsequently demonstrated for the submandibular and inferior mesenteric ganglia (Barsoum *et al.*, 1934; Emmelin and Muren, 1950). Finally, Koppanyi and his co-workers demonstrated that the sympathetic ganglia are the strategic sites for the pressor action of acetylcholine and nicotine, and that physostigmine and other anti-ChEs including organophosphorus anti-ChEs exert ganglionic actions that are consistent with the cholinergic nature of ganglionic transmission (Koppanyi *et al.*, 1936, 1947). Koppanyi (1929) also contributed to the generalization of the concept of the cholinergicity of ganglionic transmission as he evaluated the pertinent responses in several species; in these studies, Koppanyi established the concept and the discipline of comparative pharmacology. As expressed well by Volle (1966), these and additional results, all obtained in the 1930s, constitute the final proof of cholinergicity of what we would refer to today as primary autonomic ganglionic transmission:

> In these studies it was observed that (a) repetitive stimulation of the cervical sympathetic trunk of perfused superior cervical ganglia of the cat resulted in the appearance of a substance in the perfusion fluid with characteristics similar to those of ACh (Feldberg and Gaddum, 1934); (b) the ACh-like substance evoked a ganglionic response when added to the fluid perfusing the ganglion (Feldberg and Vartiainen, 1934); (c) stimulation of the postganglionic nerve or the vagus nerve did not cause the release of the material into the perfusate (Feldberg and Vartiainen, 1934); (d) ACh injected into the perfusion fluid caused ganglionic stimulation; (e) curare prevented the responses to injected ACh and nerve stimulation but did not alter the release of the ACh-like substance from the nerve endings; (f) physostigmine enhanced the responses not only to stimulation of the preganglionic nerve but also to injected ACh (J. C. Eccles, 1935a; Feldberg and Gaddum, 1934); and (g) the amount of ACh (Brown and Feldberg, 1936) and cholinesterase (Brucke, 1937) in the ganglia decreased markedly when the preganglionic nerve to the superior cervical ganglion degenerated. Collectively, these findings represent an impressive array of data to support the concept of chemical transmission in autonomic ganglia and, further, that ACh is the mediator of the process.

Besides proving the cholinergic nature of ganglionic transmission, some of these findings demonstrated that the ganglia respond nicotinically; subsequently, a muscarinic ganglionic response was described as well, both electrophysiologically and in terms of the pharmacological response (see below).

What was added later to this compelling evidence was the demonstration of synaptic vesicles in ganglionic presynaptic terminals, miniature excitatory potentials, and, finally, localization and distribution of ChEs; this distribution, which was elucidated in the classic experiments of Koelle in which he employed the histochemical methods developed by himself and Friedenwald (see Koelle, 1963), differed in parasympathetic and sympathetic ganglia but was related, in either case, to function (Koelle, 1963). The final word on the matter of ACh as primary transmitter was given by Nishi *et al.* (1967); by employing a combination of electrophysiological and electron-microscopic methods and the ultrasensitive bioassay, frog lung preparation, capable of responding to less than 10^{-18} g ACh, these authors demonstrated that single eserinized ganglion cells of *Bufo* liberated 1.6×10^{-16} g ACh per volley, a single synaptic knob releasing approximately 2–5 quanta of ACh per impulse, each quantum containing from 8 to 12 $\times$ 10^3 molecules.

Yet even these advances are not the end of the history of chemical transmission at the autonomic ganglia. The next phase of this story concerns other than ACh transmitters.

First, there is the story of catecholamines and indoleamines. Gaddum surmised in the early 1950s and subsequently demonstrated with Giarman (Gaddum and Giarman, 1956), the presence of serotonin in autonomic ganglia. This finding was subsequently confirmed in the Kurume-Loyola laboratories by means of sensitive chemical (Dun *et al.*, 1980) and high-performance liquid chromatography (Barnes, Fareed, and Karczmar, unpublished observations) methods; these authors also showed that serotonin is located postganglionically and suggested that it acts, retrograde, presynaptically (cf. Chapter 13). Physiological functions of ganglionic serotonin can only be speculated on at this time (cf. Chapters 12 and 13).

A more complex story is that of ganglionic catecholamines. The discovery of the ganglionic SIF in the ganglionic interneuron has already been mentioned. Insofar as the SIF cells correspond to chromaffin cells, their catecholamine content was known long before (see above) (see also Watzka, 1943) the definitive work on this matter by the Helsinki group (see Eranko, 1976). The proposal that a ganglionic interneuron following its muscarinic activation, liberates a catecholamine was first made by R. M. Eccles and Libet (1961); subsequently, Libet (Libet, 1970; Libet and Owman, 1974) utilized histochemical and cytospectrofluorimetric methods to relate the slow ganglionic inhibitory postsynaptic potential (IPSP) to catecholamines released from the SIF cells. Yet the results obtained by Libet may not be relevant to all species, and indeed there is a controversy today between the supporters of the disynaptic and monosynaptic nature of the IPSP; several chapters of this book relate to this matter (Chapters 3, 9, and 13).

Even more recent investigations concern amino acids, peptides, and cyclic nucleotides. As of this writing, it has been demonstrated that γ-aminobutyric acid is present in the ganglia (Waniewski and Suria, 1977) and possibly in neuroglia (Bowery and Brown, 1972); this has been confirmed more recently by microchemical methods (Bertilsson *et al.*, 1976). It was long known that peptides such as substance P and somatostatin are present in the brain or spinal cord or both (see, for example, Pernow, 1963). More recently, several peptides have been identified in the ganglia by fluorescence and immunohistochemical methods (Hökfelt *et al.*, 1977) (see also Chapters 3, 8, 12, and 13). Also in the 1970s, the presence of cyclic nucleotides was established (cf. Bloom, 1972; Greengard, 1978), almost simultaneously in the CNS and in the ganglia (George *et al.*, 1970; McAfee *et al.*, 1971), and it was shown that ACh and/or cholinomimetics and presynaptic stimulation lead to increases in cAMP and cGMP. The physiological and pharmacological meaning of nucleotides for ganglionic transmission, particularly their role as second messengers, is a subject of controversy between Greengard (cf. Greengard, 1978) and the Loyola-Kurume laboratories (see Chapters 9 and 13). Of future importance is the physiological significance and pharmacological effects of second messengers other than nucleotides, such as phosphatidylinositol; its turnover, which was shown earlier to be involved in the regulation of permeability of muscarinic and glandular membranes, may prove to be important in the membrane regulation of ganglia as well (Larrabee and Leicht, 1965; Pickard *et al.*, 1977; see Karczmar, 1981) (see also Chapter 13).

III. NEUROPHARMACOLOGY AND NEUROPHYSIOLOGY

Ganglionic spike response to presynaptic stimulation was described first by Bishop and Heinbecker (1932), Brown (1934), and J. C. Eccles (1935a,b); at this time, Eccles also stressed the excitatory and inhibitory, modulatory role of the afterpotentials (J. C. Eccles, 1936, 1937, 1944). Subsequent research by R. M. and J. C. Eccles, Libet, and Koketsu and Nishi showed that this response is only a part of the ganglionic story and established the enormous variety and plasticity of ganglionic responses.

First, J. C. Eccles (1944), by means of extracellular recordings, related the generation of the spike to the extracellular event, the negative (N) wave. Subsequently, R. M. Eccles and Libet (1961) recorded, on presynaptic stimulation, a triphasic surface response that included, besides the primary N wave, a positive (P) and a late negative (LN) wave (see also Laporte and Lorente de Nó, 1950). When evoked by repetitive presynaptic stimulation, the LN wave gave rise to prolonged repetitive postganglionic responses that were described first by Bronk (1939) and J. C. Eccles (1944); as shown subsequently, this response is also elicited by anti-ChEs (Volle, 1962; Volle and Koelle, 1961). Finally, using intracellular recording methods, Nishi and Koketsu (for references, see Koketsu, 1969; Volle, 1966; Karczmar *et al.*, 1972) demonstrated that these surface waves correspond to three synaptic responses, the fast and slow excitatory postsynaptic potentials (EPSPs) and the inhibitory postsynaptic potential (IPSP).

At the same time, J. C. and R. M. Eccles, Libet, and Nishi and Koketsu established the pharmacological characteristics of those responses. The fast EPSPs, or the N wave, were shown to be nicotinic in nature (and blocked by D-tubocurarine or large concentrations of nicotine), while the LN wave, or the slow EPSPs, and the postsynaptic discharge corresponding to this response were muscarinic in nature (and blocked by atropine). While initially it appeared that the muscarinic response *per se* does not initiate ganglionic transmission and may, rather, serve as a modulatory response (see Chapters 3, 7, 8, and 13), it was shown later that it subserves a function, at least in the pharmacological sense, since certain muscarinic agonists, administered systemically, produce a contraction of the nictitating membrane (Koppanyi, 1932) and a pressor response in experimental animals (Roszkowski, 1961; see also Waser, 1961). On the other hand, the

P wave or the slow IPSP response was shown to be sensitive to both atropine and dibenamine (R. M. Eccles and Libet, 1961) (see Chapters 3, 9, 12, and 13). These pharmacological characteristics of the P (or IPSP) response and the presence of chromaffin cells in the ganglia led to the concept of the disynaptic nature of the P response (R. M. Eccles and Libet, 1961) (see above). As already stated, this concept constitutes the subject matter of several chapters in this book.

However, the triphasic ganglionic response does not exhaust the ganglionic repertoire. In 1967, as Koketsu and Nishi (1967) studied the repetitive postganglionic discharge described earlier by Bronk (1939) (see above), they showed that certain types of repetitive presynaptic stimulation of curarized ganglia generated a postganglionic discharge that was particularly prolonged. Only the early part of this response was atropine-sensitive; it was coupled in time with the late negative (LN) wave described earlier by J. C. and R. M. Eccles (see above), while the long-lived (up to several minutes) portion of the discharge was atropine-insensitive. Therefore, Koketsu and Nishi subdivided the afterdischarge into early (EAD) and late (LAD) components (see Koketsu and Nishi, 1967). Subsequently, Nishi and Koketsu (1968) described the intracellular correlate of the LAD, the "noncholinergic, late slow" EPSP. The significance of this noncholinergic response and the responsible neurotransmitter or neurotransmitters is a matter of great interest at present, since these phenomena concern a peptidergic response; they are discussed in Chapters 3, 8, and 13.

Even with the fourth ganglionic response, the late slow EPSP [and subsequently of additional postganglionic responses (see Chapters 12 and 13)], the plasticity of ganglionic transmission is not exhausted; indeed, the nerve terminal is responsive to a number of challenges via its receptors, since it modulates ganglionic transmission by controlling ACh release. The responsiveness of the ganglionic nerve terminal was first shown by Dempsher and his associates (Dempsher and Riker, 1957); the insulting agent was, in this case, a virus. This phenomenon subsequently led (W. K. Riker, 1965) to stress nerve-terminal responsiveness as a factor in ganglionic transmission, in analogy to the role of the nerve terminal of the motoneuron proposed by Riker and Standaert (see, for example, Riker *et al.*, 1957). Koelle (1961), on the other hand, stressed the role of cholinergic receptors of the terminal in the control of ACh release, since he felt that ACh provides a positive feedback, i.e., plays a percussive role in the release process; this remains a matter of controversy (see Chapters 10, 11, and 21).

Furthermore, the nerve terminal was early (Birks and McIntosh, 1961) shown to contain adrenergic receptors. These serve as both up- and down-modulators of release, the latter effect underlying the early discovery of Marrazzi (1939) of blockade by epinephrine of ganglionic transmission. Presynaptic modulation of transmission is an important subject that was studied extensively in the Kurume-Loyola laboratories; it is discussed in Chapters 3, 10, 11, 13, and 21. It must also be stressed that the final identification of ganglionic transmitters and of their mechanism of action, and the clarification of the role of late potentials in the modulation of ganglionic excitability and transmission, depend on the understanding of the membrane characteristics of the ganglia; this subject was vigorously pursued in the Loyola-Kurume laboratories (cf. Nishi, 1974) and is discussed in several chapters of this book (see above). The variety of postsynaptic potentials and presynaptic sites as well as the responsiveness of the postsynaptic membrane are responsible for the richness of the repertoire of ganglionic pharmacology; indeed, the ganglia respond to a multiplicity of cholinomimetic drugs including anti-ChE and anticholinergic compounds as well as to many noncholinergic agents—these compounds may act on the ganglia via noncholinergic transmitter systems, as in the case of serotonergic and antiserotonergic drugs, or via mechanisms that may not be concerned with neurotransmitters, as in the case of local anesthetics. This matter is discussed in detail in Chapters 3, 10, 11, 12, and 13 of this book. Furthermore, even drugs endowed with cholinergic activities may affect the ganglia via noncholinergic mechanisms. In fact, it was proposed early that anti-ChEs of carbamate as well as of organophosphorus type exert noncholinergic effects on the ganglia; these actions were referred to as "direct" (Heymans, 1951; Koppanyi *et al.*, 1947; Koppanyi and Karczmar, 1951) (see also Chapter 13). This important direct effect of anti-ChEs was corroborated subsequently, not only with respect to the ganglia but also with respect to the neuromyal junction and to the central nervous system (see Chapter 13).

IV. UNANSWERED QUESTIONS

This perusal of the history of ganglionic research raises several questions that were unanswered by the research of the 1960s and 1970s. Some have been answered, at least in part, in the 1980s, and they will be clarified in this book; some remain unanswered and constitute the subject of future research (see also Chapter 21).

One such subject is that of the physiology and pharmacology of the spinal preganglionic neurons. This subject is addressed in Chapters 17–19. Yet even such a basic question as that of the identity of the transmitter involved in the stimulation and of a transmitter that may be involved in the inhibition of preganglionic cells remains unanswered.

Another question is that of the generation of the slow IPSP and of the role of the SIF cell; while it is hoped that Chapters 9 and 13 clarify the alternatives, the final answer—which perhaps depends on the ganglion and on the species—cannot yet be given. A related question concerns the role of cyclic nucleotides as second messengers involved in the generation of ganglionic slow potentials, the IPSP and the slow EPSP, as proposed by Greengard (1978); this question is answered in Chapters 9 and 13. Still another substance, or a metabolic cycle, the phosphatidylinositol loop, may serve as second messenger; this problem is briefly alluded to in Chapter 13.

Probably the most important questions are those that concern the control and the modulation of ganglionic transmission, and the consequences of transmitter interaction, both pre- and postsynaptic, in the ganglia. Do peptides modify the transmission via the late slow EPSP, and does this modulation constitute a ganglionic reflex that involves the sensory system (see Chapters 8, 13, and 16)? What is the physiological meaning of the slow potentials? Can they substitute for the primary transmission under certain circumstances? Do they serve as modulators of excitability or provide for effective transmission under certain conditions and block it under others (see Chapters 3, 7, and 8)? Does the "little brain" character of the ganglia make them appropriate models for CNS studies (see Chapter 21)? Finally, do the versatility of ganglionic transmission and the ready modulatory modification of transmission, provided by the slow potentials as well as by preganglionic mechanisms, constitute a time- and experience-dependent phenomenon that may serve as a model for the study of the mechanisms of cognition, learning, and memory, as suggested by Libet (1981) (see Chapter 21)? These far-reaching questions are answered, but only in part, in this book.

REFERENCES

Auerbach, L.: Fernere vorläufige Mittheilung über den Nervenapparat des Darmes. *Arch. Pathol. Anat.* **30**:457–460 (1864).

Bacq, Z. M.: *Chemical Transmission of Nerve Impulses.* Pergamon Press, Oxford (1975).

Bard, P.: The central representation of the sympathetic nervous system as indicated by certain physiologic observations. *Arch. Neurol. Psychiatry* **22**:230–246 (1929).

Barsoum, G. S., Gaddum, J. H., and Khayyal, M. A.: The liberation of a choline ester in the inferior mesenteric ganglion. *J. Physiol. (London)* **28**:9–10P (1934).

Bernard, C.: De l'influence du système nerveux grand sympathique sur la chaleur animale. *C. R. Acad. Sci.* **34**:472 (1852).

Bernard, C.: De l'influence de deux ordres de nerfs qui determinent les variations de couleur du sang veineux dans les organes glandulaires. *C. R. Acad. Sci.* **47**:245–393 (1858).

Bernard, C.: *Leçons sur les Phénomènes de la Vie Communs aux Animaux et aux Végétaus,* 404 pp. J. B. Bailliere, Paris (1878).

Bertilsson, L., Suria, A., and Costa, E.: γ-Aminobutyric acid in rat superior cervical ganglion. *Nature (London)* **260**:540–541 (1976).

Bichat, M.-F.-X.: *Anatomie Generale Appliquée à la Physiologie et à la Médicine;* 2 vols. Brosson, Gabon, Paris (1802).

Biedl, A.: Uber die Centren der Splanchnici. *Wien. Klin. Wochenschr.* **52**:915–919 (1895).
Birks, R., and McIntosh, F. C.: Acetylcholine metabolism of a sympathetic ganglion. *Can. J. Biochem. Physiol.* **39**:787–827 (1961).
Bishop, G. H., and Heinbecker, P.: A functional analysis of the cervical sympathetic nerve supply to the eye. *Am. J. Physiol.* **100**:519–532 (1932).
Bloom, F. E.: Amino acids and polypeptides in neuronal function. *Neurosci. Res. Prog. Bull.* **10**:121–251 (1972).
Boehm, R.: Uber das Vorkommen und die Wirkungen des Cholins und die Wirkungen des künstlichen Muscarins. *Arch. Exp. Pathol. Pharmakol.* **19**:87–100 (1885).
Bok, S. T.: Das Rückenmark, in: *Handbuch der mikroskopischen Anatomie des Menschen*, 4, *Das Nervensystem*, Teil 1, pp. 478–578. Springer-Verlag, Berlin, 1928.
Botar, J.: *The Autonomic Nervous System*. Akademiai Kiado, Budapest, 1966.
Bowery, N. G., and Brown, D. A.: γ-Aminobutyric acid uptake by sympathetic ganglia. *Nature (London) New Biol.* **238**:89–91 (1972).
Brazier, M. A. B.: The historical development of neurophysiology, in: *Neurophysiology* (J. Field, ed.), Section I, Vol. 1, pp. 1–58, American Physiology Society, Washington, D.C., 1959.
Brodal, A.: *Neurological Anatomy*, 3rd ed. Oxford University Press, Oxford (1981).
Bronk, D. W.: Synaptic mechanisms in sympathetic ganglia. *J. Neurophysiol.* **2**:380–401 (1939).
Brooks, McC. C.: The development of our knowledge of the autonomic nervous system, in: *Integrative Function of the Autonomic Nervous System*, (McC. C. Brooks, K. Koizumi, and A. Sato, eds.), pp. 473–496, Elsevier/North-Holland, Amsterdam (1979).
Brooks, McC. C.: Conclusions. *J. Auton. Nerv. Sys.* **6**:107–108 (1982).
Brooks, McC. C., and Seller, H.: Early and late contributions to our knowledge of autonomic nervous systems and its control made by German scientists. *J. Auton. Nerv. Sys.* **3**:105–119, (1981).
Brown, G. L.: Conduction in the cervical sympathetic. *J. Physiol (London)* **81**:228–242 (1934).
Brown, G. L., and Feldberg, W.: The action of potassium on the superior cervical ganglion of the cat. *J. Physiol. (London)* **86**:290–305 (1936).
Brown-Séquard, C.-E.: Experimental researches applied to physiology and pathology. *Med. Exam. (Philadelphia)* **8**:481–504 (1852).
Bruce, A.: Distribution of the cells in the intermedio-lateral tract of the spinal cord. *Trans. Roy. Soc.* **45**:105 (1906).
Brucke, F. Th. V.: Cholinesterase in sympathetic ganglia. *J. Physiol. (London)* **89**:429–437 (1937).
Burnstock, G.: Purinergic nerves. *Pharmacol. Rev.* **24**:509–581 (1972).
Cannon, W. B., and Rosenblueth, A.: *Autonomic Neuro-effector Systems*. Macmillan, New York, (1937).
Carpenter, F. W., and Conel, J. L.: A study of ganglion cells in the sympathetic nervous system, with special reference to intrinsic sensory neurones. *J. Comp Neurol.* **24**:269–281 (1914).
Castiglioni, A.: *A History of Medicine*, 2nd ed. Knopf, New York (1947).
Castiglioni, G.: *Italian Medicine*. Clio Medica Series, Vol. 6, Hoeber, New York (1932).
Cole, A. E., and Nicoll, R. A.: Acetylcholine mediates a slow synaptic potential in hippocampal pyramidal cells. *Science* **221**:1299–1301 (1983).
Croone, W.: *De Ratione Motus Musculorum*. Hayes Publ., London (1852).
Dail, W. G., Jr., Evans, A. P., and Eason, H. R.: The major ganglion in the pelvic plexus of the male rat: Histochemical and ultrastructural study. *Cell Tissue Res.* **159**:49–62 (1975).
Dale, H. H.: The action of certain esters and ethers of choline, and their relation to muscarine. *J. Pharmacol. Exp. Ther.* **6**:147–190 (1914).
Dale, H. H.: *Adventures in Physiology*. Wellcome Trust, London (1953).

DeGroat, W. C., Booth, A. M., and Krier, J.: Interaction between sacral parasympathetic and lumbar sympathetic inputs to pelvic ganglia, in: *Integrative Functions of the Autonomic Nervous System* (Mc. C. Brooks, K. Koizumi, and A. Sato, eds.), pp. 234–247, Elsevier/North-Holland, Amsterdam (1979).

Dempsher, J. and Riker, W. K.: The role of acetylcholine in virus-infected sympathetic ganglia. *J. Physiol. (London)* **139**:145–156 (1957).

DeRobertis, E. D. P., and Bennett, H. S.: A submicroscopic vesicular component of Schwann cells and nerve satellite cells. *Exp. Cell Res.* **6**:543–545 (1954).

DeRobertis, E. D. P., and Bennett, H. S.: Some features of the submicroscopic morphology of synapses in frog and earthworm. *J. Biophys. Biochem. Cytol.* **1**:47–58 (1955).

Descartes, R.: *Discours de la Méthode pour Bien Conduire la Raison et Chercher la Vérité dans Les Sciences*. Henry Le Gros Publ., Paris, (1949).

Dittmar, C.: Über die Lage des sogennanten Gefässcentrums in der Medulla oblongata. *Ber. Sachs. Ges. Wiss. (Math.-Phys. Cl.)* **25**:449–469 (1873).

Dixon, W. E.: On the mode of action of drugs. *Med. Mag. (London)* **16**:454–457 (1907).

Dixon, W. E., and Hammill, O.: The mode of action of specific substances with special reference to secretion. *J. Physiol. (London)* **38**:314–336 (1909).

Dogiel, A. S.: Zur Frage über den feineren Bau des sympathischen Nervensystems bei den Sügethieren. *Arch. Mikrosk. Anat.* **46**:305–344 (1895).

DuBois-Raymond, E.: Vorläufiger Abriss einer Untersuchung über den Elektromotorischen Fische. *Ann. Physik. Chem.* **51**:1–30 (1843).

Dun, N. J., Karczmar, A. G., and Ingerson, A.: A. Neurochemical and neurophysiological study of serotonin in the superior cervical ganglia. *Soc. Neurosci. Abstr.* **6**:216 (1980).

Eccles, J. C.: Synaptic transmission through a sympathetic ganglion. *J. Physiol. (London)* **81**:88P (1934).

Eccles, J. C.: Slow potential waves in the superior cervical ganglion. *J. Physiol. (London)* **85**:464–510 (1935a).

Eccles, J. C.: Facilitation and inhibition in the superior cervical ganglion. *J. Physiol. (London)* **85**:207–238 (1935b).

Eccles, J. C.: The actions of antidromic impulses in ganglion cells. *J. Physiol. (London)* **88**:1–39 (1936).

Eccles, J. C.: Synaptic and neuro-muscular transmission. *Physiol. Rev.* **17**:538–555 (1937).

Eccles, J. C.: The nature of synaptic transmission in a sympathetic ganglion. *J. Physiol. (London)* **103**:27–54 (1944).

Eccles, J. C., Fatt, P., and Koketsu, K.: Cholinergic and inhibitory synapses in a pathway from motor-axon collaterals to motoneurons. *J. Physiol. (London)* **216**:524–562 (1954).

Eccles, R. M., and Libet, B.: Origin and blockade of the synaptic responses of curarized sympathetic ganglia. *J. Physiol. (London)* **157**:484–503 (1961).

Ehrenberg, C. G.: Nothwendigkeit einer feineren mechanischen Zerlegung des Gehirns und der Nerven. *Ann. Phys. Chem.* **104**:449–472 (1833).

Elliott, T. R.: The action of adrenalin. *J. Physiol. (London)* **32**:401–467 (1905).

Emmelin, N., and MacIntosh, F. C.: The release of ACh from perfused sympathetic ganglia and skeletal muscles. *J. Physiol. (London)* **131**:447–496 (1956).

Emmelin, N., and Muren, A.: Acetylcholine release at parasympathetic synapses. *Acta Physiol. Scand.* **20**:13–32 (1950).

Eranko, O. (ed.): *SIF Cells: Structure and Function of the Small, Intensely Fluorescent Sympathetic Cells*. DHEW Publication No. 76, Washington, D. C., (1976).

Eranko, O., and Harkonen, M.: Monoamine containing small cells in the superior cervical ganglion of the rat and an organ composed of them. *Acta Physiol. Scand.* **65**:511–512 (1965).

Estienne, C.: *De Dissectione Corporis Humani*. S. Colinaus Publ., Paris (1545).

Euler, U. S. von: A specific sympathomimetic ergone in adrenergic nerve fibres (sympathin) and its relation to adrenaline and nor-adrenaline. *Acta Physiol. Scand.* **12:**73–97 (1946).

Euler, U. S. von: *Noradrenaline.* Charles C Thomas, Springfield, Illinois (1956).

Eustachio (Eustachius), B.: *Opuscula Anatomica.* V. Luchinus Publ., Venice (1564).

Feldberg, W., and Gaddum, J. H.: The chemical transmitter at synapses in a sympathetic ganglion. *J. Physiol. (London)* **81:**305–319 (1934).

Feldberg, W., and Vartiainen, A.: Further observations on the physiology and pharmacology of a sympathetic ganglion. *J. Physiol. (London)* **83:**103–128 (1934).

Flourens, P.: *Recherches expérimentales sur les propriétes et les fonctions du système nerveux dans les animaux vertébrés.* Cuerot Publ., Paris (1824).

Friedman, A.: Circumstances influencing Otto Loewi's discovery of chemical transmission in the nervous system. *Pfluegers Arch.* **325:**85–86 (1971).

Gabella, G.: *Structure of the Autonomic Nervous System.* Chapman and Hall, London (1976).

Gaddum, J. H., and Giarman, N. J.: Preliminary studies on the biosynthesis of 5-hydroxytryptamine. *Br. J. Pharmacol.* **11:**88–92 (1956).

Galen, O.: *Opus De Usu Partium Corporis Humani.* S. Colinaeus, Paris (1528).

Galvani, A.: De viribus Electricitatis in Motu Musculari. *Comm. de Bononiensi Scientiatum et Artium Inst. atque Acad. Comm.* **7:**363 (1791).

Garrison, F. H.: *An Introduction to the History of Medicine.* W. B. Saunders, Philadelphia (1929).

Gaskell, W. H.: On the structure, distribution and function of nerves which innervate the visceral and vascular systems. *J. Physiol. (London)* **7:**1–80 (1886).

Gaskell, W. H.: On the action of the muscarin upon the heart, and on electrical changes in the non-beating cardiac muscle brought about by stimulation of the inhibitory and augmentor nerves. *J. Physiol. (London)* **8:**414–415 (1887).

Gaskell, W. H.: *The Involuntary Nervous System.* Longmans, London (1916).

George, W. J., Polson, J. B., O'Toole, A. G., and Goldberg, N. D.: Elevation of guanosine 3′,5′-cyclic phosphate in rat heart after perfusion with acetylcholine. *Proc. Natl. Acad. Sci. U.S.A.* **66:**398–403 (1970).

Glisson, F.: *Tractatus de Ventriculo et Intestinis.* Henry Brome Publ., London (1677).

Goltz, R., Freksberg, A., and Gergens, E.: Über gefasserweiternde Nerven. *Arch. Gesamte Physiol.* **11:**52 (1875).

Greengard, P.: *Cyclic Nucleotides, Phosphorylated Proteins, and Neuronal Function.* Raven Press, New York (1978).

Greving, R.: Beiträge zur Anatomie des Zwischenhirns und seiner Funktion. IV. Uber den Regulationsmechanismus der vegetativen Zentren in der Zwischenhirnbasis auf Grund cytoarchitektonischer und fasersystematischer Untersuchungen. *Z. Gesamte Neurol. Psychiatr.* **99:**231–252 (1925).

Haller, von A.: *A Dissertation on the Sensible and Irritable Parts of Animals* (English translation by M. Tissot). J. Nourse, London (1755).

Hering, E., and Breuer, J.: Die Selbststeuerung der Athmung durch den Nervus vagus. *S. B. Akad. Wiss. Wien Abt. II* **57:**672–677 (1868).

Hess, W. R.: Uber die Wechselbeziehungen zwischen psychischen und vegetativen Funktionen. *Schweiz. Arch. Neurol.* **16:**36– (1925).

Hess, W. R.: *Diencephalon: Autonomic and Extrapyramidal Function.* Grune and Stratton, New York (1954).

Heymans, C.: Les substances anticholinesterasiques. *Expo. Annu. Biochim. Méd.* **12:**21–53 (1951).

Heymans, C., and Neil, E.: *Reflexogenic Areas of the Cardiovascular Center.* Churchill, London (1958).

Hirschmann, L.: Zur Lehre von der durch Arzneimittel hervorgerufenen Myosis und Mydriasis. *Arch. Gesamte Physiol. Wiss. Med.*, p. 309 (1863).

Hökfelt, T., Elfvin, L. G., Elde, R., Schultzber, M., Goldstein, M., and Luft, R.: Occurrence of somatostatin-like immunoreactivity in some peripheral sympathetic noradrenergic neurons. *Proc. Natl. Acad. Sci. U.S.A.* **74**:3587–3591 (1977).

Holmstedt, B.: The ordeal bean of old calabar: The pageant of *Physostigma venenosum* in medicine, in: *Plants in the Development of Modern Medicine* (T. Swain, ed.), pp. 303–360, Cambridge University Press., Cambridge, Massachusetts (1972).

Holmstedt, B., and Liljestrand, G.: *Readings in Pharmacology*. Pergamon Press, Oxford (1963).

Hunt, R., and Taveau, R. de M.: On the physiological action of certain cholin derivatives and new methods for detecting cholin. *Br. Med. J.* **II**:1788–1791 (1906).

Johnstone, J.: Essay on the use of the ganglions of the nerves. *Philos. Trans. Roy. Soc. (London)* **54**:177 (1764).

Karczmar, A. G.: Introduction: History of the research with anticholinesterase agents, in: *Anticholinesterase Agents* (A. G. Karczmar, ed.), pp. 1–44, Internatl. Encyclop. Pharmacol. Therap., Section 13, Vol. 1, Pergamon Press, Oxford (1970).

Karczmar, A. G.: Basic phenomena underlying novel use of cholinergic agents, anticholinesterases and precursors in neurological including peripheral and psychiatric disease, in: *Cholinergic Mechanisms* (G. Pepeu and H. Ladinsky, eds.), Advances in Behavioral Biology, Vol. 25, pp. 853–869, Plenum Press, New York (1981).

Karczmar, A. G., Nishi, S., and Blaber, L. C.: Synaptic modulations, in: *Brain and Human Behavior*, (A. G. Karczmar and J. C. Eccles, eds.), pp. 63–92, Springer-Verlag, Berlin (1972).

Karplus, J. P., and Kreidl, A.: Gehirn und Sympathicus, in: *Zwischenhirnbasis und Halssympathicus. Pfluegers Arch. Gesamte Physiol.* **129**:138–144 (1909).

Kibjakow, A. W.: Uber humorale Ubertragung der Erregung von einem Neuron auf das andere. *Pfluegers Arch. Gesamte Physiol.* **232**:432–433 (1933).

Kobert, R., and Sobot, A.: Über die Wirkung des salzsäueren Hyoscin. *Arch. Exp. Pathol. Pharmakol.* **22**:396–429 (1887).

Koelle, G. B.: A proposed dual neurohumoral role of acetylcholine: Its function at the pre- and post-synaptic sites. *Nature (London)* **190**:208–211 (1961).

Koelle, G. B.: Cytological distributions and physiological functions of cholinesterases, in: *Cholinesterases and Anticholinesterase Agents* (G. B. Koelle, ed.), Handbuch der Experimentellen Pharmakologie, Ergänzungswerk., **15**:187–298, Springer-Verlag, Berlin (1963).

Koelle, G. B.: The neurohumoral theory, in: *Nerve as a Tissue* (K. Rodahl, ed.), pp. 287–292, Haber, New York (1966).

Koketsu, K.: Cholinergic synaptic potentials and the underlying ionic mechanisms. *Fed. Proc. Fed. Am. Soc. Exp. Biol.* **28**:101–112 (1969).

Koketsu, K., and Nishi, S.: Characteristics of the slow inhibitory postsynaptic potential of bullfrog sympathetic ganglion cells. *Life Sci.* **6**:1827–1836 (1967).

Koppanyi, T.: Studies on pupillary reactions in Tetrapods. *J. Pharmacol. Exp. Ther.* **36**:179–183 (1929).

Koppanyi, T.: Studies on the synergism and antagonism of drugs. I. The non-parasympathetic antagonism between atropine and the miotic alkaloids. *J. Pharmacol. Exp. Ther.* **46**:395–405 (1932).

Koppanyi, T., and Karczmar, A. G.: Contribution to the study of the mechanism of action of cholinesterase inhibitors. *J. Pharmacol. Exp. Ther.* **101**:327–344 (1951).

Koppanyi, T., Dille, J. M., and Linegar, C. R.: The action of physostigmine on autonomic ganglia. *J. Pharmacol. Exp. Ther.* **58**:105–110 (1936).

Koppanyi, T., Karczmar, A. G., and King, T. O.: Effect of tetraethylpyrophosphate on sympathetic ganglionic activity. *Science* **106**:492–493 (1947).

Korner, P. I.: Integrative neural cardiovascular control. *Physiol. Rev.* **51**:312–367 (1971).

Kuhne, W.: *Uber die peripherischen Endorgane der motorischen Nerven*. Engelmann, Leipzig (1862).

Kuntz, A.: Histological variations in autonomic ganglia and ganglion cells associated with age and disease. *Am. J. Pathol.* **14**:783–795 (1938).

Kuntz, A., and Moseley, R. L.: An experimental analysis of the pelvic autonomic ganglia in the cat. *J. Comp. Neurol.* **64**:63–74 (1936).

Langley, J. N.: On the physiology of the salivary secretion: The influence of the chorda tympani and sympathetic nerves upon the secretion of the maxillary gland of the cat. *J. Physiol. (London)* **1**:96–103 (1878).

Langley, J. N.: On changes in serous glands during secretion. *J. Physiol. (London)* **2**:261–280 (1879).

Langley, J. N.: On the reaction of cells and of nerve endings to certain poisons, chiefly as regards the reaction of striated muscle to nicotine and to curare. *J. Physiol. (London)* **33**:374–413 (1905).

Langley, J. N.: *The Autonomic Nervous System*, Part 1. W. Heffer, Cambridge (1921).

Langley, J. N., and Anderson, H. K.: The innervation of the pelvic and adjoining viscera. Part VI. Histological and physiological observations upon the effects of section of the sacral nerves. *J. Physiol. (London)* **19**:372–384 (1895).

Langley, J. N., and Anderson, H. K.: Histological and physiological observations upon the effects of section of the sacral nerves. *J. Physiol. (London)* **19**:372–384 (1896).

Langley, J. N., and Dickinson, W. L.: On the local paralysis of peripheral ganglia, and on the connection of different classes of nerve fibers with them. *Proc. R. Soc.* **46**:423–431 (1889).

Laporte, Y., and R. Lorente de Nó, R.: Potential changes evoked in a curarized sympathetic ganglion by presynaptic volleys of impulses. *J. Cell. Comp. Physiol.* **35**:61–106 (1950).

Larrabee, M. G., and Leicht, W. S.: Metabolism of phosphatidyl inositol and other lipids in active neurones of sympathetic ganglia and other peripheral nervous tissues: The site of inositide effect. *J. Neurochem.* **12**:1–3 (1965).

Lawrentjew, B. I.: Experimentell-morphologische Studien über den feineren Bau des autonomen Nervensystems. I. Die Beteiligung des Vagus an der Herzinnervation. *Z. Mikrosk-Anat. Forsch.* **16**:383–411 (1929).

Lembeck, F., and Giere, W. *Otto Loewi*. Springer-Verlag, Heidelberg (1968).

Lennander, K. G.: Uber die Sensibilität der Bauchhohle und über lokale und allgemeine Anästhesie bei Bruch- und Bauchoperationen. *Abh. Chir.* **28**:209–223 (1901).

Libet, B.: Generation of slow inhibitory and excitatory postsynaptic potentials. *Fed. Proc. Fed. Am. Soc. Exp. Biol.* **29**:1945–1956 (1970).

Libet, B.: Slow potentials in sympathetic ganglia, in: *Electrohysiological Approaches to Human Cognitive Processing* (R. Galambos and S. A. Hillyard, eds.), pp. 236–239, Neurosciences Research Program Bulletin, Vol. 20, MIT Press Journals, Cambridge, Massachusetts (1981).

Libet, B., and Owman, C.: Concomitant changes in formaldehyde-induced fluorescence of dopamine interneurons and in slow inhibitory postsynaptic potentials of the rabbit superior cervical ganglion, induced by stimulation of the preganglionic nerve or by a muscarinic agent. *J. Physiol. (London)* **237**:635–662 (1974).

Lim, R. K. S., Wang, S. C., and Yi, C. L.: On the question of a myelencephalic sympathetic centre. VII. The depressor area: A sympatho-inhibitory center. *Clin. J. Physiol.* **13**:61–78 (1938).

Loewi, O.: Über humorale Übertragbarkeit der Herzenwirkung. *I. Pfluegers Arch. Gesamte Physiol.* **189**:239–242 (1921).

Loewi, O., and Gettwert, W.: Über die Folgen der Nebennierenextirpation. I. Mitteilung: Untersuchungen am Kaltbluter. *Pfluegers Arch. Gesamte Physiol.* **158**:29–40 (1914).

Loewi, O., and Navratil, E.: Über humorale Übertragbarkeit der Herzenwirkung. VII Mitteilung. *Pfluegers Arch. Gesamte Physiol.* **206**:123–134 (1924).

Luchsinger, B.: Neue Versuche einer Lehre von der Schweissecretion, ein Beitrag zur Physiologie der Nervencentrer. *Arch. Gesamte Physiol.* **14**:369 (1876).
Ludwig, C. T.: Neue Versuche über die Beihilfe zur Speichelabsonderung. *Z. Rat. Med. Heidelberg N.F.* **1**:254–277 (1851).
Magendie, F.: Sur le siège du mouvement et du sentiment dans la moelle éspinière. *J. Physiol. Expér. et Pathol.* **3**:153 (1823).
Marey, E. J.: *Physiologie Médicale de la Circulation du Sang.* G. Masson, ed. (1863).
Marey, E. J.: *La Circulation du Sang.* G. Masson Editeur, Paris (1881).
Marrazzi, A. S.: Electrical studies on the pharmacology of autonomic synapses, in: The action of parasympathomimetic drugs on sympathetic ganglia. *J. Pharmacol.* **65**:18–35 (1939).
Mateucci, C.: *Essai sur les Phenoménes Electriques des Animaux.* Cavilian-Goeury and Dalmont, Paris (1840).
McAfee, D. A., Schorderet, M., and Greengard, P.: Adenosine 3′,5′-monophosphate in nervous tissue: Increase associated with synaptic transmission. *Science* 171:1156–1158 (1971).
McGeer, P. L., Eccles, J. C., and McGeer, E. G.: *Molecular Neurobiology of the Mammalian Brain.* Plenum Press, New York (1978).
Meckel, J. F.: Observation anatomique avec l'examen physiologique du véritable usage des noeuds, ou ganglions des nerfs. *Mém. Acad. Roy. Sci.* **5**:84–178 (1751).
Meissner, G.: Uber die Nerven der Darmwand. *Z. Rat. Med. N.F.* **8**:364–366 (1857).
Monro, A.: *Observations on the Structure and Functions of The Nervous System*, W. Creech, Edinburgh (1783).
Nishi, S.: Ganglionic transmission, in: *The Peripheral Nervous System* (J. I. Hubbard, ed.), pp. 225–255, Plenum Press, New York (1974).
Nishi, S., and Koketsu, K.: Early and late afterdischarges of amphibian sympathetic ganglion cells. *J. Neurophysiol.* **31**:109–121 (1968).
Nishi, S., Soeda, H., and Koketsu, K.: Release of acetylcholine from sympathetic preganglionic nerve terminals. *J. Neurophysiol.* **30**:114;–134 (1967).
Norberg, K.-A., Ritzen, M., and Ungerstedt, U.: Histochemical studies on a special catecholamine-containing cell type in sympathetic ganglia. *Acta Physiol. Scand.* **67**:260–270 (1966).
North, R. A.: Electrophysiology of the enteric nervous system. *Neuroscience* **7**:315–325 (1982).
O'Malley, C. D.: *Andreas Vesalius of Brussels.* University of California Press, Berkeley (1964).
Owsjannikov, P.: Die tonischen and reflectorischen Centren des Gefassernerven. *Ber. Sachs. Gesamte Wiss. (Math.-Phys. C1.)* **21**:135–143 (1869).
Palay, S. L., and Palade, G. E.: The fine structure of neurons. *J. Biophys. Biochem. Cytol.* **1**:69–88 (1955).
Pernow, B.: Pharmacology of substance P. *Ann. N. Y. Acad. Sci.* **104**:393–402 (1963).
Petit, F.-P. de: *Memoire dans laquelle il est demontré que les nerfs intercostaux fournissent des rameaux qui portent des esprits dans les yeux.* Historical Academy of Royal Sciences, p. 1 (1727).
Pfluger, E. F. W.: *Über das Hemmungs-Nervensystem für die peristaltischen Bewegungen der Gedärme.* A. Hirschwald Publ., Berlin (1857).
Pick, J.: *The Autonomic Nervous System.* J. B. Lippincott, Philadelphia (1970).
Pickard, M. R., Hawthorne, J. N., Higashi, E., and Yamada, S.: Effects of surugatoxin and other nicotinic and muscarinic antagonists on phosphatidyl inositol metabolism in active sympathetic ganglia. *Biochem. Pharmacol.* **26**:448–450 (1977).
Poljak, S.: Über die sogenannten versprengten Ganglienzellen in der weissen Substanz des menschlichen Rückenmarks. *Arb. Neurol. Inst. Wien. Univ.* **23**:1 (1922).
Ramón y Cajal, S.: *Observaciones Microscopicas sobre las Terminaciones Nerviosas en los Musculos Voluntarios de la Rana.* Zaragoza (1881).

Ramón y Cajal, S.: Note sobre el plexo de Auerbach de los batracios. *Trab. Lab. Hist. Barcelona*, pp. 23–28 (1892).
Ramón y Cajal, S.: *Histologie du Système Nerveux de l'Homme et des Vertebrés*. A. Maloine, Paris (1911).
Ranson, S. W., and Billingsley, P. R.: Vasomotor reaction from stimulation of the floor of the fourth ventricle. *Am. J. Physiol.* **41**:85–90 (1916).
Reid, A.: *A Description of the Body of Man*. T. Coates Publ., London (1634).
Reil, J. C.: Über die Eigenschaften des Gangliensystems und sein Verhältnis zum cerebralen System. *Arch. Gesamte Physiol.* **7**:189–245 (1807).
Remak, R.: *Observationes Anatomicae et Microscopicae de Systematis Nervosi Structura*. Reimerianis, Berlin (1838).
Remak, R.: Über multipolare Ganglienzellen. *Ber. Verh. Preuss. Akad. Wiss.* pp. 26 (1854).
Riker, W. K.: Effects of tetraethylammonium chloride on electrical activities of frog sympathetic ganglion cells. *J. Pharmacol. Exp. Ther.* **145**:317–325 (1965).
Riker, W. F., Standaert, F. G., and Fujimori, H.: The motor nerve terminal as the primary focus for drug-induced facilitation of neuromuscular transmission. *J. Pharmacol. Exp. Ther.* **121**:286–312 (1957).
Rosenblueth, A.: *The Transmission of Nerve Impulses at Neuroeffector Junctions and Peripheral Synapses*. Technology Press, MIT, Cambridge; John Wiley, New York (1950).
Rosenbluth, J.: Contrast between osmium-fixed and permanganate-fixed toad spinal ganglia. *J. Cell Biol.* **16**:143–157 (1963).
Roszkowski, A. P.: An unusual type of sympathetic ganglionic stimulant. *J. Pharmacol. Exp. Ther.* **132**:156–170 (1961).
Rothschuh, K. E.: *History of Physiology*. Robert E. Kreigher, Huntington, New York (1973).
Saper, C. B., Loewy, A. D., Swanson, L. W., and Cowan, W. M.: Direct hypothalamico-autonomic connections. *Brain Res.* **117**:305–312 (1976).
Saunders, J. B. de C. M., and O'Malley, C. D.: *Andreas Vesalius De Humani Corporis Fabrica*, Basel, 1543, (English translation). Schuman, New York (1947).
Schaefer, H.: Letter to Seller. *J. Auton. Nerv. Sys.* **3**:116–119 (1981).
Schaefer, H., and Janisch, A.: *Arch. Kreislaufforsch.* **7**:260–274 (1940).
Schafer, E. A.: *Textbook of Physiology*, Macmillan, London (1900).
Schmiedeberg, O., and Koppe, R.: *Das Muscarin, das giftige Alkaloid des Fliegenpilzes*. F. C. W. Vogel, Leipzig (1869).
Smirnow, A. E. von: Die Struktur der Nervenzellen im Sympathicus der Amphibien. *Arch. Mikrosk. Anat.* **35**:409–424 (1890).
Sperling, F., and Koppanyi, T.: Histophysiologic studies on sweating. *Am. J. Anat.* **84**:335–364 (1949).
Taxi, J.: Etude de l'ultrastructure des zones synaptiques dans les ganglions sympathiques de la Grenouille. *C. R. Acad. Sci.* **252**:174–176 (1961).
Taxi, J.: Contribution a l'étude des connexions des neurons moteurs du système nerveux autonome. *Ann. Sci. Natur. Zool.* **7**:413 (1965).
Taxi, J.: Observations on the ultrastructure of the ganglionic neurons and synapses in the frog, *Rana esculenta* L., in: *The Neuron* (H. Hyden, ed.), pp. 221–254, Elsevier, Amsterdam (1967).
Volle, R. L.: The responses to ganglionic stimulating and blocking drugs of cell groups within a sympathetic ganglion. *J. Pharmacol. Exp. Ther.* **135**:54–61 (1962).
Volle, R. L.: Muscarinic and nicotinic stimulant actions at autonomic ganglia, in: *Ganglionic Blocking and Stimulating Agents* (A. G. Karczmar, ed.), pp. 1–106, Internatl. Encyclop. Pharmacol. Therap., Section 12, Vol. 1, Pergamon Press, Oxford (1966).
Volle, R. L., and Koelle, G. B.: The physiological role of acetylcholinesterase in sympathetic ganglia. *J. Pharmacol. Exp. Ther.* **133**:223–240 (1961).
Waniewski, R. A., and Suria, A.: Alterations in gamma-aminobutyric acid content in the rat superior cervical ganglion and pineal gland. *Life Sci.* **21**:1129–1141 (1977).

Waser, P. G.: Chemistry and pharmacology of muscarine, muscarone and some related compounds. *Pharmacol. Rev.* **13:**465–515 (1961).

Watzka, M.: *Die Paraganglien.* Hanbuchder Mikroskopishea Anatomie, Vol. 6, Springer-Verlag, Berlin (1943).

Weber, E., and Weber, E. H.: Experimenta, Quibus Probatur Nervos Vagos Rotatione Machinae Galvanomagneticae Irritatos, Motum Cordis Retardare et Adeo Intercipere. Annali Universali di Medicina, Seria Terza, **20:**227–318 (1845).

Willis, T.: *Cerebri Anatome, Cui Accessit Nervorum Descriptio et Usus.* J. Flesher, London (1664).

Whytt, R.: *An Essay on the Vital and Other Involuntary Motions of Animals.* G. Hamilton, J. Balfour, and P. Neill Publs., Edinburgh (1751).

Whytt, R.: *Observations on the Nature, Causes and Cure of those Disorders Which have been Commonly Called Nervous, Hypochondriac or Hysteric; to Which are Prefixed Some Remarks on the Sympathy of the Nerves.* T. Beckett Publ., Second Edition, Edinburgh (1765).

Winslow, J. B.: *Exposition Anatomique du Corps Humain.* Duprez et Dessartz, Publ., Paris (1732).

Witanowski, W. R.: Über humorale Übertragbarkeit der Herzenwirkung. VIII Mitteilung. *Pfluegers Arch. Gesamte Physiol.* **208:**694–704 (1925).

2

Anatomy, Histology, and Electron Microscopy of Sympathetic, Parasympathetic, and Enteric Neurons

GORDON M. LEES

I. INTRODUCTION

Developments in immunohistochemical and immunocytochemical techniques and methods of marking individual cells for subsequent examination in the light, fluorescence, and electron microscopes have led to a major change in the approach to the study of the morphology of the autonomic nervous system (ANS). Increasingly, morphological investigations involve the use of biochemical, immunological, and electrophysiological techniques, in addition to traditional methods. Conventionally, the neurons of the ANS are described as forming efferent nerve pathways, comprised of two neurons arranged in series and synapsing outside the central nervous system (CNS); they originate in particular parts of the CNS to innervate virtually every tissue in the body, with the major exception of skeletal muscle fibers. Such a description overlooks the possible occurrence of autonomic reflex arcs utilizing synapses entirely outside the CNS. The existence of a reflex arc concerned with the control of intestinal motility and involving the colon and inferior mesenteric ganglion was deduced by Kuntz (1940) and first proven in an elegant series of experiments by Szurszewski and his colleagues (Crowcroft *et al.*, 1971; Szur-

GORDON M. LEES • Department of Pharmacology, Marischal College, University of Aberdeen, Aberdeen, Scotland.

szewski and Weems, 1976a). Their observations have since been amply confirmed and extended to other prevertebral ganglia and to reflex responses concerned with other aspects of gastrointestinal function [for reviews, see Szurszewski (1981) and Chapters 16 and 18]. Thus, at least with respect to the autonomic innervation of the alimentary tract, integrated activity has been demonstrated as occurring (1) entirely within the wall of the gut, i.e., in the myenteric (Auerbach's) and submucous (Meissner's) plexuses, either singly or together; (2) in the prevertebral ganglia; and (3) in the CNS, particularly the brainstem and spinal cord (Kosterlitz, 1968; Szurszewski and Weems, 1976b; Davison and Grundy, 1978). Similar information has begun to be available with respect to autonomic innervation of the blood vessels (see, for example, Janig *et al.*, 1982). It is important, therefore, that there should be information about the location and morphological features as well as the physiological characteristics of the neurons involved in these reflexes.

In recent years, it has become generally accepted that the traditional division of the ANS into sympathetic and parasympathetic divisions is outmoded and that the entire nervous system should be recognized as a distinct and complex entity, in accordance with the proposal of Langley (1921), on the basis of several unique features. Another major change in concept concerns the occurrence of interneurons in autonomic ganglia. The possibility of interneurons between preganglionic and the principal ganglionic neurons (Dogiel, 1896; Lawrentjew, 1924) has been a recurring question, but ultrastructural evidence was lacking until Williams (1967) adduced evidence for the presence of small granule-containing cells in the rat superior cervical ganglion; these cells received synapses from preganglionic nerve terminals and themselves synapsed with a principal ganglion cell or its processes. This finding, together with the occurrence of small intensely fluorescent (SIF) cells, singly or in clusters and sometimes near specialized blood vessels in most sympathetic ganglia, has led to a great deal of interest in the possibility that these cells may subserve neural, neurocrine, endocrine, or paracrine functions (see below). Since the main emphasis of this chapter will be on recent advances in selected aspects of the morphology of autonomic neurons, the reader is referred to the detailed account by Gabella (1976) concerning the gross anatomy of the ANS and the morphology of autonomic neurons of many species.

II. PARASYMPATHETIC AND SYMPATHETIC GANGLIA

A. Gross Anatomical Considerations

Autonomic ganglia are relatively small, often diffused structures that may be connected by minute nerve trunks. In larger animals and man, fine dissection alone may be sufficient to locate and describe in detail the features of complex ganglia, e.g., the coeliac (Ward *et al.*, 1979) and ciliary (Sinnreich and Nathan, 1981), even for clinical purposes. In small laboratory species, however, the dissection of certain parts of the ANS (e.g., the pelvic plexus) may present considerable difficulty, particularly when the location of a ganglion is unknown, uncertain, or variable. A method of staining for cholinesterase activity applied to organs *in toto* has led to detailed descriptions of the extrinsic innervation of the abdominal organs of the rat (Baljet and Drukker, 1979). Certain ganglia, or more correctly clusters of ganglion neurons, are buried with the nerve trunks (e.g., sympathetic neurons in the hypogastric nerve or parasympathetic neurons in the oculomotor nerve); these can be identified only by cutting histological sections through the nerves.

The parasympathetic ganglia of the head and neck are often difficult to locate and in some species may be comprised of sheets of irregularly aggregated neurons (Lichtman, 1977; Gienc and Kuder, 1980). While such very small ganglia may be more difficult to dissect, they offer special experimental advantages. For example, the submandibular (parasympathetic) ganglion of the rat has proved extremely valuable for studying the mechanisms of action of the neurotransmitter acetylcholine and of ganglion-blocking agents (Rang, 1981; Gurney and Rang, 1984). Similarly, this ganglion yielded much interesting information about the development and persistence of synaptic connections (Lichtman, 1977, 1980; Purves and Lichtman, 1978). The rat submandibular ganglion consists of three to ten clusters of neurons, each containing 5–20 cells usually lying side by side; this arrangement makes them easily visualized with differential interference contrast optics. Furthermore, the size and appearance of the neurons are fairly uniform, the somata lacking an extensive dendritic tree. The horseradish peroxidase (HRP) staining method revealed very few large dendrites but numerous small projections from both the perikaryon and the initial portion of the axon, which did not branch in its course of several hundred micrometers. Electron-microscopically, the synapses of this ganglion were usually located on small projections of the soma and, electrophysiologically, the great majority of neurons had only one synaptic input, there being no evidence functionally or structurally for interneurons or small granule-containing cells (Lichtman, 1977, 1980). Interestingly, however, any one preganglionic nerve fiber synapsed with several neurons in a cluster, though not necessarily with closely adjacent cells. Thus, the ganglion provides a certain degree of divergence of autonomic outflow but no convergence, since each neuron, in the adult, is singly innervated.

A landmark in the attempt to correlate structure with function in autonomic neurons was the work of Kuffler and his colleagues (McMahan and Kuffler, 1971; Dennis *et al.*, 1971) on the ganglion cells in the amphibian heart. In addition to describing the electrophysiological characteristics of these neurons, these authors correlated, with amazing precision, the exquisite detail seen in the living neuron by means of differential interference contrast optics with the electron-microscopic features of the same neuron. They were able to compare synaptic transmission with the responses of the soma membrane elicited by focal application of acetylcholine to these parasympathetic neurons.

The patterns of innervation of these ganglion cells by both the right and the left vagus nerve have been studied in the normal and decentralized states (Roper, 1976; Courtney and Roper, 1976). It was found that when the left vagus nerve was crushed, there was sprouting from the right vagus nerve so that it provided the innervation until there was regeneration from the left side.

Morphological features of cat ciliary ganglion neurons have been correlated with electrophysiological data for the same neuron by means of intracellular injection of the fluorescent dye, procion yellow (Figure 1). Most cells were found to be innervated by two or more presynaptic fibers, B- and C-type fibers innervating exclusively B and C cells, respectively (Nishi and Christ, 1971). Wolf (1941) calculated that the ratio of preganglionic fibers to the neurons in this ganglion ranged from 1 : 1.7 to 1 : 2.6. Skok (1973) has provided detailed information about the complex conduction and synaptic pathways in this ganglion. In contrast, in the rat ciliary ganglion, which has about 200 neurons, the equivalent ratio ranges from 1 : 1 to 4 : 1 (Wigston, 1983).

B. Convergence and Divergence

From the physiological viewpoint, it is important to establish whether autonomic ganglia are simply relay stations or are capable of providing integration of synaptic activity. The degrees of divergence and convergence in ganglia have therefore been the subject of continued interest. Although it is generally accepted that the ratio of preganglionic nerves to ganglion cell bodies is higher in the parasympathetic than in the sympathetic division, the estimates of ratios show large variations, even within one division; for example, in the superior cervical ganglion of the cat, the ratio varies from about 1 : 11 to 1 : 17 (Wolf, 1941), whereas it can be up to 1 : 200 in man (Ebbesson, 1968) and the guinea pig (Purves and Wigston, 1983). However, such ratios are open to doubt, since significant numbers of postganglionic fibers may be present in the preganglionic nerve trunk (see below). Since the number of nonmyelinated preganglionic fibers usually constitutes a substantial proportion of the total population, ratios based on counts of myelinated preganglionic nerves may be seriously in error. Furthermore, although the estimation of the number of cell bodies in such a relatively small structure as a ganglion may appear to be straightforward, it is not an easy task (Smolen *et al.*, 1983). It would seem, there-

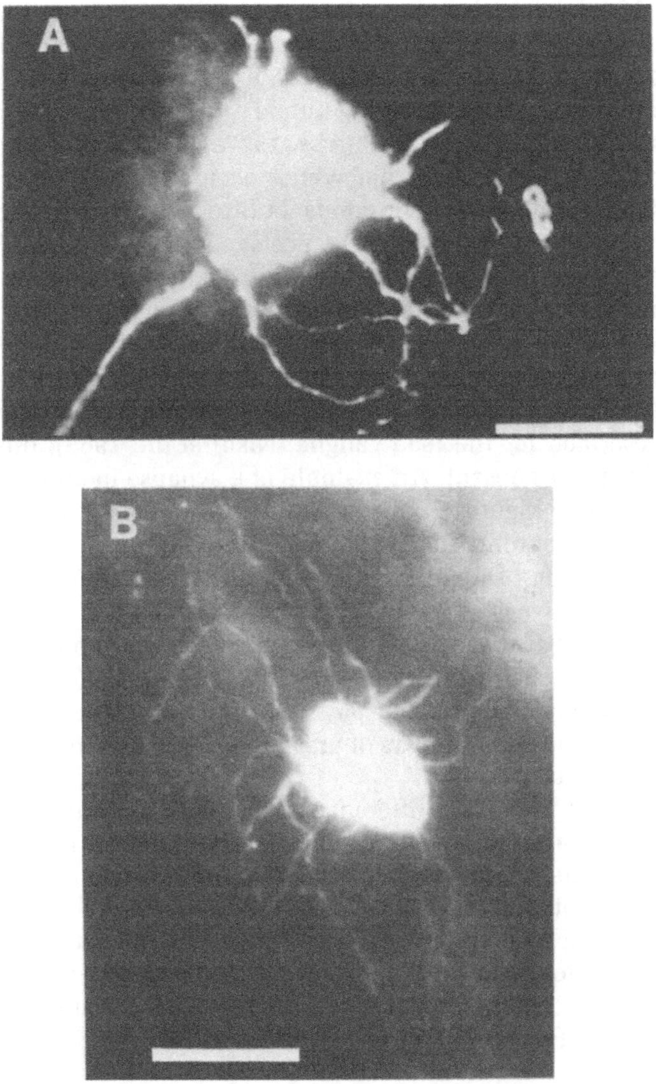

Figure 1. Morphology cat ciliary ganglion cells revealed by intracellular microintophoretic application of procion yellow. (A) Note the numerous short, tapering dendrites and one axon. Courtesy of Professor S. Nishi; reproduced from Gabella (1976) with permission. (B) Neuron with numerous fine, long processes and a few short ones. Courtesy of Professor S. Nishi. Scale bars: 50 μm.

fore, that morphological methods have too many pitfalls to provide an accurate estimate of divergence based on fiber and soma counts.

Functionally, there is also wide variation in estimates of divergence and convergence. On the basis of estimates of numbers of neurons and preganglionic axons in the rabbit superior cervical ganglion, Wallis and

North (1978) calculated that each preganglionic axon must synapse on average with 240 cells, a remarkable degree of divergence of signals; in the same ganglion, on the average, 7.5 axons synapse with one ganglion neuron, while the corresponding number for the guinea pig superior cervical ganglion is 10 (Njå and Purves, 1977a). This degree of convergence is very much higher than that which occurs in neurons of amphibian sympathetic ganglia. In conscious bullfrogs, when tonically (spontaneously) occurring fast excitatory postsynaptic potentials (EPSPs) were recorded from type B sympathetic neurons in the 9th or 10th paravertebral ganglion, it was found that 30% of cells were innervated by a single preganglionic fiber, 55% by two preganglionic fibers, and 15% had three excitatory synaptic inputs (Yoshimura *et al.*, 1979). These neurons were innervated solely by preganglionic (myelinated) B-type fibers, as previously described for the toad ganglia (Nishi *et al.*, 1965); innervation by C-type fibers was absent. An example of a synapse in an amphibian ganglion is given in Figure 2. For a new classification of sympathetic ganglion paravertebral neurons, see Dodd and Horn (1983).

The problem of identifying the controlling mechanisms in different functional pathways is an intriguing one, particularly since there seems to be, physiologically or pathophysiologically, an important proportion of transmission that may utilize slow EPSPs rather than the conventional fast EPSPs (Henderson and Ungar, 1978), but the relationship of the slow EPSP to the type and numbers of preganglionic cholinergic fibers in mammals has not been studied.

In view of the major difference in the pattern of innervation of sympathetic and parasympathetic ganglia, it would be important to know the factors that control the establishment of some synaptic connections in preference to others; for discussions of the events involved in embryonic life and during reinnervation in the adult, see Dennis and Sargent (1978), Purves and Lichtman (1978), Lichtman and Purves (1980), Hume and Purves (1981, 1983), Johnson and Purves (1981), Sargent (1983), and Forehand and Purves (1984).

C. Origin of Fibers That Form Postganglionic Nerves from Superior Cervical Ganglia

The general appearance of this ganglion in rabbits, especially of the New Zealand White strain, is extremely variable, but most commonly, the ganglion is bilobed, with a narrow isthmus between the swellings (Wallis *et al.*, 1975). Since the isthmus is often so narrow that it is almost devoid of somata (Best and Lees, unpublished observations), it is highly likely that the majority of the somata that give rise to fibers forming the internal carotide nerve lie in the distal (rostral) pole of the ganglion, while the

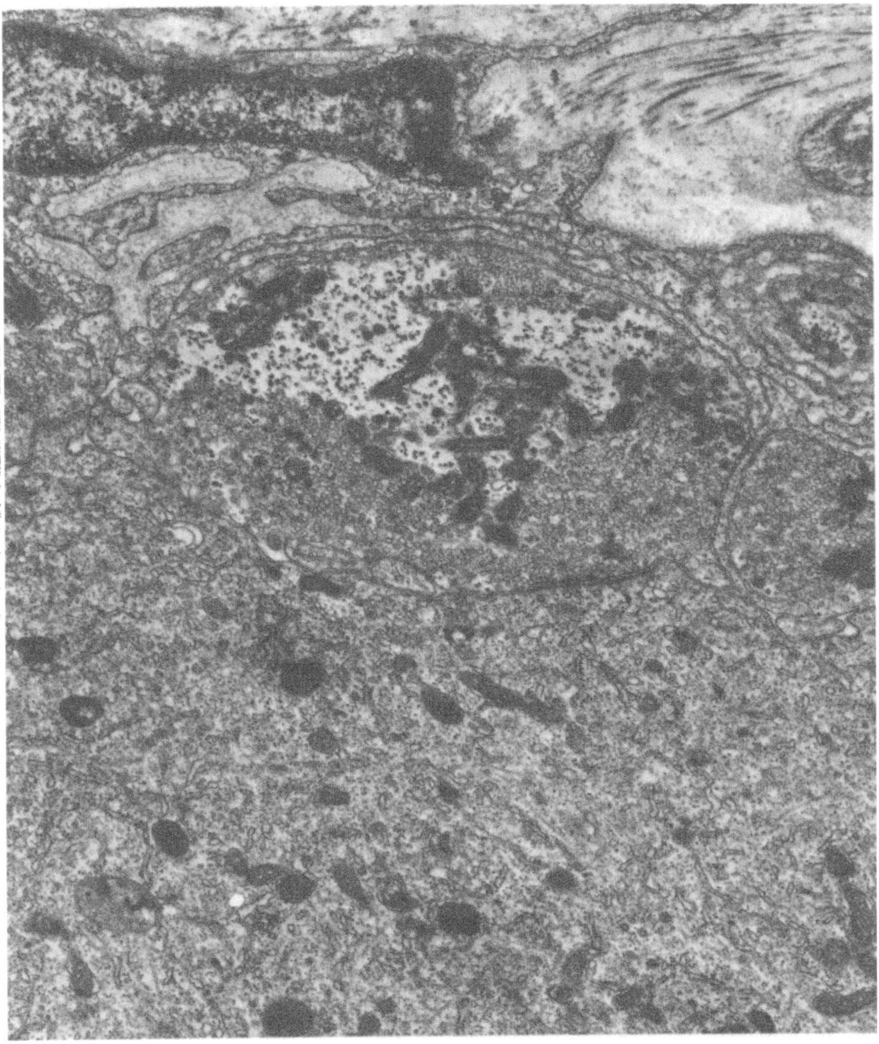

Figure 2. Abdominal sympathetic ganglion of a frog. In the center, a large nerve ending of preganglionic origin synapses on the soma of a ganglion neuron. The nerve ending is packed with small agranular vesicles, mitochondria, glycogen granules, and a few larger granular vesicles. The cell at the top is a satellite cell. Its processes are associated with the ganglion neuron and with the nerve ending. ×27,500. (Reproduced at 86%.) Courtesy of G. Gabella.

proximal swelling contains cell bodies for the fibers emerging in the external carotid nerve; electrophysiological evidence has been provided to support this contention (Kosterlitz *et al.*, 1970; Lees and Wallis, unpublished observations). When the ganglion is compact, however, a more complex arrangement may exist. It is important to establish the location of particular neurons within a ganglion to map functional connections of the neurons with the structures innervated by their axons and with spinal cord segments of origin of the corresponding preganglionic nerves (Njå and Purves, 1977a,b, 1978; Eccles and Wallis, 1976; Purves and Lichtman, 1978; Lichtman *et al.*, 1979; Dalsgaard and Elfvin, 1979; Elfvin, 1980).

Bowers and Zigmond (1979) used the HRP staining technique to investigate the location of neurons in the rat superior cervical ganglion that project into different postganglionic nerve trunks. About 45% of the total population of neurons sent processes into the external carotid nerve, and their cell bodies were located almost exclusively in the proximal (caudal) part of the ganglion. The axons of the cell bodies located near but distal to the point of origin of the external carotid nerve usually passed into the internal carotid nerve, as did all those originating from the somata in the distal (rostral) part of the ganglion; axons of approximately 30% of neurons coursed in the internal carotid nerve. In the proximal part of the ganglion, however, a population of somata was found with axons projecting into the cervical sympathetic trunk ("preganglionic nerve"). Some fibers in the external carotid nerve come from the middle and inferior cervical ganglia (Bowers and Zigmond, 1981). Similar observations had already been made for the rabbit (Douglas *et al.*, 1960) and cat superior cervical ganglia (Jacobowitz and Woodward, 1968). Not all adrenergic nerves found in the cervical sympathetic trunk at or close to the superior cervical ganglion can be assumed to originate from these cell bodies because, at least in the rabbit, many axons arising from the stellate, middle, and accessory cervical ganglia pass through the superior cervical ganglion to emerge in the external carotid nerve (Dail *et al.*, 1980).

D. Innervation of the Gallbladder

The innervation of the gallbladder has been somewhat controversial from both the histological and the functional viewpoint (see, for example, Davison *et al.*, 1978; Cai and Gabella, 1983). The occurrence of SIF cells in neonatal and adult guinea pig gallbladders has recently been investigated and a semiquantitative study made of the ganglionated plexus of gallbladder and biliary tracts (Cai and Gabella, 1983, 1984).

E. Innervation of the Urinary Bladder

The sympathetic and parasympathetic innervation of the urinary bladder and urethra has been reexamined by means of the HRP retrograde axonal transport technique. Petras and Cummings (1978) found that the axons of sympathetic postganglionic nerves from the caudal mesenteric ganglion form the bulk of the hypogastric nerves and that they course through the pelvic plexus to join axons of sympathetic neurons of the pelvic ganglion before entering the bladder. The parasympathetic preganglionic nerves arise from sacral spinal segments forming the sacral splanchnic nerves that innervate pelvic parasympathetic ganglia and project directly to the bladder to synapse with intramural ganglion cells. When HRP was injected into the bladder, reaction product was found in cranial lumbar preganglionic sympathetic and sacral preganglionic nerves. Affected cells of the cranial lumbar segment were located predominantly in the nucleus intercalatus spinalis (IC L_1–L_4); they were less numerous in the nucleus intermediolateralis thoracolumbalis pars principalis (IL_p) of the same segment. Petras and Cummings (1978) concluded that there was a direct anatomical preganglionic pathway to the urinary bladder from sympathetic preganglionic IC and IL_p neurons. As has been pointed out by Al-Khafaji *et al.*, (1980), however, care is required in interpreting experimental results such as these because of the avid uptake of HRP from the blood by some autonomic neurons. It must be added that certain specialized pelvic ganglia, such as those of the rat, may contain cholinergic and adrenergic neurons that synapse (Dail and Evna, 1978; Dail *et al.*, 1980); this possibility must be further explored (see Chapters 3, 13, and 21).

In the cat, guinea pig, and rat, the openings of the ureters and of the posterior urethra into the trigonum have a particularly rich innervation by nerves containing vasoactive intestinal polypeptide (VIP), as have the openings of the vasa deferentia and prostatic ducts (Alumets *et al.*, 1979).

Recent light- and electron-microscopic studies of human urinary bladder autonomic neurons have confirmed the existence of ganglia in the lateral walls and dome. Unusually, there was a lack of noradrenergic nerve terminals in these ganglia, in addition to a close apposition of neuronal soma membranes without specialization or satellite cell covering (Dixon *et al.*, 1983; Gilpin *et al.*, 1983).

F. Ultrastructural Considerations of Principal Sympathetic Ganglion Cells

1. Inferior Mesenteric Ganglion

In a series of papers, Elfvin (1971a–c) made a number of important observations on the ultrastructure of synapses found in the inferior mesenteric ganglion. He described the occurrence of narrow, short soma processes ("accessory dendrites") containing dense core vesicles of the kind associated with noradrenergic nerves; unlike these processes, the main dendrites did not contain such vesicles. The accessory dendrites formed specialized membrane contacts with dendrites of adjacent neurons and themselves received conventional synaptic inputs from preganglionic nerves (Elfvin, 1971a). Although the concept of interaction between dendrites of neighboring cells is not new, it is of considerable interest, especially for those working with prevertebral ganglia, since the latter are clearly not simple relay stations (Szurszewski, 1980) (see Chapters 3, 12, and 13). In addition, Elfvin (1971a) interpreted other observations as indicating the possible occurrence of axo-axonal synapses in this ganglion. If substantiated, these findings would add a further complexity to this small but highly complicated ganglion. Finally, Elfvin (1971c) reinterpreted the origin of small granules of the type that appear to be characteristically present in the interneurons (Williams, 1967; Williams and Palay, 1969; Matthews and Raisman, 1969) (see above); furthermore, he suggested that the bright, basketlike structures seen in catecholamine fluorescence preparations of this ganglion are not due to networks of noradrenergic nerve terminals but to the catecholamine content of the accessory dendrites; he thus accounted for the persistence of these fluorescent structures after complete decentralization of this ganglion. The importance of these conclusions is discussed below.

2. Superior Cervical Ganglion

In a combined electrophysiological and morphological study of rat superior cervical ganglion cells, Kondo *et al.* (1980) found that one third of neurons labeled with HRP showed varicosities in their dendrites, but not in their intraganglionic axons. Moreover, these varicose dendrites came into very close contact with adjacent nonlabeled nerve processes, a finding that was confirmed at the electron-microscopic level. The varicosities contained clusters of vesicles and were located opposite to the membranes of the other nonlabeled processes that showed a degree of specialization at that point. Kondo *et al.* (1980) suggested, therefore, that dendrodendritic and dendrosomatic synaptic connections exist in this ganglion. The functional significance of these synapses remains to be elucidated.

An additional interesting observation concerns the finding that these dendrites store considerable amounts of catecholamines, probably in clusters of small vesicles (Gabella, personal communication). Such observations are reminiscent of fluorescence histochemical findings of Jacobowitz and Woodward (1968). An example of a typical axodendritic synapse in the rat superior cervical ganglion is shown in Figure 3.

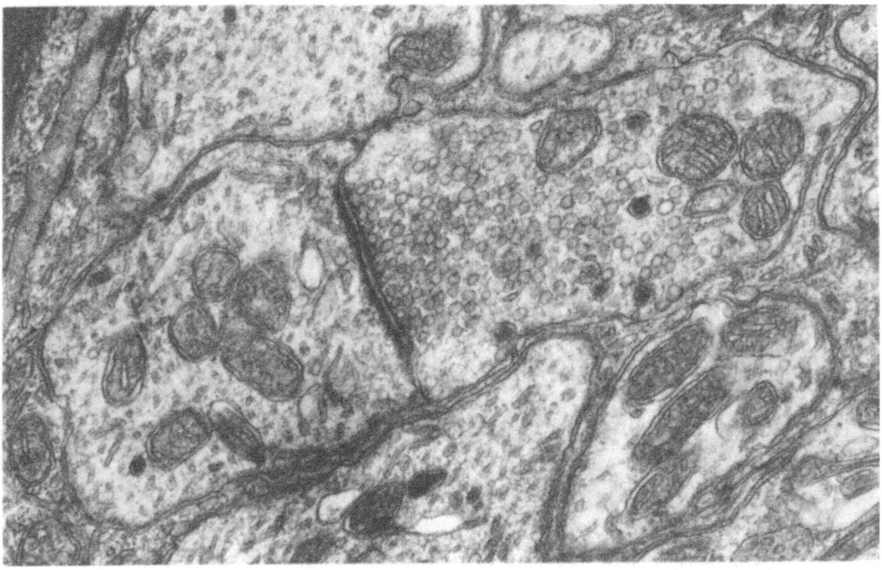

Figure 3. Electron micrograph of the superior cervical ganglion of a rat. A nerve ending of preganglionic origin synapses on a dendrite. The synaptic specializations include the postsynaptic density, the clustering of electron-lucent vesicles on the presynaptic membrane, and a slightly expanded intercellular space. Note that the nerve ending (which is presumably cholinergic) contains, in addition to small granular vesicles and mitochondria, several large granular vesicles. ×43,000. (Reproduced at 82%.) Courtesy of G. Gabella.

III. CHROMAFFIN CELLS, SMALL INTENSELY FLUORESCENT CELLS, SMALL GRANULE-CONTAINING CELLS, INTERNEURONS, PARANEURONS, AND PARAGANGLIA

In addition to ganglionic neurons, sympathetic ganglia contain several other cell types such as fibroblasts, vascular cells, mast cells, glial cells (either satellite cells, wrapping up nerve cell bodies, or Schwann cells, associated with myelinated and nonmyelinated axons), and chromaffin cells. For nearly a century, it has been known that chromaffin cells occur in sympathetic ganglia and that they are innervated by preganglionic fibers. The chromaffin reaction has low sensitivity, however, and it is not surprising that the study of these cells essentially began when they were identified specifically and accurately by the fluorescence histochemical methods for catecholamines. These cells are readily distinguished from the ganglionic neurons by their small size and intense fluorescence. Soon after, the same cell type was identified electron-microscopically by virtue of its content of granule-containing vesicles.

Classification of SIF cells is possible on the basis of their spatial location in a ganglion and their morphological features, as well as on the basis of their fluorescence histochemical characteristics; in addition, they can be classified according to their content of amine-synthesizing enzymes (tyrosine hydroxylase, L-aromatic amino acid decarboxylase, dopamine β-hydroxylase, and phenylethanolamine N-methyltransferase). In other words, for a proper appreciation of their potential physiological significance, it is important to note (1) their subcapsular or interstitial location; (2) the identity of their amine content (norepinephrine, epinephrine, dopamine, or 5-hydroxytryptamine); (3) their proximity and relationship to principal ganglionic neuron somata or processes; (4) their proximity to the microcirculation of the ganglion, particularly their closeness to fenestrated capillaries; and (5) their occurrence in clusters or as single cells (see also Chapter 13). Analysis of their electron-microscopic features is no less complicated. Added to these sources of technical difficulty are, of course, the problems of correctly interpreting the significance of the color and intensity of fluorescence before and after treatment with certain chemicals and drugs (these problems arise mainly from nonlinear relationships between concentration of fluorophore and fluorescence) and of reconstructing the original three-dimensional architecture from serial electron-microscopic sections. Not surprisingly, therefore, opinions differ as to the subtype of cell and its probable functional charcteristics or connections. It is therefore pertinent to ask whether SIF cells or small granule-containing cells release a neurotransmitter and secrete it directly or indirectly onto the neurons (neurocrine function) or secrete directly into the bloodstream or to neighboring cells (endocrine and paracrine functions, respectively) (see also Chapters 9 and 13).

Some cells of this general type have synapses but do not provide efferent ones, while others have both properties and therefore, by definition, are regarded as interneurons (Type I); some authors, however, have denied that this is a correct interpretation of the morphology and function of the SIF cells (Elfvin, 1971a,c). Type II SIF cells are closely associated with blood vessels, which immediately raises the still unanswered question—why? Related to this is the concept of the paraneuron, the criteria of which are that (1) the cell possesses neurosecretion-like and/or synaptic-vesicle-like granules; (2) the cell may produce substances identical or related to neurosecretions or neurotransmitters; and (3) the cell is "receptosecretory" in function; i.e., on receiving an adequate stimulus, it releases its secretory substance(s) as a direct consequence of the stimulus (Fujita *et al.*, 1980). This term should not be confused with "paraganglion," which refers to cells that exhibit a positive chromaffin reaction and are innervated by preganglionic nerves; these paraganglia are often situated close to abdominal sympathetic ganglia or the adrenal gland, the medullary cells of which they closely resemble.

From a functional viewpoint, there has been considerable interest in providing evidence to test the hypothesis that the slow hyperpolarization (slow IPSP) of the principal ganglion cells arises from the action of a catecholamine released from an interneuron, which itself had been excited via a preganglionic cholinergic nerve fiber (Eccles and Libet, 1961). For background information on the morphology of this controversial interloper cell and its relationship to ganglionic neurotransmission, the reader is referred to Chapters 9 and 13 and to papers and reviews by Libet and Owman (1974), Chiba and Williams (1975), Elfvin *et al.* (1975), Williams *et al.* (1975), Weight and Weitsen (1977), Chiba (1977), Eranko (1978), Kojima *et al.* (1978), Autillo-Touati (1979), Eranko *et al.* (1980), Dun (1980), Verhofstad *et al.* (1981), Black *et al.* (1982), Hadjiconstantinou *et al.* (1982), and Abe *et al.* (1983).

Evidence has been presented for the possibility that some sympathetic ganglion neurons are dopaminergic and not noradernergic (see Bell and McLachlan, 1982).

IV. ENTERIC NERVOUS SYSTEM

A. Extrinsic Nerves

Since this topic has been the subject of several recent detailed reviews (Furness and Costa, 1974; Gabella, 1976, 1979, 1981a,b), only selected aspects will be discussed in this chapter (see also Chapter 16).

The important role of the prevertebral ganglia in mediating and modulating (Hirst and McKirdy, 1974b) reflex activity in the intestine has already been mentioned. In the guinea pig, it has been shown that ganglion cells in the coeliac and superior mesenteric ganglion may receive excitatory synaptic inputs from different parts of the colon (Kreulen and Szurszewski, 1979). Afferents from the proximal colon synapse in the coeliac ganglion and those from the distal colon synapse in the inferior mesenteric ganglion, while the superior mesenteric ganglion receives approximately equal mechanosensory inputs from the proximal and distal colon. Correspondingly, the efferent noradrenergic fibers from the coeliac ganglion pass mainly to the proximal colon, whereas the outflow from the inferior mesenteric ganglion mainly provides an inhibitory innervation of the distal part of the colon. These electrophysiological observations are supported by data concerning the distribution in the appropriate ganglia of neurons retrogradely labeled *in vitro* by the HRP method (Szurszewski, 1981).

Another interesting finding concerns probable pathways between parts of prevertebral ganglia. Their presence is postulated from results of autoradiographic studies of the cat solar plexus, in which fibers could be traced from their origin in one of the ganglia of the solar plexus to their termination in a nest of endings around the somata of other principal ganglion cells in another part of the coeliac ganglion (Kelts *et al.*, 1979).

The spinal origin of preganglionic fibers projecting to the inferior mesenteric ganglion has been located by means of HRP labeling techniques (Dalsgaard and Elfvin, 1979). The preganglionic neurons were elongated or round, sometimes multipolar, and 20–40 μm in diameter. Relatively larger numbers of preganglionic fibers originating in the nucleus intercalatus of the spinal cord synapsed in the inferior mesenteric ganglion than in the superior cervical ganglion, which was used for comparison as an example of a paravertebral ganglion. Dalsgaard and Elfvin (1979) made the additional, interesting observation that the axons of neurons in nucleus intermediolateralis (pars principalis and pars funicularis) are myelinated (type B), whereas those from nucleus intercalatus are nonmyelinated (type C). This arrangement suggests that there may be zoning (distinct groupings of neurons) within the population of prevertebral ganglion cells and therefore a differential inhibitory input to parts of the intestine. The spinal arrangement is not unlike that for amphibian paravertebral sympathetic ganglia, in which C-neurons are innervated by cholinergic and noncholinergic preganglionic neurons through the VIIth and VIIIth spinal nerves, whereas B-neurons receive their cholinergic input through the IInd, IIIrd, and IVth spinal nerves (Nishi and Koketsu, 1960; Libet *et al.*, 1968; Yoshimura *et al.*, 1979; Jan *et al.*, 1980). In this connection, it is interesting to note that the myelinated preganglionic fibers innervating the intestine conduct at only 0.5–1.4 $m \cdot s^{-1}$, whereas those driving the urinary bladder (in the same nerve trunk) conduct at 8–10 $m \cdot s^{-1}$ (DeGroat and Krier, 1976).

It is generally considered that parasympathetic preganglionic nerves to the part of the alimentary tract that is located proximal to the splenic flexure emerge from somata in the medulla oblongata, the distal colon and rectum being innervated from the sacral spinal cord. With HRP labeling, it was found, however, that in the cat there is no clear line of demarcation, there being a gradual replacement of the cranial parasympathetic innervation by sacral fibers within the descending colon and upper rectum (Satomi *et al.*, 1978). The topographic representation of the stomach and duodenum was found to be in the medial aspect of the dorsal nucleus; in contrast, the intestine distal to the duodenum was innervated from the lateral aspect of the rostral dorsal vagal nucleus. It has also been shown (Lundberg *et al.*, 1978a; Ahlman *et al.*, 1978) that HRP, injected into many sites in the wall of the duodenum and first third of the jejunum of the cat and guinea pig, was retrogradely transported in the vagus nerve not only to the nodose ganglia, but also to the superior and middle cervical, stellate, and thoracic ganglia. The widespread distribution was not the result of leakage of HRP into the circulation and subsequent uptake by neurons. Efferent fibers in the abdominal vagus nerve, whether cholinergic, noradrenergic, or containing peptides (Lundberg *et al.*, 1978b; 1979a), constitute only about 10% of the total vagal fiber population (Evans and Murray, 1954; Agostini *et al.*, 1957).

The localization of the cell bodies of parasympathetic preganglionic and sympathetic neurons innervating the lower esophageal spinchter of the cat has also been investigated by means of retrograde axonal transport of HRP (Niel *et al.*, 1980). Labeled cell bodies were identified not only as expected, in the medulla, the stellate, and other thoracic ganglia, but also in the coeliac ganglion, in which they had a widespread distribution. A similar approach has been used by Elfvin and Lindh (1982) to investigate the extrinsic innervation of the guinea pig pylorus.

In the guinea pig small intestine, substance-P-containing extrinsic nerve fibers form arborizations around arterioles of the submucosa and nerve cell bodies of the submucous plexus (Furness *et al.*, 1980; Costa *et al.*, 1981). The origin of these fibers is not known, but is likely to be identical to that of the noradrenergic nerves that have a similar distribution in this tissue (Furness *et al.*, 1980).

B. Intrinsic Nerves

The complexity of the "little brains along the intestine" has been recognized for nearly a century, but it is only in the past few years that major studies have been undertaken to explore the organization of the intrinsic plexuses (see, for example, Gershon and Erde, 1981). The application of a systematic approach to the morphological arrangements of the neurons and their processes—whether by intracellular staining with fluorescent dyes (Hodgkiss and Lees, 1978, 1980, 1983; Gray *et al.*, 1980; Erde *et al.*, 1980; Bornstein *et al.*, 1983a, 1984) or by immunohistochemical techniques (Furness and Costa, 1979; Costa *et al.*, 1980c, 1981; Furness *et al.*, 1980)—has provided a possible morphological basis for known physiological events characterized by a preferred or mandatory orientation (Bayliss and Starling, 1899; Hirst and McKirdy, 1974a, 1975; Hirst *et al.*, 1975; Yokoyama and Ozaki, 1978).

The functional importance of the variation in the general architecture of the intramural plexuses in different parts of the small and large intestine (Irwin, 1931; Gabella, 1979) is not yet clear. The irregularity of the size and shape of the ganglia (Figure 4) makes it unlikely that neurons with particular electrophysiological, biochemical, pharmacological, or immunohistochemical characteristics are arranged in an obvious and regularly recurring pattern. On the other hand, it is clear that neurons do not synapse

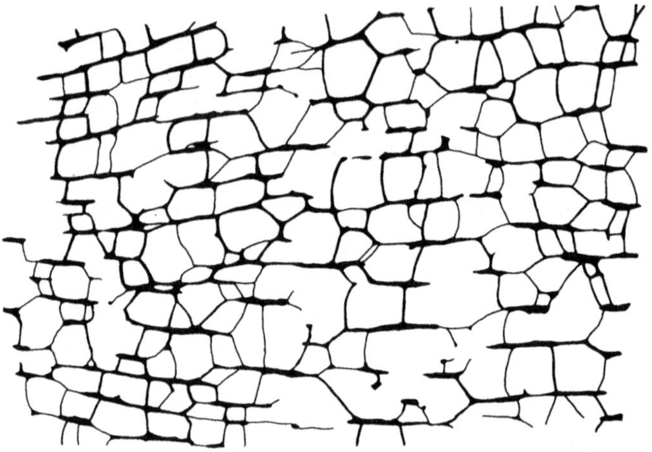

Figure 4. Drawing of the myenteric plexus of the guinea pig ileum, obtained from a stretch preparation, to illustrate the distribution of ganglia and connecting strands. ×11. (Reproduced at 65%.) Courtesy of G. Gabella.

with each other in a haphazard way, but nothing is known of the factors that influence the establishment and maintenance of synapses, particularly in the course of the maturation of the animal, during which neuronal numbers decline.

In the case of the guinea pig myenteric plexus, many neurons are sensitive to acetylcholine (Nishi and North, 1973; Hirst *et al.*, 1974), 5-hydroxytryptamine (Wood and Mayer, 1979), and substance P (Katayama *et al.*, 1979), but it is not firmly established how these neurons are interrelated (Franco *et al.*, 1979b; North *et al.*, 1980). There are two broad electrophysiological categories of neurons; S-neurons have conventional fast EPSPs, which are cholinergic in origin, whereas AH-neurons exhibit large fast EPSPs and uniquely long-lasting after-hyperpolarizations that follow the firing of soma action potentials (see also Chapter 16).

Although S-neurons do not have a uniform morphology, most of them have a soma that is irregular in appearance, being endowed with many short, thick processes; it generally exhibits one long process (Figure 5). The long process may be varicose and usually runs aborally or, if there are other long processes, both aborally and circumferentially (Hodgkiss and Lees, 1978, 1980, 1983; Bornstein *et al.*, 1983, 1984). In contrast, the other S-neurons tend to have a smooth soma with finer short processes that are distributed circumferentially. Generally, AH-neurons have a large smooth soma membrane with a few short, fine processes and several branching long processes. Even within the AH-neuron category, the morphology of substance-P-sensitive cells was widely variable. Unfortunately, however, it has not been possible to establish a completely reliable but simple classification to correlate electrophysiological and pharmacolog-

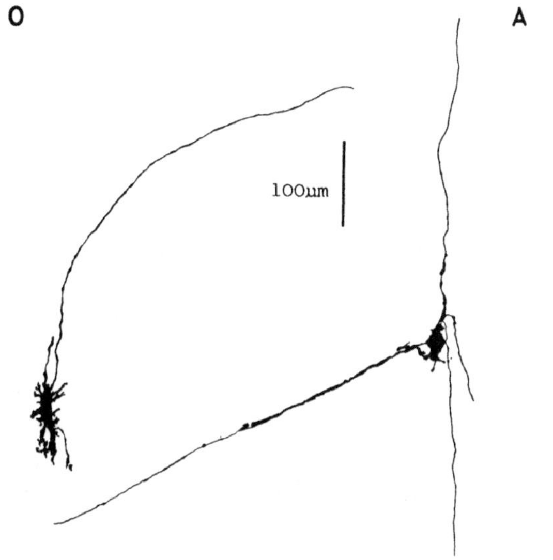

Figure 5. Morphology of guinea pig myenteric neurons revealed by intracellular microiontophoretic application of procion yellow after electrophysiological identification (tracings of photographic montages). (O) Oral; (A) aboral. *Left:* S-neuron. Note the many large, thick processes and a single, aborally directed process, which reached the adjacent ganglion. *Right:* AH-neuron. This neuron is an unusual example of the AH-type in having (1) an exceptionally long orally directed process, (2) a soma with thick processes, (3) nonbranching long processes, and (4) a soma that is no larger than that of an S-neuron (Gray, Hodgkiss, and Lees, unpublished observations).

ical properties with soma shape and distribution of long processes because there were so many exceptions to these generalizations (see also Bornstein *et al.*, 1984). A contributory factor to the variability is likely to be the degree of stretch applied to such preparations (Gabella and Trigg, 1984).

Analysis of the projection of long soma processes marked with procion yellow or lucifer yellow showed that although there were three to four times as many circumferential processes as longitudinal ones, there were at least twice as many aborally directed processes as orally directed ones (Hodgkiss and Lees, 1980, 1983; Gray and Lees, unpublished observations). These latter findings are in accord with electrophysiological data concerning long pathways in the myenteric plexus (Hirst *et al.*, 1975; Hirst and McKirdy, 1974a; Hodgkiss, 1981) and with the immunohistochemical observations of Furness and Costa. The latter authors have examined in whole mounts the location of somata and the projections of processes of guinea pig myenteric and submucous plexus neurons exhibiting immunoreactivity to substance P, VIP, somatostatin, gastrin-releasing peptide, bombesin, neuropeptide Y (NPY), cholecystokinin, enkephalins, and 5-hydroxytryptamine (Furness and Costa, 1979, 1982; Franco *et al.*, 1979a; Furness *et al.*, 1980, 1981, 1983a,b, 1984; Costa *et al.*, 1980a–c, 1981, 1982, 1984; Costa and Furness, 1982, 1983; Keast *et al.*, 1984). Similar results have also been obtained with slightly different techniques in other species (see Schultzberg *et al.*, 1980; Jessen *et al.*, 1980b; Malmfors *et al.*, 1980; Bishop *et al.*, 1982; Feher and Leranth, 1983).

In summary, the peptide-positive nerves were almost exclusively of intrinsic origin; they could be easily demonstrated, whereas visualization of their cell bodies required treatment (*in vivo*) with vinblastin or colchicine to impair intraneuronal transport of the peptide. Substance-P-like immunoreactivity was found in 3.6% of cell bodies of the myenteric plexus and in 11.3% of those of the submucous plexus of guinea pig ileum. Five complex types of projections of myenteric substance-P-positive neurons were established: (1) very short projections ending either within the same row of ganglia or within adjacent rows on both oral and aboral sides; (2) longer projections within the myenteric plexus; (3) very short projections to the circular muscle plexus and a denser, deep muscle plexus; (4) projections to the submucous plexus ganglia, forming pericellular baskets of varicose fibers; and (5) projections to the villi, though the latter are mainly innervated from the submucous plexus. The submucous plexus therefore exhibits substance-P-positive axons from three sources, namely, submucous neurons, myenteric neurons, and extrinsic fibers, the latter being distinguishable by their coarser varicosities (Costa *et al.*, 1981). The ultrastructural localization of substance-P-containing nerves in cat, guinea pig, and human small intestine has recently been described (Feher and Wenger, 1981; Llewellyn-Smith *et al.*, 1984 and personal communication).

In tissue-culture preparations, substance-P-containing nerve cell bodies were found only in the myenteric plexus of guinea pig cecum, while there endings were located principally in the submucous plexus (Jessen *et al.*, 1980a). In contrast, in the same study, cell bodies containing VIP-like immunoreactivity were found to be located mainly in the submucous plexus. It would seem that the cultured neurons may not provide a reliable model for the *in vivo* disposition because it was subsequently found that in normal small intestine, VIP neurons were common in the myenteric plexus and found only occasioanlly in the submucous plexus (Jessen *et al.*, 1980b). VIP-positive fibers were common in the myenteric plexus, often surrounding neuronal cell bodies (immunonegative for VIP), rarely seen in the longitudinal muscle, and distributed throughout the circular muscle, the innermost part of which had a rich innervation of this fiber type.

Leander *et al.* (1981) have studied the occurrence and effects of substance P, VIP, somatostatin, and enkephalin in the myenteric plexus of guinea pig taenia caeci. They found that nerve fibers displaying substance-P-, VIP-, and enkephalinlike immunoreactivity were numerous, whereas there were only a few somatostatin-positive fibers. Nerve cell bodies containing substance P or VIP were commonly seen, but there were very few enkephalin-containing somata and no somatostatin-containing cell bodies. On the basis of morphological as well as pharmacological data, these authors concluded that (1) substance P neurons are probably excitatory motor neurons, although some may act as sensory or interneurons; (2) VIP neurons are probably inhibitory motor neurons, but some may act as sensory or interneurons; (3) enkephalin neurons are interneurons or sensory neurons; and (4) somatostatin neurons probably act as interneurons in the myenteric plexus.

The finding by several workers of substance-P- and VIP-containing nerves in the deeper parts of the circular muscle are reminiscent of the dense noradrenergic and nonadrenergic innervation of the special muscle cell layer present in mammalian small intestine (Gabella, 1974), but are probably unrelated to the rich intrinsic innervation, which is likely to be afferent in nature. On the other hand, some of the extrinsic substance P fibers might be involved in producing a motor input to these special smooth muscle cells.

In the submucous plexus, VIP-positive fibers were mainly confined to the ganglia; elsewhere, apart from the lamina propria of the mucosa, the innervation was scanty and mainly associated with blood vessels. Essentially, the same observations have been made by Costa *et al.* (1980b); in their study, 45% of submucous plexus neurons of ileum stained positively for VIP, whereas only 2.5% of myenteric plexus neurons did so. Furthermore, these authors describe a rich innervation of mucosal blood vessels, a finding in keeping with the results of the recent experiments on the involvement of VIP in reflex vasodilation in the intestine (Eklund *et al.*, 1980). The discrepancies in results obtained with different experimental models and procedures should be resolved to allow assessment of the possible physiological implications of the results.

With respect to somatostatin-containing neurons and processes, 4.7% of myenteric plexus cells and 17.4% of submucous plexus cells reacted positively. In the myenteric plexus, there was a distinct aboral orientation of the processes, which arborized around other myenteric neurons, some being somatostatin-positive; Costa *et al.* (1980c) therefore postulated that these neurons might constitute a descending nervous pathway mediating descending inhibition associated with peristalsis or distension of the gut. There was no obvious orientation of the fibers from somatostatin neurons of the submucous plexus.

Two other distributions of VIP fibers are of note. Some fibers originating in the intramural ganglia project to prevertebral ganglia (Hökfelt *et al.*, 1978a). Second, in the cat, guinea pig, and rat, the musculature of the gastroesophageal junction, pylorus, and sphincter of Oddi has an unusually rich innervation by VIP fibers (Alumets *et al.*, 1979; Schultzberg *et al.*, 1980). Since VIP is thought to relax the smooth muscle associated with these nerves, this rich innervation probably accounts for the inhibitory innervation of these tissues. Substance P- and enkephalinlike immunoreactivity was found in nerves in the same areas.

From an examination of the occurrence, location, and projection of neurons containing material immunoreactive with the peptides VIP, somatostatin, substance P, and enkephalin and those containing the necessary enzymes and uptake mechanisms for handling amines (Costa and Furness, 1979), Furness and Costa (1980) have accounted for nearly 80% of submucous neurons of guinea pig ileum and concluded that most of the remainder might be cholinergic.

Four questions arise, assuming that there were no false-positive reactions among the neurons supposed to contain a particular peptide (i.e., the demonstrated immunoreactivity was not that due to some closely related peptide sequence). First, how accurate are the predicted proportions of immunoreactive somata? Second, to what extent do neurons react positively for more than one peptide? Third, can some of the peptide-containing neurons synthesize and store another neurotransmitter substance or modulator? Fourth, how may cholinergic neurons be positively identified?

The bases for the proportional representation of the richness of innervation are the counts of immunopositive somata in ganglia of myenteric and submucous plexuses and the total number of neurons in the ganglia examined. Unfortunately, however, the latter figure is merely surmised on statistical grounds, there being, on average, 43 neurons in myenteric ganglia and 8 neurons in submucous ganglia (Wilson *et al.*, 1981a). Quite apart from the vexing problem of recognizing weakly positive cells, a further complication stems from the lack of a clear definition of a ganglion; quite often, it is difficult to decide when one ganglion ends and another begins. The reader's attention is drawn to the fact that in neither type of ganglion is the frequency distribution of number of neurons normal. Unfortunately, however, since the data are incompletely stated, one can only estimate that the medians of the populations are not more than 40 and 6 and the modes are 30 (or less) and 4 or 5, respectively, for myenteric and submucous plexus ganglia. Thus, the use of means would appear to be quite inappropriate. Furthermore, the citation of percentages may give a misleading picture of the true frequency distribution of classes of neurons, particularly when one considers the possibility of two or more immunoreactive substances being colocated (see below).

With respect to the remaining questions, important information has recently become available. Although it seemed initially that few enteric neurons of fibers would react positively to two or more peptide antibodies (Schultzberg *et al.*, 1980), recent evidence indicates that such reactions are not at all uncommon in the myenteric plexus (Furness and Costa, personal communication) and a frequent phenomenon in submucous plexus neurons (Furness *et al.*, 1984), just as such a possibility has been recognized in the case of other nerves (Burnstock, 1976; Hökfelt *et al.*, 1977, 1978b; Chan-Palay *et al.*, 1978; Lundberg *et al.*, 1979b; Klein *et al.*, 1982 Olschowka and Jacobowitz, 1983; Ekblad *et al.*, 1984). For example, cholecystokinin is colocated with somatostatin and NPY with somatostatin, there being a complete overlap between these groups. Furthermore, these neurons also contain choline acetyltransferase, the enzyme essential for acetylcholine synthesis. Problems in obtaining good immunoreactivity to antibodies raised against this enzyme have so far prevented accurate assessments of the distribution of cholinergic neurons in the myenteric plexus, although a preliminary account has been given (Furness *et al.*, 1983c).

The functional significance of the tremendous variety of neuronal and nonneuronal ultrastructural features to be found in the plexuses (Cook and Burnstock, 1976a,b; Furness and Costa, 1980; Gabella, 1981b; Llewellyn-Smith *et al.*, 1981; Wilson *et al.*, 1981a,b) remains to be unraveled. The size of neurons and glial cells in the enteric ganglia of mice, rabbits, sheep, and guinea pigs has recently been examined (Gabella and Trigg, 1984). The wide range in neuronal size is remarkable, as is the capability of both myenteric and submucous neurons to hypertrophy in response to moderate intestinal obstruction of only a few weeks' duration (Gabella, 1984). The causative factors and functional significance of such changes remain to be elucidated.

C. Glial Cells

In view of the large numbers of glial cells relative to neurons (Gabella, 1972), it is surprising that they were not encountered more frequently during intracellular electrophysiological recording (Hodgkiss and Lees, 1983). Probably the explanation lies in their small size, delicacy of structure, and largely unknown electrophysiological characteristics (Figure 6); thus, the microelectrodes may have been inserted, occasionally or even frequently, into glial cells, but the fact was not recognized. In some recent experiments, in which it has been possible to visualize simultaneously myenteric ganglion cells and the morphology of one of the neurons when intracellular electrodes were used for both recording and staining (as the electrode was filled with a solution of a fluorescent dye), slight swellings have been noted as occurring around and sometimes over the surface of the neuron. The dye, however, did not leak into these swollen cell entities, but could be seen to enter one or two of these apparently extraneuronal swellings when the microelectrode was just withdrawn from the neuron while still staining with lucifer yellow applied by means of pressure ejection or passage of current. These observations are tentatively interpreted as indicating that glial cells, which surround the whole ganglion (Gabella, 1972, 1981a) (Figure 7), may be damaged during the advancement of the microelectrode while attempting to impale a neuron. From these and other experiments, it was concluded that dye-coupling between neurons and between neurons and other cell types, including glia, is rare.

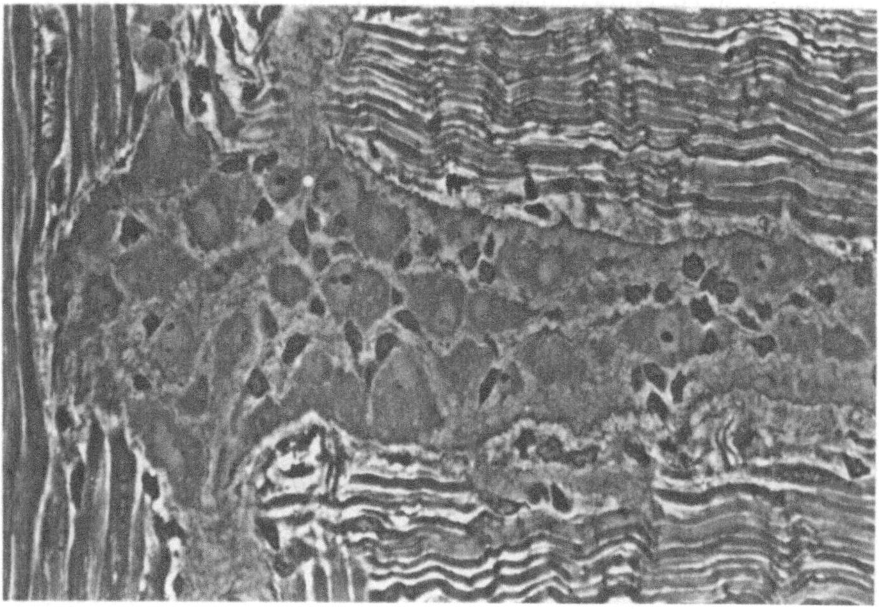

Figure 6. Tangential section of the wall of the guinea pig ileum. Araldite section stained with toluidine blue. To the left is the longitudinal muscle, to the right the circular muscle, both cut longitudinally. In the middle is a large ganglion of the myenteric plexus, with more than 30 neurons and an even greater number of glial cells. Of the latter cells, which are much smaller than the neurons, mainly the darkly stained nucleus is visible. ×675. (Reproduced at 82%.) Courtesy of G. Gabella; reproduced from Gabella (1981c) with permission.

V. PEPTIDES IN OTHER AUTONOMIC GANGLIA

The enteric nervous system is not unique in having peptide-containing nerves surrounding ganglion cell bodies; not only prevertebral but also paraverterbral sympathetic and intracardiac ganglia have such fibers, but the physiological importance of this arrangement is not fully understood. The pattern of these peptide-containing nerve fibers, especially in the prevertebral ganglia, is extremely complex. The reader is referred to

Figure 7. Ganglion of the myenteric plexus of the guinea pig ileum with part of the adjacent circular musculature (far right). In the center is a ganglion neuron with its nucleus and an extensive perikaryon that reaches the surface of the ganglion, where it lies immediately beneath the basal lamina. A densely packed neuropil formed by dendrites, axons, and glial processes lies around the nerve cell body. Around the ganglion are a basal lamina, abundant collagen fibrils and elastic fibers, and processes of interstitial cells. ×16,000. (Reproduced at 77%.) Courtesy of G. Gabella; reproduced from Gabella (1981b) with permission.

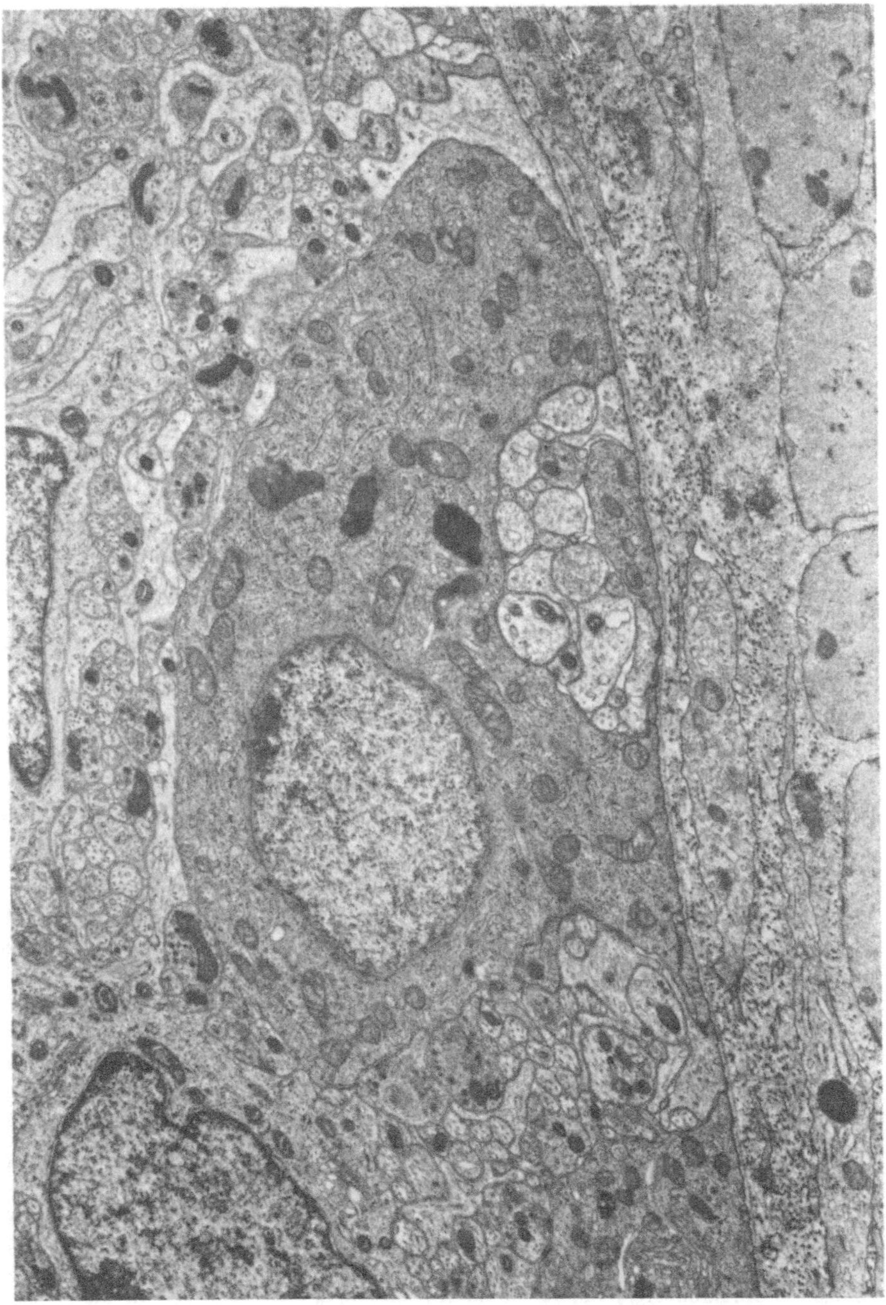

studies by Schultzberg *et al.* (1979), Hökflet *et al.* (1980), Campbell *et al.* (1982), Dalsgaard *et al.* (1982, 1983a,b), Kondo and Yui (1981, 1982), and Matthews and Cuello (1984) (see also Chapters 2 and 16).

In a detailed study of the innervation of the cat pancreas by peptidergic and noradrenergic nerves, Larsson and Rehfeld (1979) only rarely found such nerves in the vicinity of exocrine cells, the vast majority of nerves staining positively for VIP, substance P, enkephalin, and COOH-terminal gastrin/cholecystokinin being located in pancreatic ganglia, where they appared to innervate ganglion cell somata. The ganglion cells themselves did not stain for these peptides. This latter finding is in contrast with the results obtained with dog pancreas, in which there are VIP-containing cell bodies; these give off processes to islet cells, blood vessels, and the acini. Such species differences are important in interpreting responses of acinar cells to electric transmural stimulation and to the local application of peptides; since the peptides do not share a common mode of action, a response of acinar cells to noncholinergic, nonadrenergic nerve stimulation could easily be missed (Pearson *et al.*, 1981).

A peptide of a different character, namely, luteinizing-hormone-releasing hormone or a very closely related peptide, has been found to mimic the noncholinergically mediated late slow EPSP of bullfrog sympathetic ganglia, and immunopositive reactions have been found in synaptic boutons on C-neurons of the 9th and 10th paravertebral ganglia (Jan *et al.*, 1980, 1983) (see Chapters 8 and 13). Although the authors thought that the active substance must diffuse from these sites to B-neurons, to account for the late slow EPSP in these cells, they found that most immunopositive terminals were localized around B-cells and not C-cells in decentralized ganglia. An explanation for this unexpected result is awaited with interest.

ACKNOWLEDGMENT. I wish to thank Dr. G. Gabella, Department of Anatomy and Embryology, University College London, for many helpful discussions and assistance in the preparation of this chapter.

REFERENCES

Abe, H., Watanabe, H., and Yamamoto, T. Y.: Relationship between granule-containing cells and blood vessels in the rat autonomic ganglia. *Anat. Rec.* **205**:65–72 (1983).

Agostini, E., Chinnock, J. E., Daly, M. D. B., and Murray, J. G.: Functional and histological studies of the vagus nerve and its branches to the heart, lungs, and abdominal viscera in the cat. *J. Physiol. (London)* **135**:182–205 (1957).

Ahlman, B. H., Lundberg, J. M., Dahlstrom, A., Larsson, I., Pettersson, G., Kewenter, J., and Nyhus, L. M.: Evidence for innervation of the small intestine from the cervical sympathetic ganglia. *J. Surg. Res.* **24**:142–149 (1978).

Al-Khafazi, F. A. H., Anderson, P. N., Mitchell, J., and Mayor, D.: The uptake of intravenous horseradish peroxidase by the guinea-pig inferior myenteric ganglion. *J. Anat.* **130**:883–889 (1980).

Alumets, J., Schaffalitzkyde Muckadell, O., Fahrenkrug, J., Sundler, F., Hakanson, R., and Uddman, R.: A rich VIP nerve supply is characteristic of sphincters. *Nature (London)* **280**:155–156 (1979).

Autillo-Touati, A.: A cytochemical and ultrastructural study of the "S.I.F." cells in cat sympathetic ganglia. *Histochemistry* **60**:189–223 (1979).

Baljet, B., and Drukker, J.: The extrinsic innervation of the abdominal organs in the female rat. *Acta Anat.* **104**:243–267 (1979).

Bayliss, W. M., and Starling, E. H.: The movements and innervation of the small intestine. *J. Physiol. (London)* **24**:99–143 (1899).

Bell, C., and McLachlan, E. M.: Dopaminergic neurons in sympathetic ganglia of the dog. *Proc. R. Soc. London Ser. B* **215**:175–190 (1982).

Bishop, A. E., Ferri, G. L., Probert, L., Bloom, S. R., and Rolak, J.: Peptidergic nerves. *Scand. J. Gastroenterol.* (Supp.) **71**:43–59 (1982).

Black, A. C., Jr., Wamsley, J. K., Sandquist, D., and Williams, T. H.: The guinea pig further demonsrates that rodent superior cervical ganglia differ from those of other species. Exp. *Neurol.* **77**:314–323 (1982).

Bornstein, J. C., Costa, M., Furness, J. B., and Lees, G. M.: Electrophysiology and enkephalin immunoreactivity of identified myenteric plexus neurones of guinea-pig small intestine. *J. Physiol. (London)* **351**:313–325 (1984).

Bornstein, J. C., Lees, G. M., Costa, M. and Furness, J. B.: Morphology of intracellularly-stained guinea-pig myenteric neurones correlated with enkephalin immunoreactivity. *Neurosci. Lett. Suppl.* **11**:31 (1983).

Bowers, C. W., and Zigmond, R. E.: Localization of neurons in the rat superior cervical ganglion that project into different postganglionic trunks. *J. Comp. Neurol.* **185**:381–392 (1979).

Bowers, C. W., and Zigmond, R. E.: Sympathetic neurons in the lower cervical ganglia send axons through the superior cervical ganglion, *Neuroscience* **6**:1783–1791 (1981).

Burnstock, G.: Do some nerve cells release more than one transmitter? *Neuroscience* **1**:239–248 (1976).

Cai, W.-Q., and Gabella, G.: Innervation of the gall bladder and biliary pathways in the guinea-pig. *J. Anat.* **136**:97–109 (1983).

Cai, W.-Q., and Gabella, G.: Catecholamine-containing cells in the nerve plexus of the guinea pig gallbladder. *Acta Anat.* **119**:10–17 (1984).

Campbell, G., Gibbins, I. C., Morris, J. C., Furness, J. B., Costa, M., Oliver, J. R., Beardsley, A. M., and Murphy, R.: Somatostatin is contained in and released from cholinergic nerves in the heart of the toad *Bufo marinus. Neuroscience* **1**:2013–2023 (1982).

Chan-Palay, V., Jonsson, G., and Palay, S. I.: Serotonin and substance P coexist in neurons of the rat's central nervous system. *Proc. Natl. Acad. Sci. U.S.A.* **75**:1582–1586 (1978).

Chiba, T.: Monoamine-containing paraneurons in the sympathetic ganglia of mammals. *Arch. Histol. Jpn.* **40**(Suppl.):163–176 (1977).

Chiba, T., and Williams, T. H.: Histofluorescence characteristics and quantification of small intensely fluorescent (SIF) cells in sympathetic ganglia of several species. *Cell Tissue Res.* **162**:331–341 (1975).

Cook, R. D., and Burnstock, G.: The ultrastructure of Auerbach's plexus in the guinea-pig. I. Neuronal elements. *J. Neurocytol.* **5**:171–194 (1976a).

Cook, R. D., and Burnstock, G.: The ultrastructure of Auerbach's plexus in the guinea-pig. II. Non-neuronal elements. *J. Neurocytol.* **5**:195–206 (1976b).

Costa, M., and Furness, J. B.: On the possibility that an indoleamine is a neurotransmitter in the gastrointestinal tract. *Biochem. Pharmacol.* **28**:565–571 (1979).

Costa, M., and Furness, J. B.: Neuronal peptides in the intestine, in: *Regulatory Peptides of Gut and Brain. Br. Med. Bull.* **38**:247–252 (1982).

Costa, M., and Furness, J. B.: The origins, pathways and terminations of neurons with VIP-like immunoreactivity in the guinea-pig small intestine. *Neuroscience* **8**:665–676 (1983).

Costa, M., Cuello, A. C., Furness, J. B., and Franco, R.: Distribution of enteric neurons showing immunoreactivity for substance P in the guinea-pig ileum. *Neuroscience* **5**:323–331 (1980).

Costa, M., Furness, J. B., Buffa, R., and Said, S. I.: Distribution of enteric nerve cell bodies and axons showing immuno-reactivity for vasoactive intestinal polypeptide in the guinea-pig intestine. *Neuroscience* **5**:587–596 (1980).

Costa, M., Furness, J. B., Llewellyn-Smith, I. J., Davies, B., and Oliver, J.: An immunohistochemical study of the projections of somatostatin-containing neurons in the guinea-pig intestine. *Neuroscience* **5**:841–852 (1980c).

Costa, M., Furness, J. B., Llewellyn-Smith, I. J., and Cuello, A. C.: Projections of substance P-containing neurons within the guinea-pig small intestine. *Neuroscience* **6**:411–424 (1981).

Costa, M., Furness, J. B., Cuello, A. C., Verhofstad, A. A. J., Steinbusch, H. W. J., and Elde, R. P.: Neurons with 5-hydroxytryptamine-like immunoreactivity in the enteric nervous system: Their visualization and reactions to drug treatment. *Neuroscience* **7**:351–363 (1982).

Costa, M., Furness, J. B., Yanaihara, N., Yanaihara, C., and Moody, C. W.: Distribution and projections of neurones with immunoreactivity for both gastrin-releasing peptide and bombesin in the guinea-pig small intestine. *Cell Tissue Res.* **235**:285–293 (1984).

Courtney, K., and Roper, S.: Sprouting of synapses after partial denervation of frog cardiac ganglion. *Nature (London)* **259**:317–319 (1976).

Crowcroft, P. J., Holman, M. E., and Szurszewski, J. H.: Excitatory input from the distal colon to the inferior mesenteric ganglion in the guinea-pig. *J. Physiol.* **219**:443–461 (1971).

Dail, W. G., and Evan, A. P.: Ultrastructure of adrenergic terminals and SIF cells in the superior cervical ganglion of the rabbit. *Brain Res.* **148**:469–477 (1978).

Dail, W. G., Khoudary, A., Barraza, C., Murray, H. M., and Bradley, C.: The fate of adrenergic fibers which enter the superior cervical ganglion, in: Histochemistry and Cell Biology of Autonomic Neurons, SIF Cells and Paraneurons (O. Eranko, S. Soionila, H. Paivarinto, eds.), Advances in Biochemical Psychopharmacology, Vol. 25, pp. 287–297, Raven Press, New York (1980).

Dalsgaard, C.-J., and Elfvin, L.-G.: Spinal origin of preganglionic fibers projecting onto the superior cervical ganglion and of mesenteric ganglion of the guinea-pig, as demonstrated by the horseradish peroxidase technique. *Brain Res.* **172**:139–143 (1979).

Dalsgaard, C.-J., Hokfelt, T., Elfvin, L.-G., and Terenius, L.: Enkephalin containing sympathetic preganglion neurones projecting to the inferior mesenteric ganglion: Evidence from combined retrograde tracing and immunohistochemistry. *Neuroscience* **1**:2039–2050 (1982).

Dalsgaard, C.-J., Hokfelt, T., Schultzberg, M., Lundberg, J. M., Terenius, L., Dockray, G. J., and Goldstein, M.: Origin of peptide-containing fibers in the inferior mesenteric ganglion of the guinea-pig: Vasoactive intestinal polypeptide, cholecystokinin and bombesin. *Neuroscience* **9**:191–211 (1983a).

Dalsgaard, C.-J., Vincent, S. R., Hokfelt, T., Christensson, I., and Terenius, L.: Separate origins for the dynorphin and enkephalin immunoreactive fibers in the inferior mesenteric ganglion of the guinea pig. *J. Comp. Neurol.* **221**:482–489 (1983a).

Davison, J. S., and Grundy, D.: Modulation of single vagal efferent fiber discharge by gastrointestinal afferents in the rat. *J. Physiol. (London)* **284**:69–82 (1978).

Davison, J. S., Al-Hassani, M., Crowe, R., and Burnstock, G.: The non-adrenergic, inhibitory innervation of the guinea-pig gallbladder. *Pfluegers Arch.* **77**:43–49 (1978).

DeGroat, W. C., and Krier, J.: An electrophysiological study of the sacral parasympathetic pathway to the colon of the cat. *J. Physiol. (London)* **260**:425–445 (1976).

Dennis, M. J., and Sargent, P. B.: Multiple innervation of normal and re-innervated parasympathetic neurones in the frog cardiac ganglion. *J. Physiol. (London)* **281**:63–75 (1978).

Dennis, M. J., Harris, A. J., and Kuffler, S. W.: Synaptic transmission and its duplication by focally applied acetylcholine in parasympathetic neurons in the heart of the frog. *Proc. R. Soc. London Ser. B.* **177**:509–539 (1971).

Dixon, J. S., Gilpin, S. A., Gilpin, C. J., and Gosling, J. A.: Intramural ganglia of the human urinary bladder. *Br. J. Urol.* **55**:195–198 (1983).

Dodd, J., and Horn, J. P.: A reclassification of B and C neurones in the ninth and tenth paravertebral sympathetic ganglia of the bullfrog. *J. Physiol. (London)* **344**:255–269 (1983).

Dogiel, A. S.: Zwei Arten sympathischer Nervenzellen. *Anat. Anz.* **11**:679–687 (1896).

Douglas, W. W., Lywood, D. W., and Straub, R. W.: On the excitant effect of acetylcholine on structures in the preganglionic trunk of the cervical sympathetic: With a note on the anatomical complexities of the region. *J. Physiol. (London)* **153**:250–264 (1960).

Dun, N. J.: Ganglionic transmission: Electrophysiology and pharmacology. *Fed. Proc. Fed. Am. Soc. Exp. Biol.* **39**:2982–2989 (1980).

Ebbesson, S. O. E.: Quantitative studies of superior cervical sympathetic ganglia in a variety of primates including man. I. The ratio of preganglionic fibers to ganglionic neurons. *J. Morphol.* **124**(1):117–132 (1968).

Eccles, R. M., and Libet, B.: Origin and blockade of the synaptic responses of curarized sympathetic ganglia. *J. Physiol. (London)* **157**:484–503 (1961).

Eccles, R., and Wallis, D. I.: Characteristics of the sympathetic innervation of the nictitating membrane and the vasculature of the nose and tongue of the cat. *J. Neural Transm.* **39**:113–130 (1976).

Ekblad, E., Edvinsson, L., Wahlesttedt, C., Uddman, R., Hakanson, R., and Sundler, F.: Neuropeptide Y co-exists and co-operates with noradrenaline in perivascular nerve fibers. *Regulatory Peptides* **8**:225–235 (1984).

Eklund, S., Fahrenkrug, J., Jodal, M., Lundgren, O., Schaffalitzky de Muckadell, O. B., and Sjoqvist, A.: Vasoactive intestinal polypetide, 5-hydroxytryptamine and reflex hyperaemia in the small intestine of the cat. *J. Physiol. (London)* **302**:549–557 (1980).

Elfvin, L.-G.: Ultrastructural studies on the synaptology of the inferior mesenteric ganglion of the cat. I. Observations on the cell surface of the postganglionic perikarya. *J. Ultrastruct. Res.* **37**:411–425 (1971a).

Elfvin, L.-G.: Ultrastructural studies on the synaptology of the inferior mesenteric ganglion of the cat. II. Specialized serial neuronal contacts between preganglionic end fibers. *J. Ultrastruct. Res.* **37**:426–431 (1971b).

Elfvin, L.-G.: Ultrastructural studies on the synaptology of the inferior mesenteric ganglion of the cat. III. The structure and distribution of the axodendritic and dendodendritic contacts. *J. Ultrastruct. Res.* **37**:432–448 (1971c).

Elfvin, L.-G., Hökfelt, R., and Goldstein, M.: Fluorescence microscopical, immunohistochemical and ultrastructural studies on sympathetic ganglia of the guinea-pig, with special reference to the SIF cells and their catecholamine content. *J. Ultrastruct. Res.* **51**:377–396 (1975).

Elfvin, L.-G.: Morphology studies on central and peripheral connections of sympathetic ganglila, in: *Histochemistry and Cell Biology of Autonomic Neurons, SIF Cells and Paraneurons* (O. Eranko, S. Soinila, and H. Paivarinta, eds.), *Advances in Biochemical Psychopharmacology*, Vol. 25, pp. 335–340, Raven Press, New York (1980).

Elfvin, L.-G., and Lindh, B.: A study of the extrinsic innervation of the guinea pig pylorus with the horseradish peroxidase tracing technique. *J. Comp. Neurol.* **208**:317–324 (1982).

Eränkö, O.: Small intensely fluorescent (SIF) cells and nervous transmission in sympathetic ganglia. *Annu. Rev. Pharmacol. Toxicol.* **18**:417–430 (1978).

Eränkö, O., Soinila, S., and Paivarinta, H. (eds.): *Histochemistry and Cell Biology of Autonomic Neurons, SIF Cells and Paraneurons: Advances in Biochemical Psychopharmacology*, Vol. 25, Raven Press, New York (1980).

Erde, S. M., Gershon, M. D., and Wood, J. D.: Morphology of electrophysiologically identified types of myenteric neuron: Intracellular recording and injection of lucifer yellow in the guinea-pig small bowel. *Neurosci. Abstracts* **6**:274 (1980).

Evans, D. H. L., and Murray, J. G.: Histological and functional studies on the fiber composition of the vagus nerve of the rabbit. *J. Anat.* **88**:320–337 (1954).

Feher, E., and Leranth, C.: Light and electron microscopic immunocytochemical localization of vasoactive intestinal polypeptide (VIP)-like activity in the rat small intestine. *Neuroscience* **10**:97–106 (1983).

Feher, E., and Wenger, T.: Ultrastructural immunocytochemical localization of substance P in the cat small intestine. *Anat. Anz.* **150**:137–145 (1981).

Forehand, C. J., and Purves, D.: Regional innervation of rabbit ciliary ganglion cells by the terminals of preganglionic axons. *J. Neurosci.* **4**:1–12 (1984).

Franco, R., Costa, M., and Furness, J. B.: Evidence for the release of endogenous substance P from intestinal nerves. *Naunyn-Schmiedeberg's Arch. Pharmacol.* **306**:195–201 (1979a).

Franco, R., Costa, M., and Furness, J. B.: The presence of a cholinergic excitatory input to substance P neurons in the intestine. *Proc. Aust. Physiol. Pharmacol. Soc.* **10**:255P (1979b).

Fujita, T., Kobayashi, S., and Yui, R.: Paraneuron concept and its current implications, in: *Histochemistry and Cell Biology of Autonomic Neurons, SIF Cells and Paraneurons* (O. Eranko, S. Soinila, and H. Paivarinta, eds.), *Advances in Biochemical Psychopharmacology*, Vol. 25, pp. 321–325, Raven Press, New York (1980).

Furness, J. B., and Costa, M.: The adrenergic innervation of the gastrointestinal tract. *Rev. Physiol. Biochem. Pharmacol.* **69**:1–51 (1974).

Furness, J. B., and Costa, M.: Projections of intestinal neurons showing immunoreactivity for vasoactive intestinal polypeptide are consistent with these neurons being the enteric inhibitory neurons. *Neurosci. Lett.* **15**:199–204 (1979).

Furness, J. B., and Costa, M.: Types of nerves in the enteric nervous system. *Neuroscience* **5**:1–20 (1980).

Furness, J. B., and Costa, M.: Neurons with 5-hydroxytryptamine-like immunoreactivity in the enteric nervous system: Their projections in the guinea-pig small intestine. *Neuroscience* **7**:341–349 (1982).

Furness, J. B., Costa, M., Franco, R., and Llewellyn-Smith, I. H.: Neuronal peptides in the intestine: Distribution and possible functions, in: *Neural Peptides and Neuronal Communications* (E. Costa and M. Trabucchi, eds.), pp. 601–617, Raven Press, New York (1980).

Furness, J. B., Costa, M., and Walsh, J. H.: Evidence for and significance of the projection of VIP neurons from the myenteric plexus to the taenia coli in the guinea-pig. *Gastroenterology* **80**:1557–1561 (1981).

Furness, J. B., Costa, M., and Miller, R. J.: Distribution and projections of nerves with enkephalin-like immunoreactivity in the guinea-pig small intestine. *Neuroscience* **8**:653–664 (1983a).

Furness, J. B., Costa, M., Emson, P. C., Håkanson, R., Moghimzadeh, E., Sundler, F., Taylor, I. L., and Chance, R. E.: Distribution, pathways and reactions to drug treatment of nerves with neuropeptide-Y and pancreatic polypeptide-like immunoreactivity in the digestive tract. *Cell Tissue Res.* **234**:71–92 (1983b).

Furness, J. B., Costa, M., and Eckstein, F.: Neurones localized with antibodies against choline acetyltransferase in the enteric nervous system. *Neurosci. Lett.* **40**:105–109 (1983c).

Furness, J. B., Costa, M., and Keast, J. R.: Choline acetyltransferase and peptide immunoreactivity of submucous neurons in the small intestine of the guinea-pig. *Cell Tissue Res.* **237**:329–336 (1984).

Gabella, G.: Fine structure of the myenteric plexus in the guinea-pig ileum. *J. Anat.* **111**:69–97 (1972).

Gabella, G.: Special muscle cells and their innervation in the mammalian small intestine. *Cell Tissue Res.* **153**:63–77 (1974).

Gabella, G.: *Structure of the Autonomic Nervous System*. Chapman and Hall, London (1976).

Gabella, G.: Innervation of the gastrointestinal tract. *Int. Rev. Cytol.* **59**:129–193 (1979).

Gabella, G.: Structural aspects of the intramural plexuses of the gut: Appendix to Chapter 14, Holman, M. E.: The intrinsic innervation and peristaltic reflex of the small intestine, in: *Smooth Muscle: An Assessment of Current Knowledge*. (E. Bulbring, A. F. Brading, A. W. Jones, and T. Tomita, eds.), pp. 339–351, Edward Arnold, London (1981a).

Gabella, G.: Structure of muscles and nerves in the gastrointestinal tract, in: *Physiology of the Gastrointestinal Tract* (L. R. Johnson, J. Christensen, M. I. Grossman, E. Jacobson, and S. G. Schlutz, eds.), pp. 197–241, Raven Press, New York (1981b).

Gabella, G.: Ultrastructure of the nerve plexuses of the mammalian intestine: The enteric glial cells. *Neuroscience* **6**:425–436 (1981c).

Gabella, G.: Size of neurons and glial cells in the intramural ganglia of the hypertrophic intestine of the guinea-pig. *J. Neurocytol.* **13**:73–84 (1984).

Gabella, G., and Trigg, P.: Size of neurons and glial cells in the enteric ganglia of mice, guinea-pig, rababits and sheep. *J. Neurocytol.* **13**:49–71 (1984).

Gershon, M. D., and Erde, S. M.: The nervous system of the gut. *Gastroenterology* **80**:1571–1594 (1981).

Gienc, J., and Kuder, T.: Morphology and topography of the otic ganglion in guinea-pig detected with thiocholine technique. *Folia Morphol. (Warsaw)* **39**:79–85 (1980).

Gilpin, C. J., Dixon, J. S., Gilpin, S. A., and Gosling, J. A.: The fine structure of autonomic neurons in the wall of the human urinary bladder. *J. Anat.* **137**:705–713 (1983).

Gray, M. J., Hodgkiss, J. P., and Lees, G. M.: Morphological characteristics of enteric neurones revealed with different intracellular staining methods. *Neurosci. Lett. Suppl.* **5**:S193 (1980).

Gurney, A. M., and Rang, H. P.: The channel-blocking action of methonium compounds on rat submandibular ganglion cells. *Br. J. Pharmacol.* **82**:623–642 (1984).

Hadjiconstantinou, M., Potter, P. E., and Neff, N. H.: Trans-synaptic modulation via muscarinic receptors of serotonin-containing small intensely fluorescent cells of cervical ganglion. *J. Neurosci.* **2**:1836–1839 (1982).

Henderson, C. G., and Ungar, A.: Effect of cholinergic antagonists on sympathetic ganglionic transmission of vasomotor reflexes from the carotid baroreceptors and chemoreceptors of the dog. *J. Physiol. (London)* **277**:379–385 (1978).

Hirst, G. D. S., and McKirdy, H. C.: A nervous mechanism for descending inhibition in guinea-pig small intestine. *J. Physiol. (London)* **238**:129–143 (1974a).

Hirst, G. D. S., and McKirdy, H. C.: Presynaptic inhibition at mammalian peripheral synapse? *Nature (London)* **250**:430–431 (1974b).

Hirst, G. D. S., and McKirdy, H. C.: Synaptic potentials recorded from neurones of the submucous plexus of guinea-pig small intestine. *J. Physiol. (London)* **249**:269–285 (1975).

Hirst, G. D. S., Holman, M. E., and Spence, I.: Two types of neurones in the myenteric plexus of duodenum in the guinea-pig. *J. Physiol.* **236**:303–326 (1974).

Hirst, G. D. S., Holman, M. E., and McKirdy, H. C.: Two descending nerve pathways activated by distension of guinea-pig small intestine. *J. Physiol. (London)* **244**:113–127 (1975).

Hodgkiss, J. P.: Slow synaptic potentials in AH-type myenteric plexus neurons. *Pfluegers Arch.* **391**:331–333 (1981).

Hodgkiss, J. P., and Lees, G. M.: Correlated electrophysiological and morphological characteristics of myenteric plexus neurones. *J. Physiol. (London)* **285**:19–20P (1978).

Hodgkiss, J. P., and Lees, G. M.: Morphological features of guinea-pig myenteric plexus neurones, in: *Gastrointestinal Motility* (J. Christensen, ed.), pp. 111–117, Raven Press, New York (1980).

Hodgkiss, J. P., and Lees, G. M.: Morphological studies of myenteric plexus neurons of the guinea-pig ileum. *Neuroscience* **8**:593–608 (1983).

Hökflet, T., Elfvin, L.-G., Elde, R., Schultzberg, M., Goldstein, M., and Luft, R.: Occurrence of somatostatin-like immunoreactivity in some peripheral sympathetic noradrenergic neurons. *Proc. Natl. Acad. Sci. U.S.A.* **74**:3587–3591 (1977).

Hökfelt, T., Johansson, O., Ljungdahl, A., Lundberg, J. M., Schultzberg, M., Terenius, L., Goldstein, M., Elde, R., Steinbusch, H., and Verhofstad, A.: Histochemistry of transmitter interactions—neuronal coupling and coexistence of putative transmitters. *Adv. Pharmacol. Ther.* **2**:131–143 (1978a).

Hökfelt, T., Ljungdahl, A., Steinbusch, H., Verhofstad, A., Nilsson, G., Brodin, E., Pernow, B., and Goldstein, M.: Immunohistochemical evidence of substance P-like immunoreactivity in some 5-hydroxytryptamine-containing neurons in the rat central nervous system. *Neuroscience* **3**:517–538 (1978b).

Hökfelt, T., Johansson, O., Ljungdahl, A., Lundberg, J. M., and Schultzberg, M.: Peptidergic neurones. *Nature (London)* **284**:515–521 (1980).

Hume, R. I., and Purves, D.: Geometry of neonatal neurones and the regulation of synapse elimination. *Nature (London)* **295**:469–471 (1981).

Hume, R. I., and Purves, D.: Apportionment of the terminals from single preganglionic axons to target neurones in the rabbit ciliary ganglion. *J. Physiol. (London)* **388**:259–275 (1983).

Irwin, D. A.: The anatomy of Auerbach's plexus. *Am. J. Anat.* **49**:141–166 (1931).

Jacobowitz, D., and Woodward, J. K.: Adrenergic neurons in the cat superior cervical ganglion and cervical sympathetic nerve trunk: A histochemical study. *J. Pharmacol. Exp. Ther.* **162**:213–226 (1968).

Jan, L. Y., Jan, Y. N., and Brownfield, M. S.: Peptidergic transmitters in synaptic boutons on sympathetic ganglia. *Nature (London)* **280**:380–382 (1980).

Jan, Y. N., Bowers, C. W., Branton, D., Evans, L., and Jan, L. Y.: Peptides in neuronal function: Studies using frog autonomic ganglia. *Cold Spring Harbor Symp. Quant. Biol.* **48**:363–374 (1983).

Janig, W., Krauspe, R., and Wiedersatz, G.: Transmission of impulses from pre- to postganglionic vasoconstrictor and sudomotor neurons. *J. Auton. Nerv. Syst.* **6**:95–106 (1982).

Jessen, K. R., Polak, J. M., Van Noorden, S., Bloom, S. R., and Burnstock, G.: Peptide-containing neurones connect the two ganglionated plexuses of the enteric nervous system. *Nature (London)* **283**:391–393 (1980a).

Jessen, K. R., Saffrey, M. J., Van Noorden, S., Bloom, S. R., Polak, J. M., and Burnstock, G.: Immunohistochemical studies of the enteric nervous system in tissue culture and *in situ:* Localization of vasoactive intestinal polypeptide (VIP), substance-P and enkephalin immunoreactive nerves in the guinea-pig gut. *Neuroscience* **5**:1717–1735 (1980b).

Johnson, D. A., and Purves, D.: Postnatal deduction of neural unit size in the rabbit ciliary ganglion. *J. Physiol. (London)* **318**:143–159 (1981).

Katayama, Y., North, R. A., and Williams, J. T.: The action of substance P on neurons of the myenteric plexus of the guinea-pig small intestine. *Proc. R. Soc. London Ser. B* **206**:191–208 (1979).

Keast, J. R., Furness, J. B., and Costa, M.: Origins of peptide and norepinephrine nerves in the mucosa of the guinea-pig small intestine. *Gastroenterology* **86**:637–644 (1984).

Kelts, K. A., Whitlock, D. G., Ledbury, P. A., and Reese, B. A.: Postganglionic connections between sympathetic ganglia in the solar plexus of the cat demonstrated autoradiographically. *Exp. Neurol.* **63**:120–134 (1979).

Klein, R. L., Wilson, S. R., Dzielak, D. J., Yang, W. H., and Viveros, O. H.: Opioid peptides and noradrenaline co-exist in large dense-cored vesicles from sympathetic nerve. *Neuroscience* **7**:2255–2260 (1982).

Kojima, H., Anraku, S., Onogi, K., and Ito, R.: Histochemical studies on two types of cells containing catecholamine in sympathetic ganglion of the bullfrog. *Experientia* **34**:92–93 (1978).

Kondo, H., and Yui, R.: An electron microscopic study on substance P-like immunoreactive nerve fibers in the celiac ganglion of guinea pigs. *Brain Res.* **252:**142–145 (1981).

Kondo, H., and Yui, R.: An electron microscopic study on enkephalin-like immunoreactive nerve fibers in the celiac ganglion of guinea pigs. *Brain Res.* **252:**142–145 (1982).

Kondo, H., Dun, N. J., and Pappas, G. D.: A light and electron microscopic study of the rat superior cervical ganglion cells by intracellular HRP-labeling. *Brain Res.* **197:**193–199 (1980).

Kosterlitz, H. W.: Intrinsic and extrinsic nervous control of motility of the stomach and the intestines, in *Handbook of Physiology*, Section 6, *Alimentary Canal*, Vol. IV (C. F. Code, ed.), pp. 2147–2171, American Physiological Society, Washington, D.C. (1968).

Kosterlitz, H. W., Lees, G. M., and Wallis, D. I.: Synaptic potentials recorded by the sucrose-gap method from the rabbit superior cervical ganglion. *Br. J. Pharmacol.* **40:**275–293 (1970).

Kreulen, D. L., and Szurszewski, J. H.: Nerve pathways in celiac plexus of the guinea-pig. *Am. J. Physiol.* **237:**E90–E97 (1979).

Kuntz, A.: The structural organization of the inferior mesenteric ganglion. *J. Comp. Neurol.* **72:**371–382 (1940).

Langley, J. N.: *The Autonomic Nervous System*. Heffer, Cambridge (1921).

Larsson, L.-I., and Rehfeld, J. F.: Peptidergic and adrenergic innervation of pancreatic ganglia. *Scand. J. Gastroenterol.* **14:**433–437 (1979).

Lawrentjew, B.J.: Zur Morphologie des Ganglion cervical super. *Anat. Anz.* **58:**529–539 (1924).

Leander, S., Hakanson, R., and Sundler, F.: Nerves containing substance P, vasoactive intestinal polypeptide, enkephalin or somatostatin in the guinea-pig teania coli: Distribution, ultrastructure and possible functions. *Cell Tissue Res.* **215:**21–39 (1981).

Libet, B., and Owman, C.: Concomitant changes in formaldehyde-induced fluorescence of dopamine interneurones and in slow inhibitory post-synaptic potentials of the rabbit superior cervical ganglion, induced by stimulation of the preganglionic nerve or by a muscarinic agent. *J. Physiol. (London)* **237:**635–662 (1974).

Libet, B., Chichibu, S, and Tosaka, T.: Slow synaptic responses and excitability in sympathetic ganglia of the bullfrog. *J. Neurophysiol.* **31:**383–395 (1968).

Lichtman, J. W.: The reorganization of synaptic connexions in the rat submandibular ganglion during post-natal development. *J. Physiol. (London)* **273:**155–177 (1977).

Lichtman, J. W.: On the predominantly single innervation of submandibular ganglion cells in the rat. *J. Physiol. (London)* **302:**121–130 (1980).

Lichtman, J. W., and Purves, D.: The elimination of redundant preganglionic innervation to hamster sympathetic ganglion cells in early post-natal life. *J. Phsyiol. (London)* **301:**213–228 (1980).

Lichtman, J. W., Purves, D., and Yip, J. W.: On the purpose of selective innervation of guinea-pig superior cervical ganglion cells. *J. Physiol. (London)* **292:**69–84 (1979).

Llewellyn-Smith, I. J., Wilson, A. J., Furness, J. B., Costa, M., and Rush, R. A.: Ultrastructural identification of noradrenergic axons and their distribution within the enteric plexuses of the guinea-pig small intestine. *J. Neurocytol.* **10:**331–352 (1981).

Llewellyn-Smith, I. J., Furness, J. B., Murphy, R., O'Brien, P. E., and Costa, M.: Substance P-containing nerves in the human small intestine: Distribution, ultrastructure and characterization of the immunoreactive peptide. *Gastroenterology* **86:**421–435 (1984).

Lundberg, J. M., Dahlstrom, A., Larsson, I., Pettersson, G., Ahlman, H., and Kewenter, J.: Efferent innervation of the small intestine by adrenergic neurons from the cervical sympathetic and stellate ganglia, studied by retrograde transport of peroxidase. *Acta Physiol. Scand.* **104:**33–42 (1978a).

Lundberg, J. M., Hökfelt, T., Nilsson, G., Terenius, L., Rehfeld, J. F., Elde, R., and Said, S.: Peptide neurons in the vagus, splanchnic and sciatic nerves. *Acta Physiol. Scand.* **104:**499–501 (1978b).

Lundberg, J. M., Hökfelt, T., Kewenter, J., Pettersson, G., Ahlman, H., Edin, R., Dahlstrom, A., Nilsson, G., Terenius, L., Uvnäs-Wallensten, K., and Said, S.: Substance P-, VIP- and enkephalin-like immunoreactivity in the human vagus nerve. *Gastroenterology* **77**:468–471 (1979a).

Lundberg, J. M., Hökfelt, T., Schultzberg, K., Uvnäs-Wallensten, K., Kohler, C., and Said, S. I.: Occurrence of vasoactive intestinal polypeptide (VIP)-like immunoreactivity in certain cholinergic neurons of the cat: Evidence from combined immunohistochemistry and acetylcholinesterase staining. *Neuroscience* **4**:1539–1559 (1979b).

Malmfors, G., Hakanson, R., Okmian, L., and Sundler, F.: Peptidergic nerves persist after jejunal auto-transplantation: An experimental study in the piglet. *J. Pediatr. Surg.* **15**(1):53–56 (1980).

Matthews, M. R., and Cuello, A. C.: The origin and possible significance of substance P immunoreactive networks in the prevertebral ganglia and related structures in the guinea-pig. *Philos. Trans. R. Soc. London Ser. B* **306**:247–276 (1984).

Matthews, M. R., and Raisman, G.: The ultrastructure and somatic efferent synapses of small granule-containing cells in the superior cervical ganglion. *J. Anat.* **105**(2):255–282 (1969).

McMahan, U. J., and Kuffler, S. W.: Visual identification of synaptic boutons on living ganglion cells and of varicosities in postganglionic axons in the heart of the frog. *Proc. R. Soc. London Ser. B* **177**:485–508 (1971).

Niel, J. P., Gonella, J., and Roman, C.: Localisation par la technique de marquage a la peroxydase des corps cellulaires des neurones ortho et parasympathiques innervant le sphincter oesophagien inferieur du chat. *J. Physiol. (Paris)* **76**:591–599 (1980).

Nishi, S., and Christ, D. D.: Electrophysiological and anatomical properties of mammalian parasympathetic ganglion cells. Proc. XXV Int. Physiol Cong. Munich, 1249 (1971).

Nishi, S., and Koketsu, S.: Electrical properties and activities of single sympathetic neurons in frogs. *J. Cell. Comp. Physiol.* **55**:15–30 (1960).

Nishi, S., and North, R. A.: Intracellular recording from the myenteric plexus of the guinea-pig ileum. *J. Physiol. (London)* **231**:471–491 (1973).

Nishi, S., Soeda, H., and Koketsu, K.: Studies on sympathetic B and C neurons and patterns of preganglionic innervation. *J. Cell Comp. Physiol.* **66**:19–32 (1965).

Njå, A., and Purves, D.: Specific innervation of guinea-pig superior cervical ganglion cells by preganglionic fibres arising from different levels of the spinal cord. *J. Physiol. (London)* **264**:565–583 (1977a).

Njå, A., and Purves, D.: Reinnervation of guinea-pig superior cervical ganglion cells by preganglionic fibers arising from different levels of the spinal cord. *J. Physiol. (London)* **272**:633–651 (1977b).

Njå, A., and Purves, D.: Specificity of initial synaptic contacts made on guinea-pig superior cervical ganglion cells during regeneration of the cervical sympathetic trunk. *J. Physiol. (London)* **281**:45–62 (1978).

North, R. A., Henderson, G., Katayama, Y., and Johnson, S. M.: Electrophysiological evidence for presynaptic inhibition of acetylcholine release by 5-hydroxytryptamine in the enteric nervous system. *Neuroscience* **5**:581–586 (1980).

Olschowka, J. A., and Jacobowitz, D. M.: The coexistence and release of bovine pancreatic polypeptide-like immunoreactivity from noradrenergic superior cervical ganglia neurons. *Peptides* **4**:231–238 (1983).

Pearson, G. T., Davison, J. S., Collins, R. C., and Petersen, O. H.: Non-cholinergic, non-adrenergic nerves control enzyme secretion in guinea-pig pancreas. *Nature (London)* **290**:259–261 (1981).

Petras, J. M., and Cummings, J. F.: Sympathetic and parasympathetic innervation of the urinary bladder and urethra. *Brain Res.* **153**:363–369 (1978).

Purves, D., and Lichtman, J. W.: Formation and maintenance of synaptic connections in autonomic ganglia. *Physiol. Rev.* **58**:821–862 (1978).

Purves, D., and Wigston, D. J.: Neural units in the superior cervical ganglion of the guinea-pig. *J. Physiol. (London)* **334**:169–178 (1983).

Rang, H. P.: The characteristics of synaptic currents and responses to acetylcholine of rat submandibular ganglion cells. *J. Physiol. (London)* **311**:23–55 (1981).

Roper, S.: Sprouting and regeneration of synaptic terminals in the frog cardiac ganglion. *Nature (London)* **261**:148–149 (1976).

Sargent, P. B.: The number of synaptic boutons terminating on *Xenopus* cardiac ganglion cells is directly correlated with cell size. *J. Physiol. (London)* **343**:85–104 (1983).

Satomi, H., Yamamoto, T., Ise, H., and Takatama, H.: Origins of the parasympathetic preganglionic fibers to the cat intestine as demonstrated by the horseradish peroxidase method. *Brain Res.* **151**:571–578 (1978).

Schultzberg, M., Hökfelt, T., Terenius, L., Elfvin, L.-G. Lundberg, J. M. Brandt, J., Elde, R. P., and Goldstein, M.: Enkephalin immunoreactive nerve fibres and cell bodies in sympathetic ganglia of the guinea pig and rat. *Neuroscience* **4**:249–270 (1979).

Schultzberg, M., Hökfelt, T., Nilsson, L., Terenius, L., Rehfeld, J. F., Brown, M., Elde, R., Goldstein, M., and Said, S.: Distribution of peptide- and catecholamine-containing neurons in the gastro-intestinal tract of rat and guinea-pig: Immunohistochemical studies with antisera to substance P, enkephalins, somatostatin, gastrin/cholecystokinin, neurotensin and dopamine β-hydroxylase. *Neuroscience* **5**:689–744 (1980).

Sinnreich, Z., and Nathan, H.: The ciliary ganglion in man. *Anat. Anz.* **150**:287–297 (1981).

Skok, V. I.: *Physiology of Autonomic Ganglia.* Igaku Shoin, Tokyo (1973).

Smolen, A. J., Wright, L. L., and Cunningham, T. J.: Neuron numbers in the superior cervical ganglion of the rat: A critical comparison of methods for cell counting. *J. Neurocytol.* **12**:739–750 (1983).

Szurszewski, J. H.: Physiology of mammalian prevertebral ganglia. *Annu. Rev. Physiol.* **43**:53–68 (1981).

Szurszewski, J. H., and Weems, W. A.: A study of peripheral input to and its control by postganglionic neurons of the inferior mesenteric ganglion. *J. Physiol. (London)* **256**:541–556 (1976a).

Szurszewski, J. H., and Weems, W. A.: Control of gastrointestinal motililty of preverterbral ganglia, in *Physiology of Smooth Muscle* (E. Bulbring and M. F. Shuba, eds.), pp. 313–319, Raven Press, New York (1976b).

Verhofstad, A. A. J., Steinbusch, H. W. M., Penke, B., Varga, J., and Joosten, H. W. J.: Serotonin-immunoreactive cells in the superior cervical ganglion of the rat. Evidence for the existence of separate serotonin- and catecholamine-containing small ganglionic cells. *Brain Res.* **212**:39–49 (1981).

Wallis, D. I., and North, R. A.: Synaptic input to cells of the rabbit superior cervical ganglion. *Pfluegers Arch.* **374**:145–152 (1978).

Wallis, D. I., Lees, G. M., and Kosterlitz, H. W.: Recording resting and action potentials by the sucrose-gap method. *Comp. Biochem. Physiol.* **50C**:199–216 (1975).

Ward, E. M., Rorie, D. K., Nauss, L. A., and Bahn, K. C.: The coeliac ganglia in man: Normal anatomic variations. *Anesth. Analg. (Cleveland)* **58**:461–465 (1979).

Weight, F. F., and Weitsen, H. A.: Identification of small intensely fluorescent (SIF) cells as chromaffin cells in bullfrog sympathetic ganglia. *Brain Res.* **128**(2):213–226 (1977).

Wigston, D. J.: Maintenance of cholinergic neurons and synapses in the ciliary ganglion of aged rats. *J. Physiol. (London)* **344**:223–231 (1983).

Williams, T. H.: Electron microscopic evidence for an autonomic interneuron. *Nature (London)* **214**:309–310 (1967).

Williams, T. H., and Palay, S. L.: Ultrastructure of the small neurons in the superior cervical ganglion. *Brain Res.* **15**:17–34 (1969).

Williams, T. H., Black, A. C., Jr., Chiba, T., and Bhalla, R. C.: Morphology and biochemistry of small, intensely fluorescent cells of sympathetic ganglia. *Nature* **256**:315–317 (1975).

Wilson, A. J., Furness, J. B., and Costa, M.: The fine structure of the submucous plexus of the guinea-pig ileum. I. The ganglia, neurons, Schwann cells and neuropil. *J. Neurocytol.* **10**:759–784 (1981a).

Wilson, A. J., Furness, J. B., and Costa, M.: The fine structure of the submucous plexus of the guinea-pig ileum. II. Description and analysis of vesiculated nerve profiles. *J. Neurocytol* **10**:785–804 (1981b).
Wolf, G. A.: The ratio of preganglionic neurons to post-ganglionic neurons in the visceral nervous system. *J. Comp. Neurol.* **75**:235–243 (1941).
Wood, J. D., and Mayer, C. J.: Serotonergic activation of tonin-type enteric neurons in the guinea-pig small bowel. *J. Neurophysiol.* **42**:582–593 (1979).
Yokoyama, S., and Ozaki, T.: Polarity of effects of stimulation of Auerbach's plexus on longitudinal muscle. *Am. J. Physiol.* **235**(4):E345–E353 (1978).
Yoshimura, M., Higashi, H., and Nishi, S.: Multiple innervation of amphibian sympathetic ganglion cells. *Kurume Med. J.* **26**(4):381–385 (1979).

II

Transmission and Modulation in Sympathetic Ganglia and Their Neuropharmacology

3

General Concepts of Ganglionic Transmission and Modulation

K. KOKETSU and ALEXANDER G. KARCZMAR

I. INTRODUCTION

A. Historical Remarks

The complexity of ganglionic morphology was noted as early as the 18th century, and indeed Winslow used the term "little brains" to describe the ganglia (see Chapter 1). Further studies added to the perception of the heterogeneity and cytological as well as the morphological complexity of the ganglia, as described in Chapter 2. It could therefore be expected that the early investigators of ganglionic transmission and its neurotransmitters would be cognizant of the complexity of the pertinent phenomena. Yet, as described in Chapter 1, the early students of ganglionic transmission such as Langley and Dickinson (1880) as well as their immediate followers who provided the proof of the cholinergicity of this transmission essentially regarded this transmission as a one-to-one, monosynaptic relationship (Sherrington, 1906). A similar concept emerged from the early electrophysiological studies of the ganglionic response to presynaptic stimulation carried out by Bishop and Heinbecker (1932) and G. L. Brown (1934), as described in detail in Chapter 1. The actual complexity of ganglionic

K. KOKETSU • Department of Physiology, Kurume University School of Medicine, Kurume, Japan. ALEXANDER G. KARCZMAR • Department of Pharmacology and Experimental Therapeutics, Stritch School of Medicine, Loyola University Medical Center, Maywood, Illinois 60153.

events became evident only following the description by Eccles and Libet (1961) of the triphasic pattern of the postganglionic response. The implication of this demonstration was clearly posited in terms of the hypothesis of the disynaptic nature of the P response as formulated by Eccles (1952) and Eccles and Libet (1961) (see also Laporte and Lorente de Nò, 1950) (see also Chapter 1). There were several subsequent way-stations on the road to the full realization of the subtlety and multifactorial control of ganglionic transmission; these way-stations included the discovery of transmitter-sensitive receptors on the presynaptic ganglionic terminals (Birks and MacIntosh, 1961) and the demonstration of the presence in the ganglia or neuroglia or both of several nonacetylcholine bioactive substances such as amino acids, peptides, cyclic nucleotides, and phospholipids.

As a consequence, the ganglion could no longer be described as a simple synaptic relay station. In fact, the researches initiated in the late 1950s (see Chapter 1) and thereafter continued without interruption led to complex conceptualization and terminology that may appear ambiguous unless defined and described in terms of ganglionic function.

B. Scope of This Chapter

Accordingly, this chapter concerns the elucidation and definition of concepts pertaining to vertebrate ganglionic transmission, particularly transmission in the sympathetic ganglia. These concepts relate the events of sympathetic ganglionic transmission to its anatomical and microanatomical components such as preganglionic input vs. postganglionic output, neurons and interneurons, and their functional connections. These concepts also concern the receptors, both cholinergic and noncholinergic, and the components of the synaptic responses, such as intra- and extracellularly recorded potentials, as well as the interactions between the receptors and between the synaptic responses. Finally, it is important to differentiate between the synaptic and the modulatory phenomena; the pertinent definitions are adduced in this chapter, and the bioactive substances that are involved in either synaptic (transmissive) or modulatory phenomena are referred to. In exploring and defining these concepts and interactions, this chapter describes the framework of this book.

II. SYNAPTIC TRANSMISSION, ITS MODULATION, AND ITS COMPONENTS

A. Synaptic and Modulatory Events in Sympathetic Ganglia

It is clear today that ganglionic transmission is not constituted by a simple excitatory monosynaptic transmission; indeed, preganglionic signals are conveyed to the postganglionic neurons in extremely complicated ways that involve a variety of synaptic events subserving integrative functions of the sympathetic ganglia. The pertinent phenomena that were studied particularly in the last two decades involve processes of either transmission or what is referred to as modulations. Let us then explore and define the concepts of transmission and modulation and the pertinent terminology, since the latter will be employed in the subsequent chapters.

The synaptic transmission described in this book includes synaptic phenomena mediated by neurotransmitters and resulting in postsynaptic potentials such as the fast excitatory postsynaptic potential (EPSP), the slow EPSP, the late slow EPSP, and the slow inhibitory postsynaptic potential (IPSP) (Koketsu, 1969; Nishi, 1970, 1974; Dun, 1980) (see also Chapters 5–9). The synaptic modulation, on the other hand, may be defined as including a number of synaptic phenomena that occur in the ganglion without producing these particular postsynaptic potentials, yet lead to a change (or modulation) in the postsynaptic potentials resulting in a modification of their effectiveness or synaptic transmission, or both, or to a change (or modulation) of the action potential (Karczmar *et al.*, 1972). This modulation may be presynaptic in nature, since it modifies the release of bioactive substances (see Chapters 10 and 11); thus, the presynaptic modulation regulates the amounts of neurotransmitters [mainly, but not exclusively, acetylcholine (ACh)] that are released from preganglionic nerve terminals. Or it may be postsynaptic, concerning either postsynaptic potentials or action potentials or both (Chapter 12); this modulation controls the excitability or excitation of the ganglion cells.

Postsynaptic modulations may relate to changes in resting membrane potential (other than those resulting in postsynaptic potentials), or it may be reflected in the change in the duration, configuration, or patterns of action potentials [voltage-dependent currents (see Chapter 12)]. Postsynaptic modulations may also include changes in the sensitivity of the postsynaptic receptors to their neurotransmitter (see Chapter 12). It must be noted that similar mechanisms—such as changes in the resting membrane potential or conductance, or in the configuration of the action potential—are probably involved in both pre- and postsynaptic modulations (cf. Chapters 10 and 11).

It must be further clarified that the term "modulator," as used in this book, refers solely to endogenous or bioactive substances, but not to pharmacological agents or toxins. The endogenous neuromodulators may be released within the ganglion by neural elements located inside the ganglion, or they may be extrinsic and liberated in some organs or tissues located outside the ganglion; in fact, the origin and the site of release of a bioactive substance that is capable of modulation may be not always readily demonstrated, just as in the case with neurotransmitters. A hypothesis enunciated by Williams (cf. Williams and Jew, 1983) concerns what may be a special instance of modulation via extrinsic extraganglionic factors. Williams proposed that ganglionic interneurons, the small intensely fluorescent (SIF) cells (see below), particularly of the rat and the shrew, besides abutting on ganglionic neurons, release bioactive substances into the ganglionic portal system or ganglionic extracellular space, thereby influencing a large number of primary ganglionic neurons. Since SIF cells contain a number of bioactive substances (see Tables 1 and 2), it is not known which substances are involved in the process proposed by Williams (Williams and Jew, 1983).

It should be added that certain bioactive substances may generate postsynaptic potentials as well as cause modulation. For example, catecholamines may either generate the slow IPSP (perhaps via their release from the SIF cell or via other mechanisms) (Libet, 1970, 1976) (see also Chapters 9 and 13) or modify (inhibit or facilitate) the postsynaptic potentials via both pre- and postsynaptic mechanisms; we deal with modulations in this latter case (see Chapters 9 and 11). The same appears to be true for peptides that may both generate a transmissive response, the late slow EPSP (Chapter 8), and induce modulatory, particularly presynaptic but also postsynaptic, changes (see Chapters 12 and 13).

B. Neural Elements, Transmitters, and Modulators

1. Neural Elements

Several neural elements participate in ganglionic transmission and modulation. These are, first preganglionic (spinal) neurons, preganglionic nerve fibers and their endings that terminate in the ganglia; second, postganglionic neurons or cells of the autonomic ganglia (frequently referred to as efferent ganglionic neurons) (see, for example, Skok, 1976) and their fibers; and third, SIF cells or ganglionic interneurons (Kohn, 1898; Libet, 1976); (Chapters 2 and 9). While the synaptic connections of the first two components were well described for ganglia and their presence as such definitely ascertained (Chapters 1 and 2), the synaptic connections of SIF cells with respect to pre- and postganglionic neurons and their functional role remain, after much research, still ambiguous (see Chapters 2, 9, and 13).

Glial cells and related elements are also present in parasympathetic and sympathetic as well as enteric ganglia (see Chapter 2). They may be a source of modulatory effect on ganglionic transmission due to their action at either presynaptic or postsynaptic sites, via a release of such substances as serotonin or γ-aminobutyric acid (see Table 2).

Table 1. Occurrence and Localization of Peptides in Sympathetic Ganglia

Peptide[a]	Localization[b]	Sympathetic ganglia[c]	Species	References
BOM	NF	CMG	Rat	Schultzberg and Dalsgaard (1983)
CCK	NF	IMG, CMG	Guinea pig	Larsson and Rehfeld (1979)
DYN	PGN	CMG	Guinea pig, rat	Vincent *et al.* (1984)
ENK	NF	IMG, CMG, SCG	Guinea pig, rat	Schultzberg *et al.* (1979)
	PGN	SCG	Rat, guinea pig	DiGiulio *et al.* (1978), Schultzberg *et al.* (1979)
	SIF	SCG, IMG, CMG	Guinea pig, cat	Schultzberg *et al.* (1979)
	SIF	Paravertebral	Bullfrog	Kondo (unpublished)
GRH	NF	CMG	Rat	Kondo *et al.* (1983)
LH-RH	PGN	Paravertebral	Bullfrog	Y. N. Jan *et al.* (1979)
SOM	PGN	SCG, IMG, CMG	Guinea pig, rat	Hökfelt *et al.* (1977a), Leranth *et al.* (1980)
	NF	IMG	Cat	Lundberg *et al.* (1980)
SP	NF	IMG, CMG, SCG	Guinea pig, cat	Hökfelt *et al.* (1977c), Konishi *et al.* (1979), Dun and Jiang (1982)
	Interneurons (intrinsic)	SCG	Rat	Robinson *et al.* (1980)
VIP	NF	IMG, CMG, SCG	Guinea pig, rat	Hökfelt *et al.* (1977b)
	PGN	IMG, CMG	Guinea pig	Hökfelt *et al.* (1977b)

[a] (BOM) Bombesin; (CCK) cholecystokinin; (DYN) dynorphin; (ENK) enkephalin; (GRH) gastrin-releasing hormone; (LH-RH) luteinizing-hormone-releasing hormone; (SOM) somatostatin; (SP) substance P; (VIP) vasoactive intestinal peptide.
[b] (NF) Nerve fibers; (PGN) principal postganglionic neurons; (SIF) small intensely fluorescent cells.
[c] (CMG) Coeliac inferior mesenteric ganglia complex; (IMG) inferior mesenteric ganglia; (SCG) superior cervical ganglia.

Table 2. Bioactive Nonpeptide Substances Present in the Sympathetic Ganglia

Substance	Preganglion	SIF cell	Primary neurons	Glia, circulation	Selected references
Catecholamines		+	+	+	Eränkö and Harkonen (1965)
5-Hydroxytryptamine	+	+			Verhofstad *et al.* (1981)
Acetylcholine	+		+ (parasympathetic)		
Prostaglandin E_1	+?	+?	+?		Trevisani *et al.* (1982)
γ-Aminobutyric acid				+	Bertilsson *et al.* (1976)
Cyclic AMP	+		+		McAfee *et al.* (1971)
Cyclic GMP	+			+	See Greengard (1976)

2. Transmitters, Putative Transmitters, and Modulators

A number of bioactive peptide and nonpeptide substances are present in various ganglionic elements (see Tables 1 and 2). The identification of peptides is due mostly to the investigations carried out by means of histofluorescence methods by Hökfelt and his associates [for references see Tables 1 and 2 and Klingman *et al.* (1979)].

Several comments are appropriate. First, substances listed in these two tables are intrinsic to the ganglia and their elements and/or are present in the blood; thus, they can be classified as autacoids (Schäfer, 1916).

Second, some of these substances are either transmitters or putative transmitters; that is, their presence in ganglionic neural elements, their release, and their capacity to generate synaptic potentials were either demonstrated or are expected to be demonstrated soon. Here belong, of course, ACh, and putatively, certain peptides, such as substance P and LH-RH, that may be responsible for the late slow EPSP (see above and Chapter 8), and catecholamines (possibly only when present in the SIF cells) (see Chapters 9 and 13). However, rather than being released conventionally from preganglionic neurons, the peptides, such as substance P, may be released from afferent circuitry and thus constitute components of reflex systems (see Chapters 13 and 16–18).

Third, some of the substances listed in Tables 1 and 2 may serve as modulators rather than as neurotransmitters generating synaptic potentials; somatostatin or vasoactive intestinal peptide may serve as examples.

Serotonin and catecholamines may exert modulatory activities (see Chapters 12 and 13) that are related to their presence in the blood; acting in this capacity, they should be considered as modulators that are extrinsic with regard to the ganglion. As already indicated, catecholamines may therefore play several roles: They may serve as transmitters (via their release from SIF cells) or as extrinsic modulators (via their presence in the blood); finally, they may also serve as modulators via their release from primary neurons (see Chapter 12). Besides catecholamines, other substances listed in Tables 1 and 2 may serve as both transmitters and modulators; even ACh may be included in this category (see Chapter 11).

It should be added that additional autacoids may act as extrinsic (extraganglionic) modulators. These are histamine, angiotensin, and somatostatin, as well as certain purines, ATP, and others; they are not listed in Tables 1 and 2. It is of great interest that some of these substances exert very potent effects on the ganglion (Dun *et al.*, 1978) (see also Chapter 13). Of course, there is always a possibility that ganglionic presence of these substances may be demonstrated in the future.

3. Functional Connections between Neural Elements

Combined electrophysiological and morphological studies (see Nishi *et al.*, 1967, 1978; Skok, 1973, 1979) (see also Chapters 2 and 6) led to the general belief that the nicotinic postsynaptic receptors that subserve the fast EPSP are directly subsynaptic and that they respond sensitively [one nicotinic receptor needs for its activation just a few ACh molecules (cf. Chapters 5 and 6)] and instantaneously to released ACh. In contrast, the muscarinic receptor that generates the slow EPSP may be situated at a distance from the nerve endings, which would explain the slow appearance of the EPSP; however, its slowness was explained by some investigators (see Hartzell, 1981) as slowness of the binding processes that occur at the muscarinic site or the long duration of the channel-opening times (see Chapter 7).

There is also some controversy as to the location of the site that generates the late slow EPSP. The peptide responsible for this action may be released from only some preganglionic neurons and, after diffusion, activate only certain distant postganglionic neurons; this may occur in amphibia, in which, as suggested by L. Y. Jan and Y. N. Jan (1982), the late-slow-EPSP-generating peptide is released from the preganglionic C fiber and, after diffusion, acts on the postganglionic B cell. The site that generates the late slow EPSP is usually diagrammed as being distant from the presynaptic ending even when the peptide is assumed to act on the postganglionic neuron innervated by the analogous preganglionic neuron (Y. N. Jan *et al.*, 1979; L. Y. Jan and Y. N. Jan, 1982).

Finally, the slow IPSP is generated by a separate postganglionic site;

it is a moot point at this time whether the slow IPSP is generated disynaptically following the involvement of an interneuron, the SIF cell, as originally proposed by Eccles and Libet (1961) or monosynaptically (Weight and Padjen, 1973; Cole and Shinnick-Gallagher, 1984) (see Chapters 9 and 13). The disynaptic nature of the generation of the slow IPSP may explain its delayed and slow time-course; however, this time-course may also be explained by the nature of its receptor, channel, and ionic mechanisms (Cole and Shinnick-Gallagher, 1984) or by the kinetics of the transmitter binding, as was the case with the slow EPSP. In fact, the time sequence of the slow potentials is a problem in itself: The slow IPSP should not precede the other slow potentials if the disynaptic vs. monosynaptic nature of the potentials is the primary consideration with respect to the time of their occurrence; as already indicated, other factors may also be involved.

Limited evidence is currently available that could have a bearing on the pathways or circuitries that may generate other than cholinergic effects on the ganglion, whether transmissive, modulatory, or, finally, reflex effects. Certain aspects of this matter may be mentioned.

A likely additional circuitry that may affect the ganglion is afferent in nature (see Chapters 13 and 19). Thus, afferents from the colon may be involved reflexly in the release of substance P into the guinea pig inferior mesenteric ganglion and in the generation of the late slow EPSP (Elfvin and Dalsgaard, 1977; Dalsgaard *et al.*, 1982; Dun and Jiang, 1982). It is difficult to comment at this time on the pathway involved in the generation of the late slow EPSP of the amphibia, which appears to be evoked by another peptide, LH-RH (L. Y. Jan and Y. N. Jan, 1982).

Efferent autonomic neurons, i.e., the postganglionic cells, their axons or dendrites, may also constitute circuitry that controls ganglionic activity. Thus, Minota and Koketsu (1978) proposed that direct stimulation of amphibian postganglionic somata may generate via a recurrent pathway a feed-forward excitation of the preganglionic fibers, resulting in the initiation of the fast EPSP.

Dun and Minota (1981) proposed a variant of this efferent system, relating the posttetanic depolarization generated by repetitive stimulation of mammalian postganglionic neurons to the autoreceptive action of a substance released from postganglionic soma or dendrites. Still another circuitry that may also be termed as efferent may be mediated by interaction between various types of postganglionic cells. Thus, in the rat pelvic ganglion (see Chapters 15 and 21), which contains sympathetic and parasympathetic postganglionic cells, cholinergic cells are enclosed by a plexus of adrenergic nerve endings originating in the sympathetic postganglionic cell (Dail, 1976); unfortunately, there is at this time (see Owman *et al.*, 1983) insufficient evidence for relating these morphological findings to electrophysiological results that would indicate that there is an adrenergic efferent control of parasympathetic cells of the pelvic ganglion (see, however, DeGroat and Saum, 1971, 1976) (see also Chapter 21).

4. Ganglionic and Preganglionic Receptors

It appears that a number of specialized receptors, both pre- and postsynaptic, are present in the sympathetic ganglia.

The postsynaptic receptors include, first of all, those involved in ganglionic transmission; these are nicotinic, muscarinic, and peptidergic postsynaptic receptors as well as receptors responsible for the generation of the slow IPSP. The latter may represent either catecholaminergic receptors or cholinergic muscarinic receptors that differ from those responsible for the generation of the muscarinic slow EPSP and that are capable of causing an atropine-sensitive postsynaptic hyperpolarization (see Chapters 9 and 13). It should be emphasized that we deal here with specific receptors endowed with specialized chemical relationships expressed, for example, by their characteristic and differential structure-activity relationships (see Chapter 6) (Skok, 1979; Trcka, 1979; D. A. Brown, 1979).

In addition, we must distinguish postsynaptic receptors that respond to modulators and that do not generate synaptic potentials but affect the latter or the action potentials or both (see above and Chapter 12).

Presynaptic receptors are involved in presynaptic modulation (see above) of ganglionic transmission. These receptors are concerned with both facilitatory and inhibitory modulations. Altogether, several types of presynaptic receptors, including catecholaminergic, cholinergic, histaminergic, serotonergic, and enkephalinergic receptors, have been described (see Chapters 10 and 11). The role and the significance of inhibitory presynaptic catecholaminergic receptors are of particular interest, since their inhibitory function may be functionally more important than their inhibitory postsynaptic function reflected in the slow IPSP (see Chapter 9). It is of interest that, recently, it was proposed that certain spinal neurons, may release enkephalinergic inhibitory modulators abutting on cholinergic nerve terminals (DeGroat *et al.*, 1986) (see also Chapter 13).

III. INTERACTIONS AND REFLEX FUNCTIONS

Besides the pre- and postsynaptic modulation, other phenomena contribute to the moment-to-moment regulation of ganglionic transmission.

Pertinent events include reflex regulation of ganglionic transmissions, which was already referred to (Skok, 1979) (see also Chapter 17). Additional aspects of afferent ganglionic innervation are discussed in Chapters 18 and 19. On a more general level, there is a reflex control of preganglionic neurons; knowledge of this control, the transmitters involved, and preganglionic pharmacology is still in its infancy.

Still another set of phenomena related to ganglionic regulation concerns the interaction between postsynaptic potentials and their modulation by extrinsic and intrinsic bioactive substances. For example, there is an interplay between the slow IPSP and the slow EPSP that may regulate postsynaptic excitability (see Chapters 12 and 13); this phenomenon may also relate to the interaction between the IPSP and postganglionic repetitive firing that can be induced pharmacologically (cf. Nishi and Koketsu, 1968; Karczmar, 1969; Volle, 1966; Volle and Hancock, 1970) as well as, presumably, under physiological conditions of increased excitability. Similarly, catecholamines possibly generated by the SIF cell, as proposed by Libet (1970), may, even when present in concentrations insufficient to induce hyperpolarization, facilitate or potentiate the slow EPSP (Ashe and Libet, 1981); extrinsic (blood-borne) catecholamines could act similarly. Furthermore, the slow EPSP and the muscarinic agonists (when extrinsic or blood-borne) may augment the fast EPSP (Schulman and Weight, 1976) (see Chapters 7 and 21); accordingly, the catecholaminergic potentiation of the slow EPSP suggested by Libet (1970) may ultimately lead to the facilitation of the primary transmission. Finally, peptides may act as modulators of postsynaptic potentials, either directly or via second-messenger mechanisms (Hedlund *et al.*, 1986) (see also below).

Another interactive phenomenon may result from the generation in the ganglia of cyclic nucleotides (Greengard and Kebabian, 1974; Greengard, 1976) (see Chapters 9 and 13). While the suggestion that these nucleotides act as second messengers and underlie the generation of the slow IPSP and EPSP (Greengard and Kebabian, 1974; Greengard, 1976) may not be valid (see Chapters 9 and 13), the actual generation of these substances during ganglionic activity or via the action of ganglionic stimulants, or both, is well documented (Greengard, 1976); it is entirely possible that cyclic nucleotides may contribute to the control of ganglionic potentials and ganglionic excitability (cf. Hedlund *et al.*, 1985). A related concept may be that of a second-messenger or modulatory role for the components of the phosphatidylinositol cycle and phospholipid metabolism (Hokin-Neaverson, 1977; Karczmar, 1981; Dun and Karczmar, 1981) (see Chapter 13).

IV. CONCLUSIONS: SYMPATHETIC GANGLIA AS "LITTLE BRAINS"

Any single synaptic event—such as a slow potential—is a highly complex phenomenon subject to almost infinite variation and regulation. This complex regulation includes peripheral and central reflex mechanisms; ganglionic circuitry; interaction between ganglionic potentials; interplay between a number of transmitters, putative transmitters, and modulators, including cyclic nucleotides, on the level of both preganglionic and postganglionic neurons; regulatory action of extrinsic bioactive substances, including steroids (Owman *et al.*, 1983); and other effects.

Thus, the sympathetic ganglion functions like a "little brain," and ganglionic investigations of the mechanisms of synaptic transmission and of their interactions are important not only because of the interest in ganglionic function and its regulation of cardiovascular and visceral activities (see Chapter 20), but also because such studies help in the understanding of the central synaptic mechanisms. The latter aspect is stressed in Chapter 21. Indeed, as compared to the central nervous system (CNS), the sympathetic ganglion offers technical advantages for electrophysiological analysis (Nishi *et al.*, 1978): Ganglionic preparations can be readily isolated so that the extracellular environment can be easily altered, leading to convenient evaluation of ionic mechanisms involved in the generation of synaptic potentials; intracellular recording of a single ganglion-cell activity is easier and more stable, may be pursued for longer time intervals, and is more reliable than the recording of the activity of the central neurons; when amphibian sympathetic ganglion cells, devoid of dendrites, are used, the synaptic potentials, particularly of the slow variety, can be analyzed without the serious problem of their electronic decay, which is inevitable in the case of synaptic potentials of the neurons of the mammalian CNS and mammalian sympathetic ganglia because of the cable properties of their dendrites; and the ganglion can readily serve to study mechanisms and interplays concerning the action of several neurotransmitters and putative transmitters, modulators, and bioactive substances. Finally, comprehensive analysis of the ganglionic slow synaptic potentials, such as slow EPSP, slow IPSP, and late slow EPSP, is very useful for the understanding of the CNS transmission and function, since similar slow synaptic potentials are prevalent in the CNS (Krnjevic, 1974).

ACKNOWLEDGMENTS. Published and unpublished results obtained in our laboratories and referred to in this chapter were supported by NIH-NDB, Grants 00655 and 15848, by USA-AMDoD Contract 17-83-C-3133, and by a Grant-in-Aid from the Ministry of Education, Science and Culture of Japan.

REFERENCES

Ashe, J. H., and Libet, B.: Modulation of slow postsynaptic potentials by dopamine in rabbit sympathetic ganglion. *Brain Res.* **217**:93–106 (1981).

Bertilsson, L., Suria, A., and Costa, E.: γ-Aminobutyric acid in rat superior cervical ganglion. *Nature (London)* **260**:540–541 (1976).

Birks, R., and MacIntosh, F. C.: Acetylcholine metabolism of a sympathetic ganglion. *Can. J. Biochem.* **39**:787–827 (1961).

Bishop, G. H., and Heinbecker, P.: A functional analysis of the cervical sympathetic nerve supply to the eye. *Am. J. Physiol.* **100:**519–532 (1932).

Brown, D. A.: Locus and mechanism of action of ganglion blocking agents, in: *Pharmacology of Ganglionic Transmission* (D. A. Kharkevich, ed.), pp. 185–235, Springer-Verlag, Berlin (1979).

Brown, G. L.: Conduction in the cervical sympathetic. *J. Physiol. (London)* **81:**228–242 (1934).

Cole, A. E., and Shinnick-Gallagher, P.: Muscarinic inhibitory transmission in mammalian sympathetic ganglia mediated by increased potassium conductance. *Nature (London)* **307:**270–271 (1984).

Dail, W. C.: Histochemical and fine structure studies of SIF cells in the major pelvic ganglion of the rat, in: *SIF Cells: Structure and Function of Small Intensely Fluorescent Sympathetic Cells* (O. Eränkö, ed.), pp. 8–18, U.S. Government Printing Office, Washington, D. C. (1976).

Dalsgaard, C. J., Hökfelt, T., Elfvin, L. G., Skibroll, L., and Emson, P.: Substance P-containing primary sensory neurons projecting to the inferior mesenteric ganglion: Evidence from combined retrograde tracing and immunohistochemistry. *Neuroscience* **1:**647–654 (1982).

DeGroat, W. C., and Saum, W. R.: Adrenergic inhibition to mammalian parasympathetic ganglia. *Nature (London)* **213:**188–189 (1971).

DeGroat, W. C., and Saum, W. R.: Synaptic transmission in parasympathetic ganglia in the urinary bladder of the cat. *J. Physiol. (London)* **256:**137–158 (1976).

DeGroat, W. C., Kawatani, M., Booth, A. M., and Whitney, T.: Enkephalinergic modulation of cholinergic transmission in parasympathetic ganglia of the urinary bladder of the cat, in: *Dynamics of Cholinergic Function* (I. Hanin, ed.), Plenum Press, New York (1986) (in press).

Di Giulio, A. M., Yang, H.-Y. T., Lutold, B., Fratta, W., Hong, J., and Costa, E.: Characterization of enkephalin-like material extracted from sympathetic ganglia. *Neuropharmacology* **17:**989–992 (1978).

Dun, N. J.: Ganglionic transmission: Electrophysiology and pharmacology. *Fed. Proc. Fed. Am. Soc. Exp. Biol.* **39:**2982–2989 (1980).

Dun, N. J., and Jiang, Z. G.: Non-cholinergic excitatory transmission in inferior mesenteric ganglia of the guinea-pig: Possible mediation by substance P. *J. Physiol. (London)* **325:**145–159 (1982).

Dun, N. J., and Karczmar, A. G.: Multiple mechanisms in ganglionic transmission, in: *Cholinergic Synapses* (G. Pepeu and H. Ladinsky, eds.), *Advances in Behavioral Biology*, Vol. 25, pp. 109–118, Plenum Press, New York (1981).

Dun, N. J., and Minota, S.: Effects of substance P on neurones of the inferior mesenteric ganglia of the guinea-pig. *J. Physiol. (London)* **321:**259–271 (1981).

Dun, N. J., Nishi, S., and Karczmar, A. G.: An analysis of the effect of angiotensin II on mammalian ganglion cells. *J. Pharmacol. Exp. Ther.* **204:**669–675 (1978).

Eccles, R. M.: Action potentials of isolated mammalian sympathetic ganglia. *J. Physiol. (London)* **117:**181–195 (1952).

Eccles, R. M., and Libet, B.: Origin and blockade of the synaptic responses of curarized sympathetic ganglia. *J. Physiol. (London)* **157:**484–503 (1961).

Elfvin, L.-G., and Dalsgaard, C.-J.: Retrograde axonal transport of horseradish peroxidase in afferent fibers of the inferior mesenteric ganglion of the guinea pig: Identification of the cells of origin in dorsal root ganglia. *Brain Res.* **126:**149–153 (1977).

Eränkö, O., and Harkonen, M.: Monoamine-containing small cells in the superior cervical ganglion of the rat and an organ composed of them. *Acta Physiol. Scand.* **63:**511–512 (1965).

Greengard, P.: Possible role for cyclic nucleotides and phosphorylated membrane proteins in postsynaptic actions of neurotransmitters. *Nature (London)* **260:**101–108 (1976).

Greengard, P., and Kebabian, J. W.: Role of cyclic AMP in synaptic transmission in the mammalian peripheral nervous system. *Fed. Proc. Fed. Am. Soc. Exp. Biol.* **33**:1059–1067 (1974).

Hartzell, H. C.: Mechanisms of slow postsynaptic potentials. *Nature (London)* **291**:539–544 (1981).

Hedlund, B., Abens, J., Westlind, A., and Bartfai, T.: Vasoactive intestinal polypeptide (VIP)-muscarinic cholinergic interactions, in: *Dynamics of Cholinergic Function* (I. Hanin, ed.), Plenum Press, New York (1986) (in press).

Hökfelt, T., Elfvin, L.-G., Elde, R., Schultzberg, M., Goldstein, M., and Luft, R.: Occurrence of somatostatin-like immunoreactivity in some peripheral sympathetic noradrenergic neurons. *Proc. Natl. Acad. Sci. U.S.A.* **74**:3587–3591 (1977a).

Hökfelt, T., Elfvin, L.-G., Schultzberg, M., Fuxe, K., Said, S. I., Mutt, V., and Goldstein, M.: Immunohistochemical evidence of vasoactive intestinal polypeptide-containing neurons and nerve fibers in sympathetic ganglia. *Neuroscience* **2**:885–896 (1977b).

Hökfelt, T., Elfvin, L.-G., Schultzberg, M., Goldstein, M., and Nilsson, G.: On the occurrence of substance P containing fibers in sympathetic ganglia: Immunohistochemical evidence. *Brain Res.* **132**:29–41 (1977c).

Hokin-Neaverson, M.: Metabolism and role of phosphatidylinositol in acetylcholine stimulated membrane function. *Adv. Exp. Med. Biol.* **83**:429–446 (1977).

Jan, L. Y., and Jan, Y. N.: Peptidergic transmission in sympathetic ganglia of the frog. *J. Physiol. (London)* **327**:219–246 (1982).

Jan, Y. N., Jan, L. Y., and Kuffler, S. W.: A peptide as a possible transmitter in sympathetic ganglia of the frog. *Proc. Natl. Acad. Sci. U.S.A.* **76**:1501–1505 (1979).

Karczmar, A. G.: Quelques aspects de la pharmacologie des synapses cholinergiques et de sa signification centrale. *Actual. Pharmacol.* **22**:293–338 (1969).

Karczmar, A. G.: Basic phenomena underlying novel use of cholinergic agents, anticholinesterases and precursors in neurological including peripheral and psychiatric disease, in: *Cholinergic Mechanisms* (G. Pepeu and H. Ladinsky, eds.), pp. 853–869, Plenum Press, New York (1981).

Karczmar, A. G., Nishi, S., and Blaber, L. C.: Synaptic modulations, in: *Brain and Human Behavior* (A. G. Karczmar and J. C. Eccles, eds.), pp. 63–92, Springer-Verlag, Berlin (1972).

Klingman, J. D., Organisciak, D. T., and Klingman, G. I.: Ganglionic metabolism, in: *Pharmacology of Ganglionic Transmission* (D. A. Kharkevich, ed.), pp. 41–62, Springer-Verlag, Berlin (1979).

Kohn, A.: Die Paraganglien. *Arch. Mikrosk. Anat.* **62**(15):399–400 (1898).

Koketsu, K.: Cholinergic synaptic potentials and the underlying ionic mechanisms. *Fed. Proc. Fed. Am. Soc. Exp. Biol.* **28**:101–112 (1969).

Kondo, H., Iwanaga, T., and Yanaihara, N.: On the occurrence of gastrin releasing peptide (GAP) like immunoreactive fiber in the celiac ganglion of rats. *Brain Res.* **289**:326–329 (1983).

Konishi, S., Tsunoo, A., and Otsuka, M.: Substance P and non-cholinergic excitatory synaptic transmission in guinea pig sympathetic ganglia. *Proc. Jpn. Acad.* **55**:525–530 (1979).

Krnjevic, K.: Chemical nature of synaptic transmission in vertebrates. *Physiol. Rev.* **54**:418–540 (1974).

Langley, J. N., and Dickinson, W. L.: Actions of various poisons upon nerve fibers and peripheral nerve cells. *J. Physiol. (London)* **11**:509–527 (1880).

Laporte, Y., and Lorente de Nò, R.: Potential changes evoked in a curarized sympathetic ganglion by presynaptic volleys of impulses. *J. Cell. Comp. Physiol.* **35**(Suppl. 2):61–106 (1950).

Larsson, L.-I., and Rehfeld, J. F.: Localization and molecular heterogeneity of cholecystokinin in the central and peripheral nervous system. *Brain Res.* **165**:201–218 (1979).

Leranth, C., Williams, T. H., Jew, J. Y., and Arimura, A.: Immuno-electron microscopic identification of somatostatin in cells and axons of sympathetic ganglia in the guinea pig. *Cell Tissue Res.* **212:**83–89 (1980).

Libet, B.: Generation of slow inhibitory and excitatory postsynaptic potentials. *Fed. Proc. Fed. Am. Soc. Exp. Biol.* **29:**1945–1956 (1970).

Libet, B.: The SIF cell as a functional dopamine-releasing interneuron in the rabbit superior cervical ganglion, in: *SIF Cells: Structure and Function of Small Intensely Fluorescent Sympathetic Cells* (O. Eränkö, ed.), pp. 163–177, U.S. Government Printing Office, Washington, D.C. (1976).

Lundberg, J. M., Hökfelt, T., Anggard, A., Uvnas-Wallensten, K., Brimijoin, S., Brodin, E., and Fahrnekrug, J.: Peripheral peptide neurons: Distribution, axonal transport and some aspects on possible function, in: *Neural Peptides and Neuronal Communication* (E. Costa and M. Trabucchi, eds.), pp. 25–36, Raven Press, New York (1980).

McAfee, D. A., Schorderet, M., and Greengard, P.: Adenosine 3′,5′-monophosphate in nervous tissue: Increase associated with synaptic transmission. *Science* **171:**1156–1158 (1971).

Minota, S., and Koketsu, K.: Recurrent synaptic activation of the bullfrog sympathetic ganglion cells by direct intracellular stimulation. *Jpn. J. Physiol.* **28:**799–806 (1978).

Nishi, S.: Cholinergic and adrenergic receptors at sympathetic preganglionic nerve terminals. *Fed. Proc. Fed. Am. Soc. Exp. Biol.* **29:**1957–1965 (1970).

Nishi, S.: Ganglionic transmission, in: *The Peripheral Nervous System* (J. I. Hubbard, ed.), pp. 225–255, Plenum Press, New York (1974).

Nishi, S., and Koketsu, K.: Analysis of slow inhibitory postsynaptic potential of bullfrog sympathetic ganglion. *J. Neurophysiol.* **31:**717–728 (1968).

Nishi, S., Soeda, H., and Koketsu, K.: Release of acetylcholine from sympathetic preganglionic nerve terminals. *J. Neurophysiol.* **30:**114–134 (1967).

Nishi, S., Karczmar, A. G., and Dun, N. J.: Physiology and pharmacology of ganglionic synapses as models for central transmission, in: *Advances in Pharmacology and Therapeutics*, Vol. 2, *Neurotransmitters* (P. Simon, ed.), pp. 69–85, Pergamon Press, Oxford (1978).

Owman, C., Alm, P., and Sjoberg, N.-O.: Pelvic autonomic ganglia: Structure, transmitters, function and steroid influence, in: *Autonomic Ganglia* (L.-G. Elfvin, ed.), pp. 125–143, John Wiley, Chichester and New York (1983).

Robinson, S. E., Schwartz, J. P., and Costa, E.: Substance P in the superior cervical ganglion and the submaxillary gland of the rat. *Brain Res.* **182:**11–17 (1980).

Schäfer, E. A.: *The Endocrine Organs: An Introduction to the Study of Internal Secretion.* Longmans, Green, New York (1916).

Schulman, J. A., and Weight, F. F.: Synaptic transmissions: Long lasting potentiation by a postsynaptic mechanism. *Science* **194:**1437–1439 (1976).

Schultzberg, M., and Dalsgaard, C. J.: Enteric origin of bombesin immunoreactive fibres in the rat coeliac-superior mesenteric ganglion. *Brain Res.* **269:**190–195 (1983).

Schultzberg, M., Hökfelt, T., Terenius, L., Elfvin, L.-G., Lundberg, J. M., Brandt, J., Elde, R. P., and Goldstein, M.: Enkephalin immunoreactive nerve fibres and cell bodies in sympathetic ganglia of the guinea-pig and rat. *Neuroscience* **4:**249–270 (1979).

Sherrington, C. S.: *Integrative Action of the Nervous System.* Yale University Press, New Haven (1906).

Skok, V. I.: *Physiology of Autonomic Ganglia.* Igaku Shoin, Tokyo (1973).

Skok, V. I.: On the physiological role of slow inhibitory postsynaptic potential in the neurons of sympathetic ganglia, in: *Electrobiology of Nerve, Synapses and Muscle* (I. P. Reuben, D. P. Purpura, M. V. L. Bennett, and E. R. Kandel, eds.), pp. 123–128, Raven Press, New York (1976).

Skok, V. I.: Ganglionic transmission, morphology and physiology, in: *Pharmacology of Ganglionic Transmission* (D. A. Kharkevich, ed.), pp. 9–39, Springer, Verlag, Berlin (1979).

Trcka, V.: Relationship between chemical structure and ganglion-blocking activity, in: *Pharmacology of Ganglionic Transmission* (D. A. Kharkevich, ed.), pp. 123–154, Springer-Verlag, Berlin (1979).

Trevisani, A., Biondi, C., Belluzzi, O., Borasio, P. G., Capuzzo, A., Ferretti, M. E., and Perri, V.: Evidence for increased release of prostaglandins of E type in response to orthodromic stimulation in the guinea pig superior cervical ganglion. *Brain Res.* **236**:375–381 (1982).

Verhofstad, A. A. J., Steinbusch, H. W. M., Penke, B., Varga, J., and Joosten, H. W. J.: Serotonin-immunoreactive cells in the superior cervical ganglion of the rat: Evidence for the existence of separate serotonin- and catecholamine-containing small ganglionic cells. *Brain Res.* **212**:39–49 (1981).

Vincent, S. R., Dalsgaard, C.-J., Schultzberg, M., Hökfelt, T., Christensson, I., and Terenius, L.: Dynorphin-immunoreactive neurons in the autonomic nervous system. *Neuroscience* **11**:973–987 (1984).

Volle, R. L.: Muscarine and nicotinic stimulant actions at autonomic ganglia, in: *Ganglionic Blocking and Stimulating Agents* (A. G. Karczmar, ed.), pp. 1–106, International Encyclopedia of Pharmacology and Therapeutics, Section 12, Vol. 1, Pergamon Press, Oxford (1966).

Volle, R. L., and Hancock, J. C.: Transmission in sympathetic ganglia. *Fed. Proc. Fed. Am. Soc. Exp. Biol.* **29**:1913–1918 (1970).

Weight, F. F., and Padjen, A.: Slow synaptic inhibition: Evidence for synaptic inactivation of sodium conductance in sympathetic ganglion cells. *Brain Res.* **55**:219–224 (1973).

Williams, T., and Jew, J.: Monoamine connections in sympathetic ganglia, in: *Autonomic Ganglia* (L.-G. Elfvin, ed.), pp. 235–264, John Wiley, Chichester, and New York (1983).

4

Electrophysiological Properties of Sympathetic Neurons

S. NISHI

Rosamond Eccles (1955) was the first investigator to record intracellularly from rabbit superior cervical ganglion (SCG) cells; subsequently, data on the active and passive electrical properties of autonomic neurons have accumulated gradually. The specific membrane constants of ganglion cells are similar to those of somatic motor and other types of neurons. There are, however, some unique neuronal properties of the ganglion cells.

I. MEMBRANE CHARACTERISTICS AT REST

A. Resting Potential

Measured with intracellular microelectrodes, the resting membrane potential of ganglion cells usually varies between −50 and −70 mV, although the reported values range from −40 to −110 mV (Table 1). This large range is probably due to technical artifacts, such as neuronal injuries or tip potentials of microelectrodes.

Woodward *et al.* (1969) measured the electrolyte distribution in the rabbit SCG. The intracellular concentrations of Na^+ ($[Na^+]_i$), K^+ ($[K^+]_i$), and Cl^- ($[Cl^-]_i$) were reported to be 37.4, 163, and 23.5 μmoles/ml, respectively; the extracellular concentrations of Na^+ ($[Na^+]_o$), K^+ ($[K^+]_o$) and Cl^- ($[Cl^-]_o$) were found to be 133, 4.2, and 111 μmoles/ml, respectively.

S. NISHI • Department of Physiology, Kurume University School of Medicine, Kurume, Japan.

Table 1. Electrical Membrane Properties of Sympathetic Ganglion Cells[a]

Ganglion	Soma diameter (μm)	E_m (mV)	R_N (MΩ)	τ (msec)	R_m (Ωcm^2)	C_m (μF/cm^2)	Reference
Lumbar							
Frog	22–32	55–75	12–24	7.2–13	235–738	14–40	Nishi and Koketsu (1960)
	37.5–47.5	48 ± 8.8	100 ± 46	10–12		4.9	Hunt and Nelson (1965)
Pelvic							
Guinea pig	20–30	40–70	40–150	5–200	2,000–10,000	2–5	Blackman *et al.* (1969)
Thoracic							
Guinea pig	31[b] and 19[c]	50–70	19–186	6–14	1,000	10	Blackman and Purves (1969)
Superior cervical							
Rat	4.4[d]	45–90	39.4 ± 3.2	2.7 ± 0.3	1,800	1.6	Perri *et al.* (1970)
Guinea pig	5.6[d]	45–90	40.6 ± 4.3	4.4 ± 0.5	2,300	2.1	
Rabbit	<70	40–100		2.8–7.7		3–17	Erulkar and Woodward (1968)
Cat		57 ± 2.0	15.5 ± 2.2	8.5–10	500–3,200		Skok (1968)

[a] (E_m) Resting membrane potential; (R_N) total neuron resistance; (τ) membrane time constant; (R_m, C_m) resistance and capacitance for a unit area of membrane, respectively. Resting membrane potentials are negative in polarity but expressed as positive.
[b] Major diameter. [c] Minor diameter. [d] Surface area, $\times 10^{-5}$ cm^2.

The equilibrium potentials calculated from the Nernst equation for Na^+ (E_{Na}), K^+ (E_K), and Cl^- (E_{Cl}) were 33.6, −96.9, and −41.1 mV, respectively. Using these values and the experimentally obtained value of the resting potential (−68.6 mV), Skok (1973) estimated the relative membrane permeabilities at rest for Na^+, K^+, and Cl^- ($P_{Na} : P_K : P_{Cl}$) as 0.06 : 1 : 0.02.

D. A. Brown and Scholfield (1974a) reported that the $[Na^+]_i$ and $[K^+]_i$ of rat SCG neurons were 22 and 207 mM, respectively, which yielded values of +49 mV for E_{Na} ($[Na^+]_o$ = 149 mM) and −88 mV for E_K ($[K^+]_o$ = 6.6 mM). The $P_{Na} : P_K$ at a membrane potential of −70 mV (Perri *et al.*, 1970; Adams and Brown, 1973) appeared to be 0.04 : 1, assuming $P_{Cl} : P_K < 0.1$.

Among the intracellular element concentrations of rat SCG neurons recently obtained with electron-microprobe analysis (Galvan *et al.*, 1984) or ion-selective microelectrodes (Ballanyi *et al.*, 1983), the values for $[Na^+]_i$ [5 mM (Galvan *et al.*, 1984); 11 mM (Ballanyi *et al.*, 1983)] appear to be much lower than chemically measured values (see above), while the $[K^+]_i$ value (122 mM) obtained by ion-selective microelectrodes appears to be considerably lower than that (196 mM) obtained with electron-microprobe analysis. If these newly obtained values are more accurate than the earlier ones, E_{Na} is more positive than the conventional value of approximately +50 mV and the cellular potassium activity coefficient is much smaller than that in the extracellular space. The cause of these discrepancies remains to be investigated.

In the frog sympathetic ganglion, P_{Cl} appears to be small because the substitution of methylsulfate for Cl^- has no effect on the membrane potential (Blackman *et al.*, 1963). This small P_{Cl} of resting ganglion cells of mammals and amphibia is characteristically different from the P_{Cl} of the muscle membrane, which is as large as P_K (Hodgkin and Horowitz, 1959). Another difference between the ganglion cell and the muscle membrane concerns the relationship between $[K^+]_o$ and the resting potential (E_m). E_m values for ganglion cells are smaller than the E_K values expected from the Nernst equation even when $[K^+]_o$ values are greater than 10 mM. E_m can be approximated when the presumed ratio (α) between P_{Na} and P_K is as large as 0.1 in the equation provided by Blackman *et al.* (1963): $E_m = 59 \log ([K^+]_o + \alpha[Na^+]_o)/([K^+]_i + \alpha[Na^+]_i)$.

One would then expect that the replacement of Na^+ by an impermeant cation will hyperpolarize the membrane. It is puzzling, however, that only a small depolarization occurs when Na^+ is replaced with impermeant foreign cations.

B. Electric Constants

Since amphibian sympathetic ganglion cells are unipolar and devoid of dendrites (Taxi, 1961; Uchizono, 1964; Nishi *et al.*, 1967), electric constants such as the resistance and capacitance for a unit area of membrane (R_m and C_m) may be estimated from the total neuron resistance [effective or input resistance (R_N)] and the membrane time constant (τ_m) assuming the ganglion cell body is a sphere (Table 1). Our earlier measurements of membrance constants of amphibian ganglion cells (Nishi and Koketsu, 1960; Nishi *et al.*, 1965) underestimated R_N and overestimated C_m. The values recently obtained (unpublished observations) are similar to the values reported by Hunt and Nelson (1965) (Table 1).

Since mammalian autonomic neurons are multipolar and exhibit a varying number of processes (Matthews, 1983), the electric current applied intracellularly passes across the soma membrane and also the membranes of the dendritic tree and the axon. The current flowing along each of these paths is dependent on both the electrical and the geometrical properties of the neuron (Rall, 1959, 1960). It is therefore necessary to determine the geometrical features of each neuron from which passive electrical transients are recorded. Procion dye (Stretton and Kravitz, 1968) or certain other staining substances, e.g., horseradish peroxidase (HRP) (Snow *et al.*, 1976), can be iontophoretically injected into the ganglion cell cytoplasm from an intracellular recording electrode. This intracellular recording and staining method permits fairly precise measurements of electrical and geometrical parameters of single neurons.

Neurons of rabbit SCG or cat ciliary ganglion intracellularly stained with the procion yellow, M4RS (Nishi and Christ, 1971), or with HRP appear as ellipsoids, having 10–20 dendrites each, 1–3 μm in diameter (see Chapters 2 and 16). The dendrites usually have no branches and extend a few hundred micrometers from the soma. Table 2 shows the passive electrical membrane constants of these neurons obtained by intracellular recording and staining with procion-yellow- (Nishi and Christ, 1971) or HRP-containing electrodes (Uchiyama, Higashi, and Nishi, unpublished observations); these constants can be compared with those of cat spinal motoneurons measured with a similar technique (Barret and Crill, 1971). The membrane resistance for a unit area (R_m) and the dendritic-to-soma conductance ratio (ρ) of the ganglion cells were calculated from equations (1) and (2) established by Rall (1959):

$$R_m = [\,1 + \sqrt{1 + 4S/C^2D^3R_N}]^2C^2D^3R_N{}^2/4 \qquad (1)$$

$$\rho = C\,(D^{3/2}/S)\sqrt{R_m} \qquad (2)$$

where $C = (\Pi/2)\,(R_i)^{-1/2}$, and $D^{1/2} = \Sigma_j B_j\,(d_j)^{3/2}$. S, R_N, and R_i represent, respectively, the soma surface area, total neuron resistance, and specific resistivity of internal medium. B_j and d_j are the branch constant and dendritic

Table 2. Membrane Constants Obtained by Intracellular Recording and Staining Methods[a]

Neuron	Dye	Soma surface area (μm^2)	Dendrites Number	Dendrites Length (μm)	R_N ($M\Omega$)	R_m (Ωcm^2)	ρ	τ_0 (msec)	C_m ($\mu F/cm^2$)	λ (μm)	L
Superior cervical ganglion cells (rabbit)	PY	2900	14	180	32.4	4640	4.5	11.5	2.5	490	0.38
	HRP	1570	11	190	85.0	7855	6.2	10.3	1.7	753	0.27
							(5.6)[b]		(1.7)[b]		(0.74)[b]
Ciliary ganglion cells (cat)	PY	4400	15	200	18.5	3320	3.3	6.1	2.1	425	0.46
	HRP	3630	9	260	35.3	7430	4.9	9.0	1.4	959	0.29
							(5.5)[b]		(1.3)[b]		(0.90)[b]
Spinal motoneurons (cat)	PY	144,690[c]	—	—	1.9	1770–2520	9.3–12.7	5.2	2.1–2.9	—	1.4

[a] (R_N) Total neuron resistance; (R_m) resistance across a unit area of membrane; (ρ) ratio of combined dendritic input conductance to soma membrane conductance; (τ_0) membrane time constant; (C_m) capacitance for a unit area of membrane; (λ) space constant of dendrite; (L) electrotonic length of dendrite; (HRP) horseradish peroxydase; (PY) procion yellow.
[b] Values obtained by the peeling method.
[c] Surface area of soma and dendrites.

diameter associated with the *j*th dendrite of a ganglion cell, respectively. In actual calculation, the average values of geometric parameters of all the dendrites were used, instead of dealing with each dendrite separately. Hence, the expression for the combined dendritic parameter becomes $D^{3/2} = NBd^{3/2}$, where N is the number of dendrites and B and d are the averages of all B_j's and of all d_j's, respectively. The dendrites of SCG and ciliary neurons are relatively short and usually unbranched. This corresponds to the "sealed-end" case in Rall's terminal boundary conditions (Rall, 1959), and the branching constant would be

$$B = \tanh\,(l_0/\lambda_0) \tag{3}$$

where

$$\lambda_0 = \sqrt{(d/4)\,(R_m/R_i)} \tag{4}$$

where λ_0 represents the average dendritic space constant and l_0 the average length of all dendrites.

The value of B, hence the trunk parameter D as well, actually depends on R_m, so that a direct calculation of R_m from equation (1) is impossible. Therefore, the method of successive iteration was chosen for estimation of R_m, according to the following steps: First, B is set at 1, the upper limit of all B_j's, to calculate $D^{3/2}$ as equal to $NBd^{3/2}$; then R_m and λ are calculated by means of equations (1) and (4), respectively. This is then followed by recalculation of B according to equation (3). These steps were repeated until successive approximations of R_m gave almost identical results. This value of R_m was taken to be the final estimate of the membrane resistance.

For a multipolar neuron, the passive decay transients can be expressed as a sum of exponential decays:

$$V = C_0 e^{-t/\tau_0} + C_1 e^{-t/\tau_1} + C_2 e^{-t/\tau_2} + \cdots + C_n e^{-t/\tau_n} \tag{5}$$

where $\tau_0 = \tau_m$ (passive membrane time constant); τ_1, τ_2, and τ_n represent many equalizing time constants, all of which are smaller than τ_0; and C_0, C_1, C_2, and C_n are arbitrary constants (Rall, 1969). In the case of a lumped soma coupled with a cylinder that is sealed at one end, the relationship between τ_0 and τ_n was determined by Rall (1969) to be $\tau_n = \tau_0/(1 + \alpha_n^2)$, where α_n's are roots of the transcendental equation $\alpha L \cot(\alpha L) = -N\rho L/\tanh L$, in which N and L are, respectively, the number of dendrites and the dimensionless electrotonic length (l/λ). The "peeling" of semilogarithmic plots of dV/dt vs. t from the voltage transients induced by imposing a hyperpolarizing current step reveals the values of the membrane time constant (τ_0) and its coefficient (C_0), as well as C_n and τ_n (Lux and Pollen, 1966; Burke and ten Bruggencate, 1971; T. H. Brown *et al.*, 1981). With this exponential peeling technique, the values for ρ and L can be estimated without obtaining geometric parameters by the following equations (T. H. Brown *et al.*, 1981):

$$\rho = (G_N\tau_0/I)\sum_{i=0}^{\infty}(C_i/\tau_i) - 1 \tag{6}$$

$$L \simeq \pi\left(\frac{\rho/(\rho + 1)}{\tau_0/\tau_1 - 1}\right)^{1/2} \tag{7}$$

(In practice, τ components beyond $i = 1$ could not be extracted from the charging transients of ganglion cells.)

Table 2 shows that whereas the total neuron resistance (R_N) of the ganglion cells is at least 10 times that of motoneurons, their membrane resistance for a unit area (R_m) appears to be only about two to three times that of the motoneurons. The difference in R_N between ganglion cells and motoneurons can therefore be attributed to the difference in the effective surface area. The dendritic-to-soma conductance ratios (ρ) for the ganglion cells vary between 3 and 6 compared with ratios of 9 to 13 for the motoneurons. This indicates that approximately 1/4th–1/7th of the current applied to the soma of ganglion cells passes through the soma membrane, the rest of the current flowing through the dendritic and axonic membranes; in contrast in the case of the motoneurons, only 1/10th–1/14th of the current applied to the soma passes through the soma membrane, the rest flowing through the cell processes. The average space constants of the superior cervical and ciliary neuron dendrities appear to be much greater than the actual length of the dendrites. These data suggest that the synaptic inputs onto ganglion cell dendrites influence the soma membrane potential more effectively than the synaptic inputs onto motoneuron dendrities.

C. Voltage–Current Relationship

In amphibian as well as mammalian ganglion cells, the voltage–current (V–I) relationship, whether with depolarizing or hyperpolarizing potentials, is almost linear when the applied currents produce potential deflections of less than 10–15 mV from the resting potential of approximately −70 mV. When the depolarization exceeds 10–15 mV, a gradual decrease in slope conductance appears due to a rectification of the cell membrane. When the hyperpolarization of the amphibian ganglion cells exceeds 30–40 mV, the slope has a tendency to become slightly steeper than that obtained with smaller current intensities. When the membrane potential of mammalian (rabbit) SCG cells is increased to approximately

−90 to −100 mV by passing a steady inward current, the hyperpolarizing electrotonic potential becomes smaller (Figure 1). With a further increase in the membrane potential to the range of −120 to −130 mV, the hyperpolarizing potential regains its original amplitude while the depolarizing potential decreased (Figure 1). At membrane potentials higher than −150 mV, the depolarizing and hyperpolarizing potentials have almost the same amplitude, although they are often 10–20% larger than at the resting membrane potential. Thus, over a −90 to −140 mV range of membrane potentials, the electrotonic potential is markedly decreased; this constitutes the range of anomalous rectification of the SCG cells (Christ and Nishi, 1973).

A typical V-I relationship obtained in a rabbit SCG cell during passage of long depolarizing and hyperpolarizing currents of varying intensities is shown in Figure 2. The slope resistance decreases when the cell membrane is depolarized more than 10–15 mV. This rectification is due mainly to activation of a K^+ channel (D. A. Brown and Adams, 1980; Adams *et al.*, 1982) (see below). The slope resistance also decreases during the passage of a hyperpolarizing current (anomalous rectification). The amplitude of a short hyperpolarizing electrotonic potential (Figure 1, vertical bar), which is induced during delivery of continuous polarizing currents, is also shown. The amplitude of the short electrotonic potential is decreased within approximately the same range of membrane potentials that is associated with decrease of the slope resistance.

The membrane conductance for sodium and chloride does not appear to be changed significantly during the anomalous rectification, since the rectification is normal in the absence of these ions. The anomalous rec-

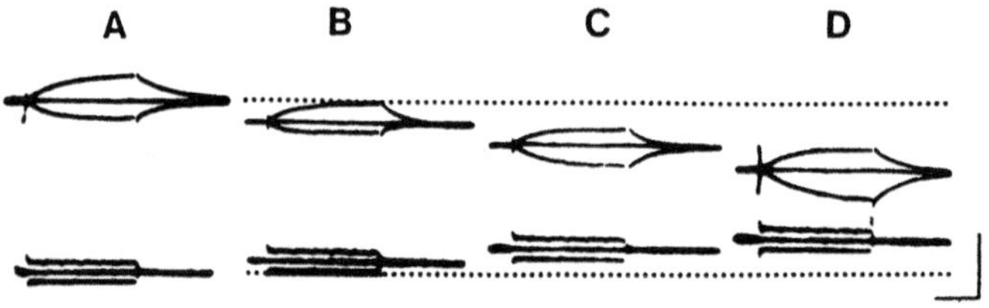

Figure 1. Depolarizing and hyperpolarizing electrotonic potentials recorded at various levels of hyperpolarization of the rabbit SCG cell membrane. Electrotonic potentials were induced by rectangular current pulses of 0.7 nA applied for 50 msec. The membrane potential was shifted by continuous anodal currents of varying intensities. Currents were passed through the recording microelectrode. (A–D) Electrotonic potentials (upper traces) were taken at −90 mV (upper dotted line), −104 mV, −118 mV, and −130 mV, respectively. The lower traces in each record are current recordings (cathodal, upward). The lower dotted line indicates the level of anodal current (0.9 nA) that shifted the resting membrane potential (−62 mV) to −90 mV. Calibrations: 40 mV (5 nA) and 20 msec. From Christ and Nishi (1973).

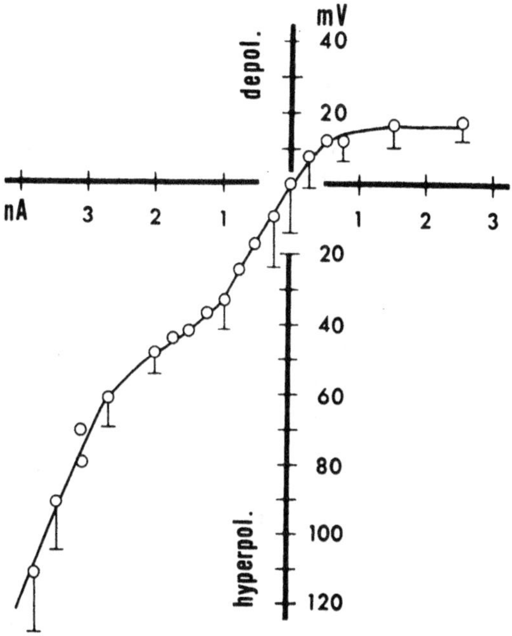

Figure 2. Steady-state V-I relationship of the rabbit SCG cell membrane obtained with a double-barreled microelectrode procedure. Steady currents were passed through the polarizing barrel and membrane potential was recorded with the recording barrel. Correction was made for the resistive coupling. Short anodal pulses (0.5 nA, 50 msec) were passed through the recording barrel. The vertical bar below each point represents the amplitude of the short hyperpolarizing electrotonic potential. The resting potential was −62 mV. From Christ and Nishi (1973).

tification is abolished by barium (10 mM) or strontium (10 mM). Calcium (10 mM) slightly reduces the anomalous rectification, whereas magnesium (10 mM) has almost no effect on the latter. Tetraethylammonium (TEA), which decreases the delayed rectification and greatly prolongs the action potential in the ganglion cells, is without effect on the anomalous rectification even at an isotonic concentration. These observations suggest that mammalian SCG cells are endowed with a K^+ channel that opens within a discrete range of hyperpolarization. This channel is depressed by Ba^{2+} or Sr^{2+} (Christ and Nishi, 1973).

The rectification of depolarizing currents is commonly seen in ganglion cells, particularly after the local response to subliminal catelectrotonic potentials is over or after the elimination of the action potential from supraliminal catelectrotonic potentials by removal of external Na^+. This rectification is not thought to be due to the conventional "delayed" rectifier described by Hodgkin and Huxley (1952a) (see below).

Brown and Adams (D. A. Brown and Adams, 1980; Adams and Brown, 1980; Adams *et al.*, 1982) demonstrated by means of experiments with voltage-clamped bullfrog sympathetic neurons the presence of a time- and voltage-dependent K^+ current that was selectively inhibited during muscarinic receptor activation; they referred to this unique current as the M current. The M current is activated at membrane potentials varying between −60 and −10 mV, with a half-maximal activation voltage at −35 mV and a minimum time constant of 150 msec at −35 mV (D. A. Brown *et al.*, 1981). The M current does not show time-dependent inactivation within its activation range. Between the potentials of −60 and −25 mV, it constitutes the only potential-sensitive outward membrane conductance of the cell.

D. A. Brown *et al.* (1981) and Adams *et al.* (1982) suggested that the primary role of the M current is to provide a braking effect on the rate of cell depolarization produced by any steady or phasic inward current that flows within potential range that exists between the resting potential and the spike threshold. Thus, the M current is expected to both reduce membrane excitability and limit the firing frequency of the neuron. More recently, Adams and Brown (Adams *et al.*, 1982) demonstrated that muscarinic agonists and depolarizing peptides such as LHRH block the M current, causing, besides depolarization, increased input resistance, reduced outward rectification, and increased excitability, as expected on the basis of their hypothesis; all these effects are expected to constitute physiologically significant aspects of the M current.

An interesting feature of this current is that it can be suppressed by synaptic processes; this will be described subsequently. Mammalian sympathetic neurons are also endowed with a K^+ channel that is similar to the M channels of amphibian sympathetic neurons (D. A. Brown *et al.*, 1980).

II. ACTIVE MEMBRANE CHARACTERISTICS

The action potential of sympathetic ganglion cells has a longer spike duration and a larger and longer after hyperpolarization than those of somatic neurons (Figure 3). Several parameters of action potentials in sympathetic and somatic neurons are shown in Table 3 (see also Nishi *et al.*, 1965). Other differences between somatic and ganglionic neurons include the difference in the time course of the recovery of excitability following an action potential (Nishi *et al.*, 1965) and the difference in excitability in modified ionic media, particularly when Na ions are replaced by divalent cations.

A. Action Potentials

1. Antidromic Action Potential

An impluse that antidromically propagates into a ganglion cell soma induces a soma action potential that frequently has a notch on its rising phase approximately 30–40 mV above the resting potential. Antidromic

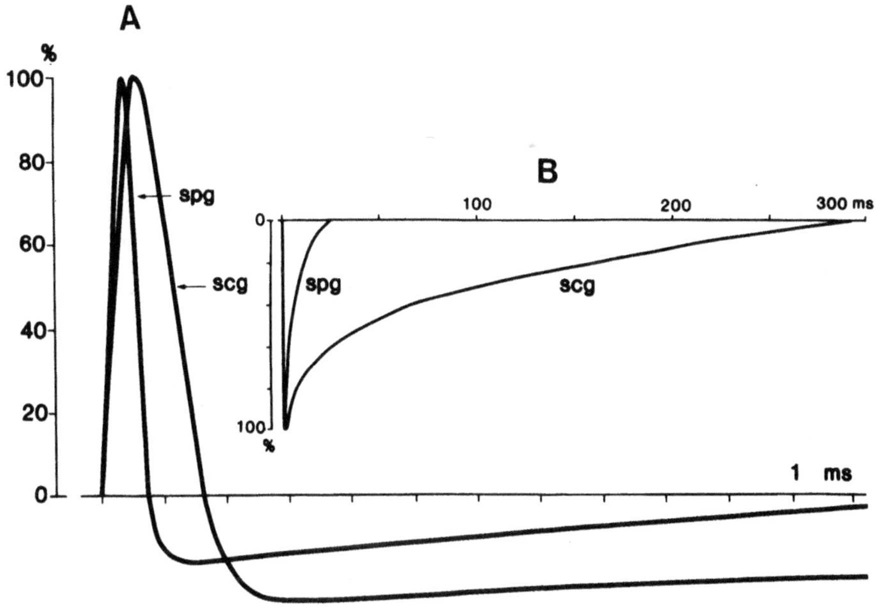

Figure 3. Comparison of typical antidromic responses of superior cervical ganglion (scg) and spinal ganglion (spg) neurons of the rabbit traced on the same time base. (A) Spike and afterpotential. (B) Afterpotential. The relative magnitude of each response is shown as a percentage; peak values were taken as 100. The horizontal lines at zero indicate the level of resting membrane potential. From Nishi (unpublished observations).

excitation at high frequencies or strong hyperpolarization of the cell membrane, or both, enhances the notch, and may even reduce the spike in an all-or-none manner, leaving a small spike (Nishi and Koketsu, 1960). The axon–soma junction, with its great expansion of the cell membrane, could be expected to impede the antidromic conduction. The notch, therefore, may be due to a delay in conduction from the axon to the soma, and the isolated small spike may be the action potential of the proximal axon, which can spread electrotonically to the soma. A full spike would be generated if the small spike attained the level of the notch. This suggests that the threshold for the soma membrane is equal to the height of the notch (Nishi and Koketsu, 1960).

In ganglion cells that have a myelinated axon (B-type neurons), the isolated small spike can be further fractionated in an all-or-none manner into a very small spike that may be attributed to the action potential of the proximal Ranvier nodes (Figure 4A). In C-type neurons, which have nonmyelinated axons, the small spike isolated from the full spike cannot be fractionated in an all-or-none manner. Instead, increasing hyperpolarization results in a gradual decrease of the amplitude of the isolated small spike (Figure 4B). This type of decrement in the isolated spike height

Table 3. Nature of Antidromic Responses in Mammalian Sympathetic and Somatic Neurons[a]

Parameter	SCG neuron	SPG neuron
Spike potential		
Height	81.9 ± 1.8 mV	103.6 ± 3.1 mV
Overshoot	23.1 ± 1.5 mV	31.9 ± 1.7 mV
Duration	1.73 ± 0.15 msec	0.69 ± 0.03 msec
Rate of rise	158.1 ± 5.6 V/sec	382.0 ± 5.9 V/sec
Rate of fall	90.5 ± 1.7 V/sec	207.5 ± 5.0 V/sec
Afterhyperpolarization		
Amplitude	25.5 ± 1.4 mV (31% of spike amplitude)	17.3 ± 1.4 mV (17% of spike amplitude)
Summit time	2.18 ± 0.20 msec	0.84 ± 0.06 msec
Decay half-time	77.0 ± 6.0 msec	5.56 ± 0.57 msec
Total duration	315.6 ± 18.3 msec	25.3 ± 2.5 msec

[a] From Nishi (unpublished observations). Values (means ± S.E.) were obtained from eight superior cervical ganglion (SCG) neurons and ten spinal ganglion (SPG) neurons of the rabbit *in vitro* at 37°C.

apparently reflects the lack of segmental myelination in the C-type neuron axon. The height of the isolated spike may be dependent on the distance between the tip of the electrode and the position where axonal blockade occurs (Nishi *et al.*, 1965).

2. Directly Activated Spike

The spike elicited by direct intracellular stimulation (direct spike) is similar in amplitude and time–course to the full-size antidromic spike. In B-type neurons of amphibian (Nishi *et al*, 1965) and mammalian [ciliary and parasympathetic (Nishi and Christ, 1971)] ganglia, however, the threshold depolarization for direct spike is 10–15 mV lower than the height of the notch on the rising phase of the antidromic spike. This difference indicates that direct intracellular stimulation initially evokes an action potential at the most excitable area of the cell membrane, equivalent to the initial segment of motoneurons (Coombs *et al*, 1957), which in turn induces the soma spike. In fact, a rapid and intense hyperpolarization applied during the rising phase of a direct spike can occasionally disclose a step on the rising phase and may even block the soma spike, leaving a small spike (Nishi and Koketsu, 1960). In C neurons of amphibian and mammalian ganglia (a majority of neurons in the latter are of the C type) the difference in excitability of the various areas of the neuron has not been clearly established (Eccles, 1963).

In amphibian B-type neurons, the firing frequency increases in proportion to the strength of applied current until it attains a maximum frequency of 40–70 action potentials per second. The majority of am-

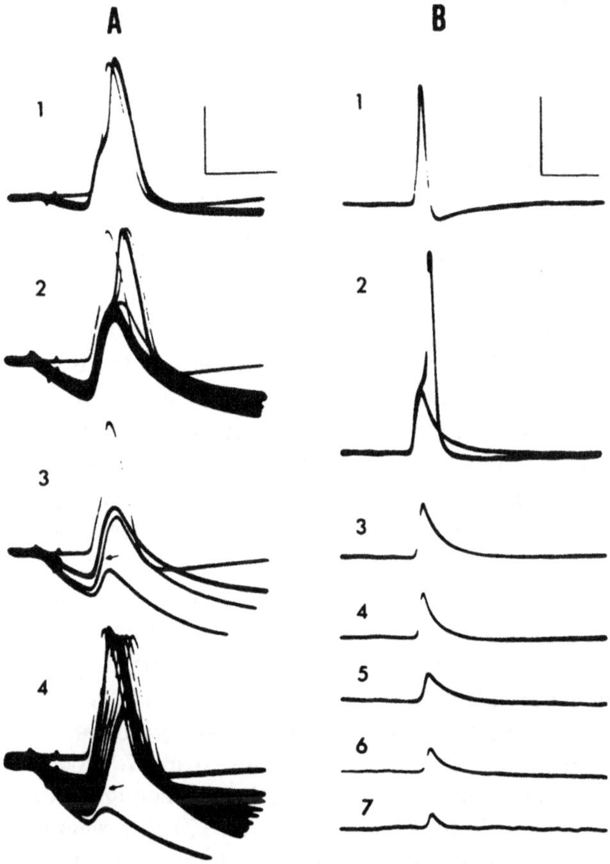

Figure 4. Separation of antidromic responses in amphibian B and C ganglion cells by hyperpolarization. (A) B-neuron responses evoked on the rising phase of hyperpolarizations of various magnitudes. In each record, the control response was always recorded before applying hyperpolarization. The intensity of applied current was increased stepwise from 1 to 3 and also changed progressively to some extent in each record. Record 4 shows the whole course of alteration of antidromic response under steeply rising hyperpolarization, the magnitude of which varied to the same extent as in records 1–3. Calibrations: 50 mV and 5 msec. From Nishi and Koketsu (1960). (B) C-neuron responses elicited at membrane potentials of −57, −100, −122, −144, −150, and −164 mV. The original resting potential was −45 mV. Calibrations: 50 mV and 20 msec. From Nishi *et al.* (1965).

phibian C-type neurons, on the other hand, develop a marked accommodation to the depolarizing current and produce only a few responses that are followed by decremental oscillations. Only a minority of C neurons are capable of firing repetitively, and the frequency of this firing amounts to only 20–40 action potentials per second. Unlike the amphibian C-type neurons, the mammaliam C-type neurons can maintain a high rate (maximum frequencies 80–170 Hz) of firing during continuous depolarization (Blackman and Purves, 1969).

3. Orthodromically Activated Spike

Comparison of the orthodromic response and the direct response recorded intracellularly from single amphibian ganglion cells reveals that the peak of the orthodromic action potential is slightly lower than the peak of the direct potential (Nishi and Koketsu, 1960). This difference is probably due to short-circuiting the neuronal membrane by the activated synaptic membrane. The threshold for the orthodromic spike is virtually identical with the threshold for the direct spike. Moreover, the orthodromic spike occasionally has a notch on its rising phase that coincides with the notch on the antidromic spike. This evidence suggests that the sites of impulse initiation by a synaptic potential and by a directly applied current are identical; namely, orthodromic and direct spikes are both initiated at the portion of the soma that has the lowest threshold. The action potential then propagates throughout the soma and down to the axon. A smilar mode of impulse generation was seen in the cat ciliary (parasympathetic) ganglion cells (Nishi and Christ, 1971). In the rabbit SCG neurons, on the other hand, the orthodromic spike seems to be generated from the ordinary soma-dendritic membrane and not from any specific area endowed with a lower threshold for excitation (Eccles, 1963).

As described in the section above, mammalian sympathetic neurons can fire at a maximum frequency of 80–170 Hz in response to direct intracellular stimulation, but they are incapable of responding at such a high frequency to repetitive orthodromic stimuli. Most of the SCG neurons, for example, can respond faithfully to preganglionic volleys applied at less than 30–35 Hz (Eccles, 1955; Skok, 1973); preganglionic volleys at higher frequencies cause intermittent postsynaptic spikes and subliminal synaptic potentials with varying amplitudes.

B. Afterpotentials

1. Afterhyperpolarization

It is a general characteristic of autonomic ganglion cells of either amphibia or mammals, that the action potential has a large and longlasting afterhyperpolarization. The amplitude of afterhyperpolarization usually amounts to 20–30% of the amplitude of the spike, and the total duration may extend to 150–300 msec (Nishi *et al.*, 1965; Skok, 1973). Both parameters are several times greater than those of spinal motoneurons and most other neurons. One might suspect that such a prominent afterhyperpolarization is due to membrane depolarization caused by microelectrode insertion, since the afterhyperpolarization increases with membrane depolarization. It is difficult to exclude this possibility, but the fact that the action potential recorded from the surface of the ganglion (Kosterlitz *et al.*, 1968; Nishi and Koketsu, 1968; Skok, 1973) also exhibits a large afterhyperpolarization supports the view that the large afterhyperpolarization is an inherent feature of these cells.

The afterhyperpolarization is associated with a decreased input resistance throughout its time course, is nullified at a potential close to E_K, and is reversed in polarity by hyperpolarizing the membrane beyond E_K (Nishi and Koketsu, 1968; Kuba and Nishi, 1976). Moreover, its amplitude and equilibrium potential are dependent on $[K^+]_o$ (Blackman *et al.*, 1963). These observations led to the hypothesis that the afterhyperpolarization is generated by a selective increase in G_K.

In rat SCG neurons, McAfee and Yarowsky (1979) observed that TEA and 4-aminopyridine, blockers of voltage-sensitive K^+ conductance changes (delayed rectifier), reduced only slightly the peak amplitude and the early phase of the afterhyperpolarization. Instead, the magnitude of the afterhyperpolarization was directly proportional to the external Ca^{2+} concentration and was antagonized by Co^{2+} and Mn^{2+} in a dose-dependent manner. They concluded that the afterhyperpolarization of rat SCG neurons results from an outward K^+ current that is Ca^{2+}- and voltage-dependent. Such a Ca^{2+}-activated G_K was described originally by Meech and Strumwasser (1970) and Meech (1972), who demonstrated that intracellular Ca^{2+} injection causes an increased G_K in *Aplysia* neurons.

The slow afterhyperpolarization of amphibian sympathetic ganglion cells is also generated by activation of the Ca^{2+}-dependent G_K (Kuba and Koketsu, 1978). Kuba *et al.* (1983) analyzed the characterisitcs of the afterhyperpolarization of bullfrog ganglion cells under various conditions that would affect either Ca^{2+} entry or intracellular Ca^{2+} release. They suggested that the origin of intracellular Ca^{2+} for the afterhyperpolarization is mainly Ca^{2+} influx during an action potential (see also McAfee *et al.*, 1981).

2. Posttetanic Hyperpolarization

A train of action potentials in a ganglion cell is followed by a prolonged hyperpolarization, the posttetanic hyperpolarization (PTH). The PTH of neuronal membranes has been ascribed to depletion of K^+ from the space immediately outside the membrane (Ritchie and Straub, 1957), activation of an electrogenic Na pump (Connelly, 1959; Straub, 1961; Nakajima and Takahashi, 1966; Rang and Ritchie, 1968; Den Hertog and Ritchie, 1969; Kuno *et al.*, 1970; Sokolove and Cooke, 1971; Koketsu, 1971), and enhancement and prolongation of G_K (Meves, 1961; Gage and Hubbard, 1964; Brodwick and Junge, 1972). Minota (1974) observed that PTH in bullfrog sympathetic ganglion cells persists in the presence of ouabain; is enhanced by TEA, which block the delayed rectifying K^+ channels; is associated with an increased conductance; and has an equilibrium potential (E_{PTH}) near E_K. Minota (1974) also found that E_{PTH} is dependent on $[K^+]_o$ but independent of $[Cl^-]_o$; moreover, PTH was abolished in either low $[Ca^{2+}]$ or high $[Mg^{2+}]$ solution. Minota (1974) hypothesized that the PTH is produced by an increased G_K that is probably triggered by an elevation of $[Ca^{2+}]_i$ (Meech and Strumwasser, 1970; Meech, 1972) that results from the calcium influx during the train of impulses. The entry of Ca^{2+} during a spike should occur, since bullfrog ganglion cells are capable of producing a Ca^{2+}-supported spike (Koketsu and Nishi, 1969). It seems likely, therefore, that the mechanisms underlying PTH and the slow afterhyperpolarization that follows a single action potential are essentially similar.

Morita *et al.* (1982) reported that in type 2 [hyperpolarizing (see Chapter 16)] myenteric neurons of the guinea pig, the calcium-activated potassium conductance declined with a single exponential time course (time constant 1.5–5 sec) following 1–6 action potentials (1–5 Hz) and with a double exponential time course (time constants about 3 and 12 sec) following 15–60 action potentials (5–20 Hz). They suggested that the time–course of this conductance increase probably reflects the free intracellular concentration of Ca^{2+} and therefore describes the calcium sequestration or extrusion process.

III. EFFECTS OF IONS AND DRUGS

A. Effects of Lithium

Li^+ can substitute completely for Na^+ in the production of action potentials in squid giant axons (Hodgkin and Katz, 1949), in frog skeletal muscle fibers (Keynes and Swan, 1959), and in rabbit cervical sympathetic fibers (Ritchie and Straub, 1957); however, it only partially substitutes for Na^+ in the case of sympathetic ganglionic transmission of the cat (Pappano and Volle, 1967). Koketsu and Yamamoto (1974) found that when action potentials evoked by direct stimulation of the bullfrog sympathetic ganglion were abolished in an Na^+-free sucrose solution, they could be restored by replacing the sucrose solution with an Na^+-free LiCl solution. However, the restoration was transient; action potentials graudally became depressed until they disappeared within 1.5–2 hr. Intracellular recording from bullfrog ganglion cells demonstrated that the resting potential (normally −60 to −65 mV) was reduced to less than −50 mV after 2 hr in the Na^+-free lithium solution. Under these conditions, no action potentials could be produced by either antidromic or orthodromic stimulation.

However, direct stimulation through an intracellular electrode could produce action potentials that had a smaller amplitude and higher threshold as compared to action potentials evoked in Ringer's solution. Some depolarized cells produced no action potentials even after strong direct stimulation. Repolarization of these cells by anodal currents restored their excitability.

When lithium ions are used to replace sodium ions in the extracellular medium, they should accumulate in the ganglion cells, since they are pumped out of the cell much more slowly than the sodium ions (Maizels, 1954; Zerahn, 1955; Keynes and Swan, 1959). This would lead to a reduction of the intracellular potassium concentration (Carmeliet, 1964; Araki *et al.*, 1965) and concomitant depolarization of the ganglion cells.

It has been reported that the membrane permeability to lithium ions is less than the permeability to sodium ions (Keynes and Swan, 1959; Armett and Ritchie, 1963). This would partially explain the smaller amplitude and higher threshold of the lithium spike. The lithium-induced depolarization that should inactivate the sodium- (lithium)-carrying system (Hodgkin and Huxley, 1952b) would enhance these changes.

B. Effects of Some Divalent Cations

Bullfrog sympathetic ganglion cells can generate action potentials in a solution in which all the NaCl has been replaced with isotonic $CaCl_2$ (Figure 5) (Koketsu and Nishi, 1969). The threshold potential for initiation of Ca^{2+} spikes is high and ranges from -15 to -5 mV. The Ca^{2+} spike reaches a peak of $+40$ to $+55$ mV and has a spike duration of 3–5 msec. The mean maximum rate of rise of Ca^{2+} spikes ($\approx$ 60 V/sec) is roughly 40% of that of Na^+ spikes. The maximum rate of rise is decreased when the resting membrane potential is lowered by a cathodal current; however,

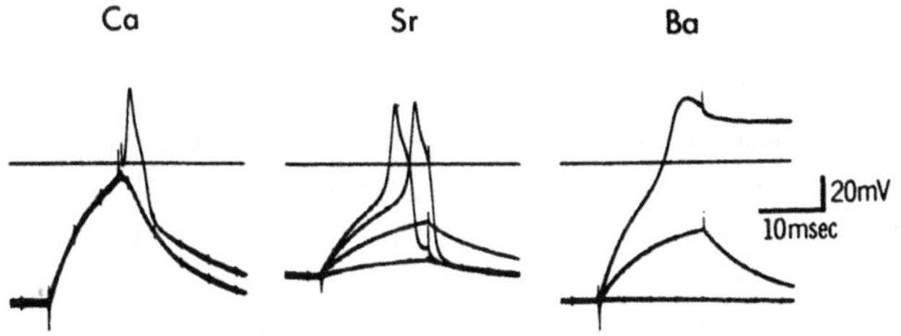

Figure 5. Action potentials of bullfrog sympathetic ganglion intracellularly recorded during perfusion with Na-free isotonic solutions of $CaCl_2$, $SrCl_2$, and $BaCl_2$. Action potentials were triggered by depolarizing current pulses applied through the cell membrane. From Koketsu and Nishi (1969).

it is not changed when the resting membrane potential is increased by an anodal current. This indicates that $P_{Ca}{}^{2+}$, like $P_{Na}{}^{+}$, is inactivated when the membrane is depolarized. The relationship between the maximum rate of rise of the Ca^{2+} spike and the resting membrane potential indicates that inactivation of the Ca-carrying system is almost completely absent when the membrane potential is maintained at a level greater than −80 mV. The peak potential of the Ca^{2+} spike is almost linearly related to the logarithmic concentration of the external Ca^{2+}, since a 10-fold increase in $[Ca^{2+}]_o$ increased the amplitude by approximately 30 mV. This suggests that the inflow of Ca^{2+} across the membrane is responsible for the production of the Ca^{2+} spike. The Ca-dependent inward current has been measured in voltage-clamped ganglion cells (Akasu and Koketsu, 1981). The Ca^{2+} spike was not appreciably affected by a concentration of tetrodotoxin (TTX) (1.6 μM) that abolishes the Na^+ spike within 5 min. The Ca^{2+} spike was also resistant to procaine (18 mM), which blocks the Na^+ spike in a few minutes. Procaine markedly prolonged the falling phase and slightly depressed the rate of rise of the Ca^{2+} spike (Koketsu and Nishi, 1969). The Ca^{2+} spike was very sensitive to the blocking action of Ca^{2+} antagonists, such as Mn^{2+}, Co^{2+}, and D-600. Mn^{2+} (10 mM) or Co^{2+} (3 mM) completely blocked the inward Ca^{2+} current in 10 min, and D-600 (50 μM) reduced the Ca^{2+} current to one half (Akasu and Koketsu, 1981).

Amphibian sympathetic neurons are also capable of producing action potentials in isotonic $SrCl_2$ or $BaCl_2$ solution (Figure 5). The Sr^{2+} spike is similar to the Ca^{2+} spike in amplitude and time course; however, the Ba^{2+} spike is a potential of long duration, since it lasts for a few seconds or more. The long-lasting nature of the Ba^{2+} spike is probably due to a suppression of the M current by Ba^{2+} (D. A. Brown and Adams, 1980), since the M current provides a braking effect on cell depolarization (see above). Like the Ca^{2+} spike, both the Sr^{2+} and the Ba^{2+} spike are very insensitive to the depressant actions of TTX and procaine. Amphibian as well as mammalian sympathetic neurons are inexcitable in an isotonic $MgCl_2$ solution.

Mammalian sympathetic neurons (rabbit SCG) can also produce a prolonged action potential in an Na^+-free $BaCl_2$ solution; however, they do not generate action potentials in an Na^+-free $CaCl_2$ or $SrCl_2$ medium (Tashiro and Nishi, 1972). McAfee and Yarowsky (1979) showed that rat SCG neurons can produce a TTX-insensitive spike that is most apparent when TEA is present. The amplitude of the TTX-insensitive spike is directly proportional to $[Ca^{2+}]_o$ and is antagonized by Co^{2+} and Mn^{2+} in a dose-dependent fashion. Thus, McAfee and Yarowsky (1979); and McAfee *et al.*, (1981) demonstrated that mammalian sympathetic neurons are also endowed with a regenerative Ca^{2+} conductance mechanism.

C. Effects of Catecholamines on Ca^{2+}-Dependent Potentials

A number of Ca^{2+}-dependent potentials and related phenomena have been described in this chapter, such as the Ca^{2+} spike and posttetanic hyperpolarization, as well as the time–course of the action potential (see above) (see also Koketsu and Nishi, 1969; Hashiguchi, 1979; Tashiro and Nishi, 1972; McAfee *et al.*, 1981). It appears that some or all of these phenomena may be regulated by α-adrenergic receptors (Minota and Koketsu, 1977; Horn and McAfee, 1980), since their agonists inhibit a number of Ca^{2+}-dependent potentials. Pharmacological analysis of these processes led McAfee and his associates (McAfee *et al.*, 1981) to propose that α_2-receptors are primarily involved, since clonidine and norepinephrine but not methoxamine acted as potent inhibitors of Ca^{2+}-dependent potentials (see also Chapters 12 and 13).

D. Hyperpolarization Caused by the Activity of an Electrogenic Sodium Pump

PTH of rabbit vagal nerve fibers appears to be induced by an electrogenic sodium pump. The activity of the electrogenic sodium pump is dependent on the external potassium concentration (Rang and Ritchie, 1968). When the external potassium concentration is zero, the PTH is very small; addition of potassium to the bathing medium causes a marked development of the PTH. Moreover, K^+ applied to the nerve a few minutes after exposure to K^+-free solution causes a hyperpolarization of the membrane, rather than the anticipated depolarization. Since the PTH cannot result from either a fall in periaxonal K^+ concentration due to active K^+ reabsorption or an increase in the potassium permeability, Rang and Ritchie (1968) suggested that the PTH is generated directly by an electrogenic sodium pump.

A similar K^+-activated hyperpolarization can be observed in bullfrog sympathetic ganglion cells as well as in their pre- and postganglionic fibers (Akasu *et al.*, 1975). In ganglia exposed to K^+-free Ringer's solution for about 60 min, addition of a 2 mM solution of potassium elicits a substantial hyperpolarization. Such hyperpolarizations with an almost constant amplitude can be induced repeatedly in a preparation for 2–3 hr provided each application of potassium is made for a few minutes at intervals of 15 min. K^+-activated hyperpolarizations are reduced by ouabain, low temperature, or replacement of sodium with lithium. All of these depress the sodium pump.

Several workers (Pascoe, 1956; D. A. Brown, 1966; Kosterlitz *et al.*, 1968, 1970; D. A. Brown *et al.*, 1969, 1972; Lees and Wallis, 1974) have shown that removal of depolarizing agents such as acetylocholine, carbachol, or choline causes a hyperpolarization of the mammalian SCG cells; this afterhyperpolarization will be termed Ch-AH. When potassium ions are omitted from the bathing solution, the amplitude of Ch-AH is never increased and is generally greatly diminished (Lees and Wallis, 1974). When 12.5 mM K^+ Krebs solution (control Krebs solution contains a 5 mM concentration of K^+) is used, the resting potential of rabbit SCG cells is close to E_K. In the presence of elevated $[K^+]_o$, the amplitude of the Ch-AH is unchanged or even increased. Thus, the Ch-AH cannot be due to an increased K^+ conductance. The amplitude of the Ch-AH is related more closely to the duration than to the amplitude of depolarization. Furthermore, a reduction in $[Na^+]_o$ results in a smaller Ch-AH. These observations are consistent with the view that the Ch-AH is related to the amount of Na^+ entering the cells during the preceding depolarization. In K^+-free solutions, acetylocholine induces a depolarization that declines slowly and is followed by a small Ch-AH. If at this time $[K^+]_o$ is raised to the control concentration level, a large, rapid Ch-AH ensues (Kosterlitz *et al.*, 1970). These results can be explained by assuming that there is an electrogenic sodium pump that requires K^+ extracellularly for its operation (Rang and Ritchie, 1968).

D. A. Brown and Scholfield (1974a) found that exposure of isolated rat SCG for a few minutes to nicotinic depolarizing agents produces a large rise in intracellular Na^+ concentration. They measured the rate of Na extrusion from ganglia loaded with ^{24}Na in the presence of a depolarizing agent and concluded that the properties of the ganglionic Na^+ pump deduced from rates of temperature-stimulated ^{24}Na extrusion support the view that the Ch-AH results from electrogenic Na^+ extrusion (D. A. Brown and Scholfield, 1974b).

E. Rhythmic Hyperpolarizations in Caffeine-Treated Neurons

Superfusion of the isolated sympathetic ganglion of the bullfrog with a solution containing caffeine (1–6 mM) often causes an initial slow hyperpolarization [initial caffeine hyperpolarization (ICH)] that is followed by a subliminal depolarization that is interrupted by rhythmic hyperpolarizations (RCHs) (Figure 6). A hyperpolarization similar to the RCHs can be triggered by an action potential or a depolarizing pulse in the presence of caffeine [evoked caffeine hyperpolarization (ECH) (Kuba and Nishi, 1976)]. All these hyperpolarizations induced by caffeine are associated with a marked reduction of the membrane resistance. Their amplitude is larger in a K^+-free solution and smaller in a high $[K^+]$ solution. Their polarity is reversed at the same membrane potential at which the afterhyperpolarization is reversed, and the reversal potential is not altered by

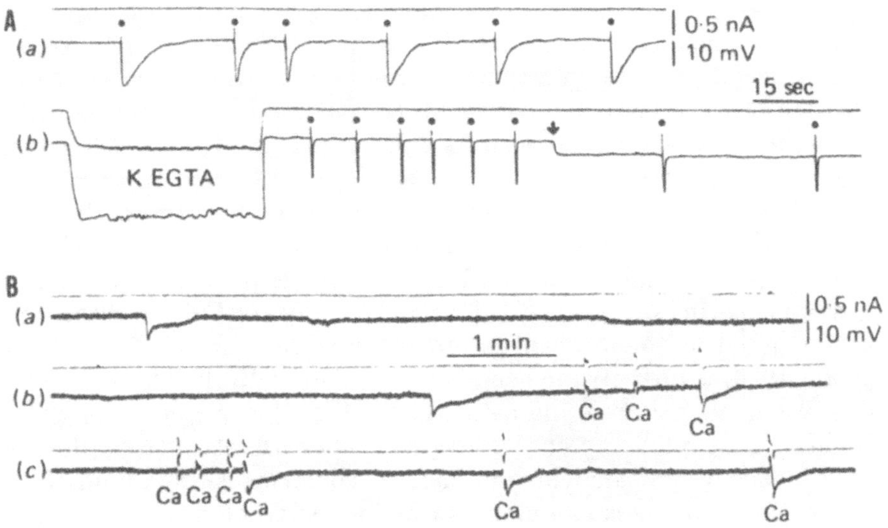

Figure 6. Suppression and activation of caffeine hyperpolarizations by intracellular injection of EGTA and Ca^{2+}. (A) Effects of the intracellular injection of EGTA into a bullfrog sympathetic neuron on its caffeine hyperpolarizations triggered by action potentials. The upper traces indicate the current for iontophoresis of EGTA, while the lower traces are membrane-potential changes. (*a*, *b*) Continuous recordings in a solution containing 10 mM caffeine. Note that only the afterhyperpolarizations of action potentials are seen after EGTA injection in (*b*). (B) Triggering of a caffeine hyperpolarization by intracellular injection of Ca^{2+}. Ca^{2+} was injected through a barrel of a double-barreled electrode filled with 0.5 M $CaCl_2$, while the membrane potential change was recorded by another barrel filled with 3 M KCl. The upper traces are the currents for Ca^{2+} injection and the lower traces show membrane-potential changes. All the recordings are continuous. The ganglion preparation was superfused with Ringer's solution containing 3 mM caffeine and 3.1 μM tetrodotoxin. From Kuba (1980).

omission of Na^+ or Cl^- from the external medium. These observations are compatible with the hypothesis that each caffeine hyperpolarization is generated by an increased K^+ permeability.

It was also demonstrated that Ca^{2+} is indispensable for the generation of caffeine hyperpolarizations and that the interval between RCHs decreases with the increase of external Ca^{2+} concentration. In view of the dependence of the hyperpolarization on external Ca^{2+} and also of the accumulated evidence that a rise in the intracellular Ca^{2+} elicits an increased K^+ permeability (Feltz *et al.*, 1972; Krnjevic and Lisiewicz, 1972; Meech, 1972; Meech and Strumwasser, 1970), it is probable that the caffeine hyperpolarizations are brought about by an elevation of $[Ca^{2+}]_i$. Caffeine does liberate Ca^{2+} from the sarcoplasmic reticulum (Weber and Herz, 1968) and increase Ca^{2+} permeability of the myoplasmic membrane (Bianchi, 1961).

If this concept is correct, lowering of $[Ca^{2+}]_i$ by intracellular injection of a Ca^{2+}-chelating agent should inhibit the production of caffeine hyperpolarizations. Indeed, iontophoretic injection of EDTA (Kuba and Nishi, 1976) or EGTA (Figure 6) depressed ICHs and abolished RCHs and ECH; (Kuba, 1980). Moreover, intracellular injection of Ca^{2+} (Figure 6) triggered the generation of a caffeine hyperpolarization before the appearance of RCHs (Kuba, 1980). This result strongly corroborates the concept that each type of caffeine hyperpolarization is associated with an elevation of $[Ca^{2+}]_i$.

The periodic appearance of RCHs may be explained in terms of a cyclic change in $[Ca^{2+}]_i$, provided that a Ca^{2+}-induced Ca^{2+}-release mechanism similar to the mechanism present in skeletal muscle fibers (Endo *et al.*, 1970) is operative in ganglion cells. Alternatively, the periodicity of RCHs may arise from the nature of the $[Ca^{2+}]_i$-dependent G_K, which could become refractory after a temporary increase. At present, there is no evidence to indicate which of these mechanisms is actually involved in the genesis of the rhythmicity of caffeine hyperpolarization.

Subcellular structures, such as the endoplasmic reticulum (DeMeis *et al.*, 1970; Diamond and Goldberg, 1971; Otsuka *et al.*, 1965; Robinson, 1968; Yoshida *et al.*, 1966), the cell membrane (Hanig *et al.*, 1972; Koketsu *et al.*, 1964), and the mitochondrion (Lust and Robinson, 1970a,b; Tjioe *et al.*, 1971; Vasington and Murphy, 1962), have been proposed as sites for storage and release of Ca^{2+}. Caffeine, however, seems to be ineffective in releasing Ca^{2+} from the myoplasmic mitochondria (Weber, 1968) and brain microsomes (Diamond and Goldberg, 1971). As described previously, an active or passive membrane depolarization can elicit an ECH, and addition or omission of external Ca^{2+} rapidly affects the caffeine hyperpolarizations. This evidence suggests that the Ca^{2+} storage sites from which caffeine liberates Ca^{2+} are the cell membrane and/or some intracellular apparatus that exists in electrical continuity with the cell membrane and readily exchanges Ca^{2+} with the extracellular medium. Electron-microscopic study of the localization of the Ca^{2+} in bullfrog sympathetic ganglion cells with the technique of Oschman and Wall (1972) revealed that the plasma membrane, subsurface cistern, and mitochondria could all be Ca^{2+} sequestration sites (Fujimoto *et al.*, 1980). Although the evidence concerning which of these sites is primarily responsible for storage and release of Ca^{2+} is still inconclusive, the subsurface cistern has been proposed as a favorable Ca^{2+} site for the production of the caffeine hyperpolarizations (Fujimoto *et al.*, 1980; Kuba, 1980). It is worthwhile to note that the actions of caffeine on mammalian sympathetic ganglion cells are essentially identical to its actions on amphibian ganglion cells (Skok *et al.*, 1978; Suzuki and Kusano, 1978). Suzuki and Kusano (1983) investigated RCHs and ECHs of the hamster submandibular ganglion and suggested that the caffeine-induced hyperpolariztaions are generated by the Ca^{2+}-activated G_K increase, which in turn is under the control of mitochondrial Ca^{2+} regulatory activity. This particular parasympathetic ganglion cell can produce, in the absence of caffeine, spontaneous rhythmic hyperpolarizations similar to the RCHs of amphibian sympathetic ganglion cells, and the mechanism underlying these spontaneous rhythmic oscillations is identical to that underlying the generation of caffeine-induced hyperpolarizations (Suzuki and Kusano, 1983).

IV. CONCLUSIONS

The advent of intracellular recording methods and their application to the studies of autonomic ganglia (Eccles, 1955) provided, over the last 30 years, much information with regard to passive and active membrane characteristics of the sympathetic neurons, and with regard to their dependence on ions, drugs, and transmitters. This information, combined with the accrued knowledge of ganglionic topography, led to a number of important implications. For example, it appears that mammalian ganglia in particular are very effective in handling, in terms of the responsiveness of the neuronal membrane, the synaptic input. At the same time, the ganglionic neurons dispose of a number of mechanisms for regulating this responsiveness and for controlling membrane excitability; several rectification processes as well as the M current (Adams *et al.*, 1982) are involved in these phenomena.

Postresponse processes, such as afterhyperpolarization and posttetanic hypopolarization, should be also listed as modulatory (see Chapters 3 and 12) in nature. It is of interest in this context that ionic interactions, particularly those between Ca^{2+} and K^{+}, underlie these modulations. Indeed, it may be speculated that Ca^{2+} storage and flux mechanisms underlie physiological functions regulating excitability of the ganglia, since the Ca^{2+}-dependent processes in particular and ionic interactions in general may be regulated by several transmitters, particularly catecholamines.

REFERENCES

Adams, P. R., and Brown, D. A.: Action of γ-aminobutyric acid (GABA) on rat sympathetic ganglion cells. *Br. J. Pharmacol.* **47**:639–640P (1973).

Adams, P. R., and Brown, D. A.: Luteineizing hormone-releasing factor and muscarinic agonists act on the same voltage-sensitive K^{+}-current in bullfrog sympathetic neurones. *Br. J. Pharmacol.* **68**:353–355 (1980).

Adams, P. R., Brown, D. A., and Constanti, D.: Pharmacological inhibition of the M-current. *J. Physiol. (London)* **332**:223–262 (1982).

Akasu, T., and Koketsu, K.: Voltage-clamp studies of a slow inward current in bullfrog sympathetic ganglion cells. *Neurosci. Lett.* **26**:259–262 (1981).

Akasu, T. Shirasawa, Y., and Koketsu, K.: The potassium-activated hyperpolarization of sympathetic ganglion cell membrane in bullfrogs. *Kurume Med. J.* **22**:177–182 (1975).

Araki, T., Ito, M., Kostyuk, P. G., Oscarsson, O., and Oshima, T.: The effects of alkaline cations on the responses of cat spinal motoneurons and their removal from the cell. *Proc. R. Soc. London Ser. B* **162**:319–332 (1965).

Armett, C. J., and Ritchie, J. M.: On the permeability of mammalian non-myelinated fibres to sodium and lithium ions. *J. Physiol. (London)* **165**:130–140 (1963).

Ballanyi, K., Grafe, P., and ten Bruggencate, G.: Intracellular free sodium and potassium, postcarbachol hyperpolarization, and extracellular potassium-undershoot in rat sympathetic neurones. *Neurosci. Lett.* **38**:275–279 (1983).

Barrett, J. N., and Crill, W. E.: Specific membrane resistivity of dye-injected cat motoneurons. *Brain Res.* **28**:556–561 (1971).

Bianchi, C. P.: Effects of caffeine on radiocalcium movement in frog sartorius. *J. Gen. Physiol.* **44**:845–858 (1961).

Blackman, J. G., and Purves, R. D.: Intracellular recordings from ganglia of the thoracic sympathetic chain of the guinea-pig. *J. Physiol. (London)* **203**:173–198 (1969).

Blackman, J. G., Ginsborg, B. L., and Ray, C.: Some effects of changes in ionic concentration on the action potential of sympathetic ganglion cells in the frog. *J. Physiol (London).* **167**:374–388 (1963).

Blackman, J. G., Crowcroft, P. J., Devine, C. E., Holman, M. E., and Yonemura, K.: Transmission from preganglionic fibres in the hypogastric nerve to peripheral ganglia of male guinea-pigs. *J. Physiol. (London)* **201**:723–743 (1969).

Brodwick, M. S., and Junge, D.: Post-stimulus hyperpolarization and slow potassium conductance increase in *Aplysia* giant neurone. *J. Physiol. (London)* **223**:549–570 (1972).

Brown, D. A.: Depolarization of normal and preganglionically denervated superior cervical ganglia by stimulant drugs. *Br. J. Pharmacol.* **26**:511–520 (1966).

Brown, D. A., and Adams, P. R.: Muscarinic suppression of a novel voltage-sensitive K^+ current in a vertebrate neurone. *Nature (London)* **283**:673–676 (1980).

Brown, D. A., and Scholfield, C. N.: Changes of intracellular sodium and potassium ion concentrations in the isolated rat superior cervical ganglia induced by depolarizing agents. *J. Physiol. (London)* **242**:307–319 (1974a).

Brown, D. A., and Scholfield, C. N.: Movements of labeled sodium ions in isolated rat superior cervical ganglia. *J. Physiol. (London)* **242**:321–351 (1974b).

Brown, D. A., Brownstein, M. J., and Scholfield, C. N.: On the nature of the drug-induced after-hyperpolarization in isolated rat ganglia. *Br. J. Pharmacol.* **37**:511P (1969).

Brown, D. A., Brownstein, M. J., and Scholfield, C. N.: Origin of the after-hyperpolarization that follows removal of depolarizing agents from the isolated superior cervical ganglion of the rat. *Br. J. Pharmacol.* **44**:651–671 (1972).

Brown, D. A., Constanti, A., and Adams, P. R.: Slow cholinergic and peptidergic transmission in sympathetic ganglia. *Fed. Proc. Fed. Am. Soc. Exp. Biol.* **40**:2625–2630 (1981).

Brown, D. A., Constanti, A., and Marsh, S.: Angiotension mimics the action of muscarinic agonists on rat sympathetic neurones. *Brain Res.* **193**:614–619 (1980).

Brown, T. H., Russel, A., Fricke, A., and Perkel, D. H.: Passive electrical constants in three classes of hippocampal neurons. *J. Neurophysiol.* **46**:812–827 (1981).

Burke, R. E., and ten Bruggencate, G.: Electrotonic characteristics of alpha motoneurones of varying size. *J. Physiol. (London)* **212**:1–20 (1971).

Carmeliet, E. E.: Influence of lithium ions on the transmembrane potential and cation content of cardiac cell. *J. Gen. Physiol.* **47**:501–530 (1964).

Christ, D. D., and Nishi, S.: Anomalous rectification of mammalian sympathetic ganglion cells. *Exp. Neurol.* **40**:806–815 (1973).

Connelly, C. M.: Recovery processes and metabolism of nerve. *Rev. Mod. Phys.* **31**:475–484 (1959).

Coombs, J. S., Curtis, D. R., and Eccles, J. C.: The generation of impulses in motoneurones. *J. Physiol. (London)* **139**:232–249 (1957).

DeMeis, L., Rubin-Altschul, B. M., and Machado, R. D.: Comparative data of Ca^{2+} transport in brain and skeletal muscle microsomes. *J. Biol. Chem.* **245**:1883–1889 (1970).

Den Hertog, A., and Ritchie, J. M.: A comparison of the effect of temperature, metabolic inhibitors and of ouabain on the electrogenic component of the sodium pump in mammalian nonmyelinated nerve fibers. *J. Physiol. (London)* **204**:523–538 (1969).

Diamond, I., and Goldberg, A. L.: Uptake and release of ^{45}Ca by brain microsomes, synaptosomes and synaptic vesicles. *J. Neurochem.* **18**:1419–1431 (1971).

Eccles, R. M.: Intracellular potentials recorded from mammalian sympathetic ganglion. *J. Physiol. (London)* **130**:572–584 (1955).

Eccles, R. M.: Orthodromic activation of single ganglion cells. *J. Physiol. (London)* **165**:387–391 (1963).

Endo, M., Tanaka, M., and Ogawa, Y.: Calcium-induced release of calcium from the sarcoplasmic reticulum of skinned muscle fibers. *Nature (London)* **228**:34–36 (1970).

Erulkar, S. D., and Woodward, J. K.,: Intracellular recording from mammalian superior cervical ganglion *in situ*. *J. Physiol. (London)* **199**:189–204 (1968).

Feltz, A., Krnjevic, K., and Lisiewicz, A.: Intracellular free Ca^{++} and membrane properties of motoneurones. *Nature (London)* **237**:179–181 (1972).

Fujimoto, S., Yamamoto, K., Kuba, K., Morita, K., and Kato, E.: Calcium localization in the sympathetic ganglion of the bullfrog and effects of caffeine. *Brain Res.* **202**:21–32 (1980).

Gage, P. W., and Hubbard, J. I.: Ionic changes responsible for post-tetanic hyperpolarization. *Nature (London)* **203**:653–654 (1964).

Galvan, M., Dorge, A., Beck, F., and Rick, R.: Intracellular electrolyte concentrations in rat sympathetic neurones measured with an electron microprobe. *Pfluegers Arch.* **400**:274–279 (1984).

Hanig, R. C., Tachiki, K. H., and Arrison, M. H.: Subcellular distribution of potassium, sodium, magnesium, calcium and chloride in cerebral cortex. *J. Neurochem.* **19**:1501–1507 (1972).

Hashiguchi, T.: The calcium-dependent components of action potentials of the rabbit superior cervical ganglion. *Tokyo Idai Z.* **37**:533–544 (1979).

Hodgkin, A. L., and Horowitz, P.: The influence of potassium and chloride ions on the membrane potential of single muscle fibers. *J. Physiol. (London)* **148**:127–160 (1959).

Hodgkin, A. L., and Huxley, A. F.: Currents carried by sodium and potassium ions through the membrane of the giant axon of *Loligo*. *J. Physiol. (London)* **116**:449–472 (1952a).

Hodgkin, A. L., and Huxley, A. F.: The dural effect of membrane potential on sodium conductance in the giant axon of *Loligo*. *J. Physiol. (London)* **116**:497–506 (1952b).

Hodgkin, A. L., and Katz, B.: The effect of sodium ions on the electrical activity of the giant axon of the squid. *J. Physiol. (London)* **108**:37–77 (1949).

Horn, J. P., and McAfee, D. A.: Alpha-adrenergic inhibition of calcium-dependent potentials in rat sympathetic neurones. *J. Physiol. (London)* **301**:191–204 (1980).

Hunt, C. C., and Nelson, P. G.: Structural and functional changes in the frog sympathetic ganglion following cutting of the presynaptic nerve fibers. *J. Physiol. (London)* **177**:1–20 (1965).

Keynes, R. D., and Swan, R. C.: The permeability of frog muscle fibers to lithium ions. *J. Physiol. (London)* **147**:626–638 (1959).

Koketsu, K.: The electrogenic sodium pump. *Adv. Biophys.* **2**:77–112 (1971).

Koketsu, K., and Nishi, S.: Calcium and action potential of bullfrog sympathetic ganglion cells. *J. Gen. Physiol.* **53**:608–623 (1969).

Koketsu, K., and Yamamoto, K.: Effects of lithium ions on electrical activity in sympathetic ganglia of the bullfrog. *Br. J. Pharmacol.* **50**:69–77 (1974).

Koketsu, K., Kitamura, R., and Tanaka, R.: Binding of calcium ions to cell membrane isolated from bullfrog skeletal muscle. *Am. J. Physiol.* **207**:509–512 (1964).

Kosterlitz, H. W., Lees, G. M., and Wallis, D. I.: Resting and action potentials recorded by the sucrose-gap method in the superior cervical ganglion of the rabbit. *J. Physiol. (London)* **195**:39–53 (1968).

Kosterlitz, H. W., Lees, G. M., and Wallis, D. I.: Further evidence for an electrogenic sodium pump in a mammalian sympathetic ganglion. *Br. J. Pharmacol.* **38**:464–465P (1970).

Krnjevic, K., and Lisiewicz, A.: Injections of calcium ions into spinal motoneurones. *J. Physiol (London)* **225**:363–390 (1972).

Kuba, K.: Release of calcium ions linked to the activation of potassium conductance in a caffeine-treated sympathetic neurone. *J. Physiol. (London)* **298**:251–269 (1980).

Kuba, K., and Koketsu, K.: Intracellular injection of calcium ions and chelating agents into the bullfrog sympathetic ganglion cells and effects of caffeine, in: *Iontophoreseis and Transmitter Mechanisms in the Mammalian Central Nervous System* (P. W. Ryall and J. S. Kelly, eds.), pp. 158–160, Elsevier/North-Holland, Amsterdam and New York (1978).

Kuba, K., and Nishi, S.: Rhythmic hyperpolarizations and depolarization of sympathetic ganglion cells induced by caffeine. *J. Neurophysiol.* **39**:547–563 (1976).

Kuba, K., Morita, K., and Nohmi, M.: Origin of calcium ions involved in the generation of a slow afterhyperpolarization in bullfrog sympathetic neurones. *Pfluegers Arch.* **399**:194–202 (1983).

Kuno, M., Miyahara, J. T., and Weakly, J. N.: Post-tetanic hyperpolarization produced by an electrogenic pump in dorsal spinocerebellar tract neurones of the cat. *J. Physiol. (London)* **210**:839–855 (1970).

Lees, G. M., and Wallis, D. I.: Hyperpolarization of rabbit superior cervical ganglion cells due to activity of an electrogenic dosium pump. *Br. J. Pharmacol.* **50**:79–94 (1974).

Lust, W. D., and Robinson, J. D.: Calcium accumulation by isolated nerve ending particles from brain. 1. The site of energy-dependent accumulation. *J. Neurobiol.* **1**:303–316 (1970a).

Lust, W. D., and Robinson, J. D.: Calcium accumulation by isolated nerve ending particles from brain. 2. Factors influencing calcium movements. *J. Neurobiol.* **1**:317–328 (1970b).

Lux, H. D., and Pollen, D. A.: Electrical constants of neurons in the motor cortex of the cat. *J. Neurophysiol.* **29**:207–220 (1966).

Maizels, E.: Active cation transport in erythrocytes. *Symp. Soc. Exp. Biol.* **8**:202– 227 (1954).

Matthews, M. R.: The ultrastructure of junctions in sympathetic ganglia of mammals, in: *Autonomic Ganglia* (L.-G. Elfvin, ed.), pp. 27–66, John Wiley, Chichester and New York (1983).

McAfee, D. A., and Yarowsky, P.: Calcium-dependent potentials in the mammalian sympathetic neurone. *J. Physiol (London)* **290**:507–523 (1979).

McAfee, D. A., Henon, B. K., Horn, J. P., and Yarowsky, P.: Calcium currents modulated by adrenergic receptors in sympathetic neurons. *Fed. Proc. Fed. Am. Soc. Exp. Biol.* **40**:2246–2249 (1981).

Meech, R. W.: Intracellular calcium injection causes increased potassium conductance in *Aplysia* nerve cells. *Comp. Biochem. Physiol.* **42A**:493–499 (1972).

Meech, R. W., and Strumwasser, F.: Intracellular calcium injection activates potassium conductance in *Aplysia* nerve cells. *Fed. Proc. Fed. Am. Soc. Exp. Biol.* **29**:834 (1970).

Meves, H.: Die Nachpotentiale isolierter markhaltiger Nervenfasern des Frosches bei tetanischer Reizung. *Pfluegers Arch. Gesamte. Physiol.* **272**:336–359 (1961).

Minota, S.: Calcium ions and the post-tetanic hyperpolarization of bullfrog sympathetic ganglion cells. *Jpn. J. Physiol.* **24**:501–512 (1974).

Minota, S., and Koketsu, K.: Effects of adrenaline on the action potential of sympthetic ganglion cells in bullfrog. *Jpn. J. Physiol.* **27**:353–366 (1977).

Morita, K., North, R. A., and Tokimasa, T.: The calcium-activated potassium conductance in guinea-pig myenteric neurones. *J. Physiol. (London)* **329**:341–354 (1982).

Nakajima, S., and Takahaski, K.: Post-tetanic hyperpolarization and electrogenic Na pump in stretch receptor neurone of crayfish. *J. Physiol. (London)* **187**:105–127 (1966).

Nishi, S., and Christ, D. D.: Electrophysiological and anatomical properties of mammalian parasympathetic ganglion cells. *Proc. Int. Union Physiol. Sci.* **9**:421 (1971).

Nishi, S., and Koketsu, K.: Electrical properties and activites of single sympathetic neurones of frogs. *J. Cell. Comp. Physiol.* **55**:15–30 (1960).

Nishi, S., and Koketsu, K.: Early and late afterdischarges of amphibian sympathetic ganglion cells. *J. Neurophysiol.* **31**:109–121 (1968).

Nishi, S., Soeda, H., and Koketsu, K.: Studies on sympathetic B and C neurons and patterns of preganglionic innervation. *J. Cell. Comp. Physiol.* **66**:19–32 (1965).

Nishi, S., Soeda, H., and Koketsu, K.: Release of acetylcholine from sympathetic preganglionic nerve terminals. *J. Neurophysiol* **30**:114–134 (1967).

Oschman, J. L., and Wall, B. J.: Calcium binding to intestinal membranes. *J. Cell Biol.* **55**:58–73 (1972).

Otsuka, M., Ohtsuki, I., and Ebashi, S.: ATP-dependent Ca binding of brain microsomes. *J. Biochem.* **58**:188–190 (1965).

Pappano, A. J., and Volle, R. L.: Actions of lithium ions in mammalian sympathetic ganglia. *J. Pharmacol. Exp. Ther.* **157**:346–355 (1967).

Pascoe, J. E.: The effects of acetylcholine and other drugs on the isolated superior cervical ganglion. *J. Physiol. (London)* **132**:242–255 (1956).

Perri, V., Sacchi, O., and Casella, C.: Electrical properties and synaptic connections of the sympathetic neurones in the rat and guinea-pig superior cervical ganglion. *Pfluegers Arch. Gesamte Physiol.* **314**:4–54 (1970).

Rall, W.: Branching dendrite trees and motoneuron membrane resistivity. *Exp. Neurol.* **1**:491–527 (1959).

Rall, W.: Membrane potential transients and membrane time constant of motoneurons. *Exp. Neurol.* **2**:503–532 (1960).

Rall, W.: Time constants and electrotonic length of membrane cylinders and neurons. *Biophys. J.* **9**:1483–1508 (1969).

Rang, H. P., and Ritchie, J. M.: On the electrogenic sodium pump in mammalian non-myelinated nerve fibers and its activation by various external cations. *J. Physiol. (London)* **196**:183–221 (1968).

Ritchie, J. M., and Straub, R. W.: The hyperpolarization which follows activity in mammalian non-medullated fibers. *J. Physiol. (London)* **136**:80–97 (1957).

Robinson, J. D.: Efflux of accumulated calcium from brain microsomes. *J. Neurochem.* **15**:1225–1235 (1968).

Skok, V. I.: The electrophysiology of cat's superior cervical sympathetic ganglion neurons. *Proc. Int. Union Physiol. Sci.* **7**:403 (1968).

Skok, V. I.: *Physiology of Autonomic Ganglia*, 77 pp. Igakushoin, Tokyo (1973).

Skok, V. I., Storch, N. N., and Nishi, S.: The effect of caffeine on the neurons of mammalian sympathetic ganglion. *Neuroscience* **3**:697–708 (1978).

Snow, P. J., Rose, P. K., and Brown, A. G.: Tracing axons and axon collaterals of spinal neurons using intracelluar injection of horseradish peroxidase. *Science* **191**:312– 313 (1976).

Sokolove, P. G., and Cooke, I. M.,: Inhibition of impulse activity in a sensory neuron by an electrogenic pump. *J. Gen. Physiol.* **57**:125–163 (1971).

Straub, R. W.: On the mechanism of post-tetanic hyperpolarization in myelinated nerve fibers from the frog. *J. Physiol. (London)* **159**:19–20p (1961).

Stretton, A. O. W., and Kravitz, E. A.: Neuronal geometry: Determination with technique of intracellular dye injection. *Science* **162**:132–134 (1968).

Suzuki, T., and Kusano, K.: Hyperpolarizing potentials induced by Ca-mediated K-con-

ductance increase in hamster submandibular ganglion cells. *J. Neurobiol.* **9**:367–392 (1978).
Suzuki, T., and Kusano, K.: Rhythmic membrane potential changes in hamster parasympathetic neurons. *J. Auton. Nerv. Syst.* **8**:213–236 (1983).
Tashiro, N., and Nishi, S.: Effects of alkali-earth cations on sympathetic ganglion cells of the rabbit. *Life Sci.* **11**:941–948 (1972).
Taxi, J.: Etude de l'ultrastructure des zones synaptiques dans les ganglions sympathiques de la grenouille. *C. R. Acad. Sci.* **252**:174–176 (1961).
Tjioe, S., Haugaard, N., and Bianchi, C. P.: The effects of psychoactive agents on calcium uptake by preparations of rat brain mitochondria. *J. Neurochem.* **18**:2171– 2178 (1971).
Uchizono, K.: On different types of synaptic vesicles in the sympathetic ganglia of amphibia. *Jpn. J. Physiol.* **14**:210–219 (1964).
Vasington, F. D., and Murphy, J. V.: Ca uptake by rat kidney mitochondria and its dependence on respiration and phosphorylation. *J. Biol. Chem.* **237**:2670–2677 (1962).
Weber, A.: The mechanism of the action of caffeine on sarcoplasmic reticulum. *J. Gen. Physiol.* **52**:760–772 (1968).
Weber, A., and Herz, R.: The relationship between caffeine contracture of intact muscle and the effect of caffeine on reticulum. *J. Gen. Physiol.* **52**:750–759 (1968).
Woodward, J. K., Bianchi, C. P., and Erulkar, S. D.: Electrolyte distribution in rabbit superior cervical ganglion. *J. Neurochem.* **16**:289–299 (1969).
Yoshida, H., Kadota, K., and Fujisawa, H.: Adenosine triphosphate dependent calcium binding of microsomes and nerve endings. *Nature (London)* **212**:291–292 (1966).
Zerahn, K.: Studies on the active transport of lithium in the isolated frog skin. *Acta Physiol. Scand.* **33**:347–358 (1955).

5

General Characteristics and Mechanisms of Nicotinic Transmission in Sympathetic Ganglia

KENJI KUBA and SHOICHI MINOTA

I. PHYSIOLOGICAL SIGNIFICANCE

The quantal release of acetylcholine (ACh) by an impulse at preganglionic terminals generates an excitatory postsynaptic potential (EPSP) at the subsynaptic membrane of a postganglionic neuron. This EPSP exhibits a rapid rise and a relatively slow decay with a time constant somewhat longer than the membrane time constant. In view of its temporal characteristics, it has been termed the fast EPSP (cf. Nishi, 1974; Kuba and Koketsu, 1978) to distinguish it from other EPSPs such as the slow EPSP and the late slow EPSP. The fast EPSP is unequivocally induced by the nicotinic action of ACh, since it is depressed by D-tubocurarine or β-erythroidine (Eccles, 1963; Blackman *et al.*, 1963) and less by atropine and is augmented by anticholinesterases.

The peak amplitude of the fast EPSP is in general large enough to generate an action potential(s) at the postganglionic neuron. Thus, the nicotinic transmission mediated by the fast EPSP is a main synaptic pathway in the sympathetic ganglia. However, the modes and efficacies of the

KENJI KUBA and SHOICHI MINOTA • Department of Physiology, Saga Medical School, Nabeshima, Saga 840-01, Japan.

nicotinic transmission in amphibian and mammalian ganglia are significantly different.

II. GENERAL CHARACTERISTICS

A. Amphibian Ganglia

Two types of postganglionic neurons, B and C, are present in frog, toad, or bullfrog sympathetic ganglia. Most of the B-type postganglionic neurons are exclusively innervated by preganglionic fibers of the B type, each cell being generally innervated by a single fiber (although sometimes by several fibers), while each C-type postganglionic neuron is innervated by several preganglionic fibers of the C type (Nishi *et al*, 1965). Each preganglionic fiber arborizes (before reaching the postganglionic cell) and terminates in many synaptic boutons at an area (close to the axon hillock) of a cell soma.

Although several branches of each presynaptic B-type fiber or several different presynaptic C-type fibers (C-type) innervate the postganglionic neurons, the release of ACh from the terminals occurs in an almost synchronized fashion, presumably because conduction velocities of each branch or fiber are similar. This in turn results in a monophasic EPSP, the amplitude of which essentially always exceeds the threshold for an action potential, setting up a condition for a one-to-one correspondence of impulse transmission. Therefore, synaptic transmission via B and C fibers belongs to the category of a "relay-type" transmission, as is the case with the skeletal neuromuscular transmission.

The fast EPSP rises to a peak within a few milliseconds and decays exponentially with a time constant that is 1.3 and 2 times longer, in the case of B and C neurons, respectively, than the cell membrane time constant [10–40 msec (see Figure 1A)] (Nishi *et al.*, 1965). This means that although most of the transmitter action ceases within a rising phase of the fast EPSP, significant residual transmitter action remains present during the decay phase; this is particularly notable in the case of C-type neurons. The synaptic delay of the fast EPSP measured from the peak of a negative deflection of an extracellularly recorded spike of the presynaptic terminal to the onset of the concurrently recorded postsynaptic current is 1.5–4 msec (Ginsborg, 1971; Kato and Kuba, 1980).

B. Mammalian Ganglia

Postsynapatic neurons of superior cervical or thoracic ganglia of mammals including rabbit, guinea pig, and rat are innervated by several pre-

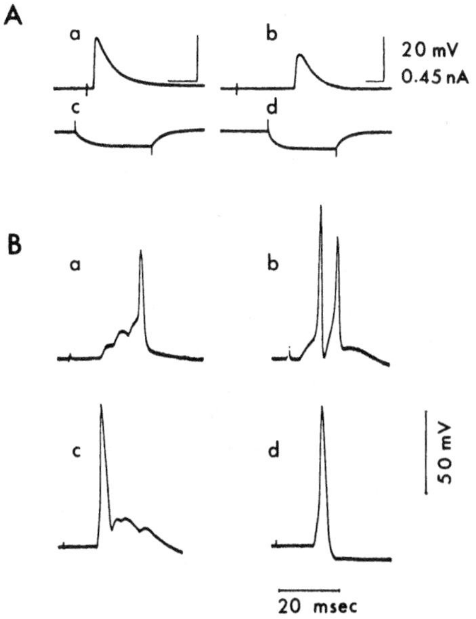

Figure 1. (A) Fast EPSPs and electrotonic potentials recorded from a B-type (a, c) and a C-type (b, d) neuron of the bullfrog sympathetic ganglion. The action potential was not generated in either type of cell. This lack of action potential is exceptional for a B-type neuron, while in the case of the C-type cell, the action potential is absent, since only a single preganglionic fiber was stimulated by a weak current pulse. From Nishi *et al.* (1965). (B) Temporal dispersion of fast EPSPs associated with action potentials. (a, b) Records obtained from different cells; (c, d) records obtained from a single cell. (d) Stimulus strength was decreased so that presumably only a single preganglionic fiber was stimulated. From Perri *et al.* (1970).

ganglionic fibers that differ in their conduction velocities. Therefore, a single stimulus of the maximum intensity generates several EPSPs of different latencies, most of which are accompanied in the postganglionic cell by single action potentials or, occasionally, by repetitive action potentials (Eccles, 1955; Erulkar and Woodward, 1968; Blackman and Purves, 1969; Perri *et al.*, 1970) (see also Figure 1B). The gradual reduction of the stimulus strength causes a stepwise decrease in the amplitude of the fast EPSP and/or disappearance of one or several of time-dispersed fast EPSPs, and eventually results in a single monophasic fast EPSP or action potential [Figure 1B(c, d)]. Thus, the nicotinic transmission in mammalian sympathetic ganglia is operated by a "convergent-type" synapse that requires coincidental activations of several preganglionic fibers for the generation of impulses at the postganglionic neuron; yet this transmission constitutes the principal and most effective pathway for ganglionic transmission. Temporal dispersal of EPSPs in response to a single preganglionic stimulus is due not only to the different velocities of the presynaptic fibers, but also to the location of synapses, most of which apparently are present at the dendrites (Elfvin, 1963) (see Section III); this is correlated with a spatial decay of the fast EPSP. Synaptic delays may also differ from one presynaptic terminal to another; the contribution of this phenomenon to the temporal dispersal of the fast EPSPs is small. The time constants of the exponential decays of the monophasic fast EPSPs in rat and guinea pig ganglion cells are 2.5 and 1.9 times longer respectively, than the time constants of their respective cell membranes (Perri *et al.*, 1970).

III. REVERSAL POTENTIAL AND IONIC MECHANISM OF THE FAST EXCITATORY POSTSYNAPTIC POTENTIAL

A. Reversal Potential

The peak amplitude of an action potential induced by preganglionic stimulation is 5–10 mV smaller than that of an action potential evoked by direct or antidromic stimulation of frog sympathetic ganglia (Nishi and Koketsu, 1960; Blackman *et al.*, 1963). This must be due to the coexistence of the fast EPSP with the action potential, since the mechanism underlying the fast EPSP is involved in the decrease of the peak of the action potential toward a zero potential level; this occurs as ACh increases the membrane permeability to ions that have an equilibrium potential less positive than the equilibrium potential for Na^+ (E_{Na}). In fact, Nishi and Koketsu (1960) succeeded in showing clear reversal of the fast EPSP of the frog sympathetic ganglion cell by a conditioning depolarizing pulse that brought the membrane potential to around −10 mV. This reversal (equilibrium) potential resembles that of the end-plate potential (Takeuchi and Takeuchi, 1959). These results clearly indicate that ion conductance of the cell membrane increases during the generation of the fast EPSP.

B. Effect of Ion Substitution

The equilibrium potential of the fast EPSP obtained by an extrapolation becomes more negative when the Na^+ or K^+ concentration of the Ringer's solution is reduced, while it shifts to a less negative (or positive) value with increasing K^+ concentration. The removal of Cl^- has no effect (Nishi and Koketsu, unpublished observation; see Koketsu, 1969; Kuba and Koketsu, 1978). Thus, the fast EPSP is generated by concomitant increases in the membrane conductances for Na^+ (G_{Na}) and K^+ (G_K).

IV. RELATIONSHIP BETWEEN THE SYNAPTIC CURRENT AND POTENTIAL

A. Time–Course of the Synaptic Current

The synaptic current [fast excitatory postsynaptic current (EPSC)] underlying the fast EPSP was first recorded by Kuba and Nishi (1971, 1979) by voltage-clamping a bullfrog sympathetic ganglion cell with two

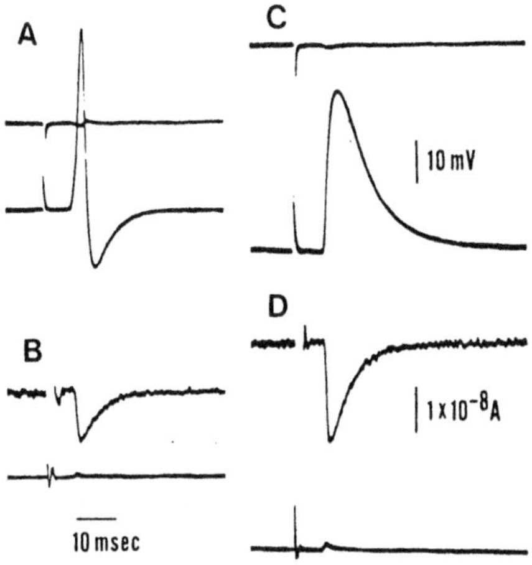

Figure 2. (A) An orthodromic action potential induced in a bullfrog sympathetic ganglion cell after insertion of two independent electrodes. (B) A fast EPSC recorded at a resting potential of −40 mV. (C, D) Fast EPSP and EPSC, respectively, recorded at −70 mV. All records were obtained from the same cell. The upper traces are membrane currents and do not indicate the zero level for membrane potential, while the lower traces are membrane potentials. From Kuba and Nishi (1979).

intracellular electrodes. Unlike other neurons of the vertebrates, the bullfrog sympathetic ganglion cell is suitable for the voltage-clamp analysis of a synaptic current, since these neurons are spheroidal and bear most of their synapses at the cell soma (although rather close to the axon hillock) (Taxi, 1961). However, in view of the small size of the cell, it is difficult to insert two independent electrodes into a single neuron. Nevertheless, successful experiments were made with cells that generated an overshoot action potential (Figure 2A).

The fast EPSC recorded at −70 mV at room temperature (Figure 2D) is an inward-directed current with a peak time of 2 msec and a half-decay time of 4.3 msec (Kuba and Nishi, 1979). Most of the falling phase follows an exponential function, except its tail part (this point will be discussed in detail later). The whole time–course of the fast EPSC is obviously faster than that of the fast EPSP recorded under the same conditions (Figure 2C, D). However, a considerable amount of the synaptic current flows during the decay phase of the synaptic potential, accounting for a longer time constant of the fast EPSP decay compared to the membrane time constant (Nishi and Koketsu, 1960).

B. Current–Voltage Relationship

Shifting the holding potential to a more negative value increases the fast EPSC amplitude and lengthens the decay phase, while shifting to a less negative value (or to a positive value) decreases the amplitude and decay constant (Figure 3A). The fast EPSC reverses its polarity at about −5 to −10 mV, as is the case with the fast EPSP. In most cells, the relationship between the peak amplitude and the membrane potential is

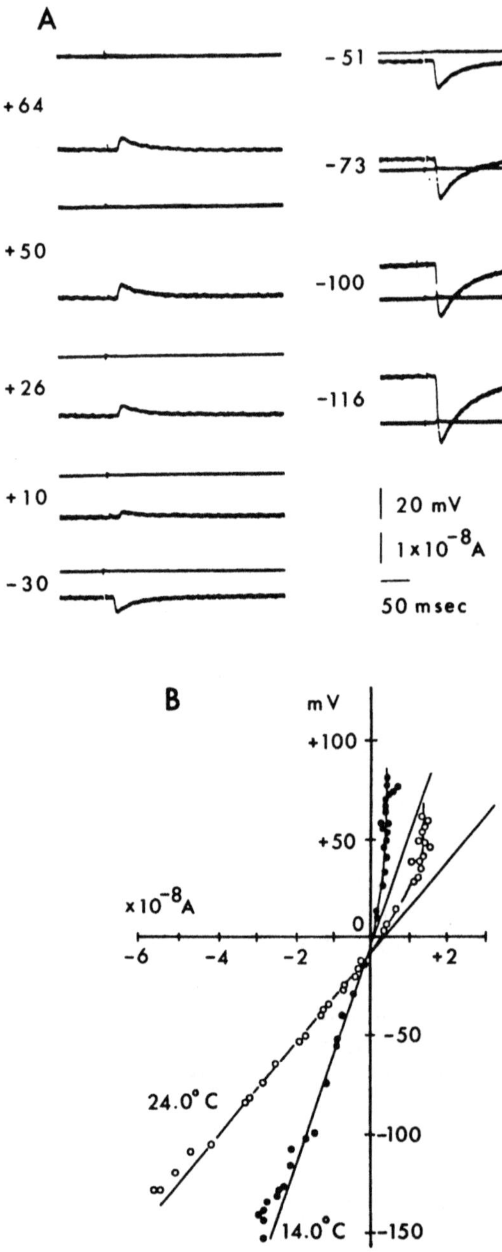

Figure 3. (A) Fast EPSCs of a bullfrog sympathetic ganglion cell evoked at various levels of membrane potential at 14°C. (B) Relationships between the peak amplitude of the fast EPSC (abscissa) and the membrane potential (ordinate) obtained at 24°C (○) and 14°C(●). From Kuba and Nishi (1979).

linear at negative membrane potentials, but it deviates from linearity at positive membrane potentials concomitant with the reduction of the synaptic conductuance (Figure 3B). This deviation cannot be explained by the voltage-dependent lifetime of the opened ion channel (to be discussed later), but must be caused by other mechanisms. Similar saturation of the synaptic current was seen in the eel electroplaque (Ruiz-Manresa and Grundfest, 1971) and in locust muscle fibers activated by γ-aminobutyric acid, in which the saturation occurred at a high negative level of membrane potential (Cull-Candy and Miledi, 1981). The conductance increase at the peak of the fast EPSC was calculated to be 13 nA/50 mV, or 260 nS.

There is almost a linear relationship between the logarithm of the half-decay time (HDT) of the EPSC and the membrane potential (Figure 4). This is expressed by the following equation:

$$\mathrm{HDT} = A \cdot \exp(V/B) \tag{1}$$

where A and B are constants.

C. Reconstruction of the Fast Excitatory Postsynaptic Potential from the Fast Excitatory Postsynaptic Current

The ionic mechanism of the fast EPSP may be simply explained by an equivalent circuit as shown in Figure 5B. What the transmitter does is turn on a switch in the circuit, causing a flow of current (shown by the arrow). The fast EPSC can be converted to the time–course of a conductance change (g_s); this conversion is used to reconstruct the fast EPSP

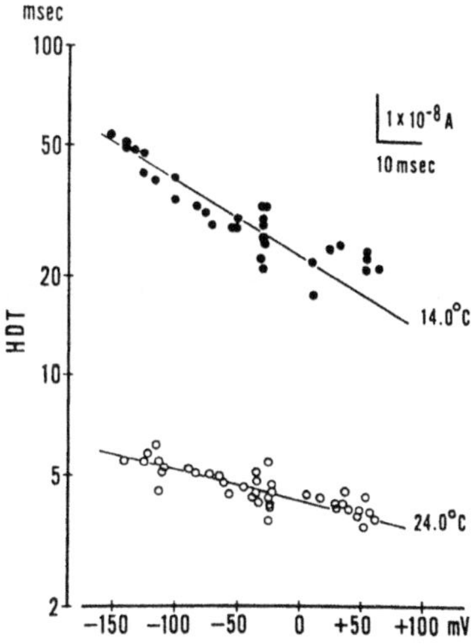

Figure 4. Relationships between the half-decay time (HDT) of the fast EPSC and the membrane potential at 24°C (○) and 14°C (●). Bullfrog sympathetic ganglion cell. *Ordinate:* HDT plotted on a logarithmic scale; this time pattern characterizes essentially the fast component of the fast EPSC decay phase. *Abscissa:* membrane potential. From Kuba and Nishi (1979).

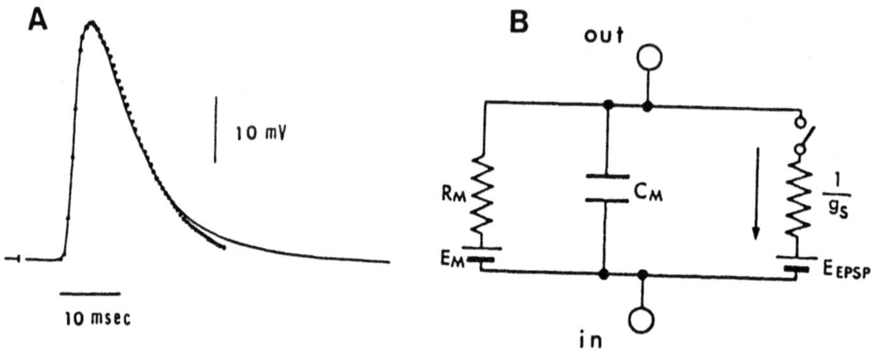

Figure 5. (A) Fast EPSP reconstructed from the fast EPSC shown in Figure 2D, using an equivalent circuit (B). The time–course of the fast EPSC was converted to that of the conductance change (g_s), which was then used for numerical integration of an equation for the equivalent electrical circuit. (B) An equivalent circuit for the nicotinic action of ACh at the subsynaptic membrane of a bullfrog sympathetic ganglion cell. (E_m, R_m, C_m) Resting membrane potential resistance, and capacitance respectively; (g_s, E_{EPSP}) conductance increase by the ACh action and its equilibrium potential, respectively.

numerically using the equivalent circuit. The reconstructed EPSP (Figure 5A, dotted curve) coincides with the experimentally recorded fast EPSP (continuous curve) except for its tail portion. This inconsistency is due to the flow of a synaptic current into the axon, this effect being particularly intense, since the synapses are located rather close to the axon hillock. Thus, even a spheroidal neuron cannot be represented by a leaky-condenser model.

D. Effects of Anticholinesterases

The decay phase of the fast EPSC is markedly prolonged in the presence of neostigmine (10 μM); the peak amplitude is either not changed or slightly augmented. Under the same conditions, the voltage sensitivity of the decay phase of the fast EPSC remained unchanged (Kuba and Nishi, 1979). MacDermott *et al.* (1980) observed that the prolonged decay phase of the fast EPSC in the presence of neostigmine or physostigmine (10 μM) exhibits a multiexponential pattern, since it may be represented by two or three functions. They ascribed this action of anticholinesterases to their effects on the ion-channel kinetics. This subject is further discussed in Chapter 13.

E. Effects of Low Temperature

Lowering of temperature from 24 to 14°C prolongs the time–course of the fast EPSC in the bullfrog sympathetic ganglion cell and reduces its amplitude to one half. The Q_{10} of the HDT measured at −60 mV in a single cell within the 14–24°C range was 4–6 (Kuba and Nishi, 1979). Pooling the results obtained with a number of cells, MacDermott *et al.* (1980) observed a Q_{10} of 3 in the same preparation. The voltage sensitivity of the decay phase of the fast EPSC appears to be increased by lowering the temperature (see Figure 4) (Kuba and Nishi, 1979), similar to the effect seen in the muscle end-plate (Magleby and Stevens, 1972). By contrast, MacDermott *et al.* (1980) reported that lowering the temperature had no effect on the voltage sensitivity of the EPSC decay phase. The reasons for these discrepancies are not known, and the effects in question should be carefully reinvestigated by minimizing possible errors, mainly caused by the application of a two-electrode voltage-clamp technique to a small cell.

V. CHARACTERISTICS OF THE ION-CHANNEL GATING

The mechanism of the ion-channel gating of the nicotinic receptor is discussed here on the basis of data obtained with amphibian sympathetic ganglia. The mechanism that obtains in mammalian ganglia, which is discussed in Chapter 6, is essentially similar to that underlying the nicotinic-receptor gating of bullfrog sympathetic ganglia, although there are some differences between these ganglia.

A. Physiological Significance of the Decay Phase of the Fast Excitatory Postsynaptic Current

That the decay phase of the fast EPSC follows in large part a single exponential function (Kuba and Nishi, 1979) can be accounted for by at least four processes: (1) the rate of ACh release from the presynaptic terminals; (2) the rate of diffusion of ACh out of the synaptic cleft; (3) the rate of hydrolysis of ACh by acetylcholinesterase; and (4) the rate of closing of the ion channel, which would include both the rate of unbinding of ACh from the receptor and the rate of a conformational change of the ion channel, no matter which is a rate limiting-step (see Magleby and Stevens, 1972). These processes may be visualized as follows:

$$\begin{array}{c} \text{Diffusion} \\ \uparrow \\ \text{Release} \rightarrow \text{ACh} + \text{R} \underset{k_2}{\overset{k_1}{\rightleftharpoons}} \text{ACh} \cdot \text{R} \underset{\alpha}{\overset{\beta}{\rightleftharpoons}} \text{ACh} \cdot \text{R}^* \\ \downarrow \\ \text{Hydrolysis} \end{array} \qquad (2)$$

where R and R* are the receptors in closing and opening conformations, respectively, and α, β, k_1 and k_2 are rate constants.

The frequency distribution of the synaptic delay, which roughly indicates the time–course of ACh release from the motor-nerve terminal (Katz and Miledi, 1965), was shown to be much shorter than the time–course of the end-plate current (cf. Magleby and Stevens, 1972). The frequency distribution of the synaptic delay in sympathetic ganglia is also shorter than the time–course of the fast EPSC. Therefore, it is highly likely that the time–course of ACh release contributes little to the decay phase of the fast EPSC. The high temperature coefficient of a fast EPSC decay seems to rule out the possibility that the rate of diffusion of ACh is the rate-limiting step. If the rate of hydrolysis of ACh by acetylcholinesterase determines the decay phase of the fast EPSC, the voltage sensitivity of the fast EPSC would disappear after the inhibition of the enzyme; however, this was not the case. Therefore, the lifetime in the synaptic cleft of ACh released from the presynaptic terminals is very short compared to the time–course of the fast EPSC; thus, ACh released by an impulse opens the ion channels almost synchronously. If so, the rate of closing of the ion channel would be reflected in the decay phase of the fast EPSC. Accordingly, the rate of closing of the ion channel (rate of conformational change of the ion channel or that of unbinding of ACh from the receptor) depends on temperature and the membrane voltage (the rate decreases with the negativity of the membrane potential).

As already mentioned, while the main part of the fast EPSC decay follows a single exponential function, the tail part deviates from a single exponential function, as shown in Figure 6. The whole decay phase of the fast EPSC is better described by a double exponential curve. The extrapolation of the slow component to the peak time of the EPSC reveals that the slow component contributes about 20% of the peak current. The time constants of the fast and slow components shown in Figure 6 are, at −55 mV, 3.2 and 13.4 msec, respectively. These characteristics resemble those of the fast EPSC of a parasympathetic ganglion cell (Rang, 1981).

B. Noise Analysis

There are several problems with applying noise analysis (see also Chapter 6 for the mammalian ganglion cell) to the nicotinic action of ACh on the sympathetic ganglion cell membrane. First, two independent intracellular electrodes having a high tip resistance must be inserted into a single ganglion cell without inflicting serious injury on the cell membrane. The membrane injury would cause a large leakage current leading to an increased thermal noise; the use of a high-tip-resistance electrode would also contribute to increased thermal noise. The second problem is the existence of the muscarinic response, which obscures the results of the analysis. The first problem may be overcome to some extent by selecting

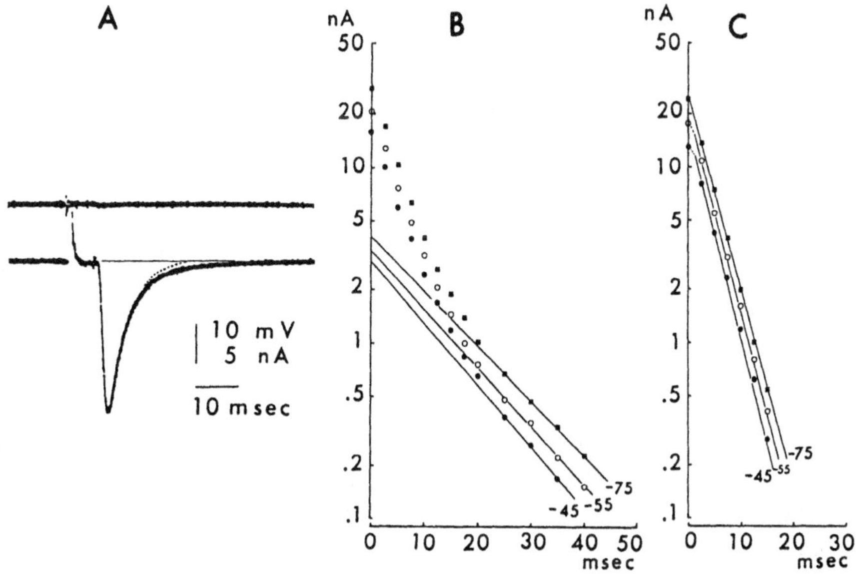

Figure 6. Double exponential decay of a fast EPSC. (A) Fast EPSC of a bullfrog sympathetic ganglion cell recorded at −55 mV. The upper and lower traces are membrane potential and current, respectively. (B) Semilogarithmic plots of the decay phase of the fast EPSCs recorded at −45, −55, and −75 mV. Straight lines are fitted for the tail part of the currents, showing a slow component. (C) Currents after subtraction of the slow component, yielding the pure fast component of the fast EPSC decay phase.

a relatively large cell, beveling the tips of the electrodes, and limiting the frequency range to be analyzed (cf. Rang, 1981). The second problem may be solved by blocking the muscarinic response with atropine. However, atropine is known to possess an inhibitory action on the ion channel gated by the nicotinic action of ACh; at relatively high concentrations (5 μM in the case of the end-plate and 10–100 μM in that of the ganglion cells), it shortens drastically the decay phase of both the end-plate current (Adler and Albuquerque, 1976) and the fast EPSC (MacDermott *et al.*, 1980). This effect would seriously affect the result of the fluctuation analysis. Fortunately, it was found that most of the depressant action of atropine disappears via desensitization during its persistent application; although the amplitude of an ACh potential (induced by iontophoresis) decreases markedly 5 min after the application of atropine, it recovers after about 60 min in the continuous presence of atropine in a perfusing solution (Minota and Kuba, 1984). Thus, the nicotinic receptor–channel complex behaves almost normally in a ganglion cell treated with atropine for a long time. However, it should be pointed out that the channel kinetics of atropine-treated and untreated neurons are quite different. Another method that can be used to suppress the muscarinic response consists of clamping the membrane potential at the equilibrium potential for K^+. This approach is based on the finding that in a certain group of bullfrog sympathetic ganglion cells, muscarinic depolarization is generated by inactivation of the K^+ conductance (Weight and Votava, 1970; Kuba and Koketsu, 1976; cf. Adams *et al.*, 1982; Adams, 1981) (see Chapter 7).

The results of the analysis of current fluctuations induced by ACh either in atropine-treated cells or in untreated cells (which are clamped at −90 mV in normal Ringer's) are illustrated in Figure 7. Figure 7A shows a membrane current induced by the perfusion of ACh in the ganglion cell; in this case, the muscarinic receptors were blocked by atropine (10 μM) for more than 1 hr. The lower records show a current fluctuation recorded with high amplification. Segments of currents before and during the action of ACh (indicated by horizontal bars in the upper record) are fed into a computer, and the net power spectral density is obtained. A typical noise spectrum of the bullfrog sympathetic ganglion cell is shown in Figure 7B. It is well described by two Lorentzian functions [as shown by equation (3)], rather than by a single Lorentzian:

$$S(f) = \frac{S_1(0)}{1 + (f/f_{1C})} + \frac{S_2(0)}{1 + (f/f_{2C})} \tag{3}$$

where $S(f)$ is the spectral density function at frequency f, $S_1(0)$ and $S_2(0)$ are the spectral density of the fast and slow components at frequency 0, respectively. At room temperature (20–24°C) and at −50 to −60 mV, the cutoff frequencies of the fast and slow components (f_{1C} and f_{2C}) are 71.2 ± 15.8 Hz (S.D., N = 8) and 6.0 ± 5.3 Hz (N = 8), respectively. These values yield time constants of 1.9 and 22.9 msec, respectively. The power spectral density obtained from the ganglion cell, which is perfused with normal Ringer's solution and clamped at −90 mV, is also fitted better by a double Lorentzian function than by a single Lorentzian function. On the other hand, the time constants of two components of the untreated cell are 2.1 and 11 msec. Thus, while the time constants of the fast component of atropine-treated and untreated cells are almost the same, the time constant of the slow component is longer in the former than in the latter case. Nevertheless, there is no doubt that the closing rate of the nicotinic channel of the bullfrog sympathetic cell is described by two exponential functions. This is consistent with the double exponential decay of a fast EPSC either in the presence or in the absence of atropine, confirming the previous conclusion that the rate of fast EPSC decay represents the rate of channel closing. Furthermore, the longer time constant of the slow component in the atropine-treated cell conforms to the finding that the second component of the fast EPSC is longer in the presence than in the absence of atropine.

The elementary conductance change (γ) of the bullfrog sympathetic ganglion treated with atropine, calculated from the mean and variance of the current fluctuation, is 7.8 ± 3.5 pS (N = 8). This is considerably smaller than the value of γ of the muscle end-plate (23–31 pS) (Anderson and Stevens,

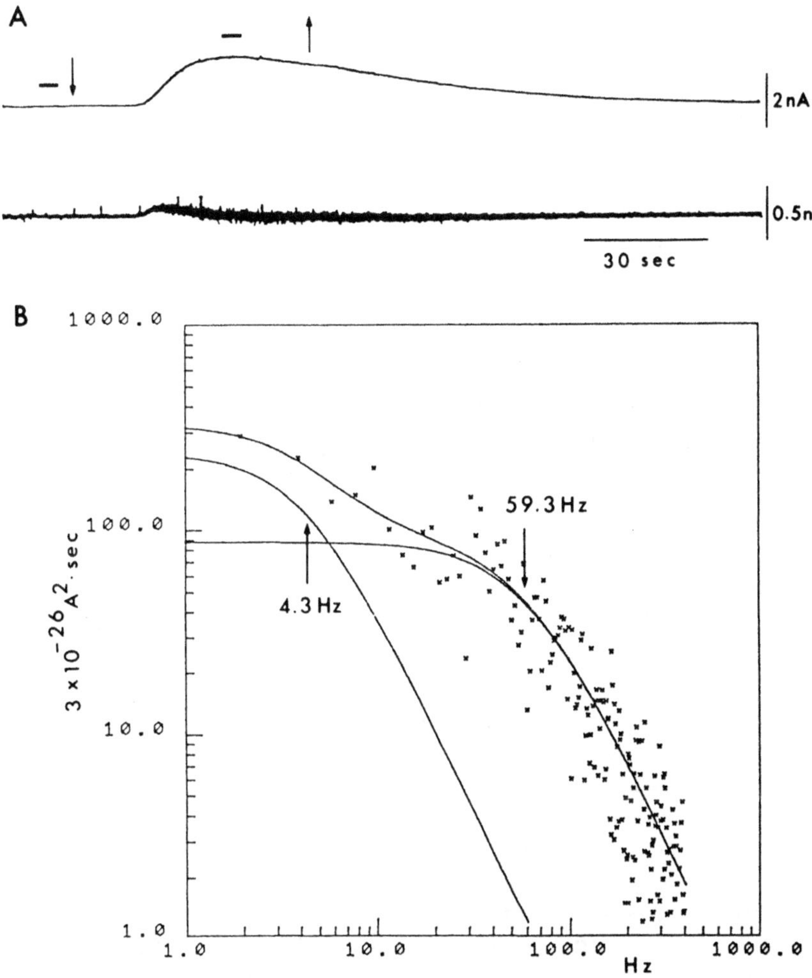

Figure 7. (A) A membrane current induced by 250 μM ACh at -59 mV in a bullfrog sympathetic ganglion cell. The upper and lower traces concern the same response, but the response shown in the upper trace was recorded with low DC amplification, that in the lower trace with high AC amplification. Segments of current indicated by horizontal bars were analyzed. (B) Power spectrum of the response to ACh after subtraction of the control spectrum. The uppermost curve shows the least-squares fit for the points in a range between 2 and 400 Hz, according to equation (3). The lower two curves show two Lorentzian components plotted separately. The arrows show cutoff frequencies. Atropine (10 μM) for 1 hr.

1973; Colquhoun, 1979) and mammalian parasympathetic (31 pS) (Rang, 1981) and sympathetic (36 pS) (Skok *et al.*, 1982; Derkach *et al.*, 1983; Skok, 1983) (see also Chapter 6) ganglion cells. The γ value of the bullfrog sympathetic ganglion cell treated with atropine apparently represents the average of the two components, whether or not the two components represent conductances of two independent channels (see below). Estimating a "mean" current for each component from their fractions in the fast EPSC decay, the elementary conductances for fast and slow components can be calculated as 8.4 ± 1.9 pS (N = 5) for $S_1(0)$ and f_{1C}, and 7.7 ± 2.1 pS for $S_2(0)$ and f_{2C}, respectively. Similar values of γ were obtained from the ganglion cell perfused with Ringer's and clamped at −90 mV. Thus, the nicotinic channels of the atropine-treated and untreated cells differ only in the time constant of the slow component. This must be ascribed to the action of atropine. Furthermore, another problem with the noise analysis of the nicotinic response is the possibility that the current fluctuations recorded in Ringer's even at −90 mV may contain a Na-current component activated by the muscarinic action of ACh (cf. Kuba and Koketsu, 1978) (see Chapter 7).

C. Two Different Types of Ion Channels or a Single Type of Ion Channel Having a Complexed Pattern of Gating

It may be possible that the slow component is caused by a response of nicotinic receptors located at a membrane remote from the synapse; this possibility, however, may be disregarded for the following reasons: The reversal potentials of the fast and slow components are the same, as evidenced by the fact that the neurally evoked EPSC at the level of the reversal potential is not biphasic. Furthermore, while the slow component follows an exponential function, this is not a resultant of several exponential functions as would be expected if it were generated by extrasynaptic receptors. Moreover, both the fast EPSC and the miniature EPSC, either in the presence or in the absence of atropine, decay with a double exponential function (however, see Selyanko *et al.*, 1979; Derkach *et al.*, 1983).

Another possibility as to the causation of the slow component is the presence for a long time period of residual ACh in the synaptic cleft. This possibility is inconsistent, however, with the demonstration of a slow component in the fluctuation analysis of the ACh-induced current.

Altogether, two mechanisms remain to be considered (cf. Rang, 1981) (see also Chapter 6): (1) There are two different groups of ion channels with a different closing rate, both of which are activated by the nicotinic action of ACh; (2) there exists only one type of ion channel that closes with two different time constants. Inconsistent with the latter mechanism is the observation that the decay phase of the miniature EPSC obviously follows a double exponential function (Colquhoun and Hawks, 1977) and the contribution of the slow component to the peak amplitude is comparable with that of the neurally induced EPSC. If the closure of the ion channel involves an intermediate closed conformation leading to two steps with comparable rate constants, it might be expected that the size of the second, slow component would depend on the amount of ACh in the synaptic cleft (cf. Rang, 1981). This argument is not strong enough to be used as a basis for distinguishing between these two possibilities. The most straightforward way to solve this problem would be to observe the frequency distribution of the duration of a single channel-current pulse (Neher and Sakman, 1976).

D. Comparison with Other Ion Channels Gated by the Nicotinic Action of Acetylcholine

The characteristics of the ion-channel gating of the nicotinic receptor in the bullfrog sympathetic ganglion cell resemble in some respects, but differ from in other respects, those of the ion channel of the motor end-plate. For instance, the fast component of the channel-closing rate of the bullfrog sympathetic ganglion cell resembles that of the end-plate. However, the second component observed in the bullfrog sympathetic ganglion cell, which is quite small, is rather similar to the slow component of the nicotinic response of the rat submandibular ganglion cell; in this neuron, about 50% of the total response is attributed to the slow component (Rang, 1981). As already mentioned, one of the important differences between the nicotinic channel of the bullfrog sympathetic ganglion cell and that of other cells is the difference in their single-channel conductance. The γ value of the former is less than half the γ values of other neurons and the muscle end-plate.

The voltage sensitivity of the closing rate constant in the amphibian sympathetic ganglion cell (Kuba and Nishi, 1979) is much smaller than that of the end-plate (Magleby and Stevens, 1972; Andersen and Stevens, 1973) and mammalian sympathetic (Selyanko *et al.*, 1979) and parasympathetic (Rang, 1981) ganglion cells. Since the membrane thickness of the bullfrog ganglion cell is not much different from that of the end-plate or of the mammalian ganglion cell, the actual voltage sensitivity of a process underlying ion translocation through the channel (which may reflect the effect on a net-dipole moment or redistribution of ions between the channel and external medium) (cf. Adams, 1981) of the bullfrog ganglion cell must be smaller than that of the end-plate and other ganglion cells.

MacDermott *et al.* (1980) reported unusually large values of the decay time constant of EPSC of bullfrog ganglion cells at positive potentials, deviating greatly from an exponential relationship. This behavior of the fast EPSC may not have been physiological, since we have occasionally encountered similar results with deteriorated cells that were possibly damaged by holding the membrane at a strongly depolarized level and by the use of low-resistance electrodes (<20 MΩ).

E. Acetylcholine-Activated Ion Channel in the Cultured Bullfrog Sympathetic Ganglion Cell

The properties of the ion channel could be clarified more directly by recording single-channel currents with a patch-clamp method (Neher and

Sakman, 1976; Hamill *et al.*, 1981) than by analyzing current fluctuations, but the existence of connective tissues and satellite cells around freshly isolated sympathetic ganglion cells of the bullfrog precludes the application of a patch-clamp method. This method can be employed, however, with respect to tissue-culture primary ganglion cells. For this purpose, paravertebral sympathetic ganglion cells were isolated from adult bullfrogs and cultured for 3–6 weeks prior to patch-clamping. Figure 8 shows single-channel currents induced by ACh [10 μM in (A), 100 μM in (B)] and records obtained under cell-patch conditions. Under these circumstances, the inward single-channel current reversed its polarity at about −10 mV (shown). The slope of the curent–voltage relationship yielded a single-channel conductance of 25 pS (Figure 9) (the values ranged from 25 to 35 pS in other cells). When the membrane was hyperpolarized, the probability of channel opening decreased with time, while membrane depolarization increased the probability, indicating the voltage dependence of the desensitization process. These characteristics differed entirely from those of nicotinically activated ion channels of the ganglion cell *in situ*. The single-channel conductance of the nicotinic channels of the cultured ganglion cells was much greater than that recorded *in situ* (8 pS); moreover, channel desensitization (Figure 8B) was much more marked in the cultured cells than under *in situ* conditions. It appears, then, that a change in the characteristics of the ion channel was produced by culturing the ganglion cell. This effect could be analogous to the effect of preganglionic denervation. Furthermore, the properties of the ion chan-

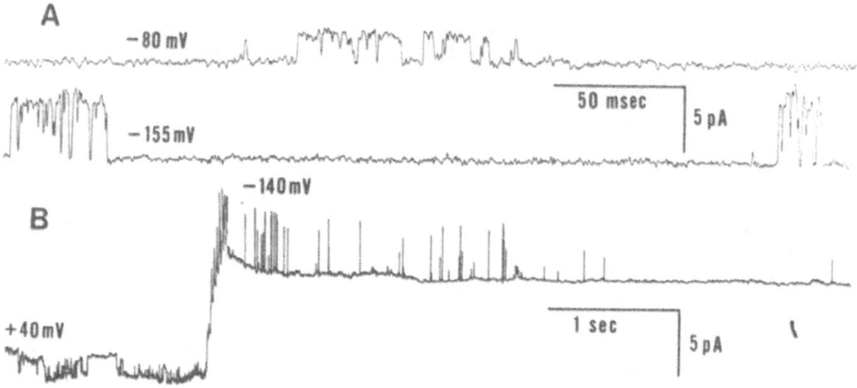

Figure 8. Single-channel currents induced by ACh [10 μM in (A), 100 μM in (B)], recorded from cultured bullfrog sympathetic ganglion cells under the cell-attached condition. The membrane potential (relative to the extracellular space) was −80 mV or −120 mV in (A) and +40 mV or −140 mV in (B). The amplitude of currents in (B) does not show faithfully the correct amplitude for the low-frequency characteristic (100 Hz) of a pen-writing recorder, especially when current pulses were short in duration.

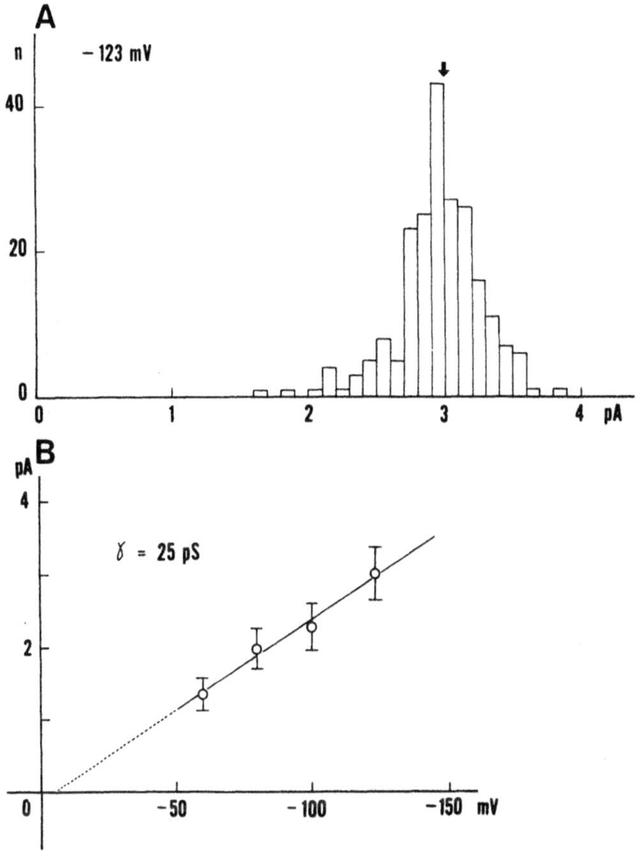

Figure 9. Frequency distribution of the amplitude of ACh-induced single-channel currents recorded at −120 mV. (B) Current–voltage relationship of ACh-activated single-channel currents of the cultured ganglion cell.

nel of the cultured ganglion cells resemble those of the intrajunctional, ACh-activated ion channels of the muscle end-plate (cf. Peper *et al.*, 1982).

VI. MODE OF ACETYLCHOLINE ACTION ON THE NICOTINIC RECEPTOR

A. Dose–Response Curve of Acetylcholine Action

A reliable dose–response curve for the nicotinic action of ACh on the sympathetic ganglion cell has not been obtained as yet. A plausible dose–response curve, however, could be obtained by means of quantitative iontophoretic applications of ACh to the ganglion cell. Iontophoresis of ACh by a current pulse applied close to a sympathetic ganglion cell produces a diphasic depolarization of the membrane, a fast phase being followed by a slow phase; only the fast phase can be recorded when the amount of iontophoresed ACh is small. The fast phase is blocked by D-tubocurarine, while the slow phase is abolished by a low concentration of atropine; thus, the two phases are generated by the nicotinic and the muscarinic action of ACh, respectively, and correspond to the fast EPSP and slow EPSP (Koketsu *et al.*, 1968a). The slow muscarinic response should not be confused with the second component of the nicotinic response as revealed in the fast EPSC or by means of noise analysis, since the muscarinic response is hundreds or thousands of times slower in its time–course than the slow component of the fast EPSC, and much more sensitive to atropine. Thus, the ganglion cell, which is equilibrated in a solution containing atropine for more than 30 min, responds only to the nicotinic action; at that time, the effect of atropine on the nicotinic response is over, due to densensitization (see above).

The peak amplitude of an ACh potential (P_{ACh}) relates to the change in the ion conductance (G_{ACh}) of the subsynaptic membrane in terms of the following equation:

$$G_{ACh} = \frac{1}{R_m}\left(\frac{P_{ACh}}{E_q - R_p - P_{ACh}}\right) \quad (4)$$

where R_m is the input resistance, E_q is the equilibrium potential, and R_p is the resting potential. Substituting experimental values for each constant, the P_{ACh} yields a G_{ACh}. G_{ACh}'s obtained with microiontophoresed ACh doses are plotted in a double-reciprocal scale for two neurons in Figure 10; the values of 1.7 and 2 were used for n, an exponent to the charge for ACh iontophoresis (cooperativity number) in the abscissa, to obtain linear relationships, for the two cells in question, between the conductance and the charge. For 23 cells, the mean value for n was 1.82 ± 0.2; thus, at least two molecules of ACh are necessary to open the ion channel.

The overall dissociation constant of the nicotinic receptor of the ganglion cell may be roughly estimated from the dose–response curve of ACh potentials using an equation that is based on point-source diffusion, steady-state kinetics of ACh–receptor interaction (at a low ACh concentration), and several additional assumptions (Peper *et al.*, 1975); the equation is as follows:

$$G_{ACh} = G_{max} \cdot (kQ/FK_m)^n \cdot (\pi 4Dt_p)^{c-1.5n} \cdot n^{-c} \cdot \exp(c-1.5n) \quad (5)$$

where t_p is the peak time of an ACh potential measured from the beginning of an iontophoretic pulse, G_{max} is the maximum conductance, k is a transport

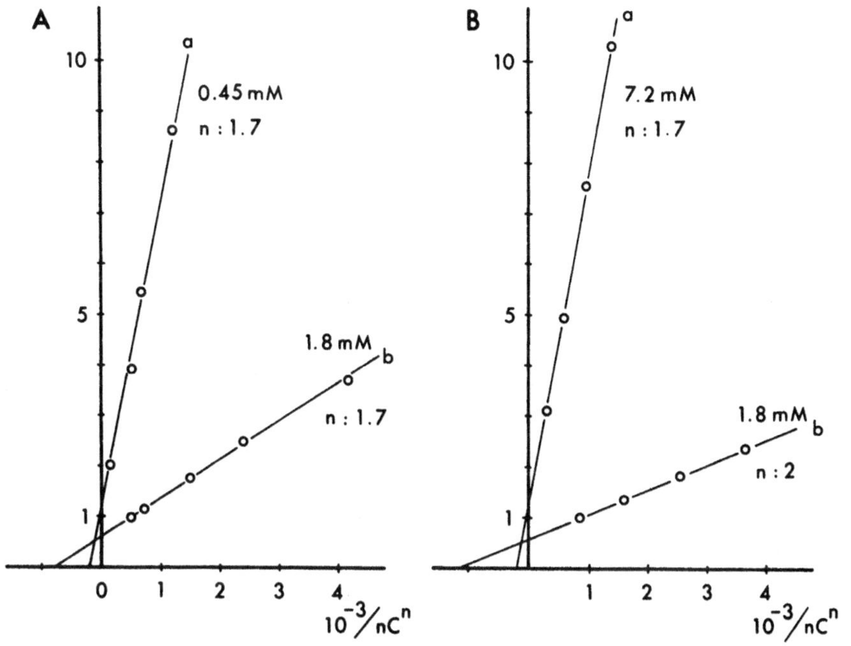

Figure 10. Dose–response curves obtained by quantitative iontophoresis of ACh in two different neurons of the bullfrog sympathetic ganglion and in different $[Ca^{2+}]_o$'s. The inverse of conductance increase at the peak of ACh potentials was plotted against the inverse of the nth power of ACh doses. (A) Effects of low $[Ca^{2+}]_o$ (0.45 mM); (B) effects of high $[Ca^{2+}]_o$ (7.2 mM).

number of ACh, Q is the charge employed for iontophoresis, D is the diffusion coefficient, K_m is an apparent dissociation constant for ACh receptor, n is a cooperativity number, and F is the Faraday constant. Finally, c is a constant determined by the pattern or geometry of the distribution of ACh receptors at the subsynaptic membrane. Since the distance between the tip of an ACh-filled electrode and the ganglion cell membrane may be assumed to be relatively long [this distance, expressed as z by Peper *et al.* (1975), may be calculated to amount to 54–75 μm when the peak time (t_p) of an ACh potential ranges from 0.45 to 0.85 sec and when D equals 11.0×10^{-6} cm$^2 \cdot$ sec^{-2}], the subsynaptic membrane of the bullfrog sympathetic ganglion cell can be considered as a point, in which case $c = 0$. (cf. Peper *et al.*, 1975).

The values for G_{max} and n may be estimated from the double-reciprocal plot of Figure 10. Assuming $k = 0.25$ (Peper *et al.*, 1975) and substituting experimental values for Q, t_p, and G_{ACh} of a small ACh potential, the apparent dissociation constant (K_m) may be readily estimated; in the case of the data illustrated in Figure 10, its value was calculated as being 100 μM. This value is much larger than that for the end-plate membrane, which is approximately

10–15 μM (Dreyer and Peper, 1975; Kuba and Takeshita, 1983), but may be comparable with that for the cultured chick ciliary ganglion cell (Ogden *et al.*, 1984).

The large value of K_m of the bullfrog ganglion cell may also be inferred from the following considerations: The amplitude of a miniature EPSC, presumably containing a single quantum of ACh, is around 200 pA at a membrane potential of −100 mV. Thus, the conductance increase induced by a quantum of ACh can be calculated as 200 pA/0.1 V = 2 nS. Since the single-channel conductance amounts to 8 pS, 250 channels must be opened by a single quantum of ACh. This value is much smaller than the value for the number of channels opened by a quantum of ACh at the end-plate membrane, which is 2000 channels. Since the vesicle hypothesis for the transmitter release mechanism should hold for both the ganglion and the myoneural junction, it is very unlikely that the amount of ACh contained in a synaptic vesicle of the preganglionic nerve terminal is different from the amount of ACh of the synaptic vesicle of the motor-nerve terminal (see Chapter 10). Thus, the small number of ion channels opened by a quantum of ACh in the ganglion cell should be accounted for by the low affinity of the ACh-receptor for ACh which is indeed indicated by the high overall dissociation constant. The possibility that a low density of ACh receptor–ion channel complexes might explain the small number of openings of the nicotinic channel was raised in the case of parasympathetic ganglion cells (Rang, 1981).

B. Pharmacological Characteristics of the Nicotinic Receptor

There are differences in the action of cholinergic agonists and antagonists on the nicotinic receptor of the ganglionic cell membrane (cf. Kuba and Koketsu, 1978). Since this subject is discussed in detail with regard to a number of agonists and antagonists in the next chapter, only the effects of erabutoxin and bungarotoxin are discussed here (see Chapter 6 for the effect of bungarotoxin).

Erabutoxin b, which blocks the ACh receptor of the end-plate membrane in a practically irreversible manner (Kato *et al.*, 1978), depresses only slightly and reversibly the nicotinic depolarization of a bullfrog sympathetic ganglion cell (Kato *et al.*, 1980). α-Bungarotoxin, which is also an irreversible inhibitor of the ACh receptor at the end-plate and the electric organ of *Torpedo* (cf. Lee, 1972) does not affect the nicotinic response in mammalian sympathetic ganglia (Chou and Lee, 1969; D. A. Brown and Fumagali, 1977). In contrast, Marshall (1981) recently reported a potent but reversible blocking action of α-bungarotoxin on the nicotinic response of an amphibian ganglion cell. Furthermore, Dun and Karczmar (1980) reported the reversible blockade of nicotinic ACh potentials by

α-bungarotoxin without any effect on the neurally induced response; this result suggests that only the extrajunctional nicotinic receptor is blocked by α-bungarotoxin. This variety of effects of neurotoxins suggests that the properties of the nicotinic receptors of sympathetic ganglion cells differ depending on the species and on whether they are junctional or extrajunctional.

VII. REGULATORY ROLE OF Ca^{2+} IN THE GATING MECHANISM OF THE NICOTINIC RECEPTOR–ION CHANNEL COMPLEX

A. Bimodal Actions of Ca^{2+}

When the Ca^{2+} concentration in Ringer's ($[Ca^{2+}]_o$) is increased from 1.8 to 7.2 mM in the presence of atropine (10 μM), the amplitude of the ACh potential of bullfrog ganglion cells decreases drastically (Figure 11). The reduction of $[Ca^{2+}]_o$ to 0.4 mM also causes a marked suppression of the ACh potential amplitude. G_{ACh}, estimated from ACh potentials by means of equation (4), decreases by 80 and 72% of the control value at 7.2 mM $[Ca^{2+}]_o$ and at 0.45 mM $[Ca^{2+}]_o$, respectively. Thus, $[Ca^{2+}]_o$ has bimodal effects on the ion conductance activated by the nicotinic action of ACh in bullfrog sympathetic ganglion cells.

B. Changes in the Kinetic Parameters

Figure 10 shows the effects of high and low $[Ca^{2+}]_o$ on the dose–response curve for the nicotinic response of the bullfrog sympathetic ganglion cell

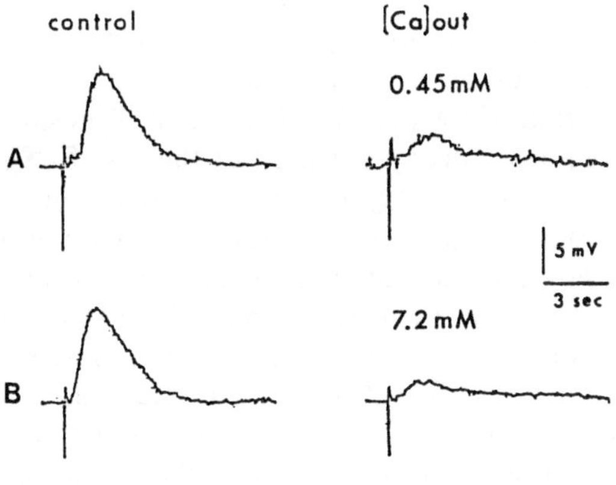

Figure 11. Effects of $[Ca^{2+}]_o$ on ACh potentials in bullfrog sympathetic ganglion cells. (A) and (B) show the effect of high and low $[Ca^{2+}]_o$, respectively.

expressed in terms of a double-reciprocal plot. Both an increase and a decrease in $[Ca^{2+}]_o$ reduced the maximum conductance increase (G_{max}) and elevated the apparent dissociation constant (K_m), while the cooperativity number (n) remained unchanged.

The decay phase of the fast EPSC was significantly lengthened at a high $[Ca^{2+}]_o$ while it was shortened at a low $[Ca^{2+}]_o$. Since an alteration in the amount of released ACh does not affect the decay phase, it may be concluded that the lifetime of the opened ion channel increases at a high $[Ca^{2+}]_o$ and decreases at a low $[Ca^{2+}]_o$, indicating a monotonic dependence of the channel lifetime on $[Ca^{2+}]_o$ (Kuba and Nishi, 1979).

These effects of Ca^{2+} on channel kinetics may be interpreted by means of equation (2′), which is a modified version of the equation (2):

$$2 \cdot \text{ACh} + \text{R} \underset{k_2}{\overset{k_1}{\rightleftharpoons}} \underset{\text{Close}}{\text{ACh}_2 \cdot \text{R}} \underset{\alpha}{\overset{\beta}{\rightleftharpoons}} \underset{\text{Open}}{\text{ACh}_2 \cdot \text{R}^*} \tag{2′}$$

At a steady state, G_{ACh} is expressed by the following equation:

$$G_{ACh} = \frac{\gamma \cdot N/(1 + \alpha/\beta)}{1 + \left\{\dfrac{k_2/k_1}{1 + \beta/\alpha}\right\}/[\text{ACh}]^2} \tag{6}$$

where γ is a single-channel conductance and N is the total number of ion channels. Assuming that essentially an equilibrium exists at the peak of the ACh potential, the G_{max} obtained from the dose–response curve for ACh potentials can be represented by the expression $\gamma \cdot N/(1 + \alpha/\beta)$, while K_m can be expressed as $(k_2/k_1)/(1 + \beta/\alpha)$.

The decrease in G_{max} at both high and low $[Ca^{2+}]_o$ signifies either a decrease in γ or an increase in α/β, assuming that N is not altered by a change in $[Ca^{2+}]_o$. The prolongation of the decay phase of the EPSC implies decreased α. Therefore, unless β is changed, a decreased G_{max} at a high $[Ca^{2+}]_o$ cannot be accounted for by an increased decay phase of the fast EPSC; presumably, however, it may be explained as resulting from a reduction of γ. Such a reduction was in fact observed at the end-plate membrane (Lewis, 1979; Kuba and Takeshita, 1983). On the other hand, the reduced decay constant of the fast EPSC at low $[Ca^{2+}]_o$ may explain the decrease in G_{max} under these conditions. An increase in K_m at high $[Ca^{2+}]_o$ may be due to a decreased α (increase in the decay phase of the fast EPSC); however, it may also result from a change in the affinity of ACh receptor (k_2/k_1) for ACh. An increase in α at low $[Ca^{2+}]_o$ is consistent with the increased K_m under the same conditions. While this interpretation is based on the model for ACh–receptor interactions illustrated under equation (2') (see above), it is obvious that alternative mechanisms for Ca^{2+} action may be considered when other models are employed. The precise values for α and for the affinity constants of the ACh receptor for ACh at different $[Ca^{2+}]_o$ are needed to construct a realistic scheme for Ca^{2+} action on the subsynaptic membrane of the sympathetic ganglion cell.

C. Possible Intracellular Action of Ca^{2+}

Since Ca^{2+} can permeate the ion channel opened by the nicotinic action of ACh at the ganglion cell membrane (Koketsu *et al.*, 1968b), a depressant effect of $[Ca^{2+}]_o$ may result from an increase in the intracellular concentration of free Ca^{2+} ($[Ca^{2+}]_i$). This may be tested by raising the $[Ca^{2+}]_i$. One of the conventional methods used to increase $[Ca^{2+}]_i$ is to apply caffeine, which is known to increase the permeability of the ganglion cell membrane to Ca^{2+} and to cause the release of Ca^{2+} from an intracellular store (Kuba and Nishi, 1976; Kuba, 1980; Morita *et al.*, 1980).

Caffeine significantly reduces the amplitudes of both the fast EPSP and the fast (nicotinic) ACh potential (Morita *et al.*, 1979). This effect depends on the $[Ca^{2+}]_o$; the higher the concentration of Ca^{2+}, the greater the depression induced by caffeine (Figure 12A). Furthermore, intracellular injection of EGTA antagonized the effect of the preceding application of caffeine (Morita *et al.*, 1979) (Figure 12B). Thus, it appears that the depressant effect of caffeine is due to increased $[Ca^{2+}]_i$. Caffeine did not affect the decay phase of the fast EPSC, indicating that $[Ca^{2+}]_o$ action on

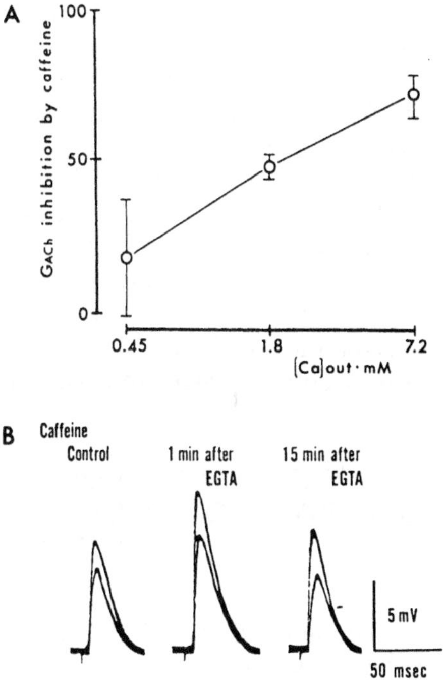

Figure 12. (A) Inhibitory action of caffeine (2 mM) on ACh-induced conductance change at different $[Ca^{2+}]_o$ in a bullfrog sympathetic ganglion cell. (B) Effects of intracellular injection of EGTA on caffeine (2 mM) depression of the fast EPSP. Fast EPSPs were recorded in low-Ca^{2+}/high-Mg^{2+} solution. The records were superimposed in each condition. EGTA was injected into a cell by a sustained current of 5 nA for 60 sec through one barrel (filled with 1 M K-citrate) of a double-barreled electrode. From Morita *et al.* (1979).

α is not mediated by $[Ca^{2+}]_i$. On the other hand, other effects of $[Ca^{2+}]_o$ (see above) could result from increased $[Ca^{2+}]_i$. It is worth noting that in the presence of caffeine, the G_{max} becomes a monotonic function of $[Ca^{2+}]_o$; for example, when the $[Ca^{2+}]_o$ is increased from 0.45 to 7.2 mM, G_{max} decreases monotonically. This implies that one of the modes of $[Ca^{2+}]_o$ action on G_{max}, which may be mediated by an increased $[Ca^{2+}]_i$, becomes dominant in the presence of caffeine. Although it is premature at this time to define the mechanism of $[Ca^{2+}]_i$, action in terms of channel kinetics, it appears that the increase in $[Ca^{2+}]_i$ affects the ion-channel-gating mechanism of the nicotinic receptor at the ganglion cell membrane. A similar idea has been proposed with respect to the effect of $[Ca^{2+}]_i$ on the end-plate membrane (Kuba, 1979).

D. Ca^{2+} Binding at the Subsynaptic Membrane

It is known that Ca^{2+} is bound to the end-plate membrane, from which it is released on activation by agonists (Csillik and Savay, 1964). Fujimoto *et al.*, (1980) demonstrated that Ca^{2+} is bound at the inner surface of the subsynaptic membrane of the bullfrog sympathetic ganglion cell. When caffeine is applied to incubation and fixation media, there is little or no Ca^{2+} at the subsynaptic membrane, indicating the release of Ca^{2+}. However, this release of Ca^{2+} may not necessarily be related to the action on the ion-channel-gating mechanism, because the removal of bound Ca^{2+} by caffeine is not consistent with the depressant effect of caffeine on the ion conductance increase induced by ACh (see above). Thus, the bound Ca^{2+} that is found at the subsynaptic membrane and released by caffeine may represent Ca^{2+} that is released and utilized for depressing the G_{ACh}. The Ca^{2+} binding responsible for the latter action may not be resolved by means of electron microscopy (procedure employed by Fujimoto *et al.*, 1980). Nevertheless, the findings of Fujimoto *et al.*, (1980) throw some light on the physiological significance of Ca^{2+} action at the subsynaptic membrane. Thus, these findings are consistent with the fact that calmodulin, a Ca-binding protein, which may be the mediator of physiological actions of Ca^{2+} in cells, is present within the postsynaptic density of the subsynaptic membrane (Grab *et al.*, 1979). Furthermore, Ca^{2+} binds to the ACh receptor isolated from *Torpedo* (Chang and Newman, 1976).

VIII. PLASTIC MODULATION OF THE NICOTINIC RECEPTOR–ION CHANNEL COMPLEX BY Ca^{2+}-DEPENDENT ACTION POTENTIALS

The quantal size of the fast EPSPs recorded from the bullfrog sympathetic ganglion cell in a low-Ca^{2+}/high-Mg^{2+} solution was enhanced for many hours when the postganglionic neuron was stimulated either intracellularly or antidromically by tetanic stimulation (20 Hz, 5 sec) in normal Ringer's solution. A similar potentiation of the nicotinic ACh potential was observed without a change in the cell input resistance after antidromic or direct tetanic stimulation (Kumamoto and Kuba, 1983). This effect may last for hours or even weeks (Bliss and Lomo, 1973). This long-lasting potentiation of the nicotinic response cannot be attributed to a persistent muscarinic action of ACh (see Chapter 7), since it was not reduced by atropine (T. H. Brown and McAfee, 1982). On the other hand, this potentiation was not seen when the tetanic stimulation was made in Ca^{2+}-free solution (Kumamoto and Kuba, 1983). These results suggest that the overall sensitivity of the nicotinic receptor–channel complex increases for a long time after a brief influx of Ca^{2+} into the postganglionic neuron during a tetanic generation of action potentials. Although the precise mechanism involved in this phenomenon is not known (whether it is mediated by a substance released from the ganglion cell soma or induced by a direct action of Ca^{2+} as it entered into the cell) (cf. Kumamoto and Kuba, 1983), the long-lasting enhancement of the sensitivity of the subsynaptic membrane of ACh should play an important role in the tonic regulation of the peripheral autonomic nervous system.

IX. DESENSITIZATION OF THE NICOTINIC RECEPTOR

There is no doubt that desensitization of the nicotinic receptor of the sympathetic ganglion cell membrane takes place when ACh or a cholinergic agonist is persistently applied (Krivoy and Willis, 1956; Ginsborg and Guerrero, 1964). Unfortunately, however, there appears to be no comprehensive study in the sympathetic ganglion cell of this important phenomenon, which is so characteristic for the end-plate. It may at least be said that the extent and rate of desensitization of the ganglion cell are not as marked as the extent and rate of desensitization of the nicotinic receptor of the motor end-plate (unpublished observation).

X. CONCLUSIONS

The fast excitatory postsynaptic potential, the fast EPSP first described in the early 1940s is the main synaptic ganglionic potential generating the primary synaptic transmission. Since the fast EPSP has been studied for so many years and its analysis is relatively easy—its amplitude is large and its time–course can be recorded quite precisely—its ionic mechanisms are today well known and relatively noncontroversial. Nor does its pharmacology present particular problems (see also Chapters 6 and 13). Yet certain phenomena remain unclear, such as the elementary events (noise) and their uni- or polychannel nature (see also Chapter 6); nor is it quite settled whether the apparent, surprising difference between the number of channels opened by a quantum of ACh at the end-plate, on the one hand, and at the ganglion, on the other, can be considered as real.

An interesting aspect of nicotinic transmission concerns its relationship to Ca^{2+}. Besides, of course, being a prerequisite for transmitter release, Ca^{2+} seems to exert a number of postsynaptic actions. Ca^{2+} appears to regulate the gating mechanism of the nicotinic receptor–channel macromolecule; it may control the channel half-life as well as affect the channel response by permeating the open channel. Furthermore, Ca^{2+} may bind to the postsynaptic receptor membrane. Some of these nicotinic, Ca^{2+}-dependent phenomena appear to underlie the phenomena of plastic regulation, as in the case of certain types of long-lasting potentiation of the nicotinic response. It is all the more striking that still another Ca^{2+}-dependent phenomenon, desensitization, that was studied extensively in the case of the neuromyal junction (see, for instance, Katz and Thesleff, 1957; Magazanik and Vyskocil, 1975; Karczmar and Ohta, 1981; Feltz and Trautman, 1982) was investigated only to a very limited extent in the case of the ganglion.

REFERENCES

Adams, P. R.: Acetylcholine receptor kinetics. *J. Membr. Biol.* **58:**161–174 (1981).

Adams, P. R., Brown, D. A., and Constanti, A.: Pharmacological inhibition of the M-current. *J. Physiol. (London)* **332:**223–262 (1982).

Adler, M., and Albuquerque, E. X.: An analysis of the action of atropine and scopolamine on the end-plate current of sartorius muscle. *J. Pharmacol. Exp. Ther.* **196:**360–372 (1976).

Anderson, C. R., and Stevens, C. F.: Voltage clamp analysis of acetylcholine produced end-plate current fluctuation at frog neuromuscular junction. *J. Physiol. (London)* **235:**655–691 (1973).

Blackman, J. G., and Purves, R. D.: Intracellular recordings from ganglia of the thoracic sympathetic chain of the guinea-pig *J. Physiol. (London)* **203:**173–198 (1969).

Blackman, J. G., Ginsborg, B. L., and Ray, C.: Synaptic transmission in the sympathetic ganglion of the frog. *J. Physiol. (London)* **167:**355–373 (1963).

Bliss, T. V. P., and Lomo, T.: Long-lasting potentiation of synaptic transmission in the dentate area of the anaesthetized rabbit following stimulation of the perforant path. *J. Physiol. (London)* **232:**331–356 (1973).

Brown, D. A., and Fumagali, L.: Dissociation of α-bungarotoxin binding and receptor block in the rat superior cervical ganglion. *Brain Res.* **129:**165–168 (1977).

Brown, T. H., and McAfee, D. A.: Long-term synaptic potentiation in the superior cervical ganglion. *Science* **215:**1411–1413 (1982).

Chang, H. W., and Newman, E.: Dynamic properties of isolated acetylcholine receptor proteins: Release of calcium ions caused by acetylcholine binding. *Proc. Natl. Acad. Sci. U.S.A.* **73**:3364–3368 (1976).

Chou, T. C., and Lee, C. Y.: Effect of whole and fractionated cobra venom on sympathetic ganglionic transmission. *Eur. J. Pharmacol.* **8**:326–330 (1969).

Colquhoun, D.: The link between drug binding and response: Theories and observations, in: *The Receptors: A Comprehensive Treatise* (R. D. O'Brien, ed.), pp. 93–141, Plenum Press, New York (1979).

Colquhoun, D., and Hawkes, A. G.: Relaxation and fluctuations of membrane currents that flow through drug-operated channels. *Proc. R. Soc. London Ser. B.* **199**:231–262 (1977).

Csillik, B., and Savay, G.: The effect of nerve degeneration and regeneration on the calcium-release in the post-junctional cytoplasm of the motor end-plate *Acta. Phys. Acad. Sci. Hung.* **26**:337–342 (1964).

Cull-Candy, S. G., and Miledi, R.: Junctional and extrajunctional membrane channels activated by GABA in locust muscle fibres. *Proc. R. Soc. London Ser. B* **211**:527–535 (1981).

Derkach, V. A., Selyanko, A. A., and Skok, V. I.: Acetylcholine-induced current fluctuations and fast excitatory post-synaptic currents in rabbit sympathetic neurons. *J. Physiol. (London)* **336**:511–526 (1983).

Dreyer, F., and Peper, K.: Density and dose-response curve of acetylcholine receptors in frog neuromuscular junction. *Nature (London)* **253**:641–643 (1975).

Dun, N. J., and Karczmar, A. G.: Blockade of ACh potentials by α-bungarotoxin in rat superior cervical ganglion. *Brain Res.* **196**:536–554 (1980).

Eccles, R. M.: Intracellular potentials recorded from a mammalian sympathetic ganglion. *J. Physiol. (London)* **130**:572–584 (1955).

Eccles, R. M.: Orthodromic activation of single ganglion cells. *J. Physiol. (London)* **165**:387–391 (1963).

Elfvin, L.-G.: The ultrastructure of the superior cervical sympathetic ganglion of the cat. II. The structure of the pre-ganglionic end fibers and the synapses as studied by serial sections. *J. Ultrastruct. Res.* **8**:441–476 (1963).

Erulkar, S. D., and Woodward, J. K.: Intracellular recording from mammalian superior cervical ganglion *in situ*. *J. Physiol. (London)* **199**:189–203 (1968).

Feltz, A., and Trautman, A.; Desensitization at the frog neuromuscular junction: A biphasic process. *J. Physiol. (London)* **322**:257–272 (1982).

Fujimoto, S., Yamamoto, K., Kuba, K., Morita, K., and Kato, E.: Calcium localization in the sympathetic ganglion cell of the bullfrog and effects of caffeine. *Brain Res.* **202**:21–32 (1980).

Ginsborg, B. L.: On the presynaptic acetylcholine receptors in sympathetic ganglia of the frog. *J. Physiol. (London)* **216**:237–246 (1971).

Ginsborg, B. L., and Guerrero, S.: On the action of depolarizing drugs on sympathetic ganglion cells of the frog. *J. Physiol. (London)* **172**:189–206 (1964).

Grab, D. J., Berzins, K., Cohen, R. S., and Siekevitz, P.: Presence of calmodulin in postsynaptic densities isolated from canine cerebral cortex. *J. Biol. Chem.* **254**:8690–8696 (1979).

Hamill, O.P., Marty, A., Neher, E., Sakman, B., and Sigworth, F. J.: Improved patch-clamp techniques for high-resolution current recording from cells and cell-free membrane patches. *Pfluegers Arch.* **391**:85–100 (1981).

Karczmar, A. G., and Ohta, Y.: Neuromyopharmacology as related to anticholinesterase action. *Fundam. Appl. Toxicol.* **1**:135–142 (1981).

Kato, E., and Kuba, K.: Inhibition of transmitter release in bullfrog sympathetic ganglia induced by γ-aminobutyric acid. *J. Physiol. (London)* **298**:271–283 (1980).

Kato, E., Kuba, K., and Koketsu, K.: Effects of erabutoxins on neuromuscular transmission in frog skeletal muscles. *J. Pharmacol.* Exp. *Ther.* **204**:446–453 (1978).

Kato, E., Kuba, K., and Koketsu, K.: Effects of erabutoxins on the cholinergic receptors of bullfrog sympathetic ganglion cells. *Brain Res.* **191**:294–298 (1980).

Katz, B., and Miledi, R.: The measurement of synaptic delay, and the time course of acetylcholine release at the neuromuscular junction. *Proc. R. Soc. London Ser. B* **161**:483–496 (1965).

Katz, B., and Thesleff, S.: A study of the "desensitization" produced by acetylcholine at the motor end-plate. *J. Physiol. (London)* **138**:63–80 (1957).

Koketsu, K.: Cholinergic synaptic potentials and the underlying ionic mechanisms. *Fed. Proc. Fed. Am. Soc. Exp. Biol.* **28**:101–112 (1969).

Koketsu, K., Nishi, S., and Soeda, H.: Acetylcholine-potential of sympathetic ganglion cell membrane. *Life Sci.* **7**:741–749 (1968a).

Koketsu, K., Nishi, S., and Soeda, H.: Calcium and acetylcholine-potential of bullfrog sympathetic ganglion cell membrane. *Life Sci.* **7**:955–963 (1968b).

Krivoy, W. A., and Willis, J. H.: Adaptation to constant concentrations of acetylcholine. *J. Pharmacol. Exp. Ther.* **116**:220–226 (1956).

Kuba, K.: Ca^{2+} and the action of acetylcholine on the subsynaptic membrane, in: *Neurobiology of Chemical Transmission* (M. Otsuka and Z. Hall, eds.), pp. 53–64, John Wiley, Chichester and New York (1979).

Kuba, K.: Release of calcium ions linked to the activation of potassium conductance in a caffeine-treated sympathetic neurone. *J. Physiol. (London)* **298**:251–269 (1980).

Kuba, K., and Koketsu, K.: Analysis of the slow excitatory postsynaptic potential in bullfrog sympathetic ganglion cells. *Jpn. J. Physiol.*. **26**:651–669 (1976).

Kuba, K., and Koketsu, K.: Synaptic events in sympathetic ganglia. *Prog. Neurobiol.* **11**:77–169 (1978).

Kuba, K., and Nishi, S.: Membrane current associated with the fast EPSP of sympathetic neurons. *Physiologist* **14**:176 (1971).

Kuba, K., and Nishi, S.: Rhythmic hyperpolarizations and depolarization of sympathetic ganglion cells induced by caffeine. *J. Neurophysiol.* **39**:547–563 (1976).

Kuba, K., and Nishi, S.: Characteristics of fast excitatory postsynaptic current in bullfrog sympathetic ganglion cells. *Pfleugers Arch.* **378**:205–212 (1979).

Kuba, K., and Takeshita, S.: On the mechanism of calcium actions on the acetylcholine receptor–channel complex at the frog end-plate membrane. *Jpn. J. Physiol.* **33**:931–944 (1983).

Kumamoto, E., and Kuba, K.: Sustained rise in ACh sensitivity of a sympathetic ganglion cell induced by postsynaptic electrical activities. *Nature (London)* **305**:145–146 (1983).

Lee, C. Y.: Chemistry and pharmacology of polypeptide toxins in snake venoms. *Annu. Rev. Pharmacol.* **121**:265–286 (1972).

Lewis, C. A.: Ion concentration dependence of the reversal potential and single channel conductance of ion channels at the frog neuromuscular junction. *J. Physiol. (London)* **286**:417–445 (1979).

MacDermott, A. B., Conner, E. A., Dionne, V. E., and Parsons, R. L.: Voltage clamp study of fast excitatory synaptic currents in bullfrog sympathetic ganglion cells. *J. Gen. Physiol.* **75**:39–60 (1980).

Magazanik, L. G., and Vyskocil, F.: The effect of temperature on desensitization kinetics at the post-synaptic membrane of the frog muscle fibre. *J. Physiol. (London)* **249**:285–300 (1975).

Magleby, K. L., and Stevens, C. F.: A quantitative description of end-plate currents. *J. Physiol. (London)* **223**:173–197 (1972).

Marshall, L. M.: Synaptic localization of α-bungarotoxin binding which blocks nicotinic transmission at frog sympathetic neurons. *Proc. Natl. Acad. Sci. U.S.A.* **78**:1948–1952 (1981).

Minota, S., and Kuba, K.: Restoration of the nicotinic receptor-channel activity from the blockade by atropine in bullfrog sympathetic ganglia. *Brain Res.* **296**:194–197 (1984).

Morita, K., Kato, E., and Kuba, K.: A possible role of intracellular Ca^{2+} in the regulation of the ACh receptor–ion channel complex of the sympathetic ganglion cell. *Kurume Med. J.* **26**:371–376 (1979).

Morita, K., Koketsu, K., and Kuba, K.: Oscillation of $[Ca^{2+}]_i$-linked K^+ conductance in bullfrog sympathetic ganglion cell is sensitive to intracellular anions. *Nature (London)* **283**:204–205 (1980).

Neher, E., and Sakman, B.: Single-channel currents recorded from membrane of denervated frog muscle fibers. *Nature (London)* **260**:779–802 (1976).

Nishi, S.: Ganglionic transmission, in: *The Peripheral Nervous System* (J. I. Hubbard, ed.), pp. 225–255, Plenum Press, New York (1974).

Nishi, S., and Koketsu, K.: Electrical properties and activities of single sympathetic neurons in frogs. *J. Cell. Comp. Physiol.* **55**:15–30 (1960).

Nishi, S., Soeda, H., and Koketsu, K.: Studies on sympathetic B and C neurons and patterns of preganglionic innervation. *J. Cell. Comp. Physiol.* **66**:19–32 (1965).

Ogden, D. C., Gray, P. T. A., Colquhoun, D., and Rang, H. P.: Kinetics of acetylcholine activated ion channels in chick ciliary ganglion neurones grown in tissue culture. *Pfluegers Arch.* **400**:44–50 (1984).

Peper, K., Dreyer, F., and Muller, K.-D.: Analysis of cooperativity of drug–receptor interaction by quantitative iontophoresis at frog motor end-plates. *Cold Spring Habor Symp. Quant. Biol.* **40**:187–192 (1975).

Peper, K., Bradley, R. J., and Dreyer, F.: The acetylcholine receptor at the neuromuscular junction. *Physiol. Rev.* **62**:1271–1340 (1982).

Perri, V., Sacchi, O., and Casella, C.: Electrical properties of the sympathetic neurons in the rat and guinea-pig superior cervical ganglion. *Pfluegers Arch. Gesarnte Physiol.* **314**:40–54 (1970).

Rang, H. P.: The characteristics of synaptic currents and responses to acetylcholine of rat submandibular ganglion cells. *J. Physiol. (London)* **311**:23–55 (1981).

Ruiz-Manresa, F., and Grundfest, H.: Synaptic electrogenesis in eel electroplaques. *J. Gen. Physiol.* **57**:71–92 (1971).

Selyanko, A. A., Derkach, V. A., and Skok, V. I.: Fast excitatory postsynaptic currents in voltage-clamped mammalian ganglion neurones. *J. Auton. Nerv. Sys.* **1**:127–137 (1979).

Skok, V. I.: Fast synaptic transmission in autonomic ganglia, in: *Autonomic Ganglia* (L. G. Elfvin, ed.), pp. 265–280, John Wiley, Chrichester and New York (1983).

Skok, V. I., Selyanko, A. A., and Derkach, V. A.: Two modes of activity of nicotinic acetylcholine receptor channels in sympathetic neurons. *Brain Res.* **238**:480–483 (1982).

Takeuchi, A., and Takeuchi, N.: Active phase of frog's end-plate potential. *J. Neurophysiol.* **22**:395–411 (1959).

Taxi, J.: Etude de l'ultrastructure des zones synaptiques dans les ganglions sympathiques de la grenouille. *C. R. Acad. Sci.* **252**:174–176 (1961).

Weight, F., and Voltava, Z.: Slow synaptic excitation in sympathetic ganglion cells: Evidence for synaptic inactivation of potassium conductance. *Science* **170**:755–758 (1970).

6

Nicotinic Receptors: Activation and Block

V. I. SKOK

I. GENERAL CHARACTERISTICS

A single nicotinic acetylcholine receptor (AChR) is an oligomeric glycoprotein molecule 6–10 nm in diameter with a molecular weight of 225,000; it is comprised of five subunits: two α-subunits and one each of β-, γ-, and δ-subunits. The subunits are assembled in a transmembrane (Raftery *et al.*, 1980; Lindstrom *et al.*, 1980) pentamer and arranged in a circle about a central pit (possibly a channel). These results were obtained in studies of AChRs of the electric organ, the most suitable object for their biochemical and molecular analysis (see Changeux *et al.*, 1976, 1984; Fambrough, 1979; O'Brien *et al.*, 1979; Raftery *et al.*, 1980; Kistler *et al.*, 1982; Zingsheim *et al.*, 1982; Conti-Tronconi *et al.*, 1982; Noda *et al.*, 1982, 1983a,b). The present concept based on considerable evidence is that the subunits of the AChR are polypeptides varying in molecular weight from 40,000 to 65,000; in fact, amino acid sequence analysis was carried out with regard to these subunits by Changeux's, Raftery's, and other groups (cf. Devillers-Thiery *et al.*, 1979, 1983; Gotti *et al.*, 1982; Wu and Raftery, 1981; Raftery *et al.*, 1983; Changeux *et al.*, 1984). It must be noted that this analysis pertains mostly to the electric organ of the eel, although some recent work concerned the mammalian AChR (Gotti *et al.*, 1982; Noda *et al.*, 1983c). While the main characteristics of the AChR may be similar in the case of the electric organ, mammalian (cf. Gotti *et al.*, 1982), and amphibian preparations including ganglia, there must be considerable difference in detail; pertinent work has not yet been carried out.

V. I. SKOK • Bogomoletz Institute of Physiology, Kiev, U.S.S.R.

Each AChR is thought to consist of two parts: a recognition component, which binds acetylcholine (ACh) and related drugs and is located in the α-subunit, and an ionic channel that opens up following the ACh-induced conformational change of the AChR molecule to permit the ion passage and is formed by α-, β-, γ-, and δ-subunits (see Lindstrom, 1980; Mishina *et al.*, 1984). Recent studies of the permeability of the end-plate channels to organic cations suggest that the open channel is a 0.65 nm × 0.65 nm square (Dwyer *et al.*, 1980). The location of the channel with respect to the AChR subunits and its precise nature are not known (see Eldefrawi *et al.*, 1978; Raftery *et al.*, 1983).

The distribution of AChRs was studied in autonomic ganglia using their binding with [^{125}I] α-bungarotoxin. The binding was inhibited by nicotinic agonists and antagonists (DeRenzis *et al.*, 1975; Fumagalli *et al.*, 1976; Chiapinelli *et al.*, 1981). It was found that AChRs developed simultaneously with the establishment of synaptic transmission through the ganglion (Gangitano *et al.*, 1978). Each neuron in a mammalian sympathetic ganglion contains about 9.2×10^5 AChRs that spread over the dendrites and are also localized at the perikaryon (Fumagalli *et al.*, 1976); they are concentrated under synaptic boutons, thus making this area most sensitive to locally applied ACh (Harris *et al.*, 1971; Skok and Selyanko, unpublished observations). It has been suggested on the basis of comparison of the current flowing through a single AChR ionic channel with the membrane current induced by a quantum of a transmitter that each quantum activates about 150 AChRs in a mammalian sympathetic ganglion neuron (Derkach *et al.*, 1983).

In contrast to AChRs of the skeletal muscle, neither the number nor the spatial distribution of nicotinic AChRs in the neurons of autonomic ganglia is affected by chronic preganglionic denervation; thus, trophic relationships between motoneuron and end-plate AChRs differ significantly from those between preganglionic neurons and postganglionic AChRs (DeRenzis *et al.*, 1975). This correlates with numerous observations that chronic preganglionic denervation does not enhance the ganglionic excitatory responses to nicotinic stimulants while enhancing their responses to muscarinic stimulants (Brown, 1966, 1969; Lukomskaya, 1969; Vicker son and Varma, 1969; Dun *et al.*, 1976). An intriguing question concerns the close similarity between kinetics of fast ACh potentials of normal and denervated ganglion cells (Kuffler *et al.*, 1971); this similarity must be contrasted with the marked difference in the kinetics of nicotinic and muscarinic responses of innervated cells (see Koketsu, 1969; Skok, 1973, 1980; Nishi, 1974; Kuba and Koketsu, 1978).

The interesting subject of the turnover of ACh receptors was studied mainly with respect to the nicotinic receptors of the *Torpedo* electric organ and tissue-cultured embryonic chick skeletal muscle cells (Fambrough,

1981). The half-life of these receptors amounted to 8–17 hr, depending on the method used. Less extensive studies were carried out with sympathetic ganglia; in an investigation concerning tissue culture of embryonic chick ganglia, the half-life of the receptors was 11 hr (Carbonetto and Fambrough, 1979).

II. RECOGNITION COMPONENT

A. Mechanisms of Acetylcholine Binding

A considerable amount of evidence as to the molecular organization of the AChR recognition component and the mechanisms of its ACh binding has been obtained from analysis of the correlation between the structure of ganglionic stimulants and blockers, on the one hand, and their effects on ganglionic AChRs, on the other. The results suggest that ACh interacts with two points of the AChR molecule, the anionic point and the esterophilic point. This notion is consistent with the finding that tetramethylammonium (TMA), which binds only to the anionic site, and carbachol, which differs from the ACh molecule only in its carbonyl group, both activate ganglionic AChRs but are, respectively, 4.6 and 7.8 times less effective than ACh (Selyanko and Skok, 1979). These results must be related to the evidence that both TMA and carbachol open the ionic channels of the AChR for a shorter time than ACh, the ester moiety of the latter serving to stabilize the open conformation of the channel (Auerbach *et al.*, 1983; Dreyer *et al.*, 1976; Ascher *et al.*, 1978).

Altogether, it may be conceptualized that the anionic point of the AChR interacts with the positively charged quaternary head-group of ACh, while the esterophilic point interacts with the carbonyl group and with the ether oxygen of ACh through their partial electric charges (see Barlow, 1955; Triggle, 1965; Anichkov, 1974). The esterophilic point attaches also to the methyl groupings via hydrophobic interactions, as evidenced by the finding that trimethylacetylcholine exhibits marked ganglionic activity (Triggle and Triggle, 1976). In general, this scheme resembles the two-point attachment scheme for the nicotinic AChR proposed by Triggle and Triggle (1976), although it emphasizes the significance of the anionic site and of the hydrophobic area.

The most probable structure for the anionic point is a carboxylate or phosphate anion (Triggle, 1965; Ilyin *et al.*, 1976). This structure is apparently identical in the case of AChRs of autonomic ganglia and skeletal muscles, since the replacement of methyl groups by ethyl groups in the ACh molecule or the increase in size of one of the TMA radicals results in essentially similar changes in the effects of the compounds in question on both types of AChRs (Barlow, 1955; Michelson and Zeimal, 1973).

It has been suggested that disulfide bonds adjacent to the anionic point are important for the activation of AChR by agonists (Karlin, 1973). According to more recent studies, reduction of disulfide bonds by dithiothreitol and the resulting generation of free SH groups changes the kinetics of gating of the AChR ionic channels of the end-plate (Terrar, 1978). Disulfide bonds may link the subunits in the oligomeric AChR molecule (O'Brien *et al.*, 1979). In the case of sympathetic ganglion neurons, the reduction of disulfide bonds markedly reduces the agonist-induced depolarization and the decrease in input resistance. Both responses to the agonist are restored by reoxidation (Brown and Kwiatkowski, 1976; Trinus and Skok, 1979).

There is evidence that binding of more than one ACh molecule is required to activate a single AChR of mammalian sympathetic ganglia (cooperativity); this evidence concerns the evaluation of the Hill number. The Hill number can be found as the slope in double logarithmic plot that relates ACh-induced depolarization of the cell membrane to the iontophoretic current employed to eject ACh; on the basis of the assumption that ACh-induced depolarization is proportional to the number of activated AChRs, the slope reflects the number of ACh molecules needed to activate the receptor (Selyanko and Skok, 1979). In sympathetic ganglia, the Hill numbers are usually greater than 1. Their mean value was 2.4, which is close to the values obtained for many other types of synaptic cholinoceptive receptors (Colquhoun, 1979).

B. Specific Pharmacological Properties of Ganglionic Acetylcholine Receptors

The effects that ACh agonists and antagonists exert on ganglionic AChRs characterize them as a unique group of receptors. Ganglionic AChRs are much more sensitive to the blocking effects of quaternary ammonium compounds, hexamethonium, and pentamethonium (Barlow, 1955; Van Rossum, 1962a,b) and of the active molluscan principle surugatoxin (Brown *et al.*, 1979; Brown, 1980) than are the AChRs of skeletal muscle end-plates, while they are much less sensitive than the latter to the excitatory effects of decamethonium and succinylcholine (Paton and Perry, 1953). It is likely that ganglionic AChRs are somewhat less specific with respect to nicotinic agonists and antagonists than nicotinic AChRs of the end-plate (see Volle, 1980).

A potent neuromuscular blocker, snake venom α-bungarotoxin, a polypeptide, does not block synaptic transmission in the sympathetic ganglion (Magazanik *et al.*, 1974; Bursztain and Gerschon, 1977) or of some neurons in parasympathetic ganglia (Luzzatto *et al.*, 1980), although it blocks their extrasynaptic AChRs (Dun and Karczmar, 1980; Chiappinelli *et al.*, 1981); this is in marked contrast to the ability of bungarotoxin to block neuromuscular transmission. On the other hand, α-bungarotoxin binds with both synaptic (Smolen, 1983) and extrasynaptic (Messing and Jonatas, 1983) AChRs of autonomic neurons; furthermore, there is a close parallelism between the ability of several drugs to protect ganglionic AChRs from being bound by α-bungarotoxin and their potency to block ganglionic transmission; thus, there is no doubt that α-bungarotoxin binds to synaptic AChRs (Chiappinelli *et al.*, 1981). Since α-bungarotoxin is devoid of channel-blocking properties (Adams, 1977; Katz and Miledi, 1978; O'Brien *et al.*, 1979) and yet blocks neuromuscular transmission, these results suggest that there are basic differences between the recognition components of the ganglionic and muscle AChRs and their relationship to the channel. Recent results indicate that α-bungarotoxin increases the duration of single-channel lifetime of sympathetic synaptic AChRs while not affecting that of the extrasynaptic AChRs and that it also partially decreases the number of AChRs that respond to activation by ACh (Selyanko, 1983). Another toxin, κ-bungarotoxin, is a potent inhibitor of transmission in avian parasympathetic ganglia (Chiappinelli, 1983).

III. IONIC CHANNEL

A. Ionic Permeability

The reversal potentials for the excitatory postsynaptic currents (EPSCs) or excitatory postsynaptic potentials (EPSPs) of neurons in autonomic ganglia of several species range from −5 to −10 mV (Kuba and Nishi, 1979; Selyanko *et al.*, 1979; MacDermott *et al.*, 1980; Rang, 1981) (see Figure 1A, B) and are similar to those for the ACh potential (Dennis *et al.*, 1971). This indicates that the ionic mechanisms for nicotinic response of synaptic and extrasynaptic AChRs are identical. Studies of the effect of changes in the ionic concentration on the EPSP reversal potential led to the conclusion that the ganglionic nicotinic AChR channels are permeable for sodium and potassium but not for chloride ions (see Nishi, 1974; Kuba and Koketsu, 1978; Selyanko and Skok, 1979) (see also Chapters 4 and 5), similarly to the channels of the AChRs of skeletal muscles. The ratio of the ACh-induced changes in sodium and potassium conductances, g_{Na}/g_K, is 1.8 (Dun *et al.*, 1976) or 2.1 (Skok and Selyanko, unpublished data), which is somewhat higher than the ratio obtained for the neuro-

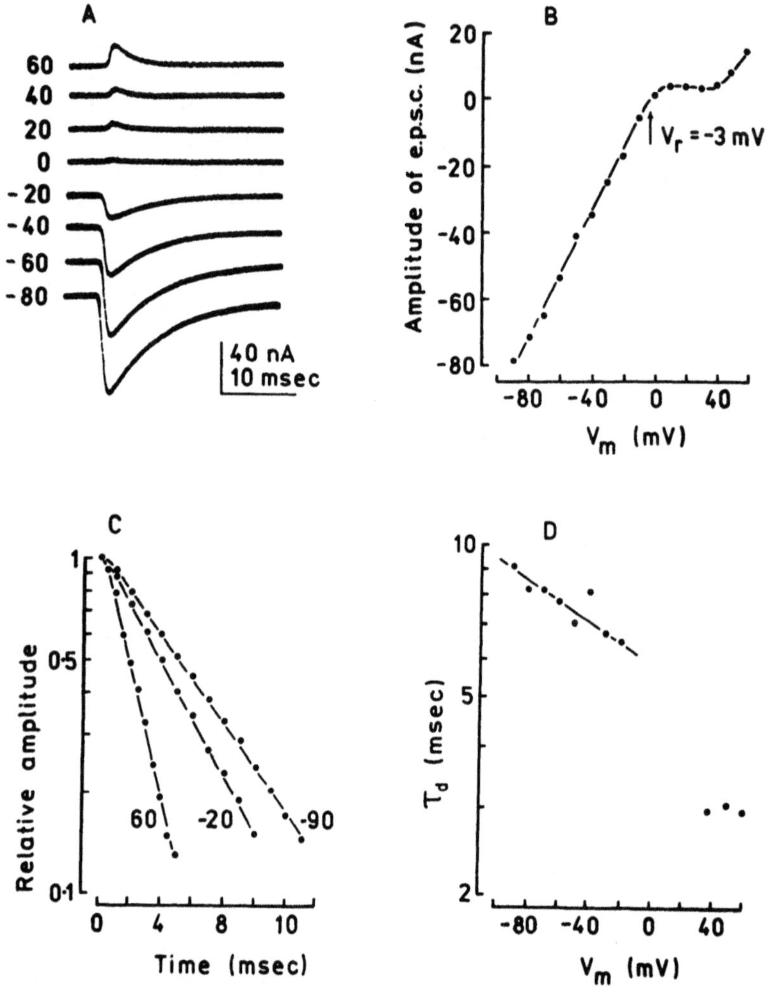

Figure 1. Characteristics of the EPSC of the rabbit superior cervical ganglion. (A) EPSCs recorded in the same cell at different levels of membrane potential (indicated in mV before each tracing). (B) Peak amplitude of the EPSC as a function of membrane potential (V_m). The reversal potential (arrow) is -3 mV. (C) Semilogarithmic plots of the decay of the EPSCs recorded at different levels of membrane potential (indicated in mV beside each plot). (D) Semilogarithmic plot of EPSC decay time constant (τ_d) as a function of membrane potential. $\tau_d = 5.8 \exp(-0.0072\ V_m)$ msec for the least-squares line fitted to the data for this cell. The data for the plots in (B–D) were obtained from records of which some are shown in (A). From Derkach *et al.* (1983).

muscular junction (cf. Takeuchi and Takeuchi, 1960; Feltz and Mallart, 1971). There is also evidence that the AChR channels in autonomic ganglion neurons are permeable to calcium ions (Pappano and Volle, 1966; Koketsu *et al.*, 1968).

B. Single-Channel Lifetime and Conductance

The mean AChR channel lifetime can be approximately estimated as the time constant τ_d of the exponential decay of the EPSC (see Katz and Miledi, 1972; Anderson and Stevens, 1973) (see also Chapter 5). The values for τ_d of the neurons of the sympathetic ganglia obtained at body tem-

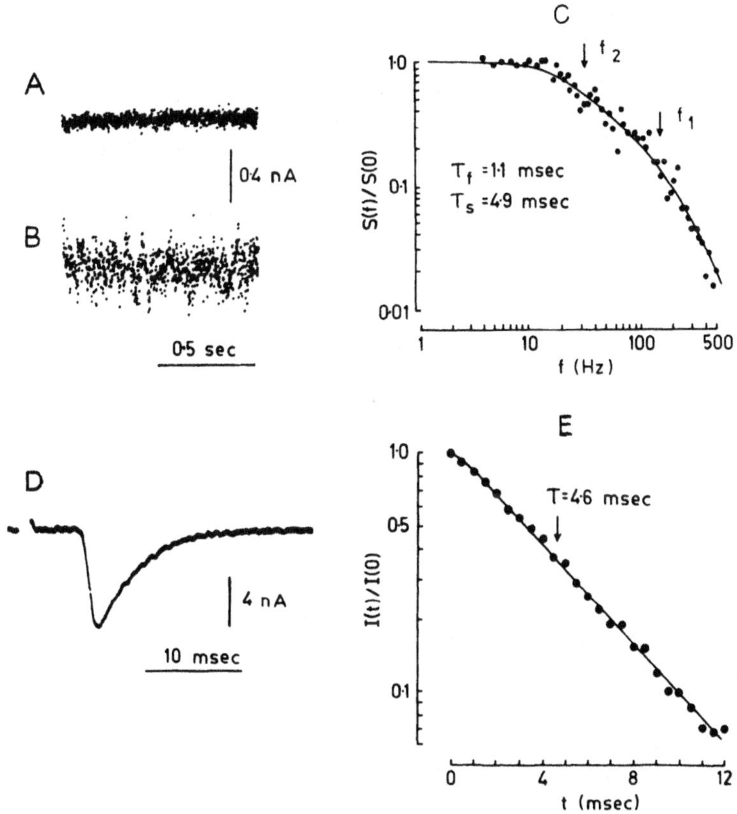

Figure 2. ACh-induced fluctuations of membrane current and EPSC recorded from the same neuron. Digitalized records of current fluctuations before (A) and during (B) the iontophoretic application of ACh (the ACh-induced steady current was 3.8 nA). The spectral density of ACh-induced current fluctuations is shown as a function of frequency (C). Spectrum (C) was obtained by subtraction of the averaged spectrum of four records similar to those shown in (A) from the averaged spectrum of four records similar to those shown in (B). The line traced in (C) represents the sum of two Lorentzian functions with the cutoff frequencies f_1 and f_2 corresponding to a mean channel lifetime τ_f = 1.1 msec (for fast-operating channels) and τ_s = 4.9 msec (for slow-operating channels), with $S_1(0)/S_2(0)$ = 0.3. The EPSC (D) evoked by a single preganglionic stimulus decays exponentially, as indicated by a straight-line relationship in the semilogarithmic plot of EPSC amplitude against time (E); the value of the EPSC decay time constant τ_d was close to the value of τ_s (4.6 msec).

perature (room temperature for poikilotherms) and at resting membrane potential amounted to about 4–6 msec (Kuba and Nishi, 1979; Selyanko *et al.*, 1979; MacDermott *et al.*, 1980) (see also Figure 2D, E). Similar τ_d values were obtained for spontaneously arising miniature EPSCs. Two-component EPSC decay was observed at room temperature in the case of the mammalian parasympathetic ganglion, suggesting contribution of two-channel populations to the EPSC (Rang, 1981); the τ_d for the fast component was about 4–6 msec, while the τ_d for the slow component was markedly longer (Rang, 1981; Rang and Gurney, 1982). The τ_d value of the mammalian sympathetic ganglia was voltage-dependent (see Figure 1C, D and Figure 4C). An e-fold increase was induced by a 140 mV hyperpolarization of the neurons (Selyanko *et al.*, 1979; Derkach *et al.*, 1983); this is close to the value obtained with respect to the nicotinic AChRs of the end-plate (Colquhoun, 1979). Much lower and much higher values for τ_d were obtained for amphibian sympathetic ganglion (Kuba and Nishi, 1979) (see Chapter 5). The Q_{10} of τ_d was equal to either 6 (Kuba and Nishi, 1979) or 3 (MacDermott *et al.*, 1980).

The analysis of membrane current fluctuations produced by a steady-state concentration of ACh (ACh noise) (see Figure 2B) provides another possibility for estimation of the mean lifetime of a single AChR channel (Katz and Miledi, 1972; Anderson and Stevens, 1973). Two kinetic components, fast and slow, corresponding to mean channel lifetimes of 1.1 and 5.0 msec, respectively, were found in the ACh noise spectrum of mammalian sympathetic ganglion neurons maintained at body temperature and at -80 mV (Skok *et al.*, 1982; Derkach *et al.*, 1983). Usually, both components were present in the same neuron (Figure 2C), but a few neurons showed the fast component only. These kinetic components were not affected by atropine (1 μM), indicating that they were due to activation of nicotinic AChRs. They were also independent of the ACh concentration, while both were shortened by hexamethonium, the slow component much more so than the fast component (Skok *et al.*, 1983). The estimate for mean channel lifetime derived from the slow kinetic component is similar to that obtained from τ_d measurements in the same neurons (Figure 2C, E). This similarity suggests that the slow component arises in synaptic AChRs.

The fast component could represent either the activity of extrasynaptic AChR population or the short-lifetime component in the multiple-burst-like activity of the channels that exhibit a slow kinetic component (for further analysis of this problem, see Chapter 5). Multiple-burst-like activity was observed in AChRs of skeletal muscles in the presence of local anesthetics (Neher and Steinbach, 1978) or desensitizing concentrations of ACh (Sakmann *et al.*, 1980). The same phenomenon seems to occur in the sympathetic neurons (Connor *et al.*, 1983). Since the fast component can be observed in some neurons in the absence of the slow component, and since it is independent of ACh concentration (Derkach *et al.*, 1983), it was suggested that the fast component arises in a distinct extrasynaptic AChR population (cf. Neher and Steinbach, 1978; Sakmann *et al.*, 1980) (see also below).

Two kinetic components of the ACh noise spectrum that closely correlate with two components of the EPSC decay and with the results of voltage-jump studies were found to characterize mammalian parasympathetic ganglion neurons kept at room temperature (Rang, 1981; Rang and Gurney, 1982).

The aforementioned values of mean channel lifetime for the AChRs of autonomic ganglia are much higher than those of mean channel lifetime of the AChRs of skeletal muscles, for which values of 1–3 and 0.3 msec were obtained for amphibian and mammalian muscles at room temperature and at body temperature, respectively (see Colquhoun, 1979). Values for channel lifetime as low as 0.7–1.2 msec were recently observed with the patch-clamp technique in cultured neurons of avian parasympathetic ganglia at 30°C (Colquhoun *et al.*, 1983).

ACh noise analysis yielded mean values for single-channel conductance of 36 and 31 pS for mammalian sympathetic (Skok *et al.*, 1982) and parasympathetic (Rang, 1981) ganglia, respectively. These values are very similar to those obtained for skeletal muscles (see Colquhoun, 1979). Somewhat higher values (41–52 pS) were found in avian parasympathetic cultured neurons (Colquhoun *et al.*, 1983).

Obviously the values for the single channel conductances reported in various laboratories are not consistent; there are also problems as to the significance of the components of the mean channel life time (Skok *et al.*, 1982; Derkach *et al.*, 1983). Recent patch-clamp studies of ACh currents carried out with respect to rat superior cervical ganglion neurons (Derkach *et al.*, 1985) seem to resolve these difficulties. In their hands, the AChR single-channel currents exhibited two-modal amplitude distribution (Figure 3). Mean single-channel conductances calculated from low- and high-amplitude currents were 26.6 pS and 71.1 pS, respectively. The low-amplitude currents were more numerous than the high-amplitude currents and appeared mostly in bursts. The most striking result was that mean duration of the burst was very close to the time constant of the EPSC decay, τ_d, and averaging of the bursts resulted in a curve with the time course similar to that of EPSC. It was concluded that the time course of EPSC in mammalian sympathetic neurons is determined by a burst of about 15 single openings, each about 1 ms long, if measured at t = 20–23°C. The number of the openings in the burst is several times higher than in the neuromuscular junction (see Derkach *et al.*, 1985).

As mentioned above, carbachol and TMA are markedly less potent than ACh in producing depolarization of sympathetic ganglion neurons. Thus, it may be suggested that these two agonists open the AChR channels for a shorter time than does ACh, similarly to what has been found in the case of skeletal muscles (Katz and Miledi, 1972; Dreyer *et al.*, 1976) and molluscan neurons (Ascher *et al.*, 1978). An increased calcium concen-

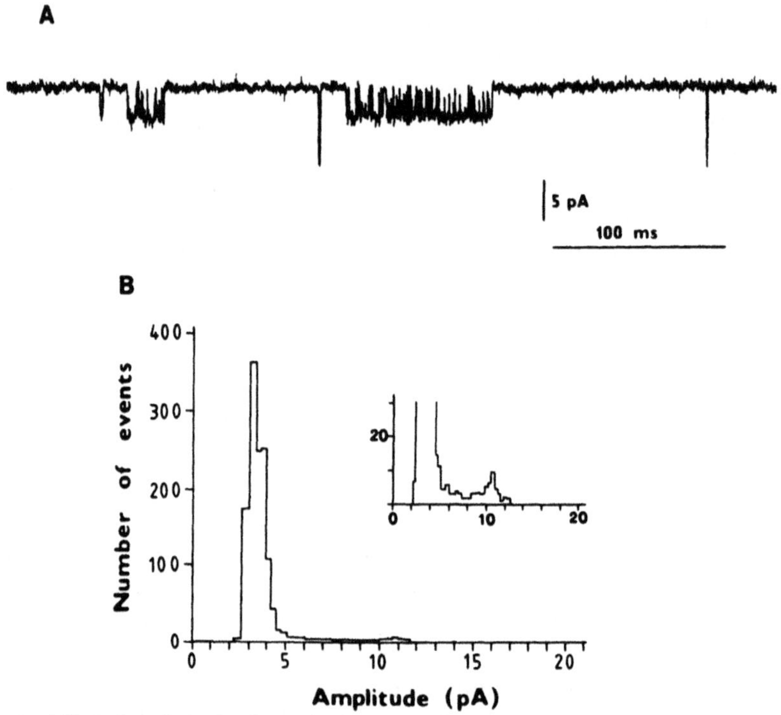

Figure 3. ACh-induced single-channel currents in the rat superior cervical ganglion neuron. (A) An example of currents recorded from cell-attached patch at −130 mV calculated membrane potential (assuming resting membrane potential was −50 mV). Inward currents are seen as downward deflections from the baseline. Pipette contained 30 μM ACh. (B) Distribution of amplitudes of single-channel currents, some of which are shown in (A). The distribution revealed two modes, with low and with high amplitudes. Inset: the same distribution but with expanded ordinate scale. Mean values of currents were 3.7 ± 0.05 pA and 10.7 ± 0.37 pA. From Derkach *et al.* (1985).

tration prolongs mean AChR channel lifetime (Kuba and Nishi, 1979; Rang, 1981), possibly due to the effect of calcium on the surface membrane changes (Cohen and Van der Kloot, 1978) or, alternatively, to the stabilizing effect of calcium on the channel in its open configuration (Ascher *et al.*, 1978).

IV. BLOCKING MECHANISMS

A. Mechanisms of the Blockade of Ganglionic Acetylcholine Receptors

According to the classic concept, there are two groups of ACh antagonists: competitive drugs that prevent activation of AChRs by ACh and depolarizing drugs that act similarly to ACh (Paton and Perry, 1953) (see also Chapters 12 and 13). Subsequently, Van Rossum (1962a,b) differentiated four groups of compounds among the ganglionic blocking substances: depolarizing, competitive, noncompetitive, and mixed ganglionic blockers. The depolarizing blockers (nicotine and 1, 1–dimethyl–4–phenyl–piperazinum, DMPP) produce AChR blockade after preliminary AChR activation. The competitive blockers (tetraethylammonium, hexa- and pentamethonium, azamethonium, and trimethaphan) compete with ACh for the binding site in the AChR. The noncompetitive blockers (chlorizondamine, trimethidine, and pentacine) bind to some other than the ACh-binding site. The mixed blockers (mecamylamine and pempidine) combine the two latter mechanisms. The competitive blockade is evidenced by a non-voltage-dependent decrease in the EPSC amplitude and, in contrast to the non-competitive blockade, is antagonized by increased doses of agonist, is not enhanced by conditional AChR activation, and does not reduce, in a voltage-sensitive manner, the AChR channel lifetime.

However, recent results show that the blockade of ganglionic AChRs by hexamethonium differs essentially from the classic scheme of competitive blockade. Figure 4 illustrates the effect of hexamethonium on the EPSC recorded from a neuron of the rabbit superior cervical ganglion. Hexamethonium reduced the amplitude and shortened the decay time of the EPSC; the decay pattern remained a one-component curve that followed a single exponential time course (Figure 3A,B). Both effects were increased by hyperpolarization, although the latter more so than the former (Figure 4C, G). Since the time constant of the EPSC decay, τ_d, is equal to the mean channel lifetime (see above), it was suggested that hexamethonium shortens mean channel lifetime, acting on the channel open by a transmitter. This notion was subsequently confirmed by means of ACh noise analysis (Skok *et al.*, 1983); both kinetic components of the ACh noise spectrum were shortened by hexamethonium, although the slow component was shortened much more markedly than the fast component (see Figure 5).

This effect of hexamethonium can be described by the following reaction scheme (see Adams, 1976; Ascher *et al.*, 1978, 1979):

$$\text{AChR} \underset{\alpha}{\overset{\beta}{\rightleftharpoons}} \text{AChR}^* + \text{B} \underset{k^*_{-\text{B}}}{\overset{k^*_{+\text{B}} \cdot X_\text{B}}{\rightleftharpoons}} \text{AChR}^*\text{B}$$

where AChR is an intermediate or nonconducting (closed) receptor–channel complex that is transformed via a voltage-dependent conformational change into the conducting (open) receptor–channel complex (AChR*); the rate constants are α and β according to a two-state model, where $\alpha = 1/\tau_d$ (Magleby and Stevens, 1972); $k^*_{+\text{B}}$ and $k^*_{-\text{B}}$ are the rate constants of binding and dissociation in the reaction between the blocker B and the open channel; and X_B is the concentration of B. When $k^*_{-\text{B}}$ is low, as when the EPSC decay is represented by a single exponential curve

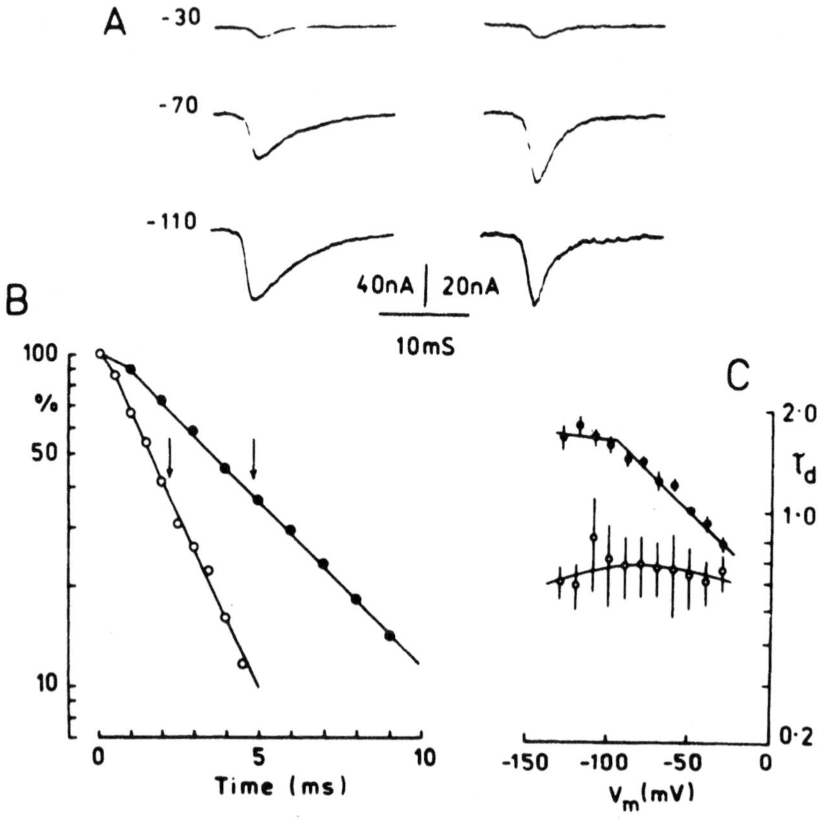

Figure 4A–C. Effects of hexamethonium (10^{-5} M) on the EPSC (A–C, F, G) and on ACh current (D, E). (A) EPSPs recorded from the neuron of the rabbit superior cervical ganglion at different membrane potential levels (indicated in mV) in normal solution (left) and in the presence of hexamethonium (right). (B) Normalized semilogarithmic plots of the EPSC decay recorded at −70 mV in normal solution (●) and in the presence of hexamethonium (○). The time constants (τ_d and τ_d) of the EPSC decay, equal to 4.8 and 2.3 msec, respectively, are indicated by arrows. (C) Semilogarithmic plot of τ_d and τ_d against membrane potential level obtained in normal solution (●) and in the presence of hexamethonium (○).

in the presence of hexamethonium, $\tau'_d = (\alpha + k^*_{+B} \cdot X_B)^{-1}$, where τ'_d is an EPSC decay time constant measured in the presence of B (see Colquhoun *et al.*, 1979). Hence, the value of k^*_{+B} can be calculated at different membrane potential levels. At −50 mV (the approximate resting membrane potential level), the mean value for k^*_{+B} in the case of hexamethonium-treated rabbit superior cervical ganglion cells is $7.3 \times 10^6\ M^{-1}sec^{-1}$. The k^*_{-B} value is 0.06 sec^{-1}, if estimated as a reciprocal of the time constant of the recovery of the second ACh response in an experiment involving paired ACh application (see Figure 3D, E) (see Skok *et al.*, 1983). The voltage sensitivity of k^*_{-B} has been clearly demonstrated for rabbit su-

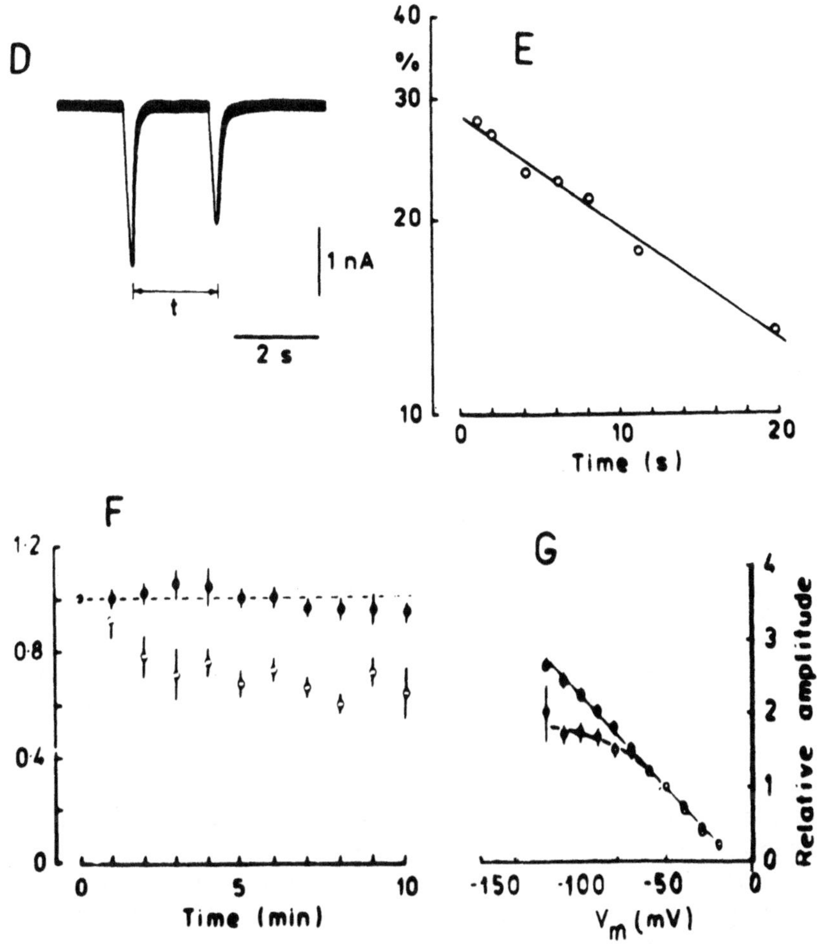

Figure 4D–G. (D) Paired ACh currents recorded at −80 mV in the presence of hexamethonium in response to two 20-msec-long iontophoretic pulses at interpulse interval *t*. (E) Semilogarithmic plot of inhibition of second ACh current as a function of interpulse interval in (D) (time constant is 25.7 sec) (Skok *et al.*, 1983). (F) Time course of hexamethonium effects on the normalized amplitude (●) and on τ_d (○) of the EPSC. Vertical bars indicate S.E. (G) Plot of the EPSC amplitude (normalized to the value at −50 mV) against membrane potential in normal solution (●) and in the presence of hexamethonium (○). From Selyanko *et al.* (1981).

perior cervical ganglion cells, with an e-fold increase of k^*_{+B} corresponding to a 67-mV hyperpolarization (Selyanko *et al.*, 1981).

It is of particular interest that the channel-blocking hexamethonium effect is observed at resting membrane potential level (about −50 mV), while the competitive effect resulting in a decrease of the EPSC amplitude is observed only in hyperpolarized cells (Figure 4C, G). Moreover, only

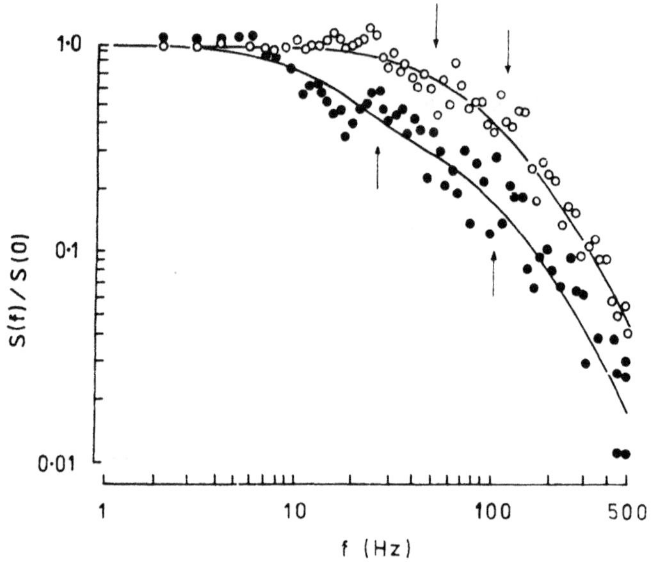

Figure 5. Spectral density of ACh-induced current fluctuations obtained as in Figure 2C, from rabbit superior cervical ganglion neuron in normal solution (●) and in the presence of 10^{-5} M hexamethonium (○). The values of channel lifetime τ_f and τ_s that correspond to the cutoff frequencies indicated by arrows were 1.5 and 10.6 msec in normal solution and 1.2 and 3.2 msec in the presence of hexamethonium. From Skok *et al.* (1983).

very high doses of hexamethonium exert the competitive effect, and at concentrations of 10^{-5} M, only the channel-blocking effect is observed (Figure 4F). It must also be noted that the shortening of the EPSC decay should result by itself in reducing the amplitude of the EPSP in the non-voltage-clamped neurons, due to the high values of the membrane time constant (Skok, 1973, 1980). This implies that the specific blocking effect of hexamethonium is due to the block of the open channel rather than to competitive blockade and that the channel-blocking site is not identical with the competitive site.

It is possible to estimate the position within the transmembrane electric field at which the blocking molecule or the agonist (channel-gating) molecule is acting by measuring the voltage sensitivity of the affinity constant $K = k^*_{+B}/k^*_{-B}$ and of $K' = \beta/\alpha$ (see Woodhull, 1973). Using this approach, it has been shown that penta-, hexa-, and heptamethonium bind at a position approximately midway (40–60%) through the membrane electric field. In contrast, the voltage sensitivity of the affinity of the transmitter molecule itself suggests that it binds to a site that is affected by only 11–14% of the transmembrane electric field, being located close to the extracellular surface (Derkach *et al.*, 1984). These results support the suggestion that the binding sites for the competitive and noncompetitive agents are not the same. Thus, it can be suggested that in addition to the classic mode of competitive ganglionic block by hexamethonium and other bisquaternary ammonium compounds (BACs), there is an additional mechanism for ganglionic AChR blockade, the channel-blocking mechanism, and that the competitive and the channel-blocking binding sites are located in different regions of the AChR. In addition, some BACs, particularly heptamethonium, may interact with the ionic channel of the ganglionic AChR in its closed conformation (Selyanko *et al.*, 1982).

The channel-blocking action of hexamethonium [first established by Blackman (1970)], as well as that of tubocurarine and decamethonium, has also been observed in mammalian parasympathetic ganglia, while surugatoxin, trimethaphan, and mecamylamine showed competitive actions (Ascher *et al.*, 1979). In mammalian sympathetic ganglia, the channel-blocking action of tubocurarine was observed only when the neurons were hyperpolarized (Selyanko *et al.*, 1981). This correlates with the finding that tubocurarine is much more potent than hexamethonium in inhibiting α-bungarotoxin binding to the neurons of autonomic ganglion (Chiapinelli *et al.*, 1981). That still another, peculiar mechanism for channel blockade may be operative in the case of the block exerted by chlorizondamine and some other noncompetitive ganglion blockers is suggested by recent results indicating that these compounds produce open-channel blockade, then appear to be "trapped" within the closed channel and need AChR reactivation by an agonist to be able to leave the channel (Lingle, 1983a, b).

It is of particular interest that the ganglion-blocking activities of BACs correlate closely with their channel-blocking activities (Skok *et al.*, 1983), as shown in Figure 6. Note that these activities relate more closely to $k^{*}_{+\mathrm{B}}$ than to the affinity constant k^{*}_{B}, which may be defined as the ratio between $k^{*}_{+\mathrm{B}}$ and $k^{*}_{-\mathrm{B}}$ (see above and Figure 6). This suggests that the open-channel blockade is the mechanism that underlies the specific ganglionic block exerted by BACs (Skok *et al.*, 1983, 1984).

B. Chemical Structure of the Binding Site for Blocking Substances

Many studies have now been carried out for several years with respect to the structure of the ganglionic AChRs as based on appropriate structure–activity evaluations of ganglionic blockers. The earlier studies of Triggle (1965) led to an emphasis on the anionic site or sites within the AChR that are needed for the potent action of the BACs; possibly two such sites are needed for the effectiveness of these bisquaternaries endowed with two cationic heads.

More recent studies (see Triggle and Triggle, 1976; Trcka, 1980a,b) led to new proposals concerning the effectiveness of the most potent ganglion blockers. All these compounds are either mono- or bifunctional;

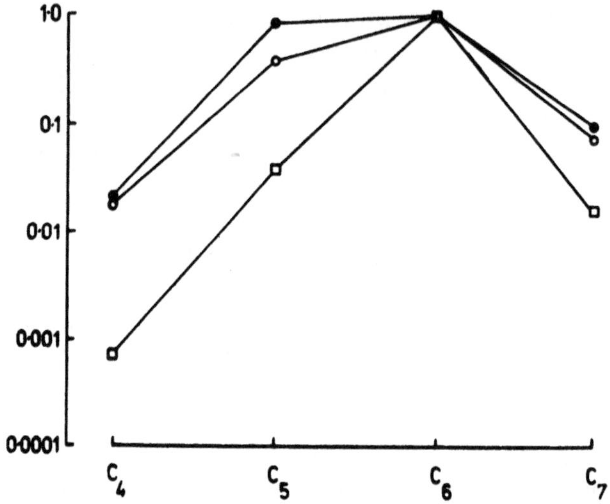

Figure 6. Relative channel-blocking activities of tetra-, penta-, hexa-, and heptamethonium (indicated as C_4, C_5, C_6, and C_7, respectively, on the abscissa) as compared with their relative ganglion-blocking activity. The channel-blocking activity was estimated as the rate constant k^*_{+B} of a blocker binding to an open channel (○) and as the affinity constant $k^*_B = k^*_{+B}/k^*_{-B}$ (□). The ganglion-blocking activity (●) was taken from Paton and Zaimis (1949) (see also Trcka, 1980a). The ganglion- and channel-blocking activities of the other compounds are normalized to those of hexamethonium. From Skok *et al.* (1983).

i.e., they interact with either one or two active points or sites within the AChR. The most potent monofunctional blockers are quaternary, tertiary, or secondary amines exhibiting one of the following three structural components (see Trcka, 1980a,b):

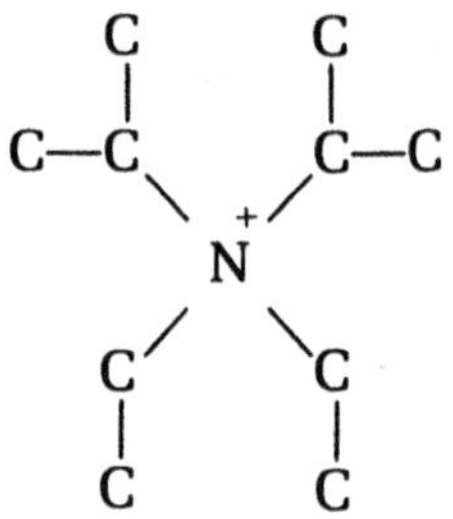

I

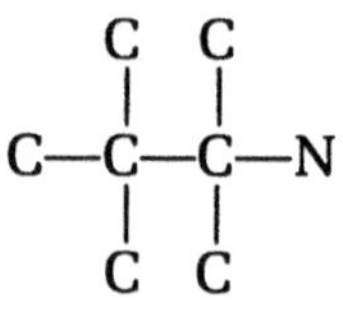

II

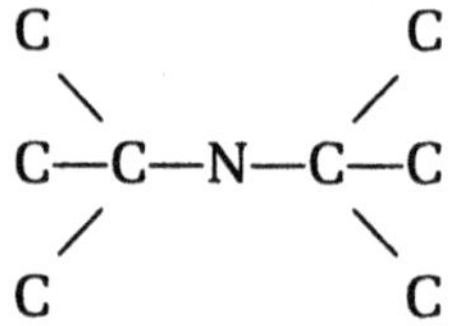

III

In the case of the bifunctional blockers, one of these structural components is connected (e.g., through a polymethylene chain) with fragment IV, the cationic group of ACh (Lukomskaya and Gmiro, 1982):

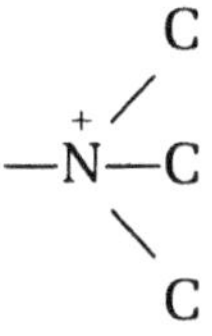

IV

These structure–activity evaluations suggest that there are probably two distinct structures within the AChR molecule that are important for blocker binding—components A and B, which bind components I, II, or III and component IV, respectively. Both structures possess an anionic point or site, most probably the carboxylate anion, since all quaternary and nonquaternary amines exist almost exclusively in the cationic form at physiological pH.

Structures A and B differ, however, in the geometry of the hydrophobic area, which is one or two intercarbon intervals longer in the case of structure A than in that of structure B. It may be speculated that structures A and B are the glutamic and aspartic acid residues, respectively, since these residues differ by one intercarbon interval.

It has long been known that the transmission-blocking potencies in the case of symmetrical BACs correlate with the distances between their two cationic groups (Barlow and Zoller, 1964; Michelson and Zeimal, 1973; Triggle and Triggle, 1976). Two potency peaks have been observed in the case of a series of ganglionic blockers that correspond to intercharge distances of 0.75–0.90 nm (for penta- and hexamethonium) and 1.91–2.11 nm (for hexa- to octadecamethonium) (see Skok, 1985). The existence of these peaks can be explained if the intercationic distance in BAC, is complementary to the interanionic distance in the binding site. The most likely model is a polypeptide chain in β conformation, with the negatively charged glutamic or aspartic acid residues appearing on one side of the β strand or β sheet, which is known to have a step of about 7 Å. On this basis, an attempt was made to estimate the number and location of the BAC-binding sites within the AChR by comparing the distance between either the glutamic or the aspartic acid residues in the electric organ AChR with the BAC intercharge distances. Two areas of most probable binding have been found in the extracellular extrabilayer portion of the AChR molecule, possibly related to the binding sites of the competitive and channel-blocking compounds (see above and Skok, 1985).

Two potency peaks were also observed in the case of the blockers of neuromuscular transmission, one corresponding to intercharge distances of 1.26–1.39 nm (decamethonium) and another identical to the second potency peak recorded for ganglionic blockers (1.91–2.11 nm) (see Skok, 1985).

In fact, the channel-blocking activity of hexamethonium has not been observed in the case of skeletal muscle (Lukomskaya *et al.*, 1981a), while some other bisquaternary ammonium derivatives exhibited this efffect (Adams and Sakmann, 1978; Lukomskaya *et al.*, 1981b).

What is the physiological function of the BAC-binding sites? It is known that calcium ions prolong channel lifetime in ganglionic AChRs, in contrast to the effect of BACs (Ascher *et al.*, 1979). Measurement of the k^*_{+B} for hexamethonium in high-calcium medium revealed a strong inhibitory effect of calcium ions on the binding of hexamethonium to the open channel; a 10-fold decrease in k^*_{+B} resulted from a 10-fold increase of calcium concentration (Selyanko *et al.*, 1985). These findings, and the structural similarity between the AChR and calcium-binding proteins (see Skok, 1984), as well as the importance of calcium ions for the normal function of the AChR (Chang and Neumann, 1976), all suggest that BACs may compete with calcium ions for the binding site in the AChR molecule.

In addition, recent results indicate that ganglionic nicotinic AChRs are under modulatory control of biogenic substances such as 5-hydroxytryptamine and certain peptides (Koketsu *et al.*, 1982; Costa *et al.*, 1983; Akasu *et al.*, 1983). The mechanism of these effects is still not clear (see Chapter 12).

V. CONCLUSIONS

The results described above suggest that the following receptor mechanisms and processes underlie ganglionic AChR activation: For activation, binding of two or more ACh molecules with each AChR is needed. Each molecule binds to two points, anionic and esterophilic, of the AChR. The binding results in the opening of the AChR ionic channel, followed by an increase in channel conductance to a mean value of about 30 pS, this increase lasting in the case of synaptic AChR for 4–6 msec with a few short channel closures; in the case of extrasynaptic AChRs, the channels stay open for about 1 msec. It is possible that the elementary openings of 1 msec duration are grouped in 4- to 6-msec bursts. An open channel is permeable for sodium and potassium ions, as well as, to a lesser extent, for calcium ions. The disulfide bonds in the AChR molecule are important for AChR activation by ACh. The selective ganglionic blockers, the bisquaternary ammonium compounds, interact with the ACh-binding site of the AChR (competitive effect) and also with a site that is in the close vicinity of the open ionic channel (channel-blocking effect). Channel-blocking activity constitutes a possible mechanism for selective ganglionic blockade, as evidenced by the close correlation between this activity and the ganglion-blocking action of ganglionic blockers. It is suggested that the BAC-binding site of the AChR normally binds calcium ions.

REFERENCES

Adams, P. R.: Drug blockade of open end-plate channels. *J. Physiol. (London)* **260:**531–552 (1976).

Adams, P. R.: Relaxation experiments using bath-applied suberyldicholine. *J. Physiol. (London)* **268:**271–289 (1977).

Adams, P. R., and Sakmann, B.: Decamethonium both opens and blocks end-plate channels. *Proc. Natl. Acad. Sci. U.S.A.* **75:**2994–2998 (1978).

Akasu, T., Kojima, M., and Koketsu, K.: Luteinizing hormone-releasing hormone modulates nicotinic ACh-receptor sensitivity in amphibian cholinergic transmission. *Brain Res.* **279:**347–351 (1983).

Anderson, A. R., and Stevens, C. F.: Voltage clamp analysis of acetylcholine produced end-plate current fluctuations at frog neuromuscular junction. *J. Physiol. (London)* **235:**655–691 (1973).

Anichkov, S. V.: *Selective Action of Mediator Drugs* (in Russian). Medicina, Leningrad (1974).

Ascher, P., Marty, A., and Neild, T. O.: Life time and elementary conductance of the channels mediating the excitatory effects of acetylcholine in *Aplysia* neurones. *J. Physiol. (London)* **278:**177–206 (1978).

Ascher, P., Large, W. A., and Rang, H. P.: Studies on the mechanism of action of acetylcholine antagonists on rat parasympathetic ganglion cells. *J. Physiol. (London)* **295:**139–170 (1979).

Auerbach, A., Del Castillo, J., Specht, P. D. and Titmus, M.: Correlation of agonist structure with acetylcholine receptor kinetics: Studies on the frog end-plate and on chick embryo muscle. *J. Physiol. (London)* **343:**551–568 (1983).

Barlow, R. B.: *Introduction to Chemical Pharmacology*. Methuen, London; John Wiley, New York (1955).

Barlow, R. B., and Zoller, A.: Some effects of long chain polymethylene bisonium salts on junctional transmission in the peripheral nervous system. *Br. J. Pharmacol.* **23:**131–150 (1964).

Blackman, J. C.: Dependence on membrane potential of the blocking action of hexamethonium at a sympathetic ganglionic synapse. *Proc. Univ. Otago Med. Sch.* **48:**4–5 (1970).

Brown, D. A.: Depolarization of normal and preganglionically denervated superior cervical ganglia by stimulant drugs. *Br. J. Pharmacol.* **26:**511–520 (1966).

Brown, D. A.: Responses of normal and denervated cat superior cervical ganglia to some stimulant compounds. *J. Physiol. (London)* **201:**225–236 (1969).

Brown, D. A.: Locus and mechanism of action of ganglion blocking agents, in: *Pharmacology of Ganglionic Transmission* (D. A. Kharkevich, ed.), pp. 185–235, Springer-Verlag, Berlin, Heidelberg, and New York (1980).

Brown, D. A., and Kwiatkowski, D.: A note on the effect of dithiothreitol (DTT) on the depolarization of isolated sympathetic ganglia by carbachol and bromo-acetylocholine. *Br. J. Pharmacol.* **56:**128–130 (1976).

Brown, D. A., Garthwaite, J., Hayashi, E., and Yamada, S.: Action of surugatoxin on nicotinic receptors in the superior cervical ganglion of the cat. *Br. J. Pharmacol.* **58:**157–159 (1979).

Bursztain, Sh., and Gerschon, M.: Discrimination between nicotinic receptors in vertebrate ganglia and skeletal muscle by α-bungarotoxin and cobra venoms. *J. Physiol.* **269:**17–31 (1977).

Carbonetto, S., and Fambrough, D. M.: Synthesis, insertion into plasma membrane and turnover of α-bungarotoxin receptors in chick sympathetic neurons. *J. Cell Biol.* **81:**555–569 (1979).

Chang, H. W., and Neumann, E.: Dynamic properties of isolated acetylcholine receptor proteins: Release of calcium ions caused by acetylcholine binding. *Proc. Natl. Acad. Sci. U.S.A.* **73:**3364–3368 (1976).

Changeux, J. P., Benedetti, L., Bourgeois, J. P., Brisson, A., Cartand, J., Devaux, P., Grunhagen, H., Moreau, M., Popot, J. L., Sorel, A., and Weher,M.: Some structural properties of the cholinergic receptor protein in its membrane environment relevant to its function as a pharmacological receptor. *Cold Spring Harbor Symp. Quant. Biol.* **40:**211–230 (1976).

Changeux, J.-P., Devillers-Thiery, A., and Chamovilli, P.: Acetylcholine receptor: An allosteric protein. *Science* **225:**1335–1345 (1984).

Chiappinelli, V. A.: Kappa-bungarotoxin: A probe for the neuronal nicotinic receptor in the avian ciliary ganglion. *Brain Res.* **277:**9–21 (1983).

Chiappinelli, V. A., Cohen, J. B., and Zigmond, R. E.: The effects of α- and β-neuro-toxins from the venoms of various snakes on transmission in autonomic ganglia. *Brain Res.* **211:**107–126 (1981).

Cohen, J., and Van der Kloot, W.: Effects of [Ca^{2+}] and [Mg^{2+}] on the decay of miniature end-plate currents. *Nature (London)* **271:**77–79 (1978).

Colquhoun, D.: The link between drug binding and response: Theories and observations, in: *The Receptors: A Comprehensive Treatise*, Vol. 1 (R. D. O'Brien, ed.), pp. 93–141, Plenum Press, New York (1979).

Colquhoun, D., Dreyer, F., and Sheridan, R. D.: The actions of tubocurarine at the neuromuscular junction. *J. Physiol. (London)* **293:**247–284 (1979).

Colquhoun, D., Gray, P. T. A., Ogden, D. C., and Rang, H. P.: Single channel ACh currents in chick ciliary ganglion cells. *J. Physiol. (London)* **334:**115–116 (1983).

Connor, E. A., Levy, S. M., and Parsons, R. L.: Kinetic analysis of atropine-induced alterations in bullfrog ganglionic fast synaptic currents. *J. Physiol. (London)* **337:**137–158 (1983).

Conti-Tronconi, B. M., Dunn, S. M., and Raftery, M. A.: Functional stability of *Torpedo* acetylcholine receptor: Effect of protease treatment. *Biochemistry* **21:**893–899 (1982).

Costa, E., Guidotti, A., Hanbauer, I., and Salani, L.: Modulation of nicotinic receptor function by opiate recognition sites highly selective for Met^5-enkephalin [Arg^6Phe^7]. *Fed. Proc. Fed. Am. Soc. Exp. Biol.* **42:**2946–2952 (1983).

Dennis, M. J., Harris, A. J., and Kuffler, S. W.: Synaptic transmission and its duplication by focally applied acetylcholine in parasympathetic neurons of the heart of the frog. *Proc. R. Soc. London Ser. B* **177:**509–539 (1971).

DeRenzis, G., Fumagalli, L., and Miani, N.: Acetylcholine receptors in sympathetic neurons. *Exp. Brain Res.* **23**(Suppl.):97 (1975).

Derkach, V. A., Selyanko, A. A., and Skok, V. I.: Acetylcholine-induced current fluctuations and fast excitatory post-synaptic currents in rabbit sympathetic neurons. *J. Physiol. (London)* **236:**511–526 (1983).

Derkach, V. A., Selyanko, A. A., and Skok, V. I.: Localization of binding sites for bisquarternary ammonium compounds in the acetylcholine receptor. *Rep. Ukr. SSR Acad. Sci.* **10:**70–72 (1984).

Derkach, V. A., North, R. A., Selyanko, A. A., and Skok, V. I.: Burst-like single-channel activity determines the duration of the synaptic current in a mammalian sympathetic neurone. *Science* (1985) (in press).

Devillers-Thiery, A., Changeux, J.-P., Paroutaud, P., and Strosberg, A. D.: The amino-terminal sequence of the 40,000 molecular weight subunit of the acetylcholine receptor protein from *Torpedo marmorata*. *FEBS Lett.* **104:**99–105 (1979).

Devillers-Thiery, A., Giraudat, J., Bentaboulet, M., and Changeux, J.-P.: Complete mRNA coding sequence of the acetylcholine binding α-subunit of *Torpedo marmorata* acetylcholine receptor: A model for the transmembrane organization of the polypeptide chain. *Proc. Natl. Acad. Sci. U.S.A.* **80:**2067–2071 (1983).

Dreyer, F., Walther, C., and Peper, K.: Junctional and extrajunctional acetylcholine receptors in normal and denervated frog muscle fibers: Noise analysis experiments with different agonists. *Pfluegers Arch.* **366:**1–9 (1976).

Dun, N. J., and Karczmar, A. G.: Blockade of ACh potentials by α-bungarotoxin in rat superior cervical ganglion cells. *Brain Res.* **196:**536–540 (1980).

Dun, N. J., Nishi, S., and Karczmar, A. G.: Alteration in nicotinic and muscarinic responses of rabbit superior cervical ganglion cells after chronic preganglionic denervation. *Neuropharmacology* **15:**211–218 (1976).

Dwyer, T. M., Adams, D. J., and Hille, B.: The permeability of the end-plate channel to organic cations in frog muscle. *J. Gen. Physiol.* **75:**469–492 (1980).

Eldefrawi, M. E., Eldefrawi, A. T., Mansour, N. A., Daly, J. W., Witkop, B., and Albuquerque, E. X.: The acetylcholine receptor and ionic channel of *Torpedo* electroplax: Binding of perhydrohistrionicotoxin to membrane and solubilized preparations. *Biochemistry* **17:** 5474–5484 (1978).

Fambrough, D. M.: Control of acetylcholine receptors in skeletal muscle. *Physiol. Rev.* **59:**165–227 (1979).

Fambrough, D. M.: Biosynthesis and turnover of nicotinic acetylcholine receptors, in: *Drug Receptors and Their Effectors* (N. J. M. Birdsall, ed.), pp. 155–164, Macmillan, London (1981).

Feltz, A., and Mallart, A.: An analysis of acetylcholine responses of junctional and extrajunctional receptors of frog muscle fibres. *J. Physiol. (London)* **218:**85–100 (1971).

Fumagalli, L., DeRenzis, G., and Miani, N.: Acetylcholine receptors: Number and distribution in intact and deafferented superior cervical ganglion of the rat. *J. Neurochem.* **27:**47–52 (1976).

Gangitano, C., Fumagalli, L., DeRenzis, G., and Sangiacomo, C. O.: α-Bungarotoxin–acetylcholine receptors in the chick ganglion during development. *Neuroscience* **3:**1101–1108 (1978).

Gotti, C., Conti-Tronconi, B. M., and Raftery, M. A.: Mammalian muscle acetylcholine receptor purification and characterization. *Biochemistry* **21:**3148–3154 (1982).

Harris, A. J., Kuffler, S. W., and Dennis, M. J.: Differential chemosensitivity of synaptic and extrasynaptic areas on the neuronal surface membrane in parasympathetic neurons of the frog, tested by microapplication of acetylcholine. *Proc. R. Soc. London Ser. B* **177:** 541–553 (1971).

Ilyin, V. A., Bregestovski, P. D., Vulfius, E. A., and Girshovich, A. S.: Identification of carboxyl groups in acetylcholine receptors of the neurones in mollusc *Lymnea stagnalis* (in Russian). *Rep. USSR Acad. Sci.* **227:**1469–1471 (1976).

Karlin, A.: Molecular interactions of the acetylcholine receptor. *Fed. Proc. Fed. Am. Soc. Exp. Biol.* **32:**1847–1853 (1973).

Katz, B., and Miledi, R.: The statistical nature of the acetylcholine potential and its molecular components. *J. Physiol. (London)* **224:**665–699 (1972).

Katz, B., and Miledi, R.: A re-examination of curare action at the motor endplate. *Proc. R. Soc. London* **203:**119–133 (1978).

Kistler, J., Stroud, R. M., Klymkowsky, M. W., Ladancette, R. A., and Flairclough, R. H.: Structure and function of an acetylcholine receptor. *Biophys. J.* **37:**371–383 (1982).

Koketsu, K.: Cholinergic synaptic potentials and the underlying ionic mechanisms. *Fed. Proc. Fed. Am. Soc. Exp. Biol.* **28:**101–112 (1969).

Koketsu, K., Nishi, S., and Soeda, H.: Calcium and acetylcholine-potential of bullfrog sympathetic ganglion cell membrane. *Life Sci.* **7:**955–963 (1968).

Koketsu, K. Akasu, T., Miyagawa, M., and Hirai, K.: Biogenic antagonists of the nicotinic receptor: Their interactions with erabutoxin. *Brain Res.* **250**:391–393 (1982).

Kuba, K., and Koketsu, K.: Synaptic events in sympathetic ganglia. Prog. *Neurobiol.* **2**(2):77–169 (1978).

Kuba, K., and Nishi, S.: Characteristics of fast excitatory postsynaptic currents in bullfrog sympathetic ganglion cells. *Pfluegers Arch.* **378**:205–212 (1979).

Kuffler, S. W., Dennis, M. J., and Harris, A. J.: The development of chemosensitivity in extrasynaptic areas of the neuronal surface after denervation of parasympathetic ganglion cells in the heart of the frog. *Fed. Proc. Fed. Am. Soc. Exp. Biol.* **28**:101–112 (1971).

Lindstrom, J.: Antibodies as probes of acetylcholine receptor structure and function, in: *Psychopharmacology and Biochemistry of Neurotransmitter Receptors* (H. Yamamura, W. Olsen, and E. Usdin, eds.), pp. 63–84, Elsevier/North-Holland, Amsterdam (1980).

Lingle, C.: Blockade of cholinergic channels by chlorisondamine on a crustacean muscle. *J. Physiol. (London)* **339**:395–417 (1983a).

Lingle, C.: Different types of blockade of crustacean acetylcholine-induced currents. *J. Physiol. (London)* **339**:419–437 (1983b).

Lukomskaya, N. Ya.: Change in cholinergic sensitivity after denervation of the superior servical ganglion of the cat (in Russian). *Zh. Evol. Biokhim. Fiziol.* **5**:65–73 (1969).

Lukomskaya, N. Ya., and Gmiro, V. E.: Study of choline receptor membrane in sympathetic ganglion by analysis of structure–activity relationship. *J. Auton. Nerv. Syst.* **6**:363–372 (1982).

Lukomskaya, N. Ya., Gmiro, V. E., and Chromov-Borisov, N. V.: Pharmacological investigation of cholinoreceptive membrane in sympathetic ganglion by analysis of structure–activity relationship, in: *Physiology of Autonomic Ganglia (Abstracts of the Symposium)* (V. Skok, ed.), pp. 61–62, Naukova Dumka, Kiev (1981a).

Lukomskaya, N. Ya., Magazanik, L. G., and Potapjeva, N. N.: Effect of bis-onium gangliolytics on kinetics of synaptic ionic channels activation, in: *Physiology of Autonomic Ganglia (Abstracts of the Symposium)* (V. Skok, ed.), pp. 62–63, Naukova Dumka, Kiev (1981b).

Luzzatto, A. C., Tronconi, B. C., Paggi, P., and Rossi, A.: Binding of *Naja Naja siamensis* α-toxin to the chick ciliary ganglion: A light-microscopy autoradiographic study. *Neuroscience* **5**:313–318 (1980).

MacDermott, A. B., Connor, E. A., Dionne, V. E., and Parsons, R. L.: Voltage clamp study of fast excitatory synaptic currents in bullfrog sympathetic ganglion cells. *J. Gen. Physiol.* **75**:39–60 (1980).

Magazanik, L. G., Ivanov, A. Ya., and Lukomskaya, N. Ya.: The effect of snake venom polypeptides on cholinoreceptors in isolated rabbit ganglia (in Russian). *Neurofiziologia* **6**:652–656 (1974).

Magleby, K. L., and Stevens, C. F.: A quantitative description of end-plate currents. *J. Physiol. (London)* **223**:173–197 (1972).

Messing, A. and Jonatas, N. K.: Extra-synaptic localization of α-bungarotoxin receptors in cultured chick ciliary ganglion neurons. *Brain Res.* **269**:172–176 (1983).

Michelson, M. Ya., and Zeimal, E. V.: *Acetylcholine: An Approach to the Molecular Mechanism of Action.* Pergamon Press, Oxford (1973).

Mishina, M., Kurosaki, T., Tobimatsu, T., Morimoto, Y., Noda, M., Yamamoto, T., Terao, M., Lindstrom, J., Takahashi, T., Kuno, M., and Numa, S.: Expression of functional acetylcholine receptor from cloned cDNAs. *Nature (London)* **307**:604–608 (1984).

Neher, E., and Steinbach, J. H.: Local anaesthetics transiently block currents through single acetylcholine-receptors channels. *J. Physiol. (London)* **277**:153–176 (1978).

Nishi, S.: Ganglionic transmission. in: *The Peripheral Nervous System* (J. I. Hubbard, ed.), pp. 225–255, Plenum Press, New York (1974).

Noda, M., Takahashi, H., Tanabe, T., Toyosato, M., Furutani, Y., Hirose, T., Asai, M., Inayama,

S., Miyata, T., and Numa, S.: Primary structure of β- and α-subunit precursors of *Torpedo californica* acetylcholine receptor deduced from cDNA sequence. *Nature (London)* **299**:793–797 (1982).

Noda, M., Takahashi, H., Tanabe, T., Toyosato, M., Furutani, Y., Hirose, T., Asai, M., Inayama, S., Miyata, T., and Numa, S.: Primary structure of β-and α-subunit precursors of *Torpedo californica* acetylcholine receptor deduced from cDNA sequence. *Nature (London)* **301**:251–255 (1983a).

Noda, M., Takahashi, H., Tanabe, T., Toyosato, M., Kikyotani, S., Furutani, Y., Hirose, T., Takashima, H., Inayama, S., Miyata, T., and Numa, S.: Structural homology of *Torpedo californica* acetylcholine receptor subunits. *Nature (London)* **302**:528–532 (1983b).

Noda, M., Furutani, Y., Takahashi, H., Toyosato, M., Tanabe, T., Shimizu, S., Kikyotani, S., Kayano, T., Hirose, T., Inayama, S., and Numa, S.: Cloning and sequence analysis of calf cDNA and human genomic DNA encoding α-subunit precursor of muscle acetylcholine receptor. *Nature (London)* **305**:818–823 (1983c).

O'Brien, R. D., Gibson, R. E., and Sumikawa, K.: The nicotine acetylcholine receptor from *Torpedo* elektroplax. *Progr. Brain Res.* **59**:279–291 (1979).

Pappano, A. J., and Volle, R. L.: Observations on the role of calcium ions in ganglionic responses to acetylcholine. *J. Pharmacol. Exp. Ther.* **152**:171–180 (1966).

Paton, W. D. M., and Perry, W. L. M.: The relationship between depolarization and block in the cat's superior cervical ganglion. *J. Physiol. (London)* **119**:43–57 (1953).

Paton, W. D. M., and Zaimis, E. J.: The pharmacological actions of polymethylene bistrimethylammonium salts. *Br. J. Pharmacol.* **4**:381–400 (1949).

Raftery, M. A., Hunkapiller, M. W., Strader, C. D., and Hood, L. E.: Acetylcholine receptor complex of homologous subunits. *Science* **208**:1454–1456 (1980).

Raftery, M. A., Dunn, S. M. J., Conti-Tronconi, B. M., Midlemas, D. S., and Crawford, R. D.: The nicotinic acetylcholine receptor: subunit structure, functional binding sites, and ion transport properties. *Cold Spring Harbor Symp. Quant. Biol.* **48**:21–33 (1983).

Rang, H. P.: The characteristics of synaptic currents and responses to acetylcholine of rat submandibular ganglion cells. *J. Physiol. (London)* **311**:23–55 (1981).

Rang, P., and Gurney, M.: Components of the synaptic current of rat submandibular ganglion cells. *J. Physiol. (London)* **78**:426–432 (1982).

Sakmann, B., Patlak, J., and Neher, E.: Single acetylcholine-activated channels show burst-kinetics in presence of desensitizing concentrations of agonist. *Nature (London)* **286**:71–73 (1980).

Selyanko, A. A.: Mechanism of the action of α-neurotoxins from snake venoms on nicotinic acetylcholine receptors in rabbit sympathetic ganglion neurons (in Russian). *Neurophysiology* **15**:377–383 (1983).

Selyanko, A. A., and Skok, V. I.: Activation of acetylcholine receptors in mammalian sympathetic ganglion neurones. *Prog. Brain Res.* **49**:241–252 (1979).

Selyanko, A. A., Derkach, V. A., and Skok, V. I.: Fast excitatory postsynaptic currents in voltage-clamped mammalian sympathetic ganglion neurones. *J. Auton. Nerv. Syst.* **1**: 127–137 (1979).

Selyanko, A. A., Derkach, V. A., and Skok, V. I.: Effects of some ganglion-blocking agents on fast excitatory postsynaptic currents in mammalian sympathetic ganglion neurones, in: *Advances in Physiological Science*, Vol. 4, *Physiology of Excitable Membranes* (J. Salanki, ed.), pp. 329–342, Pergamon Press, London; Akademiai Kiado, Budapest (1981).

Selyanko, A. A., Derkach, V. A., and Skok, V. I.: Voltage-dependent actions of short-chain polymethylene bis-trimethylammonium compounds on sympathetic ganglion neurons. *J. Auton. Nerv. Syst.* **6**:13–21 (1982).

Selyanko, A. A., Derkach, V. A., and Skok, V. I.: The effect of Ca^{2+} ions on the channel-blocking action of hexamethonium in sympathetic ganglion. *Rep. U.S.S.R. Acad. Sci.* (**284**:N1) (1985).

Skok, V. I.: *Physiology of Autonomic Ganglia.* Igaku Shoin, Tokyo (1973).

Skok, V. I.: Ganglionic transmission: Morphology and physiology, in: *Pharmacology of Ganglionic Transmission* (D. A. Kharkevich, ed.), pp. 9–38, Springer-Verlag, Berlin (1980).

Skok, V. I.: Localization and structure of the sites in the nicotinic acetylcholine receptor that bind bis-quarternary ammonium compounds. *Biol. Membr.* **2**:245–255 (1985).

Skok, V. I., Selyanko, A. A., and Derkach, V. A.: Two modes of activity of nicotinic acetylcholine receptor channels in sympathetic neurones. *Brain Res.* **238**:480–483 (1982).

Skok, V. I., Selyanko, A. A., and Derkach, V. A.: Channel-blocking activity is a possible mechanism for a selective ganglionic blockade. *Pfluegers Arch.* **398**:169–171 (1983).

Skok, V. I., Selyanko, A. A., Derkach, V. A., Gmiro, V. E., and Lukomskaya, N. Ya.: The mechanisms of ganglion-blocking action of bis-ammonium compounds (in Russian). *Neurofiziologia* **16**:54–60 (1984).

Smolen, A. I.: Specific binding of α-bungarotoxin to synaptic membranes in rat sympathetic ganglion: Computer best-fit analysis of electron microscope radioautographs. *Brain Res.* **289**:177–188 (1983).

Takeuchi, A., and Takeuchi, N.: On the permeability of the end-plate membrane during the action of transmitter. *J. Physiol. (London)* **154**:52–67 (1960).

Terrar, D. A.: Effects of dithiothreitol on end-plate currents. *J. Physiol. (London).* **276**:403–417 (1978).

Trcka, V.: Relationship between chemical structure and ganglion-blocking activity. a) Quarternary ammonium compounds, in: *Pharmacology of Ganglionic Transmission* (D. A. Kharkevich, ed.), pp. 123–153, Springer-Verlag, Berlin (1980a).

Trcka, V.: Relationship between chemical structure and ganglion-blocking activity. b) Tertiary and secondary amines, in: *Pharmacology of Ganglionic Transmission* (D. A. Kharkevich, ed.), pp. 155–184, Springer-Verlag, Berlin (1980b).

Triggle, D. J.: *Chemical Aspects of the Autonomic Nervous System.* Academic Press, London and New York (1965).

Triggle, D. J., and Triggle, C. R.: *Chemical Pharmacology of the Synapse.* Academic Press, London, New York, and San Francisco (1976).

Trinus, K. F., and Skok, V. I.: Disulfide bonds found in the cholinoreceptors of the frog sympathetic ganglion neurone (in Russian). *Neurofiziologia* **6**:593–560 (1979); also in *Neurophysiology (USSR)* **11**:445–451 (1979).

Van Rossum, J. M.: Classification and molecular pharmacology of ganglionic blocking agents. Part 1. Mechanisms of ganglionic synaptic transmission and mode of action of ganglionic stimulants. *Int. J. Neuropharmacol.* **1**:97–110 (1962a).

Van Rossum, J. M.: Classification and molecular pharmacology of ganglion blocking agents. Part II. Mode of action of competitive and non-competitive ganglion blocking agents. *Int. J. Neuropharmacol* **1**:403–421 (1962b).

Vickerson, F. H. L., and Varma, D. R.: Effects of denervation on the sensitivity of the superior cervical ganglion of the cat to acetylcholine and McN-A-343. *Can. J. Biochem. Physiol* **47**:255–259 (1969).

Volle, R. L.: Nicotinic ganglion-stimulating agents, in: *Pharmacology of Ganglionic Transmission* (D. A. Kharkevich, ed.), pp. 281–312, Springer-Verlag, Berlin (1980).

Woodhull, A. M.: Ionic Blockade of sodium channels in nerve. *J. Gen. Physiol.***61**:687–708 (1973).

Wu, W. C.-S. and Raftery, M. A.: Reconstitution of acetylcholine receptor function using purified receptor protein. *Biochemistry* **20**:694–701 (1981).

Zingsheim, H. P., Barrantes, F. J., Frank, J., Hanicke, W., and Neugebauer, D.-C.: Direct structural localization of two toxin-recognition sites on an ACh receptor protein. *Nature (London)* **299**:81–84 (1982).

7

Muscarinic Transmission

T. AKASU and K. KOKETSU

I. INTRODUCTION

It was shown early that preganglionic stimulation elicits a series of potential changes that include the late negative (LN) surface potential in the sympathetic ganglia of both turtles (Laporte and Lorente de No, 1950) and rabbits (Eccles, 1952). Eccles and Libet (Eccles and Libet, 1961; Libet, 1964) subsequently investigated the electrophysiological properties of the extracellularly recorded LN potential and revealed its unique features and muscarinic nature, and Libet (1964, 1967) applied the term "slow excitatory postsynaptic potential" (slow EPSP) to the LN potential. Nishi and Koketsu (1968) recorded by means of the sucrose-gap method a similar LN potential from bullfrog sympathetic ganglion cells. The intracellular counterpart of the LN potential is the slow EPSP; intracellular microelectrode recordings of the slow EPSP of sympathetic ganglion cells were originally carried out in frog and bullfrog paravertebral ganglia by Tosaka *et al.*, (1968) and Nishi and Koketsu (1968), respectively, and in rabbit superior cervical ganglia by Kobayashi and Libet (1968).

Preganglionic B-nerve stimulation, which activates only B neurons, seems to be enough to evoke the LN potential or the slow EPSP. Additional C-nerve stimulation, which activates both B and C neurons, does not appear to augment the LN potential in either mammalian (Eccles and Libet, 1961) or amphibian (Nishi and Koketsu, 1968) ganglia (see also Chapter 5). Therefore, only the preganglionic B neuron appears to innervate the receptor for the LN potential or the slow EPSP of postganglionic neurons

T. AKASU and K. KOKETSU • Department of Physiology, Kurume University School of Medicine, Kurume, Japan.

(Nishi and Koketsu, 1968; Tosaka *et al.*, 1968). There is, however, some controversy with regard to the types of postganglionic neurons that produce the slow EPSP. Tosaka *et al.*, (1968) reported that slow EPSPs could be recorded only from ganglionic B-type neurons, but not from the majority of C-type neurons. On the other hand, Nishi and Koketsu (1968) found the slow EPSP in neurons of both B and C type in nicotinized bullfrog sympathetic ganglia.

Eccles and Libet (1961) found that botulinum toxin, known to be a specific blocker of acetylcholine (ACh) release from preganglionic nerve terminals (Ambache, 1951), blocked the generation of LN potentials (Eccles and Libet, 1961). Low Ca^{2+}/high Mg^{2+} depressed all postganglionic responses (Libet, 1967). Furthermore, the LN potential and the slow depolarization produced by ACh or muscarinic agonists such as methacholine or bethanechol were blocked by atropine (Tosaka and Libet, 1970). Atropine also completely blocked the slow EPSP (Kobayashi and Libet, 1968; Nishi and Koketsu, 1968; Tosaka *et al.*, 1968). More direct evidence for a muscarinic action of ACh is the observation that iontophoresis of ACh to a curarized single ganglion cell produced a slow ACh depolarization (slow ACh potential) that had a time–course and latency similar to those of the slow EPSP and was blocked by atropine (Koketsu *et al.*, 1968). Conversely, the slow EPSP and the slow ACh potential were augmented and prolonged by anticholinesterase agents (Libet, 1967; Koketsu *et al.*, 1968) as were the after discharges of postganglionic fibers (Takeshige and Volle, 1962; Volle, 1962; Koketsu *et al.*, 1968) that are a concomitant of the slow potential (Volle and Hancock, 1970). The slow ACh potential could be recorded in Ca^{2+}-free isotonic Mg^{2+} solution, in which the release of any transmitter would be inhibited; this implies a direct generation of the slow ACh potential and the presynaptic nature of the slow EPSP. Altogether, it can be concluded that the slow EPSP is generated by a muscarinic action of ACh.

Recently, several investigators suggested that muscarinic receptors constitute a heterogeneous population, the precise stereochemical nature of the muscarinic receptor depending on the tissue (Birdsall *et al.*, 1976; see Vickroy *et al.*, 1984). Some *in vitro* results obtained in experiments dealing with binding characteristics of ganglionic tissue indicated that the muscarinic ganglionic receptor belongs to the M_1 category on the basis of several characteristics (Vickroy *et al.*, 1984), including high-affinity binding to the antagonist pirenzepine (Hammer *et al.*, 1980). Electrophysiological evaluation of this matter is incomplete, since Dun (unpublished data) and Ashe and Yarosh (1984) obtained diametrically opposite results—no effect vs. block, respectively—in their studies of the effect of pirenzepine action on the slow EPSP of the rabbit ganglion (see also Chapter 13).

II. NATURE OF THE SLOW EXCITATORY POSTSYNAPTIC POTENTIAL

A. Conductance Change during the Slow Excitatory Postsynaptic Potential

Libet and Kobayashi (1969) reported that there is no decrease in membrane resistance during the generation of the slow EPSP of *curarized* ganglia of both bullfrogs and rabbits. On the other hand, in *nicotinized* bullfrog sympathetic ganglia, membrane resistance appeared to be markedly increased (Kobayashi and Libet, 1970). This observation was confirmed by Weight and Votava (1970). Nishi *et al.* (1969) also observed a consistent increase in membrane resistance when the membrane was fixed at the depolarized level during the generation of the slow EPSP.

Kuba and Koketsu (1976a) carefully analyzed the changes in membrane resistance during the slow EPSP and the slow ACh potential. Contrary to the findings of Libet and Kobayashi (1969), nicotinized and curarized bullfrog sympathetic ganglion cells showed essentially the same responses (Kuba and Koketsu, 1974, 1976a). Changes in membrane resistance observed during the course of the slow ACh potential were observed to be related to the effect of membrane hyperpolarization on the amplitude of the slow EPSP of the cell in question (Kuba and Koketsu, 1976a); this amplitude was increased, unchanged, and decreased by hyperpolarization in cell types 1, 2, and 3, respectively (Figure 1). In type 1 cells, the muscarinic action of ACh decreased membrane resistance, and membrane resistance was unchanged in type 2 cells and increased in type 3 cells at membrane potentials ranging between −60 and −80 mV, while it was unchanged (type 1 cells) or increased (type 2 and type 3 cells) at the resting and depolarized levels. That a decrease in membrane conductance may occur during the slow EPSP [slow excitatory postsynaptic current (slow EPSC)] was recently confirmed by Adams and Brown (1982); they related this phenomenon to the suppression of the so-called M current (Brown and Adams, 1980) (see below). Furthermore, the relationships between the amplitude of the slow ACh potential and membrane potential level obtained from the I–V curves in the three cell types were essentially similar to those between the slow EPSP amplitude and membrane potential in these cells (Nishi *et al.*, 1969; Kuba and Koketsu, 1974; 1976a). Using a two-microelectrode voltage-clamp technique, Akasu *et al.* (1984) examined the electrophysiological properties of the slow EPSC in neurons of the bullfrog sympathetic ganglia. At least three types of slow EPSC responses were found with respect to the characteristics of their conductance changes. The type 1 slow EPSC (Figure 2) was produced by a

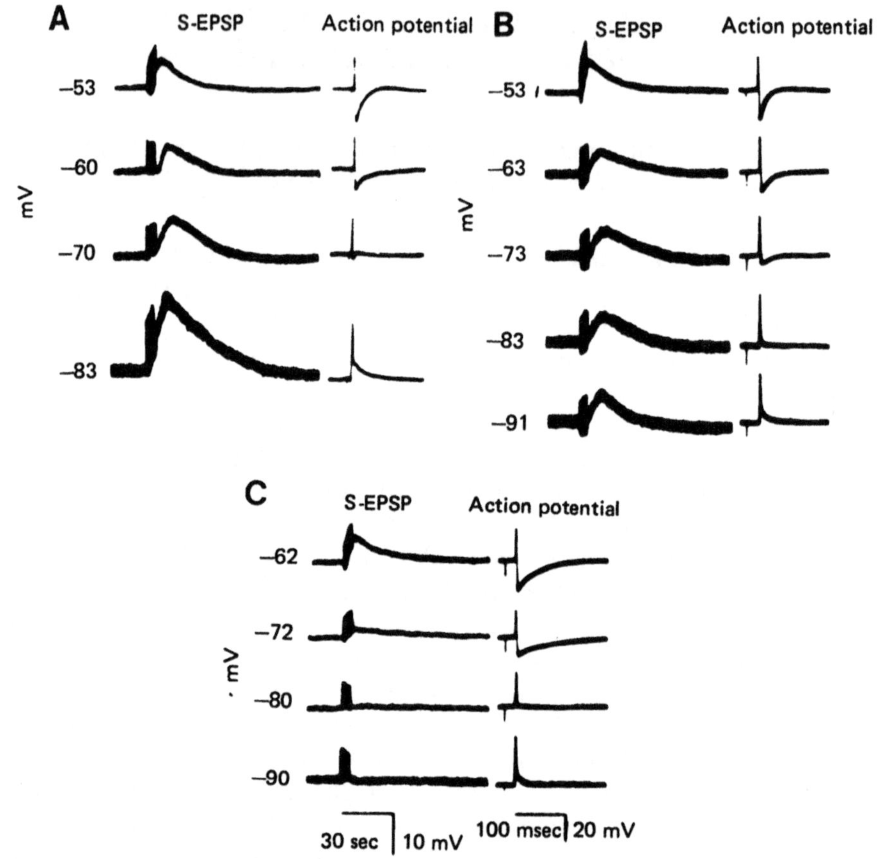

Figure 1. Effects of membrane hyperpolarization on the slow EPSP and action potentials recorded from various types of cells in the bullfrog sympathetic ganglion. The resting membrane potentials were −53 mV (A), −53 mV (B), and −62 mV (C). Action potentials were evoked by antidromic stimulations. A part of the spike of action potentials is not shown. (A) Type 1 cell; (B) type 2 cell; (C) type 3 cell. From Kuba and Koketsu (1976b).

suppression of a voltage-dependent K^+ conductance, presumably the M current (Adams and Brown, 1982). The type 2 slow EPSC, observed in about 15% of neurons tested, was associated with an increased membrane conductance, due probably to the opening of Na^+ and other cationic channels (Figure 3). The type 3 response, observed in about 20% of the neurons, was of a mixed type; the membrane conductance was increased at hyperpolarized membrane potential levels, while it was decreased at depolarizing potential levels (Figure 4). Somewhat similarly, in the case of mammalian ganglionic neurons, the conductance or resistance changes that occurred in the course of the slow EPSP or the muscarinic action of ACh or both were variable and depended on the cell; generally, the re-

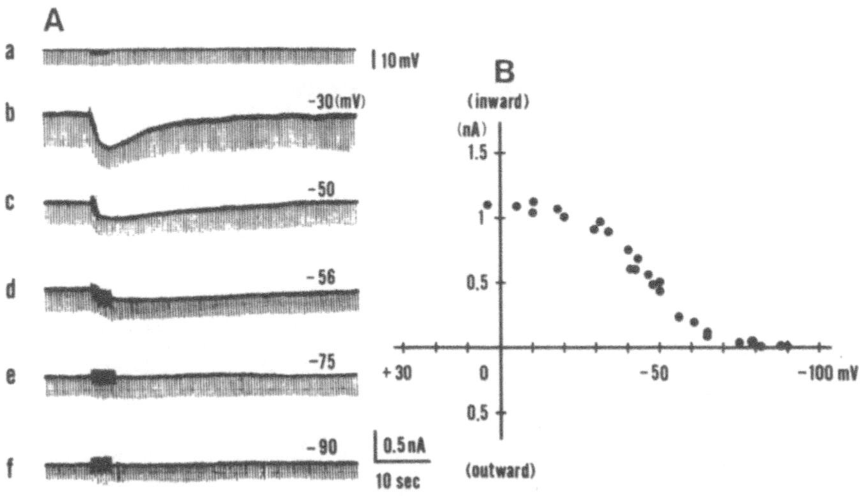

Figure 2. Relationship between the amplitude of the type 1 slow EPSC and the holding potential level. (A) Slow EPSCs in response to supramaximal preganglionic stimuli (40 Hz for 4 sec) recorded at various holding levels ranging from −30 to −90 mV. Command pulses (duration 100 msec) of 10 mV were repeated at 2 Hz (top trace). (B) Peak amplitude of each slow EPSC response (ordinate) plotted against the holding level (abscissa). From Akasu *et al.* (1984).

sistance either did not change or was increased (Dun *et al.*, 1978) (see also Nishi, 1974).

B. Effect of the Membrane Potential

Kobayashi and Libet (1968) were the first to study the effect of membrane potential on the amplitude of the slow EPSP. They reported that in curarized frog sympathetic ganglia, the amplitude of the slow EPSP was increased by depolarization and decreased by hyperpolarization of the membrane. On the other hand, in curarized rabbit superior cervical ganglia, conditioning depolarization caused a decrease in the amplitude of the slow EPSP, while it was increased by moderate hyperpolarization and then eventually decreased by large hyperpolarization (Kobayashi and Libet, 1968). In the case of nicotinized bullfrog sympathetic ganglion cells, Nishi *et al.*, (1969) found three different types of cells according to the effect of the membrane polarization on the slow EPSP. Their classification of these cells has already been briefly described; more specifically, type 1 cells showed an increase of the slow EPSP amplitude following conditioning hyperpolarization and a decrease following depolarization, conditioning hyperpolarization did not affect but depolarization slightly increased the amplitude of the slow EPSP of type 2 cells, and the slow EPSP

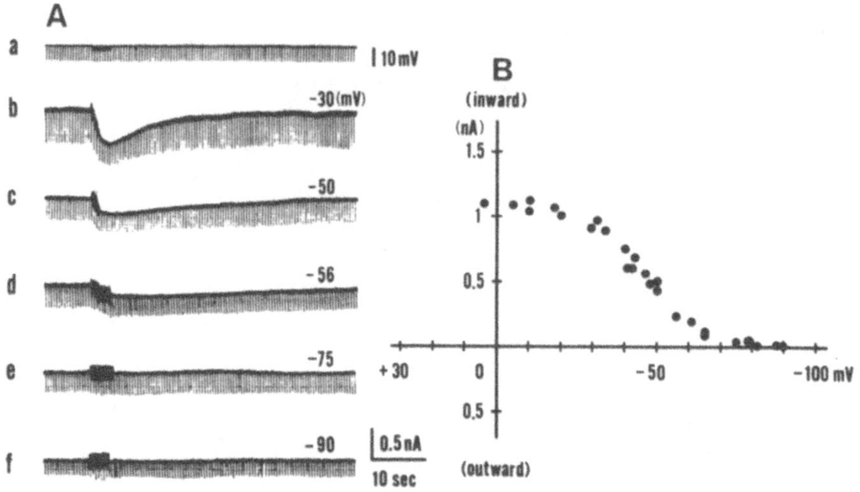

Figure 3. Relationship between the amplitude of the type 2 slow EPSC and the holding potential level. (A) Slow EPSCs elicited by a train of repetitive preganglionic stimuli (40 Hz for 5 sec) at various holding levels ranging between +5 and −80 mV. Command pulses (duration 100 msec) of 20 mV were repeated at 2 Hz (top trace). (B) Peak amplitude of each slow EPSC response (ordinate) plotted against the holding level (abscissa). From Akasu *et al.* (1984).

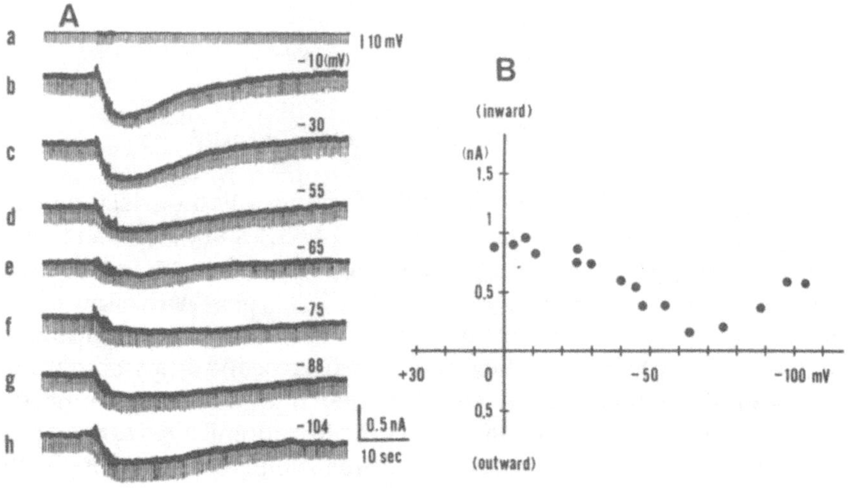

Figure 4. Relationship between the amplitude of a mixed-type response and the holding potential level. (A) Slow EPSCs elicited by a train of repetitive preganglionic stimuli (40 Hz for 4 sec) at various holding levels ranging between −10 and −104 mV. Command pulses (duration 100 msec) of 10 mV were repeated at 2 Hz (top trace). (B) Peak amplitude of each slow EPSC response (ordinate) plotted against the holding level (abscissa). From Akasu *et al.* (1984).

of type 3 cells was depressed and enhanced by hyperpolarization and depolarization, respectively. Subsequently, Kobayashi and Libet (1970) observed a depressant effect of membrane hyperpolarization on the slow ACh depolarization and the slow EPSP of nicotinized frog sympathetic ganglion cells; they also found that intense hyperpolarization reversed the initial portion of the slow EPSP. Similarly, Weight and Votava (1970) caused the reversal of the slow EPSP of nicotinized bullfrog ganglion cells by membrane hyperpolarization and also reported that the reversal potential of the polarity of the slow EPSP was similar to the E_K. However, Libet (1971) felt that the reversal of the slow EPSP at the hyperpolarized level of the membrane potential is due to its contamination by the slow inhibitory postsynaptic potential (slow IPSP), since conditioning hyperpolarization, particularly of the nicotinized frog ganglion, increased the amplitude of the IPSP. In fact, the reversed slow EPSP described by Weight and Votava (1970) was much shorter in its time–course than that observed at the resting potential level (Kobayashi and Libet, 1974). Similar partial reversal of the slow EPSC was described by Kuffler and Sejnowski (1983). However, the reversal potential of the slow EPSC was more negative than the E_K (Kuffler and Sejnowski, 1983).

Kuba and Koketsu (1974, 1976a) studied further the effect of hyperpolarization on the amplitude of the slow EPSP, in relation to the E_K, in both nicotinized and curarized bullfrog sympathetic ganglion cells (see Figure 1). As has been suggested by Nishi *et al.*(1969), three different types of cells could be characterized in terms of the effect of membrane hyperpolarization on the amplitude of the slow EPSP. Type 1 cells showed an increase in the amplitude even at the membrane potential level, which was more negative than the E_K. In the case of type 2 cells, the amplitude of the slow EPSP remained almost unchanged by hyperpolarization beyond the E_K. In the case of type 3 cells, the amplitude of the slow EPSP decreased with membrane hyperpolarization. However, Kuba and Koketsu (1974, 1976a) and Kobayashi and Libet (1974) failed to observe the reversal of the slow EPSP observed by Weight and Votava (1970). Using the voltage-clamp technique, Adams *et al.* (1982) and Adams and Brown (1982) found that the slow EPSC was completely eliminated at membrane potentials higher than -60 mV; this observation seems consistent with what Kuba and Koketsu (1974, 1976a) reported for type 3 cells.

Subsequently, Akasu *et al.* (1984) examined the voltage dependency of the slow EPSC in detail. The type 1 slow EPSC decreased at hyperpolarizing potential levels and eventually nullified its polarity at around -70 mV; thus, this EPSC was similar to the M current (Brown and Adams, 1980). No reversal of type 1 response was observed (see Figure 2). On the other hand, the type 2 response was increased by conditioning hyperpolarizations and persisted at potential levels higher than E_K (see Figure 3). At the depolarizing potential level, the type 2 response was decreased and nullified at approximately +5 mV, not exhibiting, however, any reversal of its polarity (Akasu *et al.*, 1984) (Figure 3). It was suggested that the conductance system linked with the type 2 response may be voltage-dependent. The type 3 slow EPSC showed the voltage dependence of type 1 and type 2 EPSCs, since its maximum amplitude was observed at the resting membrane potential, while conditioning hyperpolarization and depolarization depressed the amplitude of the response (see Figure 4).

C. Ionic Environment and the Slow Excitatory Postsynaptic Potential

Experiments that included the alteration of external Na^+, K^+, and Ca^{2+} concentrations provided evidence indicating that the ionic mechanism of the frog slow EPSP involves a simultaneous decrease of K^+ and an increase of Na^+ and Ca^{2+} conductance. When the K^+ concentration of Ringer's solution was increased from 2 to 10 mM, the resting potential decreased by about 10 mV and the slow ACh potential was reduced in amplitude to about 75% of control, indicating that inactivation of G_K is involved in the generation of the slow EPSP (Kuba and Koketsu, 1976a). Under these circumstances, i.e., when G_K inactivation could not play any significant role in the generation of the slow EPSP, replacement of NaCl with sucrose caused additional, marked attenuation of the remaining ACh potential. These results are consistent with the membrane-conductance increase observed during the slow EPSP and the slow ACh depolarization (Nishi *et al.*, 1969; Kuba and Koketsu, 1974, 1976a) in some types of cells since the inactivation of G_K should result in decreased rather than increased conductance, and since the increase in membrane Na^+ conductance would explain the change, in the course of these responses, of the conductance of these cells. When Ca^{2+} was completely removed from the high-K^+ Ringer's solution, the amplitude of the slow ACh potential was again further reduced (Kuba and Koketsu, 1976a); this finding suggested that the increase in the Ca^{2+} conductance is also involved in the genesis of the slow EPSP or the slow ACh potential and that Ca^{2+} serves as a charge carrier.

III. IONIC MECHANISM OF THE SLOW EXCITATORY POSTSYNAPTIC POTENTIAL

A. G_K Inactivation Hypothesis

The finding that during the generation of the slow EPSP the depolarization is accompanied by an increase in the membrane resistance suggests that the inactivation of G_K is the primary mechanism underlying the generation of the slow EPSP. On the basis of their studies of bullfrog paravertebral ganglia and of "several cells" of the rabbit superior cervical ganglion, Kobayashi and Libet (1968) discarded this possibility as the mechanism for the generation of the slow EPSP; as already pointed out above, inactivation of G_K should lead to decreased conductance, while Kobayashi and Libet (1968) could detect only a small decrease in membrane conductance in the course of the slow EPSP; indeed, at that time, these investigators were opposed to the concept that ionic movements are involved in the generation of the slow EPSP. Weight and Votava (1970) subsequently proposed that G_K inactivation underlies the generation of the slow EPSP of bullfrog sympathetic ganglion cells, since they caused a reversal of the slow EPSP at the E_K and found an increase in membrane resistance during its generation. The reversal of the slow EPSP at the E_K, however, was not reproduced by Kobayashi and Libet (1974) and Kuba and Koketsu (1974, 1976a). Nevertheless, the inactivation of G_K, which is voltage-dependent (Figure 5), seems to be a part of the mechanism for the slow EPSP (Kuba and Koketsu, 1974, 1976a), for the following reasons: (1) In the case of type 3 cells (Kuba and Koketsu, 1974, 1976a), a conditioning hyperpolarization of the membrane reduced the amplitude of the slow EPSP and the slow ACh potential; (2) in these cells, membrane resistance increased during the slow EPSP; and (3) the amplitude of the slow ACh potential was decreased in a high-K^+ solution. Adams and Brown (1982) confirmed the G_K-inactivation hypothesis in voltage-clamp experiments and concluded that the slow EPSC of bullfrog sympathetic ganglia results from a selective inhibition of the M current (see also Chapter 13).

B. Combined Mechanism of the G_K Inactivation and the G_{Na} and G_{Ca} Activations

In the case of type 1 bullfrog ganglion cells (Nishi *et al.*, 1969; Kuba and Koketsu, 1974, 1976a), the amplitudes of the slow EPSP and the slow ACh potential as well as the membrane conductance increased during conditioning hyperpolarization; furthermore, the amplitude of the slow ACh potential was depressed in Na^+-free or Ca^{2+}-free solutions. These observations indicate that increases in the G_{Na} and G_{Ca} are at least partially involved in the generation of the slow EPSP. On the other hand, it also seems certain that a decrease in the G_K plays a role in the ionic mechanism of the slow EPSP generated in the case of type 1 cells. In fact, it is likely that the three conductance changes, i.e., G_{Na}, G_K, and G_{Ca}, are all operative in the ionic mechanisms that underlie the generation of the slow EPSP, the magnitude of these three conductance changes and their contribution

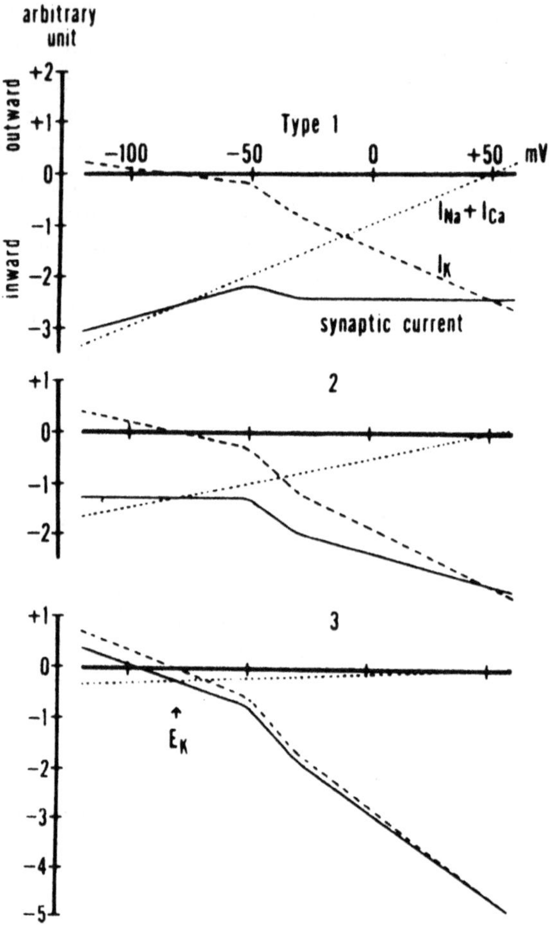

Figure 5. A hypothetical model of ionic currents for the slow EPSP of bullfrog sympathetic ganglion cells. Ordinates are membrane currents in arbitrary units. Inward and outward currents were taken as negative and positive, respectively. Abscissae represent membrane potential. Dashed lines indicate K^+ currents caused by the inactivation of the G_K. The value of the E_K was taken as −80 mV, since the E_K values of many cells were close to this value under these experimental conditions. Dotted lines indicate the sum of the Na^+ and Ca^{2+} currents caused by activation of the G_{Na} and G_{Ca}. The reversal level of this current was assumed to be +50 mV; although there is no basis for this assumption except that it would result, as expected, in positive values for the equilibrium potentials for Na^+ and Ca^{2+}, the validity of this assumption is not needed for the usefulness of this model. Continuous lines are synaptic currents for the slow EPSP. See the text. From Kuba and Koketsu (1976b).

to the generation of the slow EPSP depending on the type of cell (Kuba and Koketsu, 1974, 1976a). Kuba and Koketsu (1974, 1976a) demonstrated how different magnitudes of the increase in G_{Na} and G_{Ca} and of the decrease in G_K and the relationship between the values of these magnitudes underlie the generation of the slow EPSP and the relationship of its amplitude to the membrane potential. In the case of type 1 cells, the activation of G_{Na} or G_{Ca} or both is presumed to be greater than the G_K decrease; accordingly, the nature of the slow EPSP of this type of cell is similar to that of the fast EPSP. In the case of type 2 cells, the summed magnitude of the increases in G_{Na} and G_{Ca} is almost equal to the decrease in G_K; this could explain the lack of change in membrane resistance during the generation of the slow EPSP and the relative independence of the slow EPSP amplitude from membrane hyperpolarization. Finally, G_K inactivation seems to play a major role in the generation of the slow EPSP in the case of type 3 cells; the K^+ current involved in this inactivation is potential-dependent and exhibits rectifier characteristics (Kuba and Koketsu, 1976a). The type 3 cells in question may correspond to cells described by Weight and Votava (1970) and Brown and Adams (1980), since they proposed that only the G_K inactivation underlies the generation of the slow EPSP. Recently, Akasu *et al.* (1984) analyzed the ionic mechanism of the slow EPSC by using the voltage-clamp method. Their results led them to conclude that the type 1 slow EPSC is produced by the suppression of voltage-dependent G_K that is deactivated at -70 mV [I_M (Brown and Adams, 1980; Adams and Brown, 1982)], while the type 2 response is generated by the increase in G_{Na} and other cations; these two responses could frequently be observed in the same cells (see Figure 3). Coexistence of these two responses may be a complicating feature of the effect of the membrane potential changes on the slow EPSP.

As already pointed out, mammalian ganglionic neurons also exhibit variable changes in resistance in the course of the slow EPSP; there are only a few studies of the effect of membrane potential on the mammalian EPSP (Kobayashi and Libet, 1968) (see above). It appears likely that as in amphibian cells, several ionic and possibly nonionic mechanisms may contribute to the mammalian slow EPSP.

IV. SOME CHARACTERISTIC FEATURES OF THE SLOW EXCITATORY POSTSYNAPTIC POTENTIAL

A. Time–Course and Synaptic Delay of the Slow Excitatory Postsynaptic Potential

The LN potential as well as the slow EPSP have a latency of 100–400 msec and may last for more than 10 sec after a tetanic (30-Hz), 400-sec train of stimuli (Libet, 1967; Nishi and Koketsu, 1968). The long synaptic delay as well as the long time–course of the slow EPSP result from various factors that include (1) time for diffusion of ACh from the preganglionic terminal to the muscarinic receptor site; (2) the rates of binding and unbinding of transmitter to and from its receptor (Hartzell, 1981); (3) unknown mechanisms that mediate induction of the final membrane conductance change by the ACh-muscarinic receptor complex, such as the metabolic processes involving cyclic nucleotides (see below and Kobayashi and Tosaka, 1983); and (4) the rates of the changes in the G_{Na}, G_{Ca}, and G_K (Kuba and Koketsu, 1978).

If the muscarinic receptors that are responsible for the slow EPSP are located far from the presynaptic boutons, diffusion of ACh from presynaptic terminals to the receptor might take a long time; this could explain the slow onset and the long time–course of the slow EPSP. It should be pointed out that the slow EPSP occurs following the slow IPSP; thus, it is difficult to evaluate the real synaptic delay of the slow EPSP. However, the slow EPSP recorded after the complete elimination of the slow IPSP in Na^+-free Li^+ Ringer's solution was slow in reaching its peak (Koketsu *et al.*, 1973). Thus, the possibility that the time needed for transmitter diffusion may contribute to the time–course of the slow EPSP cannot be ruled out at present.

If it is assumed that the conductance changes of the three components take place as soon as ACh binds to the muscarinic receptor, the rates of rise and fall of the slow EPSP could be determined by the rates of increase in the local concentration of ACh close to the receptor sites and/or by the rates of binding and unbinding of ACh to and from the receptor. If the role of the diffusion factor is ruled out, the rates of binding and unbinding of ACh could be the main factors controlling the time–course of the slow EPSP (see Hartzell, 1981). In this case, the assumption must be made that the rates of ACh binding and unbinding are very slow, on the order of tens of seconds. Conversely, it is possible that the conductance changes of membrane, which occur in milliseconds in the case of the nicotinic receptor (Kuba and Koketsu, 1978; Kuba and Nishi, 1979; MacDermott *et al.*, 1980; Selyanko *et al.*, 1980), are very slow in the case of the muscarinic receptor. However, there is no direct evidence to support such a possibility at present.

It is of interest in this context that the decay time–course (falling phase) of the slow EPSC is voltage-dependent (Akasu *et al.*, 1984). The half-decay time of the slow EPSC was increased by conditioning hyperpolarization, while it was decreased by membrane depolarization, suggesting that the process that links the muscarinic receptor and the associated ionic channel might be voltage-dependent.

B. Characteristics of Conductance Change

One of the most interesting features of the ionic mechanism of the slow EPSP is the nonlinear relationship between the amplitude of the slow EPSP and the membrane potential (Kuba and Koketsu, 1974, 1976a, 1978). The nonlinearity of this relationship may be due to phenomena that concern the G_K and its rectifying nature (Kuba and Koketsu, 1974, 1976a, 1978); if the K^+ channel involved in the generation of the slow EPSP utilizes the delayed recitifier K^+ channel (Kuba and Koketsu, 1978), the net decrease of G_K caused by ACh would depend on the membrane potential and become greater at depolarized than at hyperpolarized levels of the membrane potential (Kuba and Koketsu, 1976a, 1978).

Recently, Brown and Adams (1980) found a voltage-dependent K^+ current the nature of which was different from that of the delayed rectifier K^+ current (Figure 6); since this current was not diminished by tetraethylammonium (TEA) while it was markedly decreased by muscarine, an agonist of the muscarinic receptor, they termed it the M current [I_M (Brown and Adams, 1980; Adams *et al.*, 1982)]. The M current is progressively activated between −70 and −10 mV, and does not show the time-dependent inactivation. Furthermore, the inward current relaxation (Brown and Adams, 1980; Adams *et al.*, 1982) showed an exponential time–course,

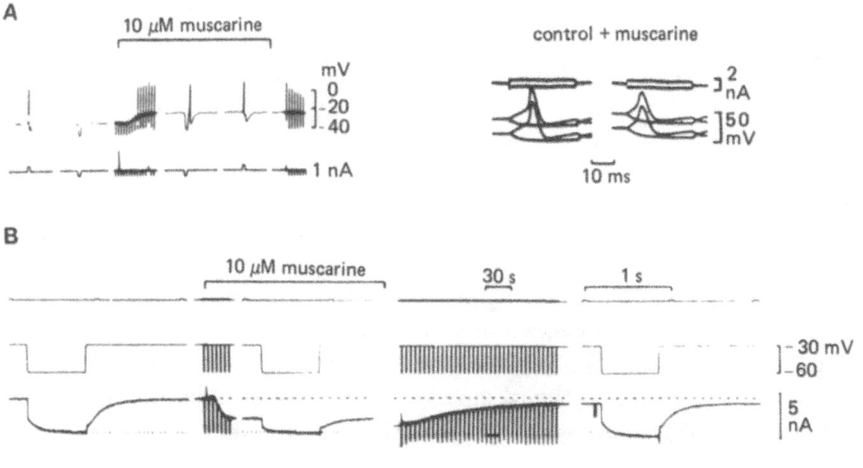

Figure 6. Effects of muscarine on a single neuron recorded under current-clamp, (*A*) and voltage-clamp conditions (*B*). The three traces on the chart-strip records read (from above downward): time (in sec), membrane voltage, and current. The recorder was periodically accelerated to display the voltage and current transients. Muscarine was applied by rapid bath-perfusion. (*A*) The oscilloscope records on the right side of the voltage response show superimposed spikes and electrotonic potentials evoked by depolarizing and hyperpolarizing current pulses, respectively. The current pulses are indicated on the top beam. Two voltage records are shown, the upper one being recorded through the current-passing electrode via an active bridge circuit. (*B*) In the voltage-clamp experiment (recorded earlier in the same cell), the membrane potential was held at −30 mV and commanded to −60 mV for 700 msec at 5-sec intervals and the induced currents recorded. From Brown and Adams (1980).

with time constants of 50 and 110 msec at membrane potentials of −60 and −30 mV, respectively. These characteristics of the M current represent rectifying properties, but are clearly different from those of the delayed rectifier K^+ current (Hodgkin and Huxley, 1952). The M current was strongly suppressed during the slow EPSC generated by preganglionic nerve stimuli. The amplitude of the slow EPSC was reduced to zero by membrane hyperpolarization to a potential at which the M channel was completely deactivated. Thus, Brown and Adams (1980) suggested that inactivation of the M channel results in the generation of the slow EPSC. They also reported that muscarine did not block the amplitude of after hyperpolarization of the action potential (Brown and Adams, 1980) (see Figure 5).

On the other hand, the observation that the muscarinic action of ACh depresses the amplitude of afterhyperpolarization (Kuba and Koketsu, 1975; 1976b) was confirmed recently (Akasu and Koketsu, 1980; Akasu, 1981). An explanation for this inconsistency between the results obtained in two laboratories does not seem to be forthcoming (Adams *et al.*, 1982). With regard to their own data, Adams *et al.* (1982) and Brown and Adams (1980) describe multiple processes that include changes in G_K and I_M and that may underlie the repolarizing phase of the spike; altogether, they feel that muscarinic inhibition of I_M should not affect spike configuration (Figure 6). A different suggestion was made by the Kurume group, who analyzed by means of the voltage-clamp technique the muscarinic effect of ACh on the G_K during the generation of the action potential (Akasu, 1981; Akasu and Koketsu, 1981a, 1982). According to their voltage-clamp analysis, both the TEA-sensitive (I_{k1}) and the TEA-insensitive (I_{k2}) voltage-dependent outward current were depressed significantly by the muscarinic action of ACh (Akasu and Koketsu, 1982). Since the TEA-insensitive outward current may be assumed to be synonymous with the delayed rectifier K^+ current (Armstrong and Binstock, 1965), and since the TEA-insensitive outward current appears to correspond to the M current [I_M (Akasu and Koketsu, 1981b)], it has been concluded that both the delayed rectifier K^+ current and the M current were depressed by the muscarinic action of ACh (Akasu, 1981; Akasu and Koketsu, 1982).

C. Role of Metabolism of Postganglionic Neurons

The slow EPSP and the slow ACh potential present some interesting metabolic features. Ouabain, which blocks the Na pump, does not seem to affect the slow potential; on the other hand, metabolic inhibitors such as dinitrophenol readily abolish it (Libet 1970; cf. Karczmar and Nishi, 1971). The relationship between these metabolic features of the slow EPSP and its ionic mechanisms remain to be explored.

McAfee and Greengard (1972) (see also Greengard, 1976) studied the effect of cyclic guanosine 3′5′-monophosphate (cGMP) and its dibutyryl derivative on rabbit superior cervical ganglion, using a sucrose-gap technique. They reported that dibutyryl cGMP produced a small, transient depolarization of postganglionic nerve cells. A similar finding was also made by Busis *et al.* (1978a,b) (see Weight *et al.*, 1974) with regard to bullfrog sympathetic ganglion cells. Weight *et al.* (1974) demonstrated that stimulation of the preganglionic nerve and application of bethanechol resulted in an increase in the cGMP level in bullfrog sympathetic ganglion cells, and this increase was prevented by atropine and absent in the low-Ca^{2+}/ high-Mg^{2+} Ringer's solution. Accordingly, these workers suggested that cGMP may mediate the slow EPSP, acting as a "second messenger." While Dun *et al.* (1977, 1978) also observed a GMP-generated depolarization of mammalian sympathetic ganglia that was similar to that described for the amphibian ganglia, the response was biphasic, cGMP-induced depolarization being followed by a hyperpolarization. Moreover, these effects were associated with reduction and increase in resistance, respectively; thus, the change in membrane resistance that occurred during the depolarizing phase of the cGMP response differed from the changes in resistance observed in the course of the muscarinic response (see also above and Chapter 13). Similarly, Hashiguchi *et al.* (1978) reported that the depolarization induced by dibutyryl cGMP was associated with conductance changes that were different from those associated with the slow EPSP, and Gallagher and Shinnick-Gallagher (1977) reported an increase in membrane conductance during depolarization induced by intracellular injection of cAMP. Furthermore, the findings of Weight *et al.* (1974) with amphibian ganglia are not uncontroversial, since many investigators could not describe any consistent potential changes in response to cGMP (Busis *et al.*, 1978a,b; Smith *et al.*, 1978; Weight *et al.*, 1978). Thus, Kuba and Koketsu (1978), Dun *et al.*(1978), Nishi *et al.* (1978), and Weight *et al.* (1979) felt that the hypothesis that cGMP serves as a "second messenger" in the slow EPSP is not tenable. To the contrary, Libet and his associates (Hashiguchi *et al.*, 1978, 1982; Libet, 1979) feel that in the case of mammalian sympathetic ganglia, the slow EPSP may be generated by cGMP; they suggest that the depolarizing action of cGMP that they observed throughout the -90 to -40 mV range of membrane potentials constitutes what they refer to as a type 2 response. According to these investigators, contrary to what they consider to be the type 1 response, the type 2 response is not dependent on the muscarine-sensitive I_M.

In 1977, Ueda and Greengard (1977) described specific proteins that serve as substrates for phosphorylating processes that underlie the second-messenger role of cyclic nucleotides. More recently, Nestler and Greengard (1982) (see also Nestler *et al.*, 1984) presented evidence suggesting that a specific protein (protein I), present in the rabbit superior cervical ganglion, is the specific substrate for pre- and postsynaptic phosphorylations and plays an important role in synaptic function. Greengard and Nestler speculated that while the presynaptic pool of protein I would be involved in transmitter release, the postsynaptic pool of this protein may concern cyclic-nucleotide-dependent postsynaptic kinases and their synaptic function. However, Greengard and his associates (Nestler and Greengard, 1982; Nestler *et al.*, 1984) do not seem at this time to relate these concepts to Greengard's earlier suggestion that cGMP generates the slow EPSP (see Chapter 13 for further discussion of the second-messenger role of cyclic nucleotides in ganglionic transmission).

V. CONCLUSIONS AND COMMENT

The slow muscarinic potential is a characteristic synaptic potential exhibited by amphibian and mammalian sympathetic ganglia. There seems to be no doubt that this potential constitutes a cholinergic response no less than the fast EPSP. What characterizes the slow excitation response in contradistinction to the fast EPSP is its unusual and complex ionic mechanism. Muscarinic inactivation of G_K, in which the block of the M current may be involved, G_{Na} and G_{Ca} effects, as well as certain, perhaps characteristic cellular metabolic energy-generating processes (Libet, 1970; see also Karczmar and Nishi, 1971), all participate in the generation of the slow EPSP. Furthermore, the significance of these various phenomena and their interaction may vary from one species to another and, in the same species, from one ganglionic neuron to another.

What is of particular importance is the physiological role of this potential. It constitutes the primary transmissive potential, and it may also control ganglionic excitability. It may then be speculated that it may facilitate and render effective the fast EPSP and, in fact, substitute for the latter under certain conditions (see Chapter 13). This modulatory effect may constitute the physiological significance of the slow EPSP; in fact, there may be an analogy between this role of the slow ganglion EPSP and the slow responses of central cholinoceptive neurons that are so characteristic for the central nervous system (cf. Chapter 21).

REFERENCES

Adams, P. R., and Brown, D. A.: Synaptic inhibition of the M-current: Slow excitatory postsynaptic potential mechanism in bullfrog sympathetic neurones. *J. Physiol. (London)* **332**:263–272 (1982).

Adams, P. R., Brown, D. A., and Constanti, A.: Pharmacological inhibition of the M-current. *J. Physiol. (London)* **332**:223–262 (1982).

Akasu, T.: Voltage clamp analysis of muscarinic effects on action potentials in sympathetic ganglion cells of bullfrogs. *Neurosci. Lett. Suppl.* **6**:59 (1981).

Akasu, T., and Koketsu, K.: Desensitization of the muscarinic receptor controlling action potentials of sympathetic ganglion cells in bullfrogs. *Life Sci.* **27**:2261–2267 (1980).

Akasu, T., and Koketsu, K.: Modulatory actions of neurotransmitters on voltage-dependent membrane currents in bullfrog sympathetic neurones. *Kurume Med. J.* **28**:345–348 (1981a).

Akasu, T., and Koketsu, K.: Voltage-clamp studies of a slow inward current in bullfrog sympathetic ganglion cells. *Neurosci. Lett.* **26**:259–262 (1981b).

Akasu, T., and Koketsu, K.: Modulation of voltage-dependent currents by muscarinic receptor in sympathetic neurones of bullfrog. *Neurosci. Lett.* **29**:41–45 (1982).

Akasu, T., Gallagher, J. P., Koketsu, K., and Shinnick-Gallagher, P.: Slow excitatory postsynaptic currents in bull-frog sympathetic neurones. *J. Physiol. (London)* **351**:583–593 (1984).

Ambache, N.: A further survey of the action of *Clostridium botulinum* toxin upon different types of autonomic nerve fibre. *J. Physiol. (London)* **113**:1–17 (1951).

Armstrong, C. M., and Binstock, L.: Anomalous rectification in the squid giant axon injected with tetraethylammonium chloride. *J. Gen. Physiol.* **48**:859–872 (1965).

Ashe, J. H., and Yarosh, C. A.: Differential and selective antagonism of the slow-IPSP and slow-EPSP by gallamine and pirenzepine in the superior cervical ganglion of the rabbit. *Neuropharmacology* **23**:1321–1329 (1984).

Birdsall, N. J. M., Burgen, A. S. V., Hammer, R., Hulme, E. C., and Stockton, J.: Pirenzepine—a ligand with original binding properties to muscarinic receptors. *J. Supramol. Struct.* **4**:367–371 (1976).

Brown, D. A., and Adams, P. R.: Muscarinic suppression of a novel voltage-sensitive K^+ current in a vertebrate neurone. *Nature (London)* **283**:673–676 (1980).

Busis, N. A., Schulman, J. A., Smith, P. A., and Weight, F. F.: Do cyclic nucleotides mediate slow postsynaptic potentials in sympathetic ganglia? *Br. J. Pharmacol.* **62**:378P–379P (1978a).

Busis, N. A., Weight, F. F., and Smith, P. A.: Synaptic potentials in sympathetic ganglia: Are they mediated by cyclic nucleotides? *Science* **200**:1079–1081 (1978b).

Dun, N. J., Kaibara, K., and Karczmar, A. G.: Direct postsynaptic membrane effect of dibutyryl cyclic GMP on mammalian sympathetic neurons. *Neuropharmacology* **16**:715–717 (1977).

Dun, N. J., Kaibara, K., and Karczmar, A. G.: Muscarinic and cGMP induced membrane potential changes: Differences in electrogenic mechanisms. *Brain Res.* **150**:658–661 (1978).

Eccles, R. M.: Action potentials of isolated mammalian sympathetic ganglia. *J. Physiol. (London)* **117**:181–195 (1952).

Eccles, R. M., and Libet, B.: Origin and blockade of the synaptic responses of curarized sympathetic ganglia. *J. Physiol. (London)* **157**:484–503 (1961).

Gallagher, J. P., and Shinnick-Gallagher, P.: Cyclic nucleotides injected intracellularly into rat superior cervical ganglion cells. *Science* **198**:851–852 (1977).

Greengard, P.: Possible role for cyclic nucleotides and phosphorylated membrane proteins in postsynaptic actions of neurotransmitters. *Nature (London)* **260**:101–108 (1976).

Hammer, R., Berrie, C. P., Birdsall, N. J. M., Burgen, A. S. V., and Hulme, E. C.: Pirenzepine distinguishes between different subclasses of muscarinic receptors. *Nature (London)* **283**:90–92 (1980).

Hartzell, H. C.: Mechanisms of slow postsynaptic potentials. *Nature (London)* **291**:539–544 (1981).

Hashiguchi, T., Ushiyama, N., Kobayashi, H., and Libet, B.: Does cyclic GMP mediate the slow excitatory synaptic potential in sympathetic ganglia? *Nature (London)* **271**:267–268 (1978).

Hashiguchi, T., Kobayashi, H., Tosaka, T., and Libet, B.: Two muscarinic depolarizing mechanisms in mammalian sympathetic neurons. *Brain Res.* **242**:378–382 (1982).

Hodgkin, A. L., and Huxley, A. F.: The dual effect of membrane potential on sodium conductance in the giant axon of the squid. *J. Physiol. (London)* **39**:715–733 (1952).

Karczmar, A. G., and Nishi, S.: The types and sites of cholinergic receptors, in: *Advances in Cytopharmacology* (F. Clementi and B. Cicciarelli, eds.), pp. 301–317, Raven Press, New York (1971).

Kobayashi, H., and Libet, B.: Generation of slow postsynaptic potentials without increases in ionic conductance. *Proc. Natl. Acad. Sci. U.S.A.* **60**:1304–1311 (1968).

Kobayashi, H., and Libet, B.: Actions of noradrenaline and acetylcholine on sympathetic ganglion cells. *J. Physiol. (London)* **208**:353–372 (1970).

Kobayashi, H., and Libet, B.: Is inactivation of potassium conductance involved in slow postsynaptic excitation of sympathetic ganglion cells? Effects of nicotine. *Life Sci.* **14**:1871–1883 (1974).

Kobayashi, H., and Tosaka, T.: Slow synaptic actions in mammalian sympathetic ganglia, with special reference to the possible roles played by cyclic mideotides, in: *Autonomic Ganglia* (L.-G. Elfvin, ed.), pp. 281–307, John Wiley, Chichester and New York (1983).

Koketsu, K., Nishi, S., and Noda, Y.: Effects of physostigmine on the after discharge and slow postsynaptic potentials of bullfrog sympathetic ganglia. *Br. J. Pharmacol.* **34**:177–188 (1968).

Koketsu, K., Shoji, T., and Nishi, S.: Slow inhibitory postsynaptic potentials of bullfrog sympathetic ganglia in sodium-free media. *Life Sci.* **13**:453–458 (1973).

Kuba, K., and Koketsu, K.: Ionic mechanism of the slow excitatory postsynaptic potential in bullfrog sympathetic ganglion cells. *Brain Res.* **81**:338–342 (1974).

Kuba, K., and Koketsu, K.: Direct control of action potentials by acetylcholine in bullfrog sympathetic ganglion cells. *Brain Res.* **89**:166–169 (1975).

Kuba, K., and Koketsu, K.: Analysis of the slow excitatory postsynaptic potential in bullfrog sympathetic ganglion cells. *Jpn. J. Physiol.* **26**:651–669 (1976a).

Kuba, K., and Koketsu, K.: The muscarinic effects of acetylcholine on the action potential of bullfrog sympathetic ganglion cells. *Jpn. J. Physiol.* **26**:703–716 (1976b).

Kuba, K., and Koketsu, K.: Synaptic events in sympathetic ganglia. *Prog. Neurobiol.* **11**:77–169 (1978).

Kuba, K., and Nishi, S.: Characteristics of fast excitatory post-synaptic current in bullfrog sympathetic ganglion cells: Effects of membrane potential, temperature and Ca ions. *Pflugers Arch. Eur. J. Physiol.* **378**:205–212 (1979).

Kuffler, S. W., and Sejnowski, T. J.: Peptidergic and muscarinic excitation at amphibian sympathetic synapses. *J. Physiol. (London)* **341**:257–278 (1983).

Laporte, Y., and Lorente de No, R.: Potential changes evoked in a curarized sympathetic ganglion by presynaptic volleys of impulses. *J. Cell. Comp. Physiol.* **35**(Suppl. 2):61–106 (1950).

Libet, B.: Slow synaptic responses and excitatory changes in sympathetic ganglia. *J. Physiol. (London)* **174**:1–25 (1964).

Libet, B.: Long latent periods and further analysis of slow synaptic responses in sympathetic ganglia. *J. Neurophysiol.* **30**:494–514 (1967).

Libet, B.: Generation of slow inhibitory and excitatory post synaptic potentials. *Fed. Proc. Fed. Am. Soc. Exp. Biol.* **29**:1945–1956 (1970).

Libet, B.: Inactivation of potassium conductance in slow post-synaptic excitation. *Science* **172**:503–504 (1971).

Libet, B.: Slow synaptic actions in ganglionic functions, in: *Interactive Function of the Autonomic Nervous System* (C. M. Brooks, K. Koizynis, and A. Sato, eds.), pp. 197–222, North-Holland, Amsterdam (1979).

Libet, B., and Kobayashi, H.: Generation of adrenergic and cholinergic potentials in sympathetic ganglion cells. *Science* **164:**1530–1532 (1969).

MacDermott, A. B., Conner, E. A., Dionne, V. E., and Parsons, R. L.: Voltage clamp study of fast excitatory synaptic currents in bullfrog sympathetic ganglion cells. *J. Gen. Physiol.* **75:**39–60 (1980).

McAfee, D. A., and Greengard, P.: Adenosine 3′, 5′-monophosphate: Electrophysiological evidence for a role in synaptic transmission. *Science* **178:**310–312 (1972).

Nestler, E. J., and Greengard, P.: Distribution of protein I and regulation of its state of phosphorylation in the rabbit superior cervical ganglion. *J. Neurosci.* **2:**1011–1023 (1982).

Nestler, E. J., Walaas, S. I., and Greengard, P.: Neuronal phosphoproteins: Physiological and clinical implications. *Science* **225:**1357–1364 (1984).

Nishi, S.: Ganglionic transmission, in: *The Peripheral Nervous System* (J. I. Hubbard, ed.) pp. 225–255, Plenum Press, New York (1974).

Nishi, S., Karczmar, A. G., and Dun, N. J.: Physiology and pharmacology of ganglionic synapses as models for central transmission, in: *Advances in Pharmacology and Therapeutics*, Vol. 2, *Neurotransmitters* (P. Simon, ed.), pp. 69–85, Pergamon Press, New York (1978).

Nishi, S., and Koketsu, K.: Early and late afterdischarges of amphibian sympathetic ganglion cells. *J. Neurophysiol.* **31:**109–121 (1968).

Nishi, S., Soeda, H., and Koketsu, K.: Unusual nature of ganglionic slow EPSP studied by a voltage-clamp method. *Life Sci.* **8:**33–42 (1969).

Selyanko, A. A., Derkach, V. A., and Skok, V. I.: Effects of some ganglion-blocking agents on fast excitatory postsynaptic currents in mammalian sympathetic ganglion neurones, in: *Physiology of Excitable Membranes*, Advances in Physiological Sciences, (J. Salanki, ed.), pp. 329–342, Pergamon Press, Budapest (1980).

Smith, P. A., Weight, F. F., Busis, N. A., and Schulman, J. A.: Cyclic nucleotides and slow post-synaptic potential generation in sympathetic ganglia of the bullfrog. *Fed. Proc. Fed. Am. Soc. Exp. Biol.* **37:**524 (1978).

Takeshige, C., and Volle, R. L.: Bimodal response of sympathetic ganglia to acetylcholine following eserine or repetitive preganglionic stimulation. *J. Pharmacol. Exp. Ther.* **138:**66–73 (1962).

Tosaka, T., and Libet, B.: Additional (adrenergic) synaptic step in sympathetic ganglia. *Fed. Proc. Fed. Am. Soc. Exp. Biol.* **29:**716 (1970).

Tosaka, T., Chichibu, S., and Libet, B.: Intracellular analysis of slow inhibitory and excitatory postsynaptic potentials in sympathetic ganglia of the frog. *J. Neurophysiol.* **31:**396–409 (1968).

Ueda, T., and Greengard, P.: Adenosine 3′:5′-monophosphate-regulated phosphoprotein system of neuronal membranes. *J. Biol. Chem.* **252:**5155–5163 (1977).

Vickroy, T. W., Watson, M., Yamamura, H. I., and Roeske, W. R.: Agonist binding to multiple muscarinic receptors. *Fed. Proc. Fed. Am. Soc. Exp. Biol.* **43:**2785–2798 (1984).

Volle, R. L.: The actions of several ganglionblocking agents on the postganglionic discharge induced by diisopropyl phosphorofluoride (DFP) in sympathetic ganglia. *J. Pharmacol. Exp. Ther.* **35:**45–53 (1962).

Volle, R. L., and Hancock, J. C.: Transmission in sympathetic ganglia. *Fed. Proc. Fed. Am. Soc. Exp. Biol.* **29:**1913–1918 (1970).

Weight, F. F., and Votava, J.: Slow synaptic excitation in sympathetic ganglion cells: Evidence for synaptic inactivation of potassium conductance. *Science* **170:**755–758 (1970).

Weight, F. F., Petzold, G., and Greengard, P.: Guanosine, 3′, 5′-monophosphate in sympathetic ganglia: Increase associated with synaptic transmission. *Science* **186**:942–944 (1974).

Weight, F. F., Smith, P. A., and Schulman, J. A.: Postsynaptic potential generation appears independent of synaptic elevation of cyclic nucleotides in sympathetic neurons. *Brain Res.* **158**:197–202 (1978).

Weight, F. F., Schulman, J. A., Smith, P. A., and Busis, N. A.: Long-lasting synaptic potentials and the modulation of synaptic transmission. *Fed. Proc. Fed. Am. Soc. Exp. Biol.* **38**:2084–2094 (1979).

8

Peptidergic Transmission

Y. KATAYAMA and S. NISHI

I. NONCHOLINERGIC EXCITATORY SYNAPTIC TRANSMISSION IN SYMPATHETIC GANGLIA

A slow depolarization resistant to cholinergic antagonists was first observed in bullfrog sympathetic ganglia by Nishi and Koketsu (1968); accordingly, it was referred to as noncholinergic excitatory transmission. Tetanic preganglionic stimulation in the presence of nicotine (or D-tubocurarine) and atropine induced, dependent on whether recorded by means of intra- or extracellular electrodes, an extremely long-lasting depolarization, or the late late negative (LLN) potential, and late afterdischarges (LADs), respectively. The relationship among the depolarization, the LLN potential, and the LAD was analyzed, and the site of origin of these responses was found to be identical (Nishi and Koketsu, 1968). When recorded intracellularly, the response exhibited a slow onset (1–5 sec after stimulation) and extremely prolonged time–course (5–10 min in duration); hence, it was called the "late slow excitatory postsynaptic potential" (EPSP) (Figure 1A). The late slow EPSP could be recorded from both B and C neurons of the bullfrog; however, the late slow EPSP could be induced only when preganglionic stimulation was strong enough to excite C fibers. This result is consistent with immunohistochemical data that demonstrate that the transmitter responsible for the late slow EPSP is present in, and released from, C fibers (L. Y. Jan *et al.*, 1980) (see Section II). Preganglionic C fibers do not innervate postganglionic B cells (Nishi

---

Y. KATAYAMA and S. NISHI • Department of Physiology, Kurume University School of Medicine, Kurume, Japan. Present address for Dr. Katayama: Department of Autonomic Physiology, Medical Research Insitute, Tokyo Medical and Dental University, Tokyo, Japan.

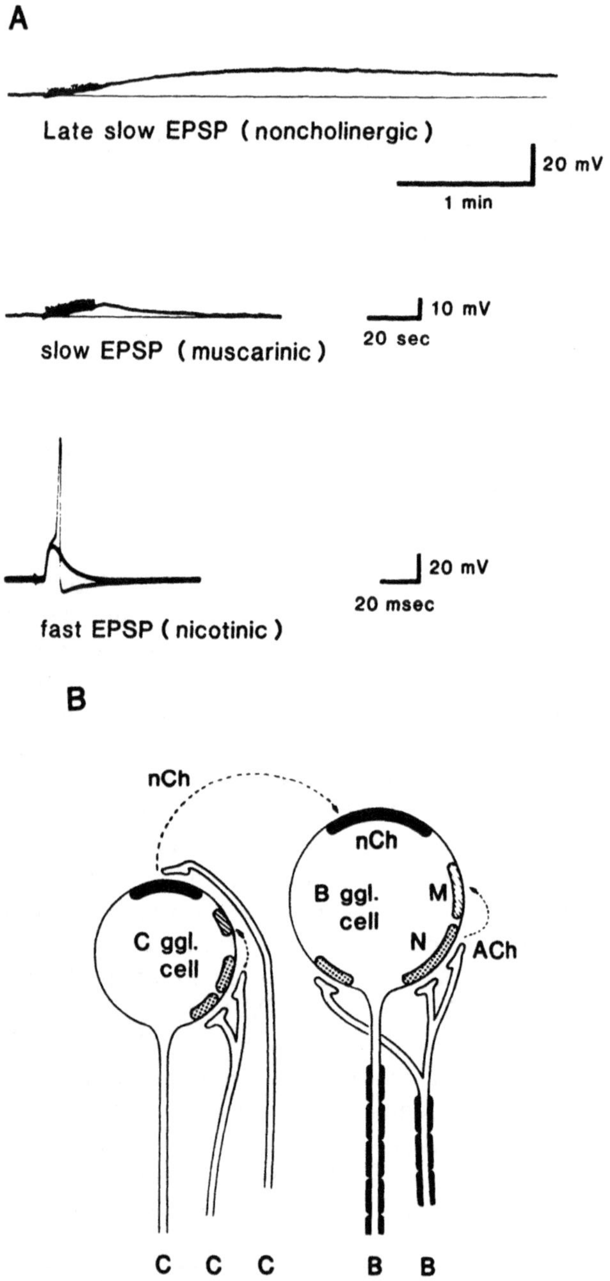

Figure 1. (A) Three types of EPSPs recorded intracellularly from bullfrog sympathetic ganglion cells. The fast EPSP (bottom), which is produced by a single preganglionic stimulation, is nicotinic in nature. The slow EPSP (middle), which is caused by tetanic stimulation, is atropine-sensitive and thus muscarinic. After nicotinization (or curarization) and atropinization of ganglion cells, the late slow EPSP (top) is produced by strong tetanic stimulation of preganglionic fibers at the 8th ganglion level. Compare the time–course of these three

et al., 1965); however, since the late slow EPSP can be generated in B cells, it was concluded that the transmitter reaches B cells by diffusion from the synapses innervated by C fibers (Figure 1B). It has not yet been demonstrated conclusively whether or not acetylcholine (ACh) and the transmitter for the late slow EPSP, presumably luteinizing-hormone-releasing hormone (LH-RH), coexist in the same C-fiber terminals, although there is some ultrastructural evidence that this is the case (L. Y. Jan and Y. N. Jan, 1982).

Noncholinergic excitatory transmission was also demonstrated for mammalian sympathetic ganglia (see Nishi, 1974; Dun, 1980, 1983; Hartzell, 1981) including the superior cervical ganglion of the dog (Chen, 1969, 1971, 1972) and cat (Alkadhi and McIsaac, 1971); the evidence included the finding that the nictitating membrane contraction caused by repetitive preganglionic stimulation was not completely suppressed by nicotinic or muscarinic antagonists. More recently, slowly developing and long-lasting depolarizations that were resistant to both nicotinic and muscarinic antagonists were recorded with a sucrose-gap method from the superior cervical ganglion of the rabbit (Ashe and Libet, 1981) and intracellularly from that of the dog (Katayama and Nishi, unpublished data).

Slowly developing depolarization was elicited in sympathetic ganglia of the guinea pig; it could be evoked in the case of the inferior mesenteric ganglion by repetitive stimulation of the hypogastric nerves (Neild, 1978; Weems and Szurszewski, 1978; Dun and Karczmar, 1979; Konishi *et al.*, 1979b; Dun and Jiang, 1982) and in that of the coeliac ganglion by stimulation of the splanchnic nerves (Dun and Ma, 1984). These depolarizations were blocked by removing Ca^{2+} from the superfusing solution, but were insensitive to nicotinic and muscarinic antagonists. This suggests that the slowly developing depolarization elicited in the inferior mesenteric and coeliac ganglion of the guinea pig is caused by a noncholinergic substance or substances released from presynaptic terminals. Another site where noncholinergic excitatory response can be generated is the enteric nervous system, since this response was noted in the case of the myenteric plexus (Katayama and North, 1978; Wood and Mayer, 1978a,b, 1979; Johnson *et al.*, 1980) and the submucous plexus (Mihara *et al.*, 1982, 1983; Surprenant, 1984) of the guinea pig.

types of EPSPs. The duration of the late slow EPSP is more than 15,000 times that of the fast EPSP. (B) Schematic illustration of the innervation pattern of ganglion cells (B and C cells) in the bullfrog sympathetic ganglion. Preganglionic B fibers innervate B cells and C fibers innervate C cells. The transmitter (ACh) for the nicotinic fast EPSP and muscarinic slow EPSP is released from B and C preganglionic fibers. The noncholinergic transmitter (nCh) for the late slow EPSP is released from C fibers, which have a higher threshold than B fibers. Because the late slow EPSP can be recorded both from B and from C cells, the late slow EPSP is brought about by the substance diffused from releasing sites, C-fiber terminals.

These noncholinergic slow depolarizations were often referred to as slow EPSPs. This nomenclature is confusing, however, and the term "slow EPSP" should be reserved for the synaptic potentials that are muscarinic in origin (see Kuba and Koketsu, 1978; Nishi *et al.*, 1978) (see also Chapters 3 and 7). To avoid confusion, the slowly developing depolarization that can be elicited in inferior mesenteric and coeliac ganglion cells and myenteric neurons will be called in this chapter the "noncholinergic slow EPSP." In the case of bullfrog sympathetic ganglia, this potential will be called the "late slow EPSP." At present, some of these EPSPs are thought to be mediated by peptides (see below).

The ionic mechanism of the noncholinergic excitatory synaptic responses will be described in Section II. The noncholinergic responses were often accompanied by a decrease in input conductance (Schulman and Weight, 1976; Katayama and Nishi, 1977; Konishi *et al.*, 1979b); sometimes input conductance was increased (Katayama and Nishi, 1977; Neild, 1978; Dun and Karczmar, 1979). The noncholinergic slow depolarization and the decrease in input conductance generated in the course of these responses may enhance the efficacy of ganglionic transmission (Weight *et al.*, 1979).

Following sustained stimulation of preganglionic fibers, the noncholinergic slow depolarization of the inferior mesenteric ganglion becomes pregressively attenuated. It was suggested that this reduction was due to presynaptic inhibition. Since the reduction was blocked by naloxone, endogenous opioids released by repetitive stimulation may have been involved. In fact, application of an enkaphalin suppressed both the nicotinic fast EPSP and the noncholinergic slow EPSP of the sympathetic ganglia (Konishi *et al.*, 1979a, 1980) (see Chapter 13).

II. INVOLVEMENT OF PEPTIDES IN EXCITATORY SYNAPTIC TRANSMISSION IN SYMPATHETIC GANGLIA

A. Peptide Candidates for a Neurotransmitter Role in the Ganglia

A number of peptides that are present in the brain have been localized in the autonomic ganglia of vertebrates by means of immunohistochemical techniques (see Chapter 3) (see also Hökfelt *et al.*, 1977a,b,c, 1980, 1984; Larsson and Rehfeld, 1979; Schultzberg *et al.*, 1979; Kondo and Yui, 1981; Kondo *et al.*, 1982; Dun, 1983); these include substance P, somatostatin, enkephalin, vasoactive intestinal polypeptide (VIP), cholecystokinin (CCK), and LH-RH (see Chapters 3 and 13). It was suggested that some of these polypeptides may serve as transmitters generating noncholinergic synaptic responses; however, the pertinent evidence is available only with respect to a few peptides such as substance P, LH-RH, and enkephalin (see below). On the other hand, although somatostatin, CCK, and VIP are present in sympathetic ganglia, there is no electrophysiological evidence as yet to support the contention that they serve as neurotransmitters or modulators or both. However, some data are available with respect to the actions of these peptides on parasympathetic and myenteric neurons; somatostatin caused a membrane hyperpolarization of feline ciliary ganglion cells (Kondo *et al.*, 1982) and exerted stimulating and inhibiting actions on guinea pig myenteric neurons (Katayama and North, 1980), while VIP increased the firing activity of guinea pig myenteric neurons (Williams and North, 1979b). Also, it was suggested that CCK-8 stimulates longitudinal muscle via stimulation of the myenteric plexus (Dockray and Hutchson, 1980).

LH-RH, substance P, and enkephalin appear to be among the most probable candidates for a transmitter role in sympathetic ganglia. The case for LH-RH was established in the course of the extensive search for noncholinergic synaptic transmitters carried out in bullfrog lumbar sympathetic ganglia by the late S. Kuffler and his colleagues (Y. N. Jan *et al.*, 1979, 1980; L. Y. Jan and Y. N. Jan, 1982; Kuffler, 1980). First, high concentrations of an LH-RH-like substance could be detected by radioimmunoassays in nerves that contain C fibers; in addition, immunohistochemical studies clearly showed that an LH-RH-like immunoreactivity was present in synaptic boutons abutting C cells (L. Y. Jan *et al.*, 1980). Second, a high-K^+ solution caused a Ca^{2+}-dependent release of an LH-RH-like substance from presynaptic fibers innervating the ganglia that could be shown to generate the late slow EPSP. Finally, electrophysiological and pharmacological investigations demonstrated that the response to LH-RH mimicked the late slow EPSP, while prolonged application of LH-RH prevented the appearance of the late slow EPSP.

Similarly, there is strong evidence for the transmitter role of substance P in the autonomic nervous system. Actually, substance P is distributed widely in both the central and the peripheral nervous system, and it is considered to be a neurotransmitter at several sites (see Nicoll *et al.*, 1980), including the enteric nervous system (Johnson *et al.*, 1981; Morita *et al.*, 1980) (see Chapter 16). As already stated, substance P was detected in pre- and paravertebral sympathetic ganglia of the guinea pig and the rat by means of immunohistochemical and radioimmunoassay methods (Hökfelt *et al.*, 1977c; Konishi *et al.*, 1979b; Robinson *et al.*, 1980). A Ca^{2+}-dependent release of an immunoreactive substance P from mesenteric ganglia was induced by high K^+. It was suggested that dorsal root ganglion cells are the origin of the fibers that release substance P in the inferior mesenteric ganglion (Konishi *et al.*, 1979b, 1980; Dun and Jiang, 1982; Tsunoo *et al.*, 1982). Two groups of investigators (Dun and Karczmar, 1979; Konishi *et al.*, 1979b) showed that substance P mimicked the noncholinergic slow EPSP in the mammalian inferior mesenteric ganglion and that the noncholinergic slow EPSP was markedly suppressed during the course of depolarization evoked by the application of substance P. The electrogenesis of substance-P-induced and synaptically evoked responses will be discussed in the next section. These results support the hypothesis that substance P is a transmitter in the inferior mesenteric ganglion of the guinea pig.

It is of interest to note that substance P also exerted an excitatory effect, similar to that exhibited by LH-RH, on the bullfrog sympathetic ganglion (Nishi *et al.*, 1980; see also L. Y. Jan and Y. N. Jan, 1982; Adams *et al.*, 1983); substance P mimicked the late slow EPSP, while the latter could be blocked by prolonged superfusion with substance P. Substance P was also found to be present in bullfrog sympathetic ganglia (L. Y. Jan and Y. N. Jan, 1982). Eledoisin-related peptide and physalaemin, which possess the same C-terminal amino acid sequence as substance P (-Gly-Leu-Met), also exerted similar excitatory actions on the ganglion (Figure

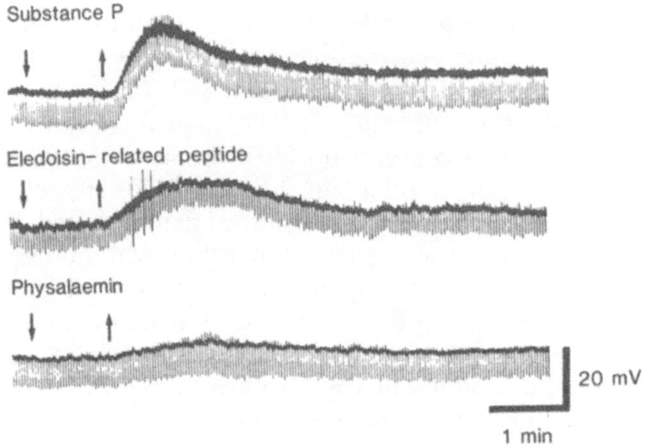

Figure 2. Depolarization of the soma membrane of 9th sympathetic ganglion cells of the bullfrog caused by not only LH-RH, but also other peptides, substance P, and eledoisin-related peptide and physalaemin, which have the same C-terminal structure (-Gly-Leu-Met) as substance P. During the period represented by the arrows, the superfusing solution contained peptides at a concentration of 50 μM. There was a time delay of 1 min before the onset of the response due to the time characteristics of the superfusing system. The depolarizations were associated wtih an increase in input resistance. Met-enkephalin, neurotensin, thyrotropin-releasing hormone, angiotensin II, and bradykinin at the same concentration had no excitatory effect on the neurons of this ganglion.

2). Furthermore, there was cross-desensitization between substance P and LH-RH (see, however, L. Y. Jan and Y. N. Jan, 1982). Usually, the potency of LH-RH action on the ganglion was much higher than that of substance P. Although serotonin is thought to be the main mediator for the noncholinergic slow EPSP in the coeliac ganglion of the guinea pig, substance P is found to cause a depolarization in a portion of coeliac ganglion cells; accordingly, substance P is considered to be a possible candidate mediating the noncholinergic slow EPSP (Dun *et al.*, 1984).

Enkephalin is still another of the peptides listed above as present in the ganglion and possibly active as a ganglionic neurotransmitter. Enkephalin is an important neuropeptide that appears to be a neurotransmitter not only in the peripheral but also in the central nervous system (see North, 1979). Immunohistochemical studies demonstrated the presence of enkephalin in rat (Schultzberg *et al.*, 1979) and bullfrog sympathetic ganglia (Kondo and Yui, 1981). In the case of the inferior mesenteric ganglion of the guinea pig, the primary site of action of enkephalin was the presynaptic terminal (Konishi *et al.*, 1979a, 1980), while in that of the lumbar sympathetic ganglion of the bullfrog, its action was thought to be both pre- and postsynaptic (Wouters and Van den Bercken, 1979, 1980) (see, however, Kondo and Yui, 1981). The pre- and postsynaptic actions of enkephalin were almost always inhibitory in the case of these sympathetic ganglia as well as in that of parasympathetic ganglia [e.g., ciliary ganglia of cat (Katayama and Nishi, 1981, 1984)] and of myenteric plexus (North and Williams, 1976; Williams and North, 1979a; North *et al.*, 1979). The actions of enkephalin will be described in detail in Chapters 11 and 13.

The case of two other peptides, angiotensin II and bradykinin, is more complicated. In 1965, Lewis and Reit (1965) reported that two endogenous peptides, angiotensin II and bradykinin, are potent ganglionic stimulants; indeed, Lewis and Reit (1965) could induce contraction of the cat nictitating membrane by applying these peptides to the superior cervical ganglion. Further electrophysiological analysis of the action of angiotensin II was carried out with regard to neurons of the superior cervical ganglion of the cat (Dun *et al.*, 1978), the superior cervical ganglion of the rat (Brown *et al.*, 1980; Constanti and Brown, 1981), and the ciliary ganglion and dorsal root ganglion of the cat (Dun *et al.*, 1978). Angiotensin II exerted a stimulating action due to increased permeability of the postsynaptic membrane to Na^+ in the case of the cat superior cervical ganglion and to suppression of K^+ conductance in that of the rat superior cervical ganglion. This peptide also depolarized inferior mesenteric ganglion cells of the guinea pig; the ionic mechanism of this action is unknown (Tsunoo *et al.*, 1982). Bradykinin did not stimulate amphibian sympathetic ganglia, as shown by means of intracellular recording (Y. N. Jan *et al.*, 1979; Nishi *et al.*, 1980). It should be mentioned that angiotensin II immunoreactivity is known to occur in the spinal ganglion (Fuxe *et al.*, 1976) (see Tsunoo *et al.*, 1982). Furthermore, these peptides may conceivably affect ganglionic transmission via their presence in the blood vessels. Recently, neuromodulating actions of peptides in the sympathetic ganglia were described; LH-RH affects transmitter release and modulates the generation of action potentials and synaptic potentials in bullfrog sympathetic ganglion cells (Akasu *et al.*, 1983) (see also Chapter 12).

B. Electrophysiological Characteristics of Noncholinergic Excitatory Postsynaptic Potentials and Peptide-Induced Depolarization

To establish a particular peptide as a neurotransmitter, it must be shown that the electrogenesis of peptide-induced and synaptically evoked responses is identical. Many pertinent electrophysiological studies were carried out with regard particularly to those peptides the presence of which in sympathetic ganglia could be demonstrated. Since substance P was proposed as a transmitter for the noncholinergic slow EPSP of the guinea pig sympathetic ganglion, and LH-RH or an LH-RH-like substance as a transmitter for the late slow EPSP of the bullfrog lumbar sympathetic ganglion, this section will concentrate on noncholinergic EPSPs of these ganglia and on substance-P- and LH-RH-induced responses at these sites.

The noncholinergic slow EPSPs of inferior mesenteric ganglion cells of the guinea pig were usually associated with a decrease in neuronal input resistance (Neild, 1978). This may suggest that increased Na^+ conductance is involved. However, it was subsequently found that only in the case of EPSPs larger than 15 mV is input resistance decreased, while small noncholinergic slow EPSPs are usually accompanied by an increased resistance (Konishi *et al.*, 1979b; Tsunoo *et al.*, 1982). Thus, the primary action of the transmitter may be to decrease the resting K^+ conductance.

It was reported that substance-P-induced depolarization is associated with either an increase or a decrease in input membrane resistance (Dun and Karczmar, 1979; Konishi *et al.*, 1979b; Dun and Jiang, 1982). These initial observations were confirmed by experiments in which the manual clamp method was employed (Minota *et al.*, 1981; Dun and Minota, 1981). The incidence of both types of responses was found to be equal. Substance-P-induced depolarization was increased by membrane hyperpolarization and almost abolished in Na^+-free solution; sometimes, the response was nearly abolished at E_K. Interestingly, substance P often caused a biphasic response in which an initial conductance increase was followed by a conductance decrease. Dun and Minota (1981) therefore concluded that substance-P-induced depolarization was generated by simultaneous increase in Na^+ and decrease in K^+ permeability. Thus, several mechanisms seem to be involved in substance-P-induced response of the inferior mesenteric ganglion. However, the relationship between the Na^+ permeability increase and the K^+ permeability decrease was not clear. In this context, it is important that both substance-P-induced depolarization and the noncholinergic slow EPSP of the myenteric plexus of the guinea pig ileum were both associated with a decrease in membrane conductance due to inactivation of K^+ conductance (Katayama *et al.*, 1979; Johnson *et al.*, 1980). In sum, there is no simple mechanism that would explain the ionic mechanism of substance-P-induced response and of the noncholinergic slow EPSP of guinea pig mesenteric ganglion cells.

The ionic mechanism of the noncholinergic response of bullfrog ganglia also appears to be complex. Nishi (1973, 1974) reported that the late slow EPSP of sympathetic ganglion cells of the bullfrog was associated with decreased membrane resistance that was due to increased membrane permeabilities to Na^+ and K^+. Later, a different type of late slow EPSP was observed; it was accompanied by increased membrane resistance (Schulman and Weight, 1976; Katayama and Nishi, 1977; Y. N. Jan *et al.*, 1980). The two responses were called Type II and Type I late slow EPSP, respectively; decreased membrane permeability (Type I response) was encountered more frequently than increased permeability (Type II response). Since the Type I late slow EPSP was occasionally reversed at approximately −90 mV, the Type I response is probably due to inactivation of K^+ conductance (Katayama and Nishi, 1977). Furthermore, the late slow EPSP associated with decreased membrane conductance at the resting potential (Type I response) became much larger in size in some cells at E_K, while it almost disappeared in a few cells at this membrane potential (Y. N. Jan *et al.*, 1980). These results indicate that several mechanisms may be involved in the generation of the late slow EPSP. It should be emphasized that the late slow EPSP and the LH-RH-induced response of bullfrog sympathetic ganglia exhibited similar conductance changes and voltage dependency (Y. N. Jan *et al.*, 1980).

Voltage-clamp analysis of the synaptic current and of the current generated by membrane-active substances including transmitters and related molecules constitutes a powerful tool for the study of synaptic responses. The late slow excitatory postsynaptic current (late slow EPSC) and the LH-RH-induced current were evaluated in this particular way; furthermore, the change in membrane conductance accompanying these currents was estimated by the voltage-jump method of applying command pulses. Of particular interest was the distinctive voltage-dependent K^+ current that was observed in bullfrog sympathetic ganglion cells (Brown and Adams, 1980) (see Chapters 7, 12, and 13). In the present context, the point is that this current was selectively suppressed not only by muscarinic agonists (and was therefore referred to as the "M current"), but also by LH-RH (Adams and Brown, 1980) (see Constanti and Brown, 1981; Adams *et al.*, 1983). In fact, suppression of the M current seems to be involved in the electrogenesis of the Type I late slow EPSP. However, a simple decrease in the voltage-sensitive K^+ conductance, i.e., the closing of the M channel, cannot completely explain the behavior of the late slow EPSC, especially its behavior when the membrane is strongly hyperpolarized. Indeed, when the membrane was hyperpolarized, the late slow EPSC, depending on the cell, was either reversed, the reversal potential being near E_K, or not reversed, although in the case of either cell, the membrane conductance was decreased when the membrane was clamped near the resting potential (Sejnowski, 1982).

Our recent studies in which the voltage-clamp method was employed (Katayama *et al.*, 1981; Katayama and Nishi, 1982) (see Kuffler and Sejnowski, 1983) have disclosed that clamped bullfrog ganglion cells can be classed into three groups according to the characteristics of their late slow EPSC and LH-RH-induced current. In the first group of cells, both currents were associated with a *decreased* membrane conductance irrespective of the holding potentials (Type I response). The value of the reversal potential of both currents was approximately −90 mv; this indicated that both currents are partially due to the inactivation of the voltage-insensitive K^+ conductance. The currents could be still detectable when the holding level was more negative than −60 mV, at which level M channels are inactivated (Brown and Adams, 1980). The mechanism that leads to the Type I response may include two components: suppression of the M current and suppression of the voltage-independent K^+ current. It could actually be shown that the former contributes to the generation of the late slow EPSP mainly at depolarized and the latter mainly at hyperpolarized levels of the membrane potential.

On the other hand, the second group of cells, which was small in number, produced a late slow EPSC and an LH-RH-induced current that were associated with an *increased* membrane conductance at every holding potential between −30 and −130 mV (Type II response). The value of the reversal potential of these responses was approximately −30 mV; therefore, the Type II response appeared to be brought about by an increase in Na^+ and K^+ conductances. The LH-RH-induced response of these cells was markedly reduced although still observable in Na^+-free solution. The response that could still be observed in the Na^+-free medium exhibited characteristics of the Type I response; that is, it was associated with a decreased conductance (see Figure 5 in Katayama and Nishi, 1982).

Finally, in the third group of neurons (Figure 3), the late slow EPSC and the LH-RH response were associated with a decreased conductance when the holding potential was low (above −40 to −50 mV); its amplitude decreased as the holding potential approached −50 to −60 mV. To the contrary, at hyperpolarized levels of membrane potentials (below

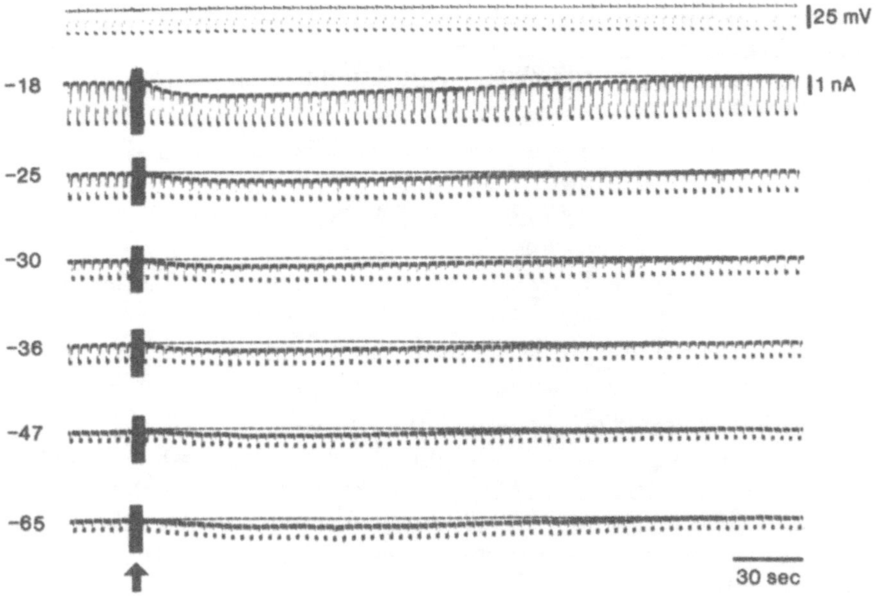

Figure 3. Late slow EPSC recorded from a voltage-clamped neuron in the bullfrog sympathetic ganglion. The late slow EPSC was triggered by supramaximal preganglionic tetanic stimulation (10 Hz for 5 sec, indicated by the arrow). Commanding pulses (25 mV for 1 sec, repeated at 0.25 Hz) were given so as to estimate a membrane conductance by the jumping method. The clamped levels are indicated at left. When the clamped level was more positive than −50 mV, the response was associated with a decreased conductance. As the holding level was shifted toward −50 mV, the response became small. On the other hand, when the holding level was −65 mV, the response was accompanied by an increased conductance and its amplitude was larger than that of the response obtained at −50 mV.

−50 mV), the conductance and the amplitude of the late slow EPSC and of the LH-RH current increased. Thus, the amplitude of the late slow EPSC and of the LH-RH response of these cells was at its lowest when the holding potential ranged betwen −50 and −60 mV; accordingly, these cells had no reversal potential. Although in these cells the LH-RH-induced response was only slightly affected by total removal of Na^+ for the perfusing solution when the cell membrane was depolarized, at the hyperpolarized level of the resting membrane potential, the response was almost eliminated in Na^+-free solution. Interestingly, the diminished response in Na^+-free solution was associated with a decreased conductance even at the hyperpolarized level of the membrane potential, whereas the response observed in normal Ringer's solution at the same hyperpolarized membrane potential was accompanied by an increased conductance; it is important that this diminished response was sometimes successfully reversed at approximately −90 mV. This indicated that the response observed in Na^+-free medium may represent the Type I response and depend on the Type I mechanism. Furthermore, a change in Cl^- concentration of the superfusing solution had no significant effect on the late slow EPSC or the LH-RH-induced current.

These results suggest that the late slow EPSC and the LH-RH-induced response may both be due to activation of Na^+ and K^+ conductances (Type II mechanism) or suppression of K^+ conductance (Type I mechanism) or both. The Type I and II mechanisms may dominate at the depolarized and hyperpolarized levels of the membrane potential, respectively. It was concluded, therefore, that bullfrog ganglion cells may be classed into three groups on the basis of the ionic mechanisms of their synaptically evoked and LH-RH-induced responses: (1) A large number of ganglion cells appear to exhibit responses that depend mainly on depression of voltage-dependent and voltage-independent K^+ currents (Type I mechanism) and, to a lesser degree, on increase of Na^+ and K^+ currents (Type II mechanism). (2) Another group of cells exhibited responses that depended only on the Type I mechanism. (3) Finally, a small group of cells exhibited responses dependent mainly on the Type II mechanism and, to a lesser extent, on the Type I mechanism. It is premature to assume that these various mechanisms reflect the existence of several types of receptors, since specific antagonists that could separate these receptors are not available.

III. MODULATION OF NONCHOLINERGIC EXCITATORY TRANSMISSION AND PEPTIDE-INDUCED EXCITATION BY CATECHOLAMINES AND CYCLIC NUCLEOTIDES

One of the prominent characteristics of the late slow EPSP of the bullfrog sympathetic ganglion is its extremely long-lasting time–course. Several factors may contribute to this long duration of the late slow EPSP and of the other slow PSPs such as the muscarinic slow EPSP, the noncholinergic slow EPSP, and also the slow inhibitory PSP (IPSP) (see Chapter 9) (Kuba and Koketsu, 1978; Hartzell, 1981): (1) the slow course of presynaptic events involved in transmitter release; (2) the time needed for diffusion of the transmitter from the nerve terminal to its receptor sites; (3) the slow kinetics of receptor–transmitter binding; and (4) the duration of intracellular events that generate the change in membrane properties, such as events involved in the action of the second messengers.

These latter processes may include those involving cyclic AMP (cAMP), which was proposed as a second messenger for the slow EPSP in sympathetic ganglia (McAfee and Greengard, 1972; Greengard, 1976) (but see

Kuba and Koketsu, 1978; Nishi *et al.*, 1978; Hartzell, 1981; Dun and Karczmar, 1978) (see Chapters 9 and 13); it must be remembered in this context that cAMP is widely known to be a second messenger in peptide actions (Labrie *et al.*, 1975). A second messenger may also be involved as a chemical mediator in the production of the late slow EPSP or the action of LH-RH or both. In this section, we will describe our recent studies focusing on the relationship between cAMP and the late slow EPSP or the LH-RH-induced response or both (Ushijima *et al.*, 1980; Nishi and Katayama, 1981); the modulation of synaptic transmission by second messengers will be reviewed in Chapters 9 and 13).

The noncholinergic LADs and the late slow EPSP of the bullfrog sympathetic ganglion were markedly enhanced and prolonged by superfusion of catecholamines at concentrations between 5 and 50 μM, while these drugs exerted no effects on the frequency of spontaneous activity or on the properties of the resting membrane. Even a short application (3 min) of catecholamines caused an enhancement of the LAD and the late slow EPSP that persisted for 1 hr (Figure 4A). The order of potency with regard to this augmentation was: isoprenaline > epinephrine > norepinephrine and dopamine. The facilitatory effect was prevented by propranolol (1 μM), but not by phentolamine (10 μM). There is some evidence that dopamine causes a long-lasting augmentation of the muscarinic slow EPSP in rabbit superior cervical ganglion cells, and it was suggested that these effects were brought about by an elevation of intrasomatic cAMP (Libet *et al.*, 1975; Kobayashi *et al.*, 1978; Libet, 1979). However, in the bullfrog sympathetic ganglion, the early afterdischarge and the slow EPSP were not significantly affected by any of the catecholamines.

Dibutyryl cAMP (1–50 μM) mimicked the actions of catecholamines on the LAD and the late slow EPSP of the bullfrog sympathetic ganglia without affecting spontaneous activity or the resting properties of the membrane of the ganglion cells. On the other hand, cAMP had no effect on the LAD or the late slow EPSP. This suggests that dibutyryl cAMP, which penetrates cell membranes better than cAMP (see Greengard, 1976), could enter the cytoplasm of pre- or postsynaptic elements or both, thereupon enhancing the LAD and the late slow EPSP.

The effects of catecholamines and dibutyryl cAMP on the LH-RH-induced response were similar to their action on the late slow EPSP; these effects were still present in Ca^{2+}-free/high-Mg^{++} solution. This suggests that the site of action of the catecholamines and dibutyryl cAMP is mainly postsynaptic. In this context, intracellular injection of cAMP into ganglion cells produced an obvious and persistent augmentation of the late slow EPSP (Figure 4B). However, dibutyryl cGMP applied by superfusion and cGMP applied by intracellular injection did not change the LAD, the late slow EPSP, or other electrical activities of bullfrog ganglion cells.

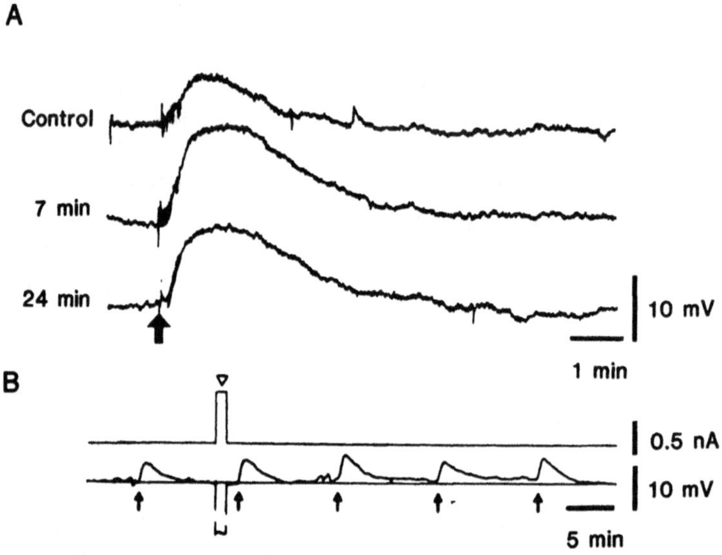

Figure 4. (A) Catecholamine-induced augmentation and prolongation of the late slow EPSP. The late slow EPSP was triggered by tetanic stimulation (10 Hz for 20 sec, indicated by the arrow). The records show the late slow EPSP before (control) superfusion and 7 and 24 min after the end of a 3-min superfusion with isoprenaline (50 μM). Isoprenaline itself had no effect on the resting membrane potential. The augmentation and prolongation of the late slow EPSP in this cell became more marked after a 7-min washing and gradually diminished after a 24-min washing. (B) Intracellular injection of cAMP augmented the late slow EPSP without changing the membrane potential. The upper trace shows the current for electrophoretic injection of cAMP [0.9 nA for 1 min (△)]. The lower trace is the intracellular record of the late slow EPSP that was evoked by tetanic stimulation (10 Hz for 10 sec, indicated by arrows). Augmentation and prolongation of the late slow EPSP appeared immediately after the cAMP injection and lasted for more than 30 min.

These observations are in agreement with the reports that adrenoceptor agonists are capable of elevating cAMP in several preparations (cf. Muir and Templeton, 1976; Horn and McAfee, 1977; Greengard, 1976; E. J. Nestler *et al.*, 1984). It may be hypothesized that activation of these receptors and resulting generation of cAMP may modulate noncholinergic excitatory transmission. However, it has not been determined whether LH-RH can elevate somatic cAMP in sympathetic ganglia. The presynaptic mechanism also appears plausible, if catecholamines and cAMP increase the release of transmitter that generates the late slow EPSP. Indeed, pretreatment with epinephrine and dibutyryl cAMP caused a sustained facilitation of quantal release of acetylcholine for 2 hr in the bullfrog sympathetic ganglion (Kuba *et al.*, 1981), and this presynaptic action was proposed to be linked with a cAMP system that leads to transmitter release (see Klein and Kandel, 1978; E. J. Nestler and Greengard, 1982).

The persistent modulatory actions of these drugs at either pre- or postsynaptic sites provide a possible model for neuronal plasticity, since they seem to be capable of facilitating synaptic transmission for long time periods. This modulatory action may be important physiologically in the case of the bullfrog sympathetic ganglion, since it contains catecholamine-containing small intensely fluorescent cells (Kojima *et al.*, 1978) [see Libet (1979) and Chapter 9], which may release catecholamines in response to preganglionic nerve activity. It must be remembered, however, that since catecholamines and cAMP or dibutyryl cAMP did not affect spontaneous ganglionic activity or the properties of the resting membrane of the bullfrog sympathetic ganglion (Nishi and Katayama, 1981) (see also Kuba and Koketsu, 1978), cAMP does not appear to be directly involved in the generation of the late slow EPSP by the noncholinergic transmitter.

IV. SUMMARY

Noncholinergic EPSPs can be evoked in amphibian and mammalian sympathetic ganglia. They have a very slow time–course; thus, they were named the "late slow EPSP" in the case of the lumbar sympathetic ganglion of the bullfrog and the "Noncholinergic slow EPSP" in the case of the inferior mesenteric ganglion of the guinea pig. Several peptides may be considered as possible neurontransmitters generating the noncholinergic EPSPs. Recent investigations suggest that LH-RH or an LH-RH-like peptide is the transmitter for the late slow EPSP and that substance P is the transmitter for the noncholinergic slow EPSP. Several ionic mechanisms appear to be involved in the generation of the noncholinergic EPSPs and the peptide-induced responses. Voltage-clamp analysis of the late slow EPSP and the LH-RH-induced response in the sympathetic ganglion of the bullfrog demonstrated that at least two different mechanisms are involved in generating these potentials, an increase in Na^+ and K^+ conductances and a decrease in K^+ conductance. A complex ionic mechanism that may include increase in Na^+ and decrease in K^+ permeability seems to underlie the noncholinergic EPSP of the mammalian inferior mesenteric ganglion.

A short superfusion with catecholamines and dibutyryl cAMP caused a sustained augmentation of the late slow EPSP and the LH-RH-induced response in bullfrog ganglia. This provides a model for neuronal plasticity that may be linked with the intracellular cAMP and that is activated by endogenous catecholamines.

The functional significance of the excitatory peptidergic transmission system in the central nervous system (CNS) is not clear. In this context, since the substance-P-containing fibers in the inferior mesenteric ganglia may actually be the collateral branches of primary sensory neurons of the dorsal root ganglia (Dalsgaard *et al.*, 1982; Matthews and Cuello, 1982), the noncholinergic excitatory potential may represent the transmission of sensory signals from the gastrointestinal tract to these ganglia; this provides a mechanism for a peripheral reflex regulation of gastrointestinal activity (Dun and Jiang, 1982; King and Szurszewski, 1984). The slowly developing peptidergic process may affect spatial and temporal summation by interacting with cholinergic or catecholaminergic processes in the ganglia. Detailed analysis of these interactions in the CNS is technically difficult; thus, the peripheral autonomic ganglia constitute a most useful model analysis of the ionic mechanism and physiological functions of the peptides.

REFERENCES

Adams, P. R., and Brown, D. A.: Luteinizing hormone-releasing factor and muscarinic agonists act on the same voltage-sensitive K^+ current in bullfrog sympathetic neurones. *Br. J. Pharmacol.* **68**:353–355 (1980).

Adams, P. R., Brown, D. A., and Jones, S. W.: Substance P inhibits the M-current in bullfrog sympathetic neurones. *Br. J. Pharmacol.* **79**:330–333 (1983).

Akasu, T., Kojim, M., and Koketsu, K.: Luteinizing hormone-releasing hormone modulates nicotinic ACh-receptor sensitivity in amphibian cholinergic transmission. *Brain Res.* **279**:347–351 (1983).

Alkadhi, K. A., and McIsaac, R. J.: Non-nicotinic ganglionic transmission during partial ganglionic blockade with chlorisondamine. *Fed. Proc. Fed. Am. Soc. Exp. Biol.* **30**:655 (1971).

Ashe, J. H., and Libet, B.: Orthodromic production of non-cholinergic slow depolarizing response in the superior cervical ganglion of the rabbit. *J. Physiol. (London)* **320**:333–346 (1981).

Brown, D. A., and Adams, P. R.: Muscarinic suppression of a novel voltage-dependent K-current in a vertebrate neurone. *Nature (London)* **283**:673–676 (1980).

Brown, D. A., Constanti, A., and Marsh, S.: Angiotensin mimics the action of muscarinic agonists on rat sympathetic neurones. *Brain Res.* **193**:614–619 (1980).

Chen, S. S.: Late contraction of nictitating membrane of the dog. *Am. J. Physiol.* **217**:1205–1210 (1969).

Chen, S. S.: Transmission in superior cervical ganglion of the dog after cholinergic suppression. *Am. J. Physiol.* **221**:209–213 (1971).

Chen, S. S.: Late discharges in dog's sympathetic ganglia. *Can. J. Physiol. Pharmacol.* **50**:263–269 (1972).

Constanti, A., and Brown, D. A.: M-current in voltage-clamped mammalian sympathetic neurones. *Neurosci. Lett.* **24**:289–294 (1981).

Dalsgaard, C. J., Hökfelt, T., Elfvin, L.-G., Skirboll, L., and Emson, P.: Substance P-containing primary sensory neurons projecting to the inferior mesenteric ganglion: Evidence from combined retrograde tracing and immunohistochemistry. *Neuroscience* **7**:647–654 (1982).

Dockray, G. J., and Hutchson, J. B.: Cholecystokinin octapeptide in guinea pig ileum myenteric plexus: Localization and biological action. *J. Physiol. (London)* **300**:28p–29p (1980).

Dun, N. J.: Ganglionic transmission: Electrophysiology and pharmacology. *Fed. Proc. Fed. Am. Soc. Exp. Biol.* **39**:2982–2989 (1980).

Dun, N. J.: Peptide hormones and transmission in sympathetic ganglia, in: *Autonomic Ganglia* (L.-G. Elfvin, ed.), pp. 341–366, Wiley, New York (1983).

Dun, N. J., and Jiang, Z. G.: Noncholinergic excitatory transmission in the inferior mesenteric ganglia of the guinea pig: Possible mediation by substance P. *J. Physiol. (London)* **325**:145–159 (1982).

Dun, N. J., and Karczmar, A. G.: Involvement of an interneuron in the generation of the slow inhibitory potential in mammalian sympathetic ganglia. *Proc. Natl. Acad. Sci. U.S.A.* **75**:4029–4032 (1978).

Dun, N. J., and Karczmar, A. G.: Actions of substance P on sympathetic neurons. *Neuropharmacology* **18**:215–218 (1979).

Dun, N. J., and Ma, R. C.: Slow non-cholinergic excitatory potentials in neurones of the guinea pig coeliac ganglia. *J. Physiol. (London)* **351**:47–60 (1984).

Dun, N. J., and Minota, S.: Effects of substance P on neurones of the inferior mesenteric ganglia of the guinea pig. *J. Physiol. (London)* **321**:259–271 (1981).

Dun, N. J., Nishi, S., and Karczmar, A. G.: An analysis of the effect of angiotensin II on mammalian ganglion cells. *J. Pharmacol.* Exp. *Ther.* **204**:669–675 (1978).

Dun, N. J., Kiraly, M., and Ma, R. C.: Evidence for a serotonin-mediated slow excitatory potential in the guinea pig coeliac ganglia. *J. Physiol. (London)* **351**:61–76 (1984).

Fuxe, K., Ganten, D., Hökfelt, T., and Bolme, P.: Immunohistochemical evidence for the existence of angiotensin II-containing nerve terminals in the brain and spinal cord in the rat. *Neurosci. Lett.* **2**:229–234 (1976).

Greengard, P.: Possible role for cyclic nucleotides and phosphorylated membrane proteins in postsynaptic actions of neurotransmitters. *Nature (London)* **260**:101–108 (1976).

Hartzell, H. C.: Mechanisms of slow postsynaptic potentials. *Nature (London)* **291**:101–108 (1981).

Hökfelt, T., Elfvin, L.-G., Elde, R., Schulzberg, M., Goldstein, M., and Luft, R.: Occurrence of somatostatin-like immunoreactivity in some peripheral sympathetic noradrenergic neurons. *Proc. Natl. Acad. Sci. U.S.A.* **74**:3587–3591 (1977a).

Hökfelt, T., Elfvin, L.-G., Schultzberg, M., Fuxe, K., Said, S. I., Mutt, V., and Goldstein, M.: Immunohistochemical evidence of vasoactive intestinal polypeptide-containing neurons and nerve fibers in sympathetic ganglia. *Neuroscience* **2**:885–896 (1977b).

Hökfelt, T., Elvin, L.-G., Schulzberg, M., Goldstein, M., and Nilsson, G.: On the occurrence of substance P-containing fibers in sympathetic ganglia: Immunohistochemical evidence. *Brain Res.* **132**:29–41 (1977c).

Hökfelt, T., Johansson, O., Ljungdahl, A., Lundberg, J. M., and Schultzberg, M.: Peptidergic neurones. *Nature (London)* **284**:515–521 (1980).

Hökfelt, T., Johanson, O., and Goldstein, M.: Chemical anatomy of the brain. *Science* **225**:1326–1334 (1984).

Horn, J. P., and McAfee, D. A.: Modulation of cyclic nucleotide levels in peripheral nerve without effect on resting or compound action potentials. *J. Physiol. (London)* **269**:753–766 (1977).

Jan, L. Y., and Jan, Y. N.: Peptidergic transmission in sympathetic ganglia of the frog. *J. Physiol. (London)* **327**:219–246 (1982).

Jan, L. Y., Jan, Y. N., and Brownfield, M. S.: Peptidergic transmitters in synaptic boutons of sympathetic ganglia. *Nature (London)* **288**:380–382 (1980).

Jan, Y. N., Jan, L. Y., and Kuffler, S. W.: A peptide as a possible transmitter in sympathetic ganglia of the frog. *Proc. Natl. Acad. Sci. U.S.A.* **76**:1501–1505 (1979).

Jan, Y. N., Jan, L. Y., and Kuffler, S. W.: Further evidence for peptidergic transmission in sympathetic ganglia. *Proc. Natl. Acad. Sci. U.S.A.* **77**:5008– 5012 (1980).

Johnson, S. M., Katayama, Y., and North, R. A.: Slow synaptic potentials in neurones of the myenteric plexus. *J. Physiol. (London)* **310**:505–516 (1980).

Johnson, S. M., Katayama, Y., Morita, K., and North, R. A.: Mediators of the slow synaptic potentials in the myenteric plexus of the guinea pig ileum. *J. Physiol. (London)* **320**:175–186 (1981).

Katayama, Y., and Nishi, S.: The ionic mechanism of the late slow EPSP in amphibian sympathetic ganglion cells. *Proc. Int. Union Physiol. Sci.* **13**:171 (1977).

Katayama, Y., and Nishi, S.: Actions of enkephalins on single neurons in ciliary ganglia. in: *Advances in Endogenous and Exogenous Opioids* (H. Takagi and E. J. Simon, eds.), pp. 205–207, Kohdansha, Tokyo (1981).

Katayama, Y., and Nishi, S.: Voltage-clamp analysis of peptidergic slow depolarizations in bullfrog sympathetic ganglion cells. *J. Physiol. (London)* **333**:305–313 (1982).

Katayama, Y., and Nishi, S.: Sites and mechanisms of actions of enkephalin in the feline parasympathetic ganglion. *J. Physiol. (London)* **351**:111–121 (1984).

Katayama, Y., and North, R. A.: Does substance P mediate slow synaptic excitation within the myenteric plexus? *Nature (London)* **274**:387–388 (1978).

Katayama, Y., and North, R. A.: The action of somatostatin on neurones of the myenteric plexus of the guinea pig ileum. *J. Physiol. (London)* **303**:315–323 (1980).

Katayama, Y., North, R. A., and Williams, J. T.: The action of substance P on neurons of the myenteric plexus of the guinea pig small intestine. *Proc. R. Soc. London* **206**:191–208 (1979).

Katayama, Y., Inokuchi, H., and Nishi, S.: Voltage-clamp study of the late slow excitatory postsynaptic current in bullfrog sympathetic ganglion cells. *Neurosci. Lett.* **6**(Suppl.):S63 (1981).

King, B. F., and Szurszewski, J. H.: Mechanoreceptor pathways from the distal colon to the autonomic nervous system in the guinea pig. *J. Physiol. (London)* **350**:93–107 (1984).

Klein, M., and Kandel, E. R.: Presynaptic modulation of voltage-dependent Ca^{2+} current: Mechanism for behavioral sensitization in *Aplysia californica*. *Proc. Natl. Acad. Sci. U.S.A.* **75**:35L2–35L6 (1978).

Kobayashi, H., Hashiguchi, T., and Ushiyama, N.: Post-synaptic modulation of excitatory process in sympathetic ganglia by cyclic AMP. *Nature (London)* **271**:268–270 (1978).

Kojima, H., Anraku, S., Onogi, K., and Ito, R.: Histochemical studies on two types of cells containing catecholamine in sympathetic ganglia of the bullfrog. *Experientia* **34**:92–93 (1978).

Kondo, H., and Yui, R.: Enkephalin-like immunoreactivity in the SIF cells of sympathetic ganglia of frogs. *Biomed. Res.* **2**:338–340 (1981).

Kondo, H., Katayama, Y., and Yui, R.: On the occurrence and physiological effect of somatostatin in the ciliary ganglion of cats. *Brain Res.* **247**:141–144 (1982).

Konishi, S., Tsunoo, A., and Otsuka, M.: Enkephalins presynaptically inhibit cholinergic transmission in sympathetic ganglia. *Nature (London)* **282**:515–516 (1979a).

Konishi, S., Tsunoo, A., and Otsuka, M.: Substance P and noncholinergic excitatory synaptic transmission in guinea pig sympathetic ganglia. *Proc. Jpn. Acad. Ser. B* **55**:525–530 (1979b).

Konishi, S., Tsunoo, A., Yanaihara, N., and Otsuka, M.: Peptidergic excitatory and inhibitory synapses in mammalian sympathetic ganglia: Roles of substance P and enkephalin. *Biomed. Res.* **1**:528–536 (1980).

Kuba, K., and Koketsu, K.: Synaptic events in sympathetic ganglia. *Prog. Neurobiol.* **11**:77–169 (1978).

Kuba, K., Kato, E., Kumamoto, E., Koketsu, K., and Hirai, K.: Sustained potentiation of transmitter by adrenaline and dibutyryl cyclic AMP in sympathetic ganglia. *Nature (London)* **291**:654–656 (1981).

Kuffler, S. W.: Slow synaptic responses in autonomic ganglia and the pursuit of a peptidergic transmitter. *J. Exp. Biol.* **89**:257–286 (1980).

Kuffler, S. W., and Sejnowski, T. J.: Peptidergic and muscarinic excitation at amphibian sympathetic synapses. *J. Physiol. (London)* **341**:257–278 (1983).

Labrie, F., Borgeat, P., Lemay, A., Lemaire, S., Barden, N., Drouin, J., Jolicoeur, P., and Belanger, A.: Role of cyclic AMP in the action of hypothalamic regulatory hormones. in: *Adv. Cyclic Nucleotide Res.* **5**:787–801 (1975).

Larsson, L. I., and Rehfeld, I. F.: Localization and molecular heterogeneity of cholecystokinin in the central and peripheral nervous system. *Brain Res.* **165**:201–218 (1979).

Lewis, G. P., and Reit, E.: The action of angiotensin and bradykinin on the superior cervical ganglion of the cat. *J. Physiol. (London)* **197**:538–553 (1965).

Libet, B.: Slow postsynaptic actions in ganglionic functions, in: *Integrative Functions of the Autonomic Nervous System* (C. McC. Brooks, K. Koizumi, and A. Sato, eds.), pp. 197–222, University of Tokyo Press, Tokyo (1979).

Libet, B., Kobayashi, H., and Tanaka, T.: Synaptic coupling into the production and storage of a neuronal memory trace. *Nature (London)* **258**:155–157 (1975).

Matthews, M. R., and Cuello, A. C.: Substance P-immunoreactive peripheral branches of sensory neurons innervate guinea pig sympathetic neurons. *Proc. Natl. Acad. Sci. U.S.A.* **79**:1668–1672 (1982).

McAfee, D. A., and Greengard, P.: Adenosine 3′,5′-monophosphate: Electrophysiological evidence for a role in synaptic transmission. *Science* **178**:310–312 (1972).

Mihara, S., Katayama, Y., and Nishi, S.: The action of substance P on neurons in the submucous plexus of the guinea pig cecum. *Neurosci. Lett.* **9**(Suppl.):S91 (1982).

Mihara, S., Katayama, Y., and Nishi, S.: Slow synaptic potentials and chemosensitivities of neurons in the isolated submucous plexus of the guinea pig cecum. *Proc. Int. Union Physiol. Sci.* **15**:454 (1983).

Minota, S., Dun, N. J., and Karczmar, A. G.: Substance P-induced depolarization in sympathetic neurons: Not simple K^+-inactivation. *Brain Res.* **216**:224–228 (1981).

Morita, K., North, R. A., and Katayama, Y.: Evidence that substance P is a neurotransmitter in the myenteric plexus. *Nature (London)* **287**:151–152 (1980).

Muir, T. C., and Templeton, D.: The role of cyclic 3′,5′-adenosine monophosphate (cyclic AMP) in the ability of sympathetic nerve stimulation to enhance growth and secretion in rat salivary glands *in vivo*. *J. Physiol. (London)* **259**:47–61 (1976).

Neild, T. O.: Slowly-developing depolarization of neurones in the guinea pig inferior mesenteric ganglion following repetitive stimulation of the preganglionic nerves. *Brain Res.* **140**:231–239 (1978).

Nestler, E. J., and Greengard, P.: Distribution of protein I and regulation of its state of phosphorylation in the rabbit superior cervical ganglion. *J. Neurosci.* **2**:1011–1023 (1982).

Nestler, E. J., Walaas, S. I., and Greengard, P.: Neuronal phosphoproteins. *Science* **225**:1357–1364 (1984).

Nicoll, R. A., Schenker, C., and Leeman, S. E.: Substance P as a transmitter candidate. *Annu. Rev. Neurosci.* **3**:227–268 (1980).

Nishi, S.: Electrogenesis of muscarinic and noncholinergic slow EPSP's of amphibian sympathetic ganglion cells (in Russia), in: *Interneuronal Transmission in the Autonomic Nervous System* (P. Kostyuk, ed.), pp. 112–135, Nankova Dumka, Kiev (1973).

Nishi, S.: Ganglionic transmission, in: *The Peripheral Nervous System* (J. I. Hubbard, ed.), pp. 225–255, Plenum Press, New York (1974).

Nishi, S., and Katayama, Y.: The non-cholinergic excitatory transmission in sympathetic ganglia, in: Advances in Physiological Sciences, Vol. 4, *Physiology of Excitable Membranes* (J. Salanki, ed.), pp. 322–327, Pergamon Press, New York (1981).

Nishi, S., and Koketsu, K.: Early and late after-discharges of amphibian sympathetic ganglion cells. *J. Neurophysiol.* **31**:109–121 (1968).

Nishi, S., Soeda, H., and Koketsu, K.: Studies on sympathetic B and C neurons and patterns of preganglionic innervation. *J. Cell. Comp. Physiol.* **66**:19–21 (1965).

Nishi, S., Karczmar, A. G., and Dun, N. J.: Physiology and pharmacology of ganglionic synapses as models for central transmission, in: Advances in Pharmacology and Physiology, Vol. 2, *Neurotransmitters* (P. Simon, ed.), pp. 69–85, Pergamon Press, Oxford and New York (1978).

Nishi, S., Katayama, Y., Nakamura, J., and Ushijima, H.: A candidate substance for the chemical transmitter mediating the late slow EPSP in bullfrog sympathetic ganglia. *Biomed. Res.* **1**(Suppl.):144–148 (1980).

North, R. A.: Opiates, opioid peptides and single neurons. *Life Sci.* **24**:1527–1546 (1979).

North, R. A., and Williams, J. T.: Enkephalin inhibits firings of myenteric neurons. *Nature (London)* **264**:460–461 (1976).

North, R. A., Katayama, Y., and Williams, J. T.: On the mechanism and site of action of enkephalin on single myenteric neurons. *Brain Res.* **165**:67–77 (1979).

Robinson, S. E., Schwartz, J. P., and Costa, E.: Substance P in the superior cervical ganglion and the submaxillary gland of the rat. *Brain Res.* **182**:11–17 (1980).

Schulman, J. A., and Weight, F. F.: Synaptic transmission: Long-lasting potentiation by a postsynaptic mechanism. *Science* **194**:1437–1439 (1976).

Schultzberg, M., Hökfelt, T., Terenius, L., Elfvin, L.-G., Lundberg, J. M., Brandt, J., Elde, R. P., and Goldstein, M.: Enkephalin immunoreactive nerve fibers and cell bodies in sympathetic ganglia of the guinea pig and rat. *Neuroscience* **4**:249–270 (1979).

Sejnowksi, T. J.: Peptidergic synaptic transmission in sympathetic ganglia. Fed. Proc. Fed. Am. Soc. Exp. Biol. **41**:2923–2928 (1982).

Surprenant, A.: Slow excitatory synaptic potentials recorded from neurones of guinea pig submucous plexus. *J. Physiol. (London)* **351**:343–362 (1984).

Tsunoo, A., Konishi, S., and Otsuka, M.: Substance P as an excitatory transmitter of primary afferent neurons in guinea pig sympathetic ganglia. *Neuroscience* **7**:2025–2037 (1982).

Ushijima, H., Katayama, Y., and Nishi, S.: The effect of catecholamines on the slow ganglionic transmission. *J. Physiol. Soc. Jpn.* **42**:327 (1980).

Weems, W. A., and Szurszewski, J. H.: An intracellular analysis of some intrinsic factors controlling neural output from inferior mesenteric ganglion of guinea pigs. *J. Neurophysiol.* **41**:305–321 (1978).

Weight, F. F., Schulman, J. A., Smith, P. A., and Busis, N. A.: Long-lasting synaptic potentials and modulation of synaptic transmission. *Fed. Proc. Fed. Am. Soc. Exp. Biol.* **38**:2084–2094 (1979).

Williams, J. T., and North, R. A.: Effects of endorphins on single myenteric neurons. *Brain Res.* **165**:57–65 (1979a).

Williams, J. T., and North, R. A.: Vasoactive intestinal polypeptide excites neurons of the myenteric plexus. *Brain Res.* **175**:174–177 (1979b).

Wood, J. D., and Mayer, C. J.: Intracellular study of electrical activity of Auerbach's plexus in guinea pig small intestine. *Pfluegers Arch.* **374**:265–275 (1978a).

Wood, J. D., and Mayer, C. J.: Slow synaptic excitation mediated by serotonin in Auerbach's plexus. *Nature (London)* **276**:836–837 (1978b).

Wood, J. D., and Mayer, C. J.: Intracellular study on tonic-type enteric neurons in guinea pig small intestine. *J. Neurophysiol.* **42**:569–581 (1979).

Wouters, W., and Van den Bercken, J.: Hyperpolarization and depression of slow synaptic inhibition by enkephalin in frog sympathetic ganglion. *Nature (London)* **277**:53–54 (1979).

Wouters, W., and Van den Berken, J.: Effect of met-enkephalin on slow synaptic inhibition in frog sympathetic ganglion. *Neuropharmacology* **19**:237–243 (1980).

9

Inhibitory Transmission: Slow Inhibitory Postsynaptic Potential

K. KOKETSU

I. INTRODUCTION

J. C. Eccles (1943) was the first to show that a slow surface positive (P) potential could be recorded, when preganglionic nerves were stimulated, from mammalian sympathetic ganglia treated with curare. This P potential was further investigated in turtle sympathetic ganglia by Laporte and Lorente de No (1950). In 1961, R. M. Eccles and Libet (1961) made a systematic analysis of the synaptic mediation of the P potential recorded from mammalian sympathetic ganglia. They suggested that the P potential might be produced by the action of a catecholamine released from chromaffin cells [subsequently referred to as small intensely fluorescent (SIF) cells (see below and Chapters 12 and 13)] when these neurons are stimulated by acetylcholine (ACh) liberated from preganglionic nerve terminals. The P potential was subsequently recorded intracellularly as a hyperpolarizing response of ganglion cells (Tosaka *et al.*, 1968). Thus, the

K. KOKETSU • Department of Physiology, Kurume University School of Medicine, Kurume, Japan.

P potential can be referred to as the slow inhibitory postsynaptic potential (slow IPSP).*

Since the work of R. M. Eccles and Libet (1961), a great number of experiments have been carried out in many laboratories to determine the neurotransmitter directly responsible for the generation of the slow IPSP and also to clarify its ionic mechanism. However, there are many inconsistencies among experimental results that have been obtained in different laboratories, as well as many disagreements with regard to the interpretation of the data.

There are a number of reasons for these discrepancies; they concern the difficulties inherent in the analysis of the nature of the slow IPSP. These reasons are as follows:

1. Although the fast excitatory postsynaptic potential (fast EPSP) is completely eliminated in the presence of nicotinic blockers that are used in studies of the slow IPSP, the slow IPSP is inevitably superimposed on the slow EPSP, which may be produced before or almost simultaneously with the slow IPSP. Unfortunately, it is impossible to selectively block the slow EPSP, leaving only the slow IPSP. This is indeed an extremely disappointing situation with respect to the analysis of the slow IPSP, and particularly of its ionic mechanism.

2. It is often very difficult to record the slow IPSP by means of an intracellular microelectrode as compared with its extracellular recording. In fact, the size of the slow IPSP recorded intracellularly is usually too small to make a reliable analysis of its nature. This suggests that the site of receptors responsible for the generation of the slow IPSP is remote from the cell body; presumably, these receptors are localized in the membrane of initial axonal segments (IAS) of ganglion cells.

3. In the case of amphibian and mammalian sympathetic ganglia, the results from this laboratory indicate that the slow IPSP is generated in both ganglionic B and C cells (cf. Kuba and Koketsu, 1978). Some investigators (e.g., Libet and Tosaka, 1970), however, believe that the slow IPSP in amphibian sympathetic ganglia is produced only in ganglionic C cells.

4. The neurotransmitter that is directly responsible for the generation of the slow IPSP is not fully identified as yet. Furthermore, the ionic mechanism of the slow IPSP appears to be quite complex and consists of a number of mechanisms. If this is the case, experimental results obtained for the analysis of the slow IPSP must be carefully interpreted.

*Generally, the terms IPSP and the P potential are used when one wishes to refer to the inhibitory events recorded intra- and extracellularly, respectively. In this chapter, however, only the term IPSP will be used, but pertinent description will make it clear which recording method was employed in each case.

5. The nature of the slow IPSP recorded from curarized ganglia is somehow different from that of the slow IPSP recorded from nicotinized ganglia. This is due to different modes of action on the ganglion cell of these two nicotinic blockers.

II. RECORDING OF THE SLOW INHIBITORY POSTSYNAPTIC POTENTIAL

The slow IPSP can be easily recorded extracellularly from sympathetic ganglia. A most convenient and reliable method for the extracellular recording of the slow IPSP is the sucrose-gap method (Koketsu and Nishi, 1967; Kosterlitz *et al.*, 1968). To record the slow IPSP, the fast EPSP must be completely eliminated. In most experiments, nicotine or D-tubocurarine (D-TC) has been used to block the fast EPSP in both amphibian and mammalian sympathetic ganglia. Although it is not impossible to record the slow IPSP with a single preganglionic volley, repetitive stimulation (10–50/sec) is generally required to obtain a clear slow IPSP, as shown in the case of an amphibian preparation in Figure 1. This figure shows that the amplitude of the slow IPSP produced by preganglionic B plus C nerve stimulation is much larger than that produced by preganglionic B stimulation alone; it also illustrates the generalization that the slow IPSP is always superimposed on the slow EPSP. The amplitude of the slow IPSP of bullfrog sympathetic ganglia recorded by the sucrose-gap method in the presence of nicotine was much larger than that recorded in the presence of D-TC.

Although the slow IPSP can be recorded intracellularly from am-

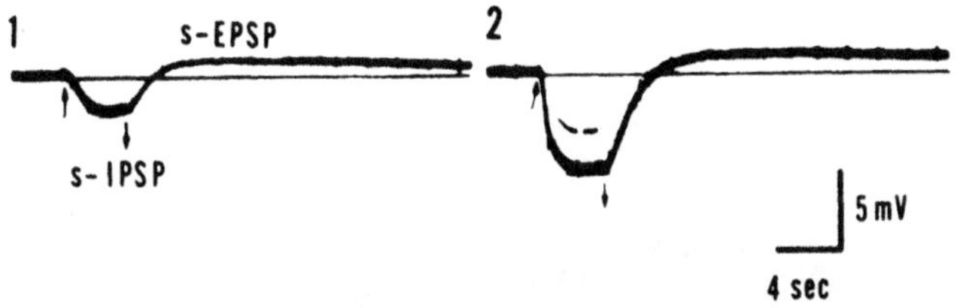

Figure 1. Slow IPSP followed by slow EPSP recorded from a nicotinized bullfrog ganglion by the sucrose-gap method. Records (1) and (2) were generated by trains of repetitive (30/sec) preganglionic B-nerve and B-plus-C-nerve stimulations, respectively. Arrows indicate the duration of stimulation.

phibian or mammalian sympathetic ganglion cells, its amplitude is generally very small (see, however, Dodd and Horn, 1983). As in the case of the extracellularly recorded slow IPSP or the IPSP recorded by means of the sucrose-gap method, or both, the amplitude of the slow IPSP recorded intracellularly from curarized ganglia was much smaller than that recorded from nicotinized ganglia. Intracellular recording of the slow IPSP from curarized amphibian B neurons is almost impossible (see Kuba and Koketsu, 1976), whereas a slow IPSP of small size can be recorded from curarized amphibian C cells (Tosaka *et al.*, 1968); this difference between B and C cells may be due to the difference in the sizes of these two types of neurons. To record intracellularly the slow IPSP from either B or C cells of nicotinized amphibian ganglia is relatively easy [see, however, Libet (1970) and below], although its amplitude is still small (Koketsu and Nakamura, 1975); (see Figure 2).

Libet and his collaborators reported that although the slow IPSP of mammalian sympathetic ganglia can be recorded intracellularly from both B and C cells, the slow IPSP of amphibian sympathetic ganglia treated with either D-TC or dihydro-β-erythroidine (DHE) can be recorded intracellularly only from C cells (Libet, 1970), and they concluded that the slow IPSP of amphibian sympathetic ganglia is produced only in C cells. They have also claimed that the slow EPSP is produced only from B cells (see Libet, 1970). On the other hand, Koketsu and Nakamura (1975) presented evidence indicating that the slow IPSP of amphibian sympathetic ganglion cells can be recorded from both B and C cells.

It must be emphasized that it is more difficult to record the slow IPSP by means of intracellular microelectrodes inserted into the soma of the ganglion cells than to record it extracellularly by the sucrose-gap method. The most reasonable explanation for this difference is that the receptors responsible for the generation of the slow IPSP are located at the membrane of the IAS of a ganglion cell. If they are, spatial decrement of the slow IPSP would account for the small amplitude of an intracellularly recorded slow IPSP; on the other hand, a full-size potential would be recorded by the sucrose-gap method even if it were generated at the IAS.

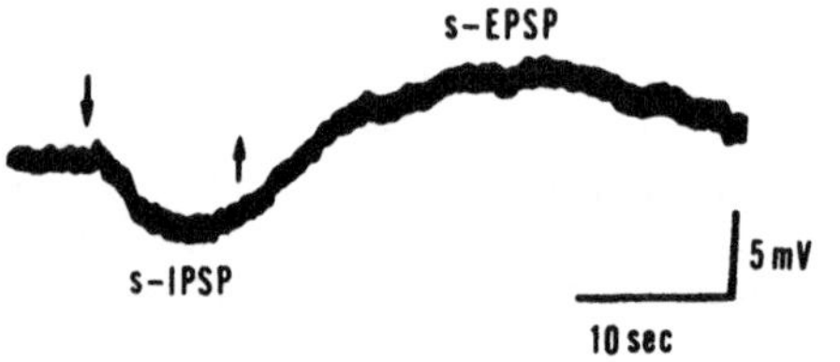

Figure 2. Slow IPSP followed by slow EPSP recorded intracellularly from a nicotinized bullfrog ganglion (B cell). These potentials were produced by a train of repetitive (30/sec) preganglionic B-nerve stimulations.

III. SYNAPTIC MEDIATION OF THE SLOW INHIBITORY POSTSYNAPTIC POTENTIAL

Neither the synaptic pathway nor the neurotransmitter involved in generation of the slow IPSP in sympathetic ganglia is entirely clarified as yet. The possibility that the synaptic pathways for the slow IPSP of amphibian and mammalian sympathetic ganglia may be different seems to be too speculative to be accepted. It is generally proposed that the slow IPSP is elicited by a direct action of either ACh or a catecholamine. ACh is undoubtedly released from preganglionic nerve terminals, but there are two possible sources for catecholamines, namely, SIF cells (chromaffin cells); (see Chapters 12 and 13) and the ganglion cells.

On the basis of the results of a pharmacological analysis, R. M. Eccles and Libet (1961) proposed a disynaptic mediation of the slow IPSP. They proposed the following sequence of events: Chromaffin cells of the ganglion are stimulated via their muscarinic receptors by ACh released from preganglionic nerve terminals; following this stimulus, the chromaffin cells release adrenergic substances that stimulate the adrenergic postganglionic receptors of ganglion cells and thereby generate the slow IPSP of these cells. On the other hand, the possibility that the slow IPSP may be generated by a direct muscarinic action of ACh was suggested by others (Volle and Hancock, 1970; Weight and Padjen, 1973a,b; Weight and Weitsen, 1977; Horn and Dodd, 1981; Weight, 1983; Dodd and Horn, 1983; Cole and Shinnick-Gallagher, 1984).

A. Disynaptic Mediation

When ACh is applied directly to the ganglion in the presence of either nicotine or D-TC, the extra- or intracellular recording demonstrates that the ganglion cells generate a hyperpolarizing response followed by a depolarizing response. These responses are caused by the muscarinic actions of ACh and correspond to the slow IPSP and the slow EPSP, respectively; both these potential changes are usually completely wiped out by the action of atropine. If these responses are generated by a direct action of ACh, they can be expected to be present after the preparations are soaked in the low-Ca^{2+}/high-Mg^{2+} solution that arrests the release of all transmitters. Alternatively, if these responses are generated via the action of a transmitter that is released by ACh, they would be eliminated in the low-Ca^{2+}/high-Mg^{2+} solution. Libet and his collaborators tested these alternatives in mammalian sympathetic ganglia (Libet, 1970). They found that when a ganglion was treated with a low-Ca^{2+}/high-Mg^{2+} solution, ACh or the muscarinic drug bethanechol generated only a depolarizing response, and no hyperpolarizing response. This result suggested that the depolarizing but not the hyperpolarizing response to ACh is generated by a direct action of ACh, the slow IPSP being then produced by a disynaptic pathway.

If chromaffin cells in ganglia play the role of internuncial neurons, the neurotransmitter that is directly responsible for the generation of the slow IPSP must be a catecholamine, as originally suggested by R. M. Eccles and Libet (1961). This concept was supported by their experimental finding that the slow IPSP of mammalian sympathetic ganglia is depressed by α-adrenergic blockers, such as dibenamine. In subsequent studies, however, the action of adrenergic blockers on the slow IPSP of both mammalian and amphibian sympathetic ganglia was not clear-cut (see, for example, Volle and Hancock, 1970); (see also Chapter 13). Further supportive findings were provided more recently by means of intracellular techniques, since membrane hyperpolarization that exhibited pharmacological characteristics similar to those of the IPSP was elicited by iontophoresis of dopamine onto noncurarized ganglion cells and ACh onto curarized cells of isolated rabbit superior cervical ganglion; furthermore, ACh-induced hyperpolarization was blocked in low-Ca^{2+}/high-Mg^{2+} solution, as well as by tetrodotoxin (TTX) (Dun *et al.*, 1977; Dun and Karczmar, 1978; cf. Dun, 1980).

Additional experimental evidence supporting the notion of the existence of an adrenergic step in the synaptic pathway of the slow IPSP was presented by Libet and Tosaka (1970), who demonstrated that prolonged treatment of isolated mammalian sympathetic ganglia with a muscarinic agent, such as bethanechol, resulted in a selective, strong depression of the slow IPSP. They related this depression to the depletion of catecholamines stored in chromaffin cells; they appeared to demonstrate this depletion histochemically. Furthermore, they showed that after the slow IPSP of the ganglia was blocked by treatment with bethanechol, it could be almost fully restored following the incubation of the ganglia in a solution containing dopamine. Incubation with noradrenaline produced only a slight restoration of the IPSP; no restoration of the IPSP was obtained following incubation with adrenaline. On the basis of these experimental results, Libet and Tosaka (1970) concluded that the neurotransmitter that is directly responsible for the generation of the slow IPSP of mammalian sympathetic ganglion is dopamine (see also Kobayashi and Tosaka, 1983).

The disynaptic mediation of the slow IPSP seems to be supported by histochemical and electron-microscopic studies of SIF cells. Libet and Owman (1974) have reported, on the basis of electrophysiological and histofluorescent studies, that SIF cells of the rabbit sympathetic ganglion contained dopamine that was released when the preganglionic nerve was stimulated, and they have provided strong evidence that the slow IPSP of the rabbit sympathetic ganglion was generated by dopamine released from SIF cells. The possibility that SIF cells play the role of internuncial neurons is also supported by electron-microscopic studies that have demonstrated that chromaffin cells receive afferent innervation from preganglionic neurons and give rise to afferent synapses with postganglionic neurons (cf. Kuba and Koketsu, 1978) (see Chapter 2).

Disynaptic mediation of the slow IPSP was also postulated for amphibian sympathetic ganglia; again in this case, it was proposed that SIF cells function as internuncial neurons that release an adrenergic substance involved in generating the slow IPSP (Tosaka *et al.*, 1968). No conclusive experimental evidence to support this concept, however, has been provided by physiological or pharmacological analysis of the slow IPSP in amphibian ganglia. It must be also pointed out that bullfrog sympathetic ganglia seem to lack efferent synapses (Weight and Weitsen, 1977) (see also below).

Altogether, Libet's group assembled impressive evidence, particularly for the mammalian ganglion, in support of the early hypothesis of R. M. Eccles and Libet (1961) that the slow IPSP is generated by a disynaptic mechanism that involves the SIF cell.

B. Monosynaptic Mediation

There are several weaknesses in the hypothesis of disynaptic mediation of the IPSP, since several results are not consistent with Libet's scheme or lead to alternative explanations for Libet's findings. For instance, the depression of the IPSP induced by bethanechol may be not as selective as supposed by Libet and Tosaka (1970) and not related to catecholamine depletion; rather, it seemed to include the slow EPSP as well (cf. Dun, 1980; Dun and Karczmar, 1978). Thus, the effect of bethanechol may not depend on the depletion of catecholamines of the SIF cell but on desensitization of muscarinic receptors that would include those that generate the slow EPSP as well as those that generate the slow IPSP, whatever the location of the latter receptors may be (see below). In addition, when several mammalian species are compared with each other and with the bullfrog, it becomes apparent that the occurrence of SIF cells and their interneuronal nature, on the one hand, and the slow IPSP, on the other, are in general poorly correlated (Dun and Karczmar, 1980). In this context, certain data support the notion of the monosynaptic nature of the IPSP.

Indeed, it is of interest that in the case of amphibian sympathetic ganglia, the concept that a catecholamine is the neurotransmitter generating the slow IPSP has been challenged by Weight and his co-workers. Weight and Padjen (1973a,b) found that in the case of nicotinized bullfrog sympathetic ganglia, the muscarinic hyperpolarization induced by a direct application of ACh was still present in the low-Ca^{2+} Ringer's solution. Furthermore, in their hands, the slow IPSP was not significantly depressed by the action of adrenergic blockers, contrary to the findings of Libet (1970) reviewed earlier. On the basis of these observations, they concluded that the slow IPSP of amphibian sympathetic ganglia is generated by a direct muscarinic action of ACh that is released from preganglionic nerves. This proposal was supported by their histochemical and ultrastructural studies of SIF cells of bullfrog ganglia (Weight and Weitsen, 1977). According to these studies, SIF cells of amphibian sympathetic ganglia are chromaffin cells that are virtually identical to those present in the adrenal gland and that do not exhibit morphological characteristics of interneurons, such as the efferent and afferent synaptic connections (cf. Kondo, 1977) (see also Chapter 13).

Libet and Kobayashi (1974) reexamined the results obtained by Weight and Padjen (1973a,b) and made extensive analysis of muscarinic hyperpolarizing responses initiated by direct application of ACh bullfrog sympathetic ganglia treated with either nicotine or DHE. In curarized ganglia, treatment with a low-Ca^{2+}/high-Mg^{2+} solution abolished the hyperpolarizing component of the response to ACh and to methacholine and enlarged the depolarizing response, whereas in nicotinized ganglia, similar treatment with a low-Ca^{2+}/high-Mg^{2+} solution had little effect on the hyperpolarizing response to ACh, indicating that in the presence of nicotine, ACh elicits a hyperpolarization by a direct action on the ganglion cells, as reported by Weight and Padjen (1973a,b). On the basis of these results, Libet (Libet and Kobayashi, 1974) concluded that in amphibian as in mammalian ganglia, the synaptic pathway for the slow IPSP includes the SIF cells, which release a catecholamine; the latter acts directly on the ganglion cell to elicit the hyperpolarizing response, viz., the slow IPSP. Libet further proposed that nicotine alters cellular conditions so that an additional, direct hyperpolarizing action of ACh on the ganglion cell becomes manifest. In connection with this proposal, Kobayashi and Libet (1974) reported that the slow IPSP can be recorded from both B and C cells when amphibian sympathetic ganglia are treated with nicotine, although under physiological conditions or in the case of curarized ganglia, the slow IPSP can be generated only from C cells.

However, Cole and Shinnick-Gallagher (1984) presented direct evidence indicating a monosynaptic mediation of the slow IPSP in mammalian sympathetic ganglia. They found that ACh-induced hyperpolarization of rabbit superior cervical ganglia persisted in a Ca^{2+}-free solution when the synaptic response, the slow IPSP, was blocked. In Ca^{2+}-free/high-Mg^{2+}, EGTA (1 mM)-containing medium, the slow IPSP was quickly abolished by 5 min of perfusion with this solution, presumably because the release of neurotransmitter that mediates the slow IPSP was blocked in this medium; in contrast, ACh-response hyperpolarization persisted at a time when the slow IPSP was blocked. TTX (1 μM) blocked the mediation of the slow IPSP, while it did not affect the ACh hyperpolarization obtained from the same cells. Furthermore, in their previous experiments, Cole and Shinnick-Gallagher (1980), showed that the slow IPSP of rabbit superior cervical ganglia was not significantly blocked by catecholamine antagonists that blocked the catecholamine-induced hyperpolarization. Thus, they concluded that the slow IPSP in mammalian sympathetic ganglia is due to the monosynaptic activation of muscarinic receptors located on the postganglionic neurons.

It must be emphasized, again, that the slow IPSP as well as the hyperpolarizing response elicited by a direct action of exogenous ACh on either nicotinized or curarized ganglia are effectively blocked by the action of atropine; furthermore, these hyperpolarized potential changes always precede the slow EPSP or ACh depolarizing response generated by the direct muscarinic action of ACh. These and other facts discussed above, particularly with respect to amphibian ganglia, have made it difficult to believe that the slow IPSP or the ACh hyperpolarizing response is generated by a transmitter released from an interneuron activated by a direct muscarinic action of ACh. It must be added that experimental tests carried out in low-Ca^{2+}/high-Mg^{2+} solution may not always provide a conclusive answer, because the change in Ca^{2+} and Mg^{2+} concentration may affect the sensitivity of receptors generating the muscarinic hyperpolarizing response. Furthermore, as discussed subsequently, it may be speculated that the slow IPSP is generated by combined direct actions of adrenergic substances and ACh (see also Chapters 2, 12, and 13). The nature—di- or monosynaptic—of the mammalian IPSP is less clear, although strong evidence that favors the monosynaptic character of the mammalian IPSP was presented by the Gallaghers and their associates (Gallagher *et al.*, 1980; Cole and Shinnick-Gallagher, 1980, 1984); (see also Weight, 1983; Kobayashi and Tosaka, 1983). Furthermore, the effectiveness of catecholaminergic, including dopaminergic, blocks, whether in mammals (R. M. Eccles and Libet, 1961) or in amphibia, cannot be readily used in favor of the hypothesis of the disynaptic nature of the IPSP in light of the lack of specificity of the action of catecholaminergic blockers that can depress muscarinic responses as well.

IV. UNUSUAL SLOW INHIBITORY POSTSYNAPTIC POTENTIALS AND RELATED RESPONSES

The question arises as to whether or not only one neurotransmitter is directly responsible for producing the slow IPSP. In other words, it is possible that the slow IPSP consists of multiple components that may be generated by the combined action of ACh and catecholamines; additional substances may be involved. In fact, ganglion cells may produce slow hyperpolarizing responses that are similar to the IPSP in appearance, but different from it with regard to the mode of their activation via preganglionic nerve stimulation and other characteristics (Koketsu, 1969; Kobayashi and Libet, 1974; for a review, see Weight, 1983).

A. Unusual Acetylcholine Responses

Ganglionic responses produced by preganglionic B-nerve stimulation are usually completely eliminated in the presence of nicotine and atropine (Koketsu, 1969). In some ganglia, however, a small, slow hyperpolarizing response remained present under these experimental conditions (Kobayashi and Libet, 1974; Koketsu and Yamamoto, 1975). This hyperpolarizing response was highly resistant to both nicotinic and muscarinic blockers. A similar hyperpolarizing response was detected when high concentrations of ACh were applied directly to these particular preparations; these hyperpolarizing responses to ACh were not blocked, but rather were enhanced, in a low-Ca^{2+}/high-Mg^{2+} solution containing both nicotine and atropine, whereas those produced by preganglionic B-nerve stimulation were completely eliminated in this solution (Koketsu and Yamamoto, 1975). This indicates that the hyperpolarizing response in question is generated by a direct action of ACh released from preganglionic nerve terminals. This unusual ACh hyperpolarization was augmented in the presence of an anticholinesterase, but it could not be initiated by either carbachol or bethanechol, suggesting that the response is produced by a specific action of ACh that is neither nicotinic nor muscarinic. Since this hyperpolarization was augmented during a conditioning hyperpolarization and easily eliminated by ouabain or by replacing the external Na^+ by Li^+, this unusual ACh hyperpolarization may be produced by an electrogenic Na^+ pump activated by this unique action of ACh (Koketsu and Yamamoto, 1975). Indeed, the possibility that ACh may augment the Na^+ pump has been suggested with respect to aplysian neurons (Pinsker and Kandel, 1969), snail parietal ganglia (Kerkut *et al.*, 1969a,b), and skeletal muscle (Dockry *et al.*, 1966).

B. Synaptically Induced K^+-Activated Hyperpolarization

It has been reported by Tosaka and Tanaka (1973) that rabbit sympathetic ganglion cells are hyperpolarized when repetitive stimuli are applied to preganglionic nerves in a Ringer's solution containing no K^+

and when cholinergic transmission is completely blocked by D-TC and atropine. They analyzed the nature of this hyperpolarizing response and suggested that it might be produced by an electrogenic Na^+ pump, when the latter, following its depression in K^+-free solution, is suddenly activated by extracellular K^+ released from preganglionic nerve terminals by the action potentials. In fact, the membrane of sympathetic ganglion cells maintained in a K^+-free perfusion solution is hyperpolarized by an addition of K^+ to the external solution, and this K^+-activated hyperpolarization appears to be generated by activation of an electrogenic Na^+ pump (Akasu and Koketsu, 1976a,b); furthermore, the physiological and pharmacological characteristics of such K^+-activated hyperpolarization are similar to those of hyperpolarization induced by K^+ released by preganglionic nerve activation in a K^+-free solution.

Figure 3 illustrates synaptically induced K^+-activated hyperpolarization induced in K^+-free solution by applying a train of repetitive preganglionic nerve stimuli. The exquisite sensitivity of this hyperpolari-

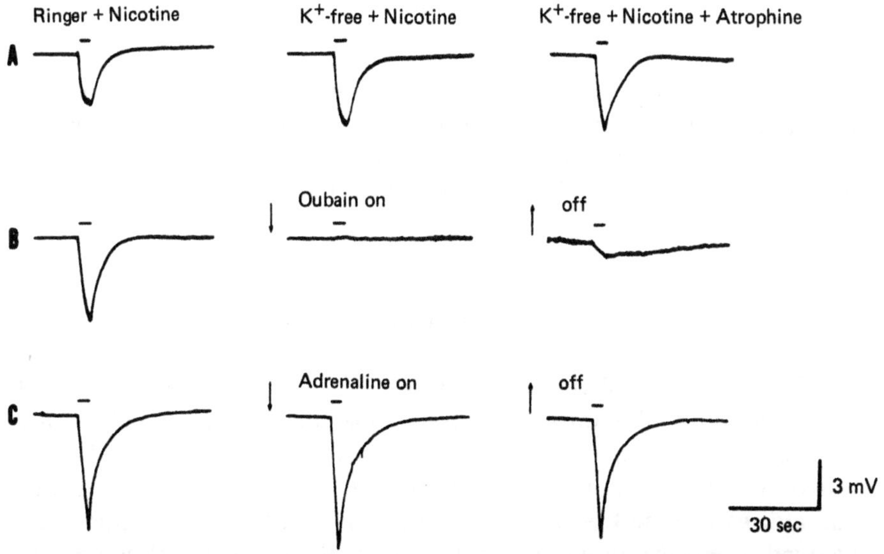

Figure 3. Synaptically induced K^+-activated hyperpolarizations recorded from bullfrog ganglia. (A) A slow IPSP recorded from a nicotinized ganglion was enhanced in K^+-free solution. A large part of this enhanced slow IPSP was still observed after addition of atropine, disclosing synaptically induced K^+-activated hyperpolarization. (B) Synaptically induced K^+-activated hyperpolarization was completely abolished by ouabain (1×10^{-6} M) and partially restored after withdrawal of ouabain. (C) Synaptically induced K^+-activated hyperpolarization was enhanced in the presence of epinephrine (0.1 mM). All potential changes were induced by trains of repetitive preganglionic B-nerve stimulations (marked by horizontal lines). Downward and upward arrows indicate addition (on) and withdrawal (off) of a drug, respectively. From Akasu *et al.* (1978).

zation to ouabain is also shown; this latter finding supports the concept that this hyperpolarization is generated by the activation of an electrogenic Na^+ pump. It must be added that the amplitude of the hyperpolarization was increased by a moderate conditioning hyperpolarization and depressed by a strong conditioning hyperpolarization (Akasu *et al.*, 1978). All these aspects of the hyperpolarization in question are quite similar to those of the slow IPSP.

A question arises as to whether or not synaptically induced K^+-activated hyperpolarization participates in the formation of the slow IPSP generated in normal K^+-containing solutions. It may be possible that the activity of the electrogenic Na^+ pump of the ganglion cells is accelerated when the extracellular K^+ concentration in the immediate vicinity of the cell membrane is raised to more than 2 mM by the generation of action potentials at preganglionic nerve terminals; at this time, it appears possible that the K^+-activated hyperpolarization may contribute to the slow IPSP.

V. NATURE OF THE SLOW INHIBITORY POSTSYNAPTIC POTENTIAL

The slow IPSP can be more readily recorded in the presence of a nicotinic blocker than in the presence of a curaremimetic blocker (see above); there should be no fundamental differences in the ionic mechanisms of the slow IPSP, however, whether recorded from nicotinized or curarized ganglia.

A. Effect of Ouabain

The effect of ouabain on the slow IPSP has been examined and discussed in many laboratories for the purpose of clarifying whether or not the generation of the slow IPSP is directly associated with active Na^+ transport. Unfortunately, the experimental results obtained in different laboratories have not been always consistent. The disagreements may be resolved, however, by taking into consideration the experimental materials and conditions. In this respect, it is important to consider whether the experiments in question are carried out on ganglia treated with nicotine or with curare.

It was reported (Nishi and Koketsu, 1967; Koketsu and Nakamura, 1975) that in nicotinized ganglia, the slow IPSP was markedly decreased or abolished 14–30 min following the application of ouabain, 10^{-6} M. The slow IPSP was selectively depressed, while other synaptic potentials, including the fast and the slow EPSP, remained unchanged in the presence of ouabain. Under these conditions, the inhibitory effect of the slow IPSP on afterdischarges of ganglion cells was eliminated. In some preparations, the slow IPSP was relatively resistant to the action of ouabain even when it was applied for more than 30 min (Koketsu and Nakamura, 1975) (see Figure 4).

Nicotinic transmission can be completely blocked when a ganglion is perfused with a solution containing 1.4 mM D-TC for more than 1 hr; it was reported that under these circumstances, the effects of ouabain on the slow IPSP were insignificant in either mammalian or amphibian ganglia (Kobayashi and Libet, 1968; Libet, 1970). In partial agreement with these findings, Akasu and Koketsu (1976c) reported that in curarized bullfrog ganglia, the amplitude of the slow IPSP was depressed only 30% by ouabain, 10^{-6} M, even following long-term application.

B. Effect of K^+

In both nicotinized and curarized ganglia, the amplitude of the slow IPSP was depressed when the external K^+ concentration was increased from 2 to 10 mM; it was also depressed when the K^+ concentration was decreased from 2 to 0.2 mM (Nishi and Koketsu, 1968). However, changes in the slow IPSP caused by lowering the external K^+ concentration are complex and depend on the protocol. For example, rapid replacement of normal Ringer's solution by a K^+-free solution, i.e., rapid application of a K^+-free Ringer's solution to the ganglia, results in a rapid and persistent increase of the amplitude of the slow IPSP (Smith and Weight, 1977a,b) (see Figure 3). It must be remembered that the ganglion cell generates K^+-activated hyperpolarization in K^+-free Ringer's solution following repetitive preganglionic nerve stimulation [synaptically induced K^+-activated hyperpolarization (see above)]. Thus, the slow IPSP recorded in a K^+-free solution may include the synaptically induced K^+-activated hyperpolarization.

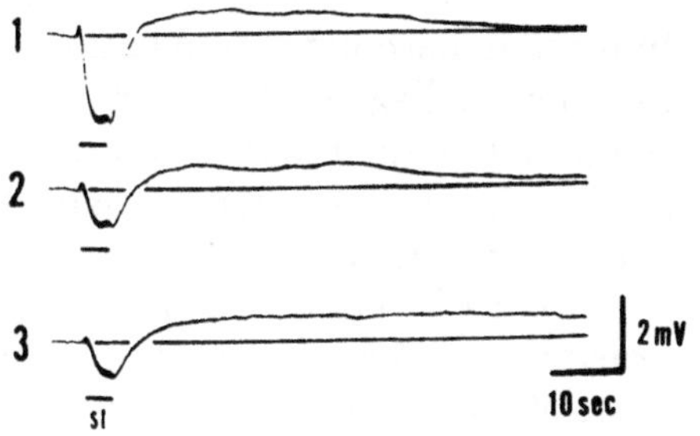

Figure 4. Effect of ouabain (1×10^{-6} M) on slow IPSP and slow EPSP recorded from a nicotinized bullfrog sympathetic ganglion by the sucrose-gap method. The slow IPSP and the slow EPSP were produced by trains of repetitive (30/sec) preganglionic B-nerve stimulations (st). Records (1), (2), and (3) were obtained, respectively, before, 15, and 30 min after an application of ouabain. Note the depression by ouabain of the slow IPSP while the slow EPSP remained unchanged. From Koketsu and Nakamura (1975).

The slow IPSP recorded from nicotinized ganglia in the presence of ouabain is also increased when the external K^+ is removed (Lees and Wallis, 1974; Smith and Weight, 1977a,b). The slow IPSP recorded under these conditions may represent the actual slow IPSP, since synaptically induced K^+-activated hyperpolarization should be completely eliminated in the presence of ouabain. The results obtained by Lees and Wallis (1974) and Smith and Weight (1977a,b) were confirmed with regard to both nicotinized and curarized ganglia maintained in K^+-free solution containing ouabain (Akasu *et al.*, 1978; Akasu and Koketsu, 1983). It was suggested on the basis of these results that the slow IPSP includes a component that is resistant to the action of ouabain and is augmented when the external K^+ is omitted (Akasu *et al.*, 1978; Akasu and Koketsu, 1983).

C. Membrane Conductance Changes

It was reported that the slow IPSP and catecholamine-induced hyperpolarization of mammalian sympathetic ganglia were generated without changes in membrane conductance (Tosaka *et al.*, 1968; Libet and Kobayashi, 1969; Kobayashi and Libet, 1970). In the pertinent experiments, intracellular methods were used to measure membrane potential changes due to the slow IPSP and catecholamines as well as the membrane conductance (evaluated in terms of intracellularly recorded potential response to constant anodal current pulses). Weight and Padjen (1973a,b) reported that membrane resistance increased during the development of the slow IPSP of the C cells of nicotinized bullfrog ganglia. On the other hand, Nishi (1979) found that in mammalian sympathetic ganglion cells, membrane resistance decreased during the hyperpolarizing response evoked by iontophoresis of ACh, which presumably corresponded to the slow IPSP, whereas Ivanov and Skok (1980) described two types of slow IPSPs occurring in different neurons of mammalian sympathetic ganglia; the slow IPSP of one type was concomitant with a decrease in cell membrane resistance, while the IPSP of another type was not followed by any change in cell membrane resistance (Figure 5). Recently, a decrease in membrane resistance was demonstrated during generation of slow IPSP and ACh-induced hyperpolarization in the superior ganglion of the rabbit (Cole and Shinnick-Gallagher, 1984). Similarly, Horn and Dodd (1981) (see also Dodd and Horn, 1983) have reported that membrane resistance was decreased during the hyperpolarizing responses induced in the bullfrog sympathetic ganglion by iontophoretic application of ACh to C cells.

It must be emphasized that accurate measurement of the membrane resistance changes during slow IPSP is difficult because of rectifying properties of the ganglion-cell membrane activated near the resting membrane potential level. However, Akasu and Koketsu (1983) could measure di-

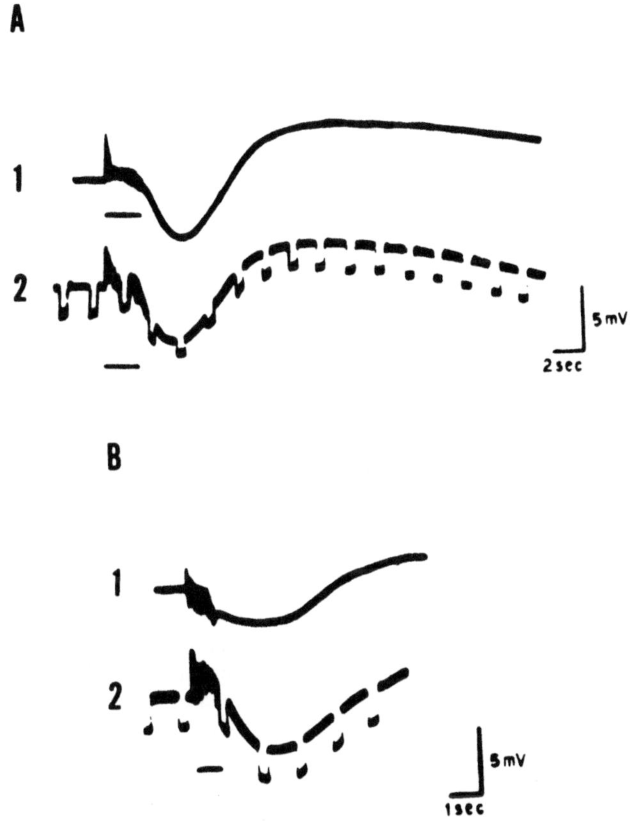

Figure 5. Slow IPSPs of two types (A, B) evoked in two neurons by short trains of repetitive orthodromic stimuli [10 Hz, 2 sec for (A), 10 Hz, 1 sec for (B)]. The intracellular response (1) and changes in input resistance that follow the response (2) are shown in both (A) and (B). Oscilloscope recording. From Ivanov and Skok (1980).

rectly by means of the voltage-clamp procedure the conductance change concomitant with the bullfrog slow inhibitory postsynaptic current (IPSC); they found that membrane conductance increases during the slow IPSC (Figure 6).

D. Effects of the Membrane Potential Level on the Slow Inhibitory Postsynaptic Potential

An interesting and important aspect of the slow IPSP concerns the effect of membrane potential level on the amplitude of the slow IPSP. When studies were carried out with nicotinized bullfrog sympathetic ganglia, and when the membrane of the ganglion cells was moderately hy-

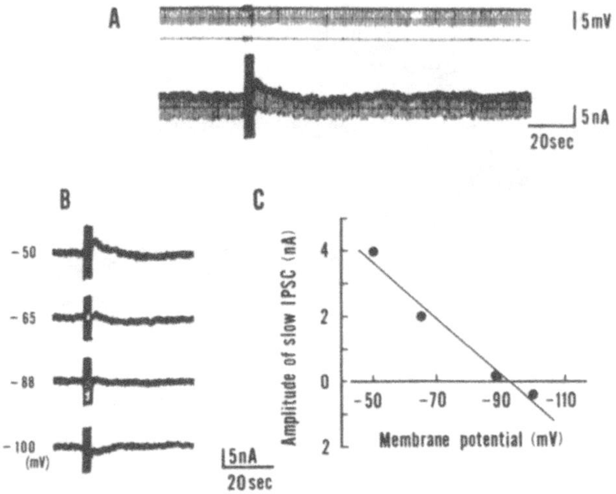

Figure 6. Conductance change and the reversal potential during generation of the slow IPSC. (A) The upper trace shows the membrane potential with hyperpolarizing command potentials (100-msec duration). The lower trace shows membrane currents evoked by a train of repetitive (30/sec for 4 sec) preganglionic stimulations. The holding membrane potential of this ganglion cell was −65 mV. (B) Slow IPSCs obtained at holding potential levels (indicated at left of each record). (C) Relationship between amplitudes of slow IPSC and holding membrane potential. Points in this graph were obtained from records shown in (B). From Akasu and Koketsu (1983).

perpolarized by extrinsic anodal currents, the amplitude of the slow IPSP was markedly increased; when the resting membrane was hyperpolarized beyond a certain potential level, the slow IPSP was inevitably and consistently depressed (Koketsu and Nishi, 1967; Nishi and Koketsu, 1968) (see Figure 7). This critical potential level ranged between −75 and −80 mV, the resting membrane potential being approximately −70 mV; an IPSP of approximately original size was recorded at −90 mV, and it disappeared when the resting membrane was strongly hyperpolarized beyond approximately −110 mV.

The results differed when bullfrog sympathetic ganglia were treated with D-TC (Akasu and Koketsu, 1976c). The amplitude of the slow IPSP recorded in many of these preparations was consistently reduced during conditioning hyperpolarization, although some preparations showed an increased size of the slow IPSP when the cells were moderately hyperpolarized. The slow IPSP was nullified at a certain potential level and eventually reversed its polarity when a strong conditioning hyperpolarization exceeded this potential level (Akasu and Koketsu, 1976c). Raising the external K^+ concentration to 10 mM decreased the amplitude of the

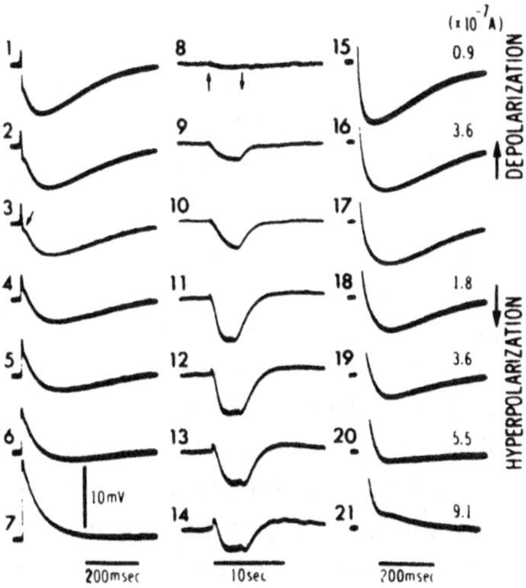

Figure 7. Changes in the slow IPSP and afterhyperpolarization of a ganglion by conditioning depolarization and hyperpolarization. (1–7) Afterhyperpolarization of orthodromic responses [the arrow in record (3) indicates the hump due to the presence of the EPSP]. (8–14) Slow IPSPs. (15–21) Afterhyperpolarization of directly induced ganglionic responses. Spikes of orthodromic and directly induced responses are out of frame. Orthodromic responses were induced by single stimuli applied to preganglionic B fibers during perfusion of the ganglion with Ringer's solution. Slow IPSPs and direct responses were recorded approximately 40 min after nicotine sulfate (0.12 mM) was added to the perfusate. Slow IPSPs were elicited by trains of repetitive stimulations (10/sec for 4 sec) of preganglionic B plus C fibers [upward and downward arrows in (8) indicate initiation and cessation of preganglionic stimulation]. Direct responses were induced by supramaximal stimulating currents (10 msec in duration) delivered through the bridge. Records in the third horizontal row, (3), (10), and (17), were taken without applying external currents, whereas those above and below this row were obtained during application of cathodal and anodal currents, respectively. Intensities of applied currents were identical for all records of any single row; they are indicated above the right-hand record. Bullfrog ganglion. From Nishi and Koketsu (1968).

slow IPSP, and the slow IPSP was reversed at a potential level much closer to the resting membrane potential established in normal Ringer's solution than to the potential level of the reversal obtained in the normal Ringer's solution (Akasu and Koketsu, 1976c). The amplitude of the slow IPSP was consistently depressed when conditioning depolarization was applied in both normal and high-K^+ solutions.

This difference between the effects of membrane potential on the slow IPSPs recorded from nicotinized ganglia and from curarized ganglia seemed to be related to the difference in their responses observed during conditioning hyperpolarization. This is suggested by the fact that the properties of the slow IPSP recorded from nicotinized ganglia in the presence of ouabain were very similar to those of curarized ganglia: After the slow IPSP of nicotinized ganglia was depressed by ouabain (10^{-6} M) by more than 50%, changes in the amplitude of the slow IPSP during conditioning hyperpolarization resembled those observed in curarized ganglia, since the amplitude of the slow IPSP was progressively reduced and the IPSP finally reversed in the course of progressive hyperpolarization of the resting membrane potential.

Reversal of the polarity of the slow IPSP recorded by intracellular microelectrodes at a hyperpolarized potential level of curarized cells was demonstrated in the case of mammalian sympathetic ganglion cells (Ivanov and Skok, 1980) and bullfrog sympathetic ganglion C cells (Horn and Dodd, 1981; Dodd and Horn, 1983). In the case of bullfrog sympathetic ganglion cells, the slow IPSP amplitude increased to a peak value with progressive hyperpolarization; it then became smaller and, finally, the polarity of the IPSP became reversed. The slow IPSP of these cells reversed at approximately −100 mV; their spike afterpotential reversed at a similar potential (Horn and Dodd, 1981; Dodd and Horn, 1983).

The reversal of the slow IPSC was also demonstrated in bullfrog sympathetic ganglia (Akasu and Koketsu, 1983). Figure 6 shows an example of the results obtained from a ganglion cell treated with ouabain (1 μM) so that the ouabain-sensitive component of the slow IPSP was abolished. The membrane potential was initially held at −65 mV. As illustrated in Figure 6, the slow IPSC was decreased and eventually reversed at a hyperpolarizing potential level of −93 mV; this reversal potential is close to the equilibrium potential of K^+.

The reversal of the polarity of the slow IPSC during conditioning hyperpolarization is an important phenomenon, since this property of the slow IPSP seems to be an important key to understanding its ionic mechanism (Akasu and Koketsu, 1983).

E. Ionic Mechanism of the Slow Inhibitory Postsynaptic Potential

1. Electrogenic Na^+ Pump Hypothesis

Nishi and Koketsu (1967, 1968) have suggested that the slow IPSP of sympathetic ganglion cells may be generated by an activation of the electrogenic Na^+ pump. This hypothesis was based on the observation that the slow IPSP recorded in the presence of nicotine was very sensitive to ouabain (Nishi and Koketsu, 1967, 1968; Koketsu and Nakamura, 1975); (see above) as well as to Na^+ loading (Nishi and Koketsu, 1968) or Na^+ depletion (Koketsu *et al.*, 1973). It is known, however, that the slow IPSP recorded from curarized ganglia is not sensitive to ouabain (Kobayashi and Libet, 1968; Libet, 1970). Furthermore, Smith and Weight (1977a,b) reported that the slow IPSP of nicotinized amphibian ganglia was depressed by only 10–40% within 1 hr of the application of ouabain, 10^{-6} M. Subsequently, it was suggested that the slow IPSP recorded from either nicotinized or curarized amphibian ganglia includes two components, viz., a component that is sensitive to ouabain and associated with the electrogenic Na^+ pump and another component that is insensitive to ouabain and not associated with the electrogenic Na^+ pump (Akasu *et al.*, 1978).

2. G_{Na}-Inactivation Hypothesis

Weight and Padjen (1973a,b) hypothesized that the slow IPSP of the sympathetic ganglion cells is produced by reduced inactivation of the resting membrane permeability to Na^+, since the reduction of the G_{Na} would lead to a hyperpolarization generated by resting currents of other ions, mainly K^+. This hypothesis was based on their observation that the slow IPSP recorded intracellularly from C cells of nicotinized amphibian ganglia was accompanied by a reduction of the membrane conductance.

3. G_K-Activation Hypothesis

As already alluded to above, Koketsu and his collaborators (Akasu and Koketsu, 1976c, 1983; Koketsu and Akasu, 1978; Kuba and Koketsu, 1978; Akasu *et al.*, 1978) suggested that the slow IPSP of the bullfrog sympathetic ganglion cell may consist of two different components, an ouabain-insensitive component that may be generated by a K^+ conductance increase and an ouabain-sensitive component that may be associated with the electrogenic Na^+ pump. This hypothesis is based particularly on the observation that the slow IPSP recorded in the presence of ouabain reversed its polarity at a potential level similar to the K^+ equilibrium potential. It is important to note in this context that ACh-induced hyperpolarization, which mimics the slow IPSP in mammalian sympathetic ganglia, was accompanied by an increase of the membrane conductance (Nishi, 1979); furthermore, this hyperpolarization reversed its polarity at a potential level (-95 mV) similar to the K^+ equilibrium potential, suggesting, again, that the slow IPSP is generated by an increase in K^+ conductance (Nishi, 1979). Similarly, Ivanov and Skok (1980) reported that the slow IPSP of mammalian sympathetic ganglia is accompanied by an increase of membrane conductance and that it reversed its polarity at a potential level that was near the K^+ equilibrium potential. The G_K activation hypothesis was also supported by experiments carried out by Horn and Dodd (1981) (see also Dodd and Horn, 1983). In their hands, the slow IPSP recorded from C cells of curarized bullfrog sympathetic ganglia was accompanied with an increase of membrane conductance and it was reversed at about -100 mV; furthermore, the reversal potential was insensitive to changes in the Cl^- gradient. Finally, the increase in conductance was blocked by Ba^{2+}, an antagonist of processes underlying changes in K^+ conductance (Dodd and Horn, 1983).

VI. SECOND MESSENGERS

On the basis of the data obtained by Greengard and his associates (see Greengard, 1976), Greengard suggested that the generation of the slow IPSP is due to cyclic AMP acting as a second messenger following its activation by dopamine. This matter is under critical review at present (Dun, 1980; see also Nishi *et al.*, 1978; Kobayashi and Tosaka, 1983), and it is discussed in detail in Chapters 12 and 13.

VII. CONCLUSIONS

Jointly with the slow EPSP and the late slow EPSP, the slow IPSP illustrates the important concept of the pluripotential nature of the synaptic ganglionic response. It is not surprising, then, that several speculations were initiated with respect to the slow potentials in an attempt to consider these responses as convenient models for the study of higher brain function and its regulation (see Chapter 21).

The nature of the slow IPSP is controversial. Thus, while the hypothesis of disynaptic nature of the IPSP constitutes one of the classic hypotheses in the area of neurosciences, this proposal cannot be considered as proven or demonstrated; in fact, it is entirely possible that both disynaptic and monosynaptic mechanisms are involved in the generation of the IPSP, or that the actual mechanism is species- and ganglion-dependent, or both. Several technical difficulties render the solution of this controversy not easy to achieve. These difficulties are also inherent in studies of the ionic mechanism or mechanisms that may be involved in its generation, and this mechanism is not as yet agreed on. Indeed, while three separate hypotheses—G_{Na} inactivation, electrogenic Na^+ pump activation, and G_K activation—were proposed to explain the IPSP, there is a tendency at this time to combine these mechanisms (see, for example, Weight, 1983; see also Hartzell, 1981; Hille, 1984). It is of interest that a similar complex mechanism consisting of muscarinically generated increased G_K and decreased Ca^{2+} and Na^+ currents may underlie the cholinergic hyperpolarization occurring at the atrial muscle (Brown, 1982; Pappano and Inoue, 1984).

It must be added that the IPSP response may not be pure; it may include several mechanisms or components or both, and such responses as "unusual" ACh hyperpolarization, K^+-activated hyperpolarization, and perhaps other responses may either be components of the slow IPSP or constitute additional potentials that are independent of the classic IPSP. In fact, it is tempting to speculate that perhaps a still-unknown transmitter or modulator may be involved in the generation of the slow IPSP.

REFERENCES

Akasu, T., and Koketsu, K.: Adrenaline and the electrogenic sodium pump in *Rana catesbeiana* sympathetic ganglion cells. *Experientia* **32**:57–59 (1976a).

Akasu, T., and Koketsu, K.: The effect of adrenaline on the K^+-activated hyperpolarization of the sympathetic ganglion cell membrane in bullfrogs. *Jpn. J. Physiol.* **26**:289–301 (1976b).

Akasu, T., and Koketsu, K.: Ionic mechanisms of the slow inhibitory postsynaptic potential of sympathetic ganglion cells in bullfrogs. *Kurume Med. J.* **23**:183–188 (1976c).

Akasu, T., and Koketsu, K.: Electrogenesis of the slow inhibitory postsynaptic potential in bullfrog sympathetic ganglia. *Jpn. J. Physiol.* **33**:279–300 (1983).

Akasu, T., Omura, H., and Koketsu, K.: Roles of electrogenic Na^+ pump and K^+ conductance in the slow inhibitory postsynaptic potential of bullfrog sympathetic ganglion cells. *Life Sci.* **23**:2405–2410 (1978).

Brown, H. F.: Electrophysiology of the sinoatrial node. *Physiol. Rev.* **62**:505–530 (1982).

Cole, A. E., and Shinnick-Gallagher, P.: Alpha-adrenoceptor and dopamine receptor antagonists do not block the slow inhibitory postsynaptic potential in sympathetic ganglia. *Brain Res.***187**:226–230 (1980).

Cole, A. E., and Shinnick-Gallagher, P.: Muscarinic inhibitory transmission in mammalian sympathetic ganglia mediated by increased potassium conductance. *Nature (London)* **307**:270–271 (1984).

Dockry, M., Kernan, R. P., and Tangney, A.: Active transport of sodium and potassium in mammalian skeletal muscle and its modification by nerve and by cholinergic and adrenergic agents. *J. Physiol. (London)* **186**:187–200 (1966).

Dodd, J., and Horn, J. P.: Muscarinic inhibition of sympathetic neurones in the bullfrog. *J. Physiol. (London)* **334**:271–291 (1983).

Dun, N. J.: Ganglionic transmission: Electrophysiology and pharmacology. *Fed. Proc. Fed. Am. Soc. Exp. Biol.* **39**:2982–2989 (1980).

Dun, N. J., and Karczmar, A. G.: Involvement of an interneuron in the generation of the slow inhibitory postsynaptic potential in mammalian sympathetic ganglia. *Proc. Natl. Acad. Sci. U.S.A.* **75**:4029–4032 (1978).

Dun, N. J., and Karczmar, A. G.: A comparative study of the pharmacological properties of the positive potential recorded from the superior cervical ganglia of several species. *J. Pharmacol. Exp. Ther.* **215**:455–460 (1980).

Dun, N. J., Kaibara, K., and Karczmar, A. G.: Dopamine and adenosine 3′,5′-monophosphate responses of single mammalian sympathetic neurons. *Science* **197**:778–780 (1977).

Eccles, J. C.: Synaptic potentials and transmission in sympathetic ganglion. *J. Physiol. (London)* **101**:465–483 (1943).

Eccles, R. M., and Libet, B.: Origin and blockade of the synaptic responses of curarized sympathetic ganglia. *J. Physiol. (London)* **157**:484–503 (1961).

Gallagher, J. P., Shinnick-Gallagher, P., Cole, A. E., Griffith, W. H., III, and Williams, B. J.: Current hypotheses for the slow inhibitory postsynaptic potential in sympathetic ganglia. *Fed. Proc. Fed. Am. Soc. Exp. Biol.* **34**:3009–3015 (1980).

Greengard, P.: Possible role for cyclic nucleotides and phosphorylated membrane proteins in postsynaptic actions of neurotransmitters. *Nature (London)* **260**:101–108 (1976).

Hartzell, H. C.: Mechanisms of slow postsynaptic potentials. *Nature (London)* **291**:539–544 (1981).

Hille, B.: *Ionic Channels of Excitable Membranes.* Sinauer, Sunderland, Massachusetts (1984).

Horn, J. P., and Dodd, J.: Monosynaptic muscarinic activation of K^+ conductance underlies the slow inhibitory postsynaptic potential in sympathetic ganglia. *Nature (London)* **292**:625–627 (1981).

Ivanov, A. Ya, and Skok, V. I.: Slow inhibitory postsynaptic potentials and hyperpolarization evoked by noradrenaline in the neurones of mammalian sympathetic ganglion. *J. Auton. Nerv. Syst.* **1**:255–263 (1980).

Kerkut, G. A., Brown, L. C., and Walker, R. J.: Postsynaptic stimulation of the electrogenic sodium pump. *Life Sci.* **8**:297–300 (1969a).

Kerkut, G. A., Brown, L. C., and Walker, R. J.: Cholinergic IPSP by stimulation of the electrogenic sodium pump. *Nature (London)* **223**:864–865 (1969b).

Kobayashi, H., and Libet, B.: Generation of slow postsynaptic potentials without increases in ionic conductance. *Proc. Natl. Acad. Sci. U.S.A.* **60**:1304–1311 (1968).

Kobayashi, H., and Libet, B.: Actions of noradrenaline and acetylcholine on sympathetic ganglion cells. *J. Physiol. (London)* **208**:353–372 (1970).

Kobayashi, H., and Libet, B.: Is inactivation of potassium conductance involved in slow postsynaptic excitation of sympathetic ganglion cells? Effects of nicotine. *Life Sci.* **14**:1871–1883 (1974).

Kobayashi, H., and Tosaka, T.: Slow synaptic actions in mammalian sympathetic ganglia, with special reference to the possible role plated by cyclic nucleotides, in: *Autonomic Ganglia* (L.-G. Elfvin, ed.), pp. 281–307, John Wiley, Chichester and New York (1983).

Koketsu, K.: Cholinergic synaptic potentials and the underlying ionic mechanisms. *Fed. Proc. Fed. Am. Soc. Exp. Biol.* **28**:101–112 (1969).

Koketsu, K., and Akasu, T.: A unique potential component of the slow IPSP of bullfrog sympathetic ganglion cells. *IBRO News* **6**:4 (1978).

Koketsu, K., and Nakamura, M.: Some analyses of the slow inhibitory postsynaptic potential in bullfrog sympathetic ganglia. *Kurume Med. J.* **22**:5–11 (1975).

Koketsu, K., and Nishi, S.: Characteristics of the slow inhibitory postsynaptic potential of bullfrog sympathetic ganglion cells. *Life Sci.* **6**:1827–1836 (1967).

Koketsu, K., and Yamamoto, K.: Unusual cholinergic response of bullfrog sympathetic ganglion cells. *Eur. J. Pharmacol.* **31**:281–286 (1975).

Koketsu, K., Shoji, T., and Nishi, S.: Slow inhibitory postsynaptic potentials of bullfrog sympathetic ganglia in sodium-free media. *Life Sci.* **13**:453–458 (1973).

Kondo, H.: Innervation of SIF cells in the superior cervical and nodose ganglia: An ultrastructural study with serial sections. *Biol. Cell.* **30**:253–264 (1977).

Kosterlitz, H. W., Lees, G. M., and Wallis, D. I.: Resting and action potentials recorded by the sucrose-gap method in the superior cervical ganglion of the rabbit. *J. Physiol. (London)* **195**:39–53 (1968).

Kuba, K., and Koketsu, K.: Analysis of the slow excitatory postsynaptic potential in bullfrog sympathetic ganglion cells. *Jpn. J. Physiol.* **26**:651–669 (1976).

Kuba, K., and Koketsu, K.: Synaptic events in sympathetic ganglia. *Prog. Neurobiol.* **11**:77–169 (1978).

Laporte, Y., and Lorente de No, R.: Potential changes evoked in a curarized sympathetic ganglion by presynaptic volleys of impulses. *J. Cell. Comp. Physiol.* **35**(Suppl. 2):61–106 (1950).

Lees, G. M., and Wallis, D. I.: Hyperpolarization of rabbit superior cervical ganglion cells due to activity of an electrogenic sodium pump. *Br. J. Pharmacol.* **50**:79–93 (1974).

Libet, B.: Generation of slow inhibitory and excitatory postsynaptic potentials. *Fed. Proc. Fed. Am. Soc. Exp. Biol.* **29**:1945–1956 (1970).

Libet, B., and Kobayashi, H.: Generation of adrenergic and cholinergic potentials in sympathetic ganglion cells. *Science* **164**:1530–1532 (1969).

Libet, B., and Kobayashi, H.: Adrenergic mediation of slow inhibitory postsynaptic potential in sympathetic ganglia of the frog. *J. Neurophysiol.* **37**:805–814 (1974).

Libet, B., and Owman, C.: Concomitant changes in formaldehyde-induced fluorescence of dopamine interneurones and in slow inhibitory postsynaptic potentials of the rabbit superior cervical ganglion, induced by stimulation of the preganglionic nerve or by a muscarinic agent. *J. Physiol. (London)* **237**:635–662 (1974).

Libet, B., and Tosaka, T.: Dopamine as a synaptic transmitter and modulator in sympathetic ganglia: A different mode of synaptic action. *Proc. Natl. Acad. Sci. U.S.A.* **67**:667–673 (1970).

Nishi, S.: The catecholamine-mediated inhibition in ganglionic transmission, in: *Integrative Functions of the Autonomic Nervous System* (C. McC. Brooks, K. Koizumi, and A. Sato, eds.), pp. 223–233, University of Tokyo Press (1979).

Nishi, S., and Koketsu, K.: Origin of ganglionic inhibitory postsynaptic potential. *Life Sci.* **6**:2049–2055 (1967).

Nishi, S., and Koketsu, K.: Analysis of slow inhibitory postsynaptic potential of bullfrog sympathetic ganglion. *J. Neurophysiol.* **31**:717–728 (1968).

Nishi, S., Karczmar, A. G., and Dun, N. J.: Physiology and pharmacology of ganglionic synapses as models for central transmission, in: *Advances in Pharmacology and Therapeutics*, Vol. 2 (P. Simon, ed.), pp. 69–85, Pergamon Press, Cambridge (1978).

Pappano, A. J., and Inoue, D.: Development of different electrophysiological mechanisms for muscarinic inhibition of atria and ventricles. *Fed. Proc. Fed. Am. Soc. Exp. Biol.* **43**:2607–2612 (1984).

Pinsker, H., and Kandel, E. R.: Synaptic activation of an electrogenic sodium pump. *Science* **163**:931–935 (1969).

Smith, P. A., and Weight, F. F.: Role of electrogenic sodium pump in slow synaptic inhibition is re-evaluated. *Nature (London)* **267**:68–70 (1977a).

Smith, P. A., and Weight, F. F.: A comparison of the effects of ouabain and potassium-free Ringer on the electrogenic sodium pump and slow synaptic inhibition in bullfrog sympathetic ganglia. *Br. J. Pharmacol.* **59**:495P (1977b).

Tosaka, T., and Tanaka, T.: Long lasting hyperpolarization following EPSPs in atropinized superior cervical sympathetic ganglia of rabbit (in Japanese). *J. Physiol. Soc. Jpn.* **35**:482 (1973).

Tosaka, T., Chichibu, S., and Libet, B.: Intracellular analysis of slow inhibitory and excitatory postsynaptic potentials in sympathetic ganglia of the frog. *J. Neurophysiol.* **31**:396–409 (1968).

Volle, R. L., and Hancock, J. C.: Transmission in sympathetic ganglia. *Fed. Proc. Fed. Am. Soc. Exp. Biol.* **29**:1913–1918 (1970).

Weight, F. F.: Synaptic mechanisms in amphibian sympathetic ganglia, in: *Autonomic Ganglia* (L.-G. Elfvin, ed.), pp. 309–344, John Wiley, Chichester and New York (1983).

Weight, F. F., and Padjen, A.: Slow synaptic inhibition: Evidence for synaptic inactivation of sodium conductance in sympathetic ganglion cells. *Brain Res.* **55**:219–224 (1973a).

Weight, F. F., and Padjen, A.: Acetylcholine and slow synaptic inhibition in frog sympathetic ganglion cells. *Brain Res.* **55**:225–228 (1973b).

Weight, F. F., and Weitsen, H. A.: Identification of small intensely fluorescent (SIF) cells as chromaffin cells in bullfrog sympathetic ganglia. *Brain Res.* **128**:213–226 (1977).

10

Presynaptic Modulation

The Mechanism and Regulation of Transmitter Liberation in Sympathetic Ganglia

KENJI KUBA and SHOICHI MINOTA

I. INTRODUCTION

The release of the chemical transmitter acetylcholine (ACh) from preganglionic nerve terminals is measured either by quantifying in the presence of an anticholinesterase the amount of ACh released in the perfusate of a ganglion (Kibjakow, 1933; Feldberg and Gaddum, 1934) or indirectly by recording intracellularly either the fast or the miniature excitatory postsynaptic potential (fast EPSP and mEPSP, respectively) from the postsynaptic membrane (Eccles, 1955; Nishi and Koketsu, 1960). The former method measures the absolute amount of ACh released from preganglionic terminals; both bioassay (Nishi and Koketsu, 1960; Nishi *et al.*, 1967) and chemical techniques were used in the measurement of ACh release. It must be noted that the bioassay employed by Nishi and Koketsu (amphibian lung preparation) was sensitive enough, in combination with certain neuroanatomical and electrophysiological evaluations, to yield dependable values of ACh release per impulse per synaptic knob (6000–8000

KENJI KUBA and SHOICHI MINOTA • Department of Physiology, Saga Medical School, Nabeshima, Saga 840–01, Japan.

molecules); (see also below). However, it is not known which fraction of ACh is released in a quantal manner, since ACh is also liberated in a nonquantal manner (cf. Katz and Miledi, 1977).

On the other hand, the indirect method, which is based on the established relationship between the quantal release of ACh and the resultant response of the subsynaptic membrane, appears to be effective and convenient for the analysis of the ACh-release mechanism involved in ganglionic synaptic transmission at the single-cell level.

The mechanism of ACh release in sympathetic ganglia has been generally understood as being parallel to the mechanism described for motor-nerve terminals (cf. Katz, 1969). To be released during synaptic transmission, ACh is stored in the synaptic vesicles in the preganglionic nerve terminal. The arrival of an impulse causes an influx of Ca^{2+} into the nerve terminal, increasing the level of intracellular free Ca^{2+} ($[Ca^{2+}]_i$) in the nerve terminal. This causes the synaptic vesicle to fuse with the presynaptic terminal membrane and subsequently to release ACh into the synaptic cleft by an exocytotic mechanism. This vesicle hypothesis (Del-Castillo and Katz, 1954; cf. Katz, 1969; Zimmermann, 1979; Ceccarelli and Hurlbut, 1980) has recently been challenged by several workers who described biochemical and morphological evidence that is inconsistent with the vesicle hypothesis (Tauc, 1979; Dunant and Israel, 1979; Marchbanks, 1978). They envisage that chemical transmitters are released directly from the cytoplasm or from some structures located close to, or at, the presynaptic membrane through the presynaptic terminal and its membrane. Although the vesicle hypothesis, in our opinion, seems superior to the latter hypothesis, the experimental evidence available does not allow a distinction between the two concepts in the case of sympathetic ganglia. However, there is no doubt that the release of ACh responsible for ganglionic synaptic transmission, whether from the cytoplasm or the vesicles, takes place in a quantal mode and depends on the $[Ca^{2+}]_i$ of the preganglionic nerve terminals.

One point that cannot be ignored in the evaluation of transmitter release mechanisms is the pattern of presynaptic ganglionic innervation, which differs notably from that of the neuromuscular junction. As already described in Chapters 2 and 5, the preganglionic axon of amphibian ganglia arborizes before reaching a postganglionic neuron soma, and the boutons of all these branches make synaptic contacts with the soma of a single postganglionic cell. On the other hand, in mammalian ganglia, the postganglionic neuron is innervated by several different preganglionic axons, so that a postsynaptic potential recorded intracellularly is the result of a summation of postsynaptic responses induced by several different presynaptic fibers. In contrast, each motor nerve branches out to innervate several muscle fibers (muscle cells); i.e., there is no convergence in the case of the neuromuscular junction, while there is both divergence and convergence and/or summation in that of ganglion neurons. The characteristics of the synaptic ganglionic connections make it difficult to interpret the results of pertinent experiments. This is particularly true in the case of statistical analysis of the quantal release of ACh.

Keeping these problems in mind, we will review in this chapter the mechanism of ACh release in sympathetic ganglia. Some of this review will concern mechanisms referred to briefly in Chapter 11. However, Chapter 11 focuses mainly on the modulation of the processes of ACh release, this modulation being only briefly referred to in this chapter with respect to the effect on ACh release of γ-aminobutyric acid (GABA) and the catecholamines; on the other hand, this chapter addresses the physiology and molecular mechanisms of ACh release and the differences between its release from the ganglia and from motor nerve terminals.

II. SYNTHESIS, STORAGE, AND TURNOVER OF ACETYLCHOLINE

The turnover of ACh in preganglionic nerve terminals has been studied mainly by assaying, under a variety of conditions, the amount of ACh released in the perfusate and relating it to ACh content remaining present in the whole ganglion. It was implicitly assumed that the amount of ACh measured in the whole ganglion represents that contained in the presynaptic nerve terminals.

A classic study (G. L. Brown and Feldberg, 1936) introduced the concept that ACh is located in the preganglionic terminals in two compartments; one compartment contains "storage" ACh that is not readily released on preganglionic stimulation, while the other contains ACh available for release. This idea is based on the finding that the total ACh in the ganglion is unaltered or only slightly decreased after repetitive preganglionic stimulation, although the amount of ACh released per impulse is reduced following such stimulation. The analysis of this problem by means of anticholinesterases and hemicholinium (HC-3), an inhibitor of choline uptake at the presynaptic membrane, led to the suggestion that there are three compartments for the storage of ACh in preganglionic nerve terminals: stationary, depot, and surplus ACh compartments (Birks and McIntosh, 1961). The stationary ACh cannot be depleted by prolonged preganglionic stimulation even in the presence of HC-3, so that it may not be involved in synaptic transmission, while the function of surplus ACh is observable only when ganglionic cholinesterase is inhibited. On the other hand, depot ACh is a compartment of ACh that appears to play a physiological role in synaptic transmission; depot ACh is assumed to be localized in the synaptic vesicles, and it is available for release. This depot compartment is further subdivided into two compartments; one constitutes a smaller pool of ACh that is readily available for release, while the other is a larger pool of ACh that is employed to replenish the former pool. This suggestion is of interest in terms of the morphological findings by Zimmermann and Denston (1977); (cf. Whittaker, 1979) with respect to the fragmented nerve terminals of the electric organ of *Torpedo*. They found two different classes of synaptic vesicles, larger and smaller vesicles, the latter being 25% smaller in diameter than the former. These observations may serve as a satisfactory answer to the arguments raised against the vesicular hypothesis (cf. Dunant and Israel, 1979; Whittaker, 1979).

Despite this interesting coincidence between the physiological findings of Birks and McIntosch (1961) and the morphological findings of Zimmermann and Denston (1977), subsequent studies did not support the two-compartment hypothesis for depot ACh. Bennett and McLachlan (1972a) observed a single exponential decline of the amplitude of the fast EPSP during the course of repetitive stimulation when the synthesis of ACh was blocked by HC-3. This observation favors the idea of a single ACh store in the nerve terminals, unless ACh of the two compartments is in a state of equilibrium. Birks and Fitch (1974) also argued against the idea of two different compartments for depot ACh on the basis of their findings of two different patterns of time–courses of ACh output at two different stimulation frequencies, 4 and 20 Hz.

Collier (1969) suggested that newly synthesized ACh is equilibrated with a smaller fraction of ACh stores and released preferentially to the preformed ACh. A similar suggestion was also made for other synapses (see Zimmermann, 1979). This idea conforms to the existence of a population of smaller synaptic vesicles; newly synthesized ACh appears to be taken up by these smaller vesicles (Zimmermann and Denston, 1977). However, Birks and Fitch (1974) observed that the percentage decrease in the ACh store evoked by repetitive stimulation in the presence of HC-3 was almost the same following and in the absence of conditioning stimulation; furthermore, in the absence of HC-3, repetitive stimulation augmented ACh stores. These and other results of Birks (Birks, 1974; Birks and Fitch, 1974) suggest that newly synthesized ACh is rapidly mixed with preformed ACh.

The findings described above led to the suggestion that ACh is released from a cytoplasmic pool. Accordingly, the depot ACh previously described by Birks and McIntosch (1961) would correspond to a cytoplasmic pool of ACh. This is apparently inconsistent with the mode of quantal release of ACh, as revealed by the existence of mEPSPs and of the quantal fluctuation of the fast EPSP amplitude. This problem is explained by assuming in the hypothesis of cytoplasmic ACh release the presence of a "vesicate" (Tauc, 1979) or "operator" (Dunant and Israel, 1979). Unfortunately, the morphological evidence for these assumptions that would be as concrete as that for synaptic vesicles in the case of the vesicular hypothesis is not available as yet. On the other hand, the effect of HC-3 on the quantal size of the fast EPSP does not favor the vesicle hypothesis (McLachlan, 1975b). The unaltered shape of the "gamma distribution" (McLachlan, 1975b) of the mEPSP amplitudes with a progressive reduction in their mean amplitude during the repetitive stimulation in the presence of HC-3 can be explained only by assuming a rather free exchange of ACh between the synaptic vesicles and the cytoplasm. These and related arguments against the vesicular hypothesis were summarized recently by Tauc (1982). While Tauc feels strongly that "it is now doubtful that . . . exocytosis of synaptic vesicles . . . is responsible for the bulk of transmitter release," the alternatives that he proposes constitute, as he states, "a hypothetical portrait" only. Thus, it cannot be concluded at this time to what extent either of these two mechanisms is operative in preganglionic nerve terminals. It should be added that much of the pertinent information concerns the *Torpedo* electric organ rather than the ganglion (e.g., Dunant *et al.*, 1981; Marchbanks, 1981).

It seems established that ACh is synthesized in the cytoplasm, but not in the synaptic vesicles (Fonnum, 1968). Furthermore, ACh synthesis in preganglionic nerve terminals depends on the uptake of choline at the preganglionic terminal membrane, which results from the hydrolysis of released ACh (Bennett and McLachlan, 1972b). Undoubtedly, the rate of ACh synthesis is controlled by the rate of ACh release. This is supported by the findings that the fast EPSP amplitude is well maintained during repetitive stimulation (Bennett and McLachlan, 1972b) and that total ACh increases by 70% in the ganglion after repetitive presynaptic stimulation (20 Hz for 60 min) and a subsequent resting period (20 min); (Birks, 1977). Interestingly, the rate of ACh synthesis activated by release was inversely related to the amount of ACh store, indicating the existence of a regulatory role for the ACh store (Birks, 1977). Subsequent experiments of Birks (1978) have shown that the activation of ACh synthesis depends not only on the total number of stimuli applied to the preganglionic nerves, but also on the pattern of impulse generation; stimulation with intermittent trains of impulses was found to be much more effective than continuous stimulation, the total number of impulses needed to produce a given increase in ACh store being much smaller with the former pattern of stimulation than with the latter. Birks (1978) suggested that this mode of acceleration of ACh synthesis may be related to the choline-uptake mechanism, which apparently depends on the Na^+ gradient across the membrane; this gradient is altered by repetitive firing of the membrane and returns to normal by means of the Na^+-pump activity.

III. QUANTAL RELEASE OF ACETYLCHOLINE

A. Spontaneous Release

1. Existence of mEPSPs

The spontaneous release of ACh in a quantal manner can be seen with an intracellular recording technique; small depolarizing potentials of the amphibian sympathetic ganglion cell, which occur without a preganglionic stimulation, can be thus recorded (Nishi and Koketsu, 1960). These spontaneous potentials are blocked by curare and are similar in their time–course to nerve-evoked fast EPSPs. The frequency of appearance of these potentials increased drastically with an increase in the K^+ concentration of a perfusing solution, while their amplitude was reduced under these conditions (Blackman *et al.*, 1963a). Furthermore, in amphibian ganglia, they are increased in both amplitude and time–course by the inhibition of cholinesterase and increased and decreased in amplitude by conditioning hyperpolarization and depolarization, respectively, of the membrane (Kuba and Minota, unpublished observations). Thus, the spontaneous potentials must be caused by the nicotinic action of ACh that is released spontaneously in a quantal manner from the nerve terminals. Accordingly, they were termed miniature EPSPs (mEPSPs) in analogy with the miniature end-plate potential (mEPP) of the motor end-plate. Similar mEPSPs were found in the case of several other autonomic ganglia, including pelvic ganglia (Blackman *et al.*, 1969), superior cervical ganglia (Sacchi and Perri, 1971), thoracic ganglia (Blackman and Purves, 1969), and hypogastric ganglia of the guinea pig (Bornstein, 1978).

2. Amplitude Distribution of mEPSPs

The frequency function of the amplitude of mEPSPs in most autonomic ganglia is somewhat skewed and better described by a gamma distribution rather than by a normal distribution (McLachlan, 1975a). It was suggested that this characteristic of mEPSP distribution is due to the dependence of the release probability of a single quantum on the quantal size in such a way that quanta are normally released if they contain more than a specific, limiting amount of ACh (McLachlan, 1975b). According to the vesicle hypothesis, it is generally conceived that one quantum corresponds to the amount of ACh contained in a single synaptic vesicle present in a preganglionic nerve terminal [see above and Taxi (1961) and Yamamoto (1963)]. Nishi *et al.* (1967) estimated the number of ACh molecules contained in a single quantum to be 8000; this estimate was based on direct measurement of the ACh released in the perfusate of the ganglion and on calculation of the total number of quanta comprised in the fast EPSPs generated for the same period of time.

Miniature EPSPs of exceptionally large amplitude have been observed in amphibian sympathetic ganglion cells (Blackman *et al.*, 1963a; Kuba and Koketsu, 1978; Kuba and Minota, unpublished observations) and other autonomic ganglion cells (Bornstein, 1978; Martin and Pilar, 1964; Blackman *et al.*, 1969). These large mEPSPs are known to consist of multiquantal units (Bornstein, 1978), which are presumably the result of concomitant spontaneous release of several quanta due to quantal interaction, as will be discussed below.

3. Mode of Appearance of mEPSPs

The original analysis by Blackman *et al.* (1963a) indicated that mEPSPs of the frog sympathetic ganglion cell appear in a random fashion and that the interval distribution of their appearance is well described by a Poisson distribution. However, later investigations revealed that mEPSPs occasionally appeared in a bursting or clustered pattern, indicating interactions among quanta. Several tests for the applicability of Poisson distribution in hypogastric ganglia failed to prove the independence of mEPSPs (Bornstein, 1978). The mode of bursting of mEPSPs in the bullfrog sympathetic ganglion cell is more drastic than that found in other cells (Figure 1): The frequency of mEPSPs rises sporadically from a low level (a few mEPSPs per min) to a very high level that is more than 100 times higher than the low, resting level. The duration of bursts of mEPSPs is widely variable, ranging from several seconds to tens of minutes; similarly, the interval between bursts is not constant, ranging from several tens of minutes to several hours. Bursting can even be seen in a low-Ca^{2+}/high-Mg^{2+} solution (Kuba, Kumamoto, Minota, and Nohmi, unpublished observations). This clustering of mEPSPs does not seem to result from injury of the presynaptic nerve terminals by a recording electrode, since they occasionally occur after a quiescent period of low frequency lasting sometimes for more than one hour, and appear repetitively.

There are several possible mechanisms for the bursting of mEPSPs. It could not be due to spontaneous repetitive firings of the presynaptic terminals, since tetrodotoxin had no effect on bursting (Bornstein, 1978). The observation that the bursting of mEPSPs can be recorded, by means of an extracellular electrode, from a focal site at the presynaptic terminals of the neuromuscular junction (Cohen *et al.*, 1974; Bennett and Pettigrew,

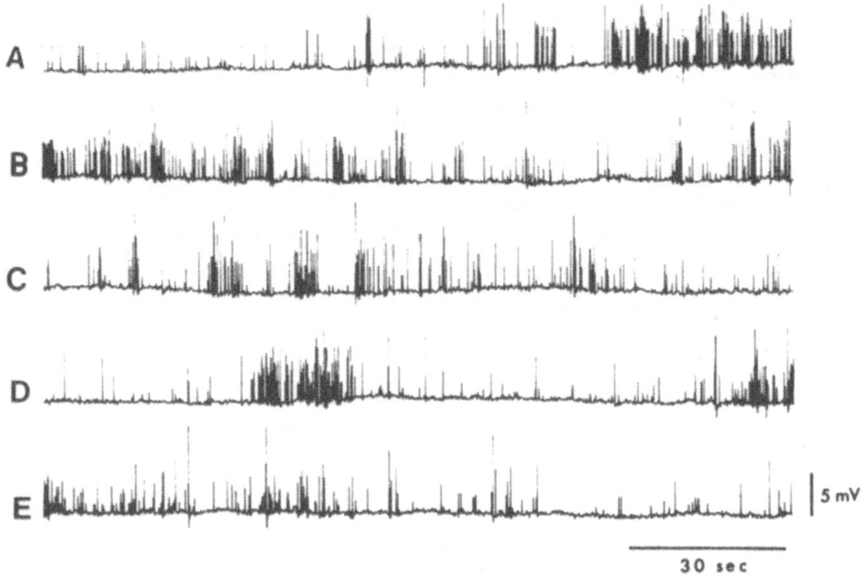

Figure 1. Miniature EPSPs recorded from a bullfrog sympathetic ganglion cell maintained in normal Ringer's solution. Note the bursting appearance of the mEPSPs. All the records (A–E) are continuous. From E. Kumamoto, K. Kuba, S. Minota, and M. Nhomi (unpublished observations).

1975) suggests that there are some interactions between individual release sites, but not among branches or varicosities. Bornstein (1978) suggested that the partial-activation hypothesis (cf. Hubbard, 1973) would account for the bursting of mEPSPs seen in the hypogastric ganglion. This hypothesis assumes that a site (X) essential for the activation of transmitter release is bound by several Ca^{2+} ions (Ca_1X, Ca_2X, Ca_3X); (see Section III.B). Following the preceding activation of ACh release, some of the sites remain in a partially activated state (Ca_1X or Ca_2X). Then, although the $[Ca^{2+}]_i$ in the unstimulated terminals is low, some Ca^{2+} in the terminals may bind to Ca_1X or Ca_2X and activate the ACh release. This idea is compatible with a statistical model proposed by Bornstein (1978) that is a modified version of a branching-Poisson model (Cohen *et al.*, 1974). This model assumes a primary, independent process that is involved in the release mechanism and a secondary process that depends on the primary process and exhibits a different time constant. Although this model describes well the bursting of mEPSPs in mammalian ganglia, it may not be appropriate for the bursting seen in bullfrog sympathetic ganglia, which occurs in a more intensive manner than that noticed in mammalian ganglia (see Figure 1).

Another interesting hypothesis is that there may be a spontaneous intracellular release of Ca^{2+} in the presynaptic nerve terminal. This Ca^{2+} release would occur in a regenerative manner from a Ca^{2+} reservoir, as seen in the postganglionic neuron (Kuba, 1980). This idea may conform to the appearance of a massive bursting of mEPSPs even in a low-Ca^{2+}/high-Mg^{2+} solution. It is to be noted that the fast EPSP amplitude was unchanged or only slightly increased during a bursting of mEPSPs. This observation may be accounted for if one assumes that the bursting of mEPSPs is a phenomenon that occurs at a single presynaptic bouton or varicosity, while the evoked release of ACh takes place at many terminals that are not in a bursting condition. The influx of Ca^{2+} by an action potential at the terminal where the bursting is occurring would have a small effect on the probability of ACh release, since the latter is already increased by the flux of intracellular Ca^{2+} release evoked by the action potential.

If the vesicular hypothesis is accepted, it may be assumed in turn that there might be some cooperative interaction between the exocytotic openings of synaptic vesicles at a specific site of the terminal membrane. Finally, it is also possible that the oscillation of the ACh concentration in the nerve terminal may occur as it does in the electric organ (Dunant *et al.*, 1977); this mechanism may be involved in the bursting of mEPSPs, if we accept the cytoplasmic ACh-release hypothesis. Altogether, although the mechanism for bursting is of great interest, the direct evidence that may support one or another hypothesis has not been made available as yet.

4. Effects of Ions and Tonicity

Effects of changes in extracellular ions on the frequency of mEPSPs in sympathetic ganglia are essentially similar to their effects on mEPP frequency at the motor end-plate. An increase in K^+ concentration ($[K^+]_o$) in the perfusing solution to less than 10 mM has little effect on the mEPSP frequency, but an increase to more than 10 mM causes a steep logarithmic rise in the frequency, indicating the existence of a threshold $[K^+]_o$ (Beume and Pott, 1978). Since the effect of $[K^+]_o$ must be due to an increased $[Ca^{2+}]_i$ in the nerve terminals, a depolarization induced by 10 mM $[K^+]_o$ would constitute the threshold for raising the Ca^{2+} permeability of the terminal membrane.

The effects of the extracellular Ca^{2+} concentration ($[Ca^{2+}]_o$) on mEPSP frequency depend on the $[K^+]_o$. At a normal $[K^+]_o$, the mEPSP frequency rises with an increase in $[Ca^{2+}]_o$. However, at a high $[K^+]_o$ of more than 15 mM, changes in the $[Ca^{2+}]_o$ produce diphasic effects; raising the $[Ca^{2+}]_o$ up to 2 mM increases, while a further increase in the $[Ca^{2+}]_o$ decreases, the mEPSP frequency (Beume and Pott, 1978). This mode of Ca^{2+} actions at a high $[K^+]_o$ is exactly similar to that prevailing at the motor end-plate (Gage and Quastel, 1966; Matthews and Wickelgren, 1977; Madden and Van der Kloot, 1978; Ohta and Kuba, 1980). A pertinent mechanism proposed by many investigators (Matthews and Wickelgren, 1977; Madden and Van der Kloot, 1978; Beume and Pott, 1978) is described as the surface potential hypothesis. According to this hypothesis, increasing $[Ca^{2+}]_o$ would reduce the surface potential of the membrane by screening the negative charges of the outer surface of the membrane; this would eventually raise the potential gradient within the terminal membrane. This would be effective in increasing mEPSP frequency when the membrane potential is in a range of the activation of the voltage-dependent Ca^{2+} conductance, e.g., when the membrane is depolarized by a high $[K^+]_o$ of more than 10 mM; it would not be effective at a normal $[K^+]_o$. Thus, $[Ca^{2+}]_o$ may have two opposite effects at a high $[K^+]_o$; one would lead to an increase of the electromotive force, while the other would reduce the Ca^{2+} conductance by increasing the voltage gradient in the membrane. Another mechanism that may also be involved, however, is as follows: When the $[Ca^{2+}]_i$ is preset at a relatively high level by raising the $[K^+]_o$, a further rise in $[Ca^{2+}]_i$ results in a reduction of the $[Ca^{2+}]_i$, presumably by the activation of the Ca^{2+}-buffering system (Ohta and Kuba, 1980). This hypothesis is based on the observation that the various procedures used to raise $[Ca^{2+}]_i$, such as the application of caffeine or tetanic nerve stimulation, decreased the mEPP frequency at a high $[K^+]_o$, similarly to the effect of increased $[Ca^{2+}]_o$.

It was reported that the frequency of mEPSPs in the guinea pig superior cervical ganglion cell is unaffected, at a normal $[K^+]_o$, by the replacement of total Ca^{2+} with equimolar Sr^{2+} or Ba^{2+}, while the evoked release of ACh is depressed by Sr^{2+} and blocked by Ba^{2+} (McLachlan, 1977). It is not known whether external Ca^{2+} of the nerve terminals, which is responsible for the occurrence of mEPSPs, can be substituted by Sr^{2+} or Ba^{2+} and whether these ions are as effective as Ca^{2+} for the activation of spontaneous ACh release. Interestingly, Ba^{2+} and Sr^{2+} accelerated a posttetanic rise in mEPSP frequency (McLachlan, 1977); (see below).

Bornstein (1979) observed a marked and steady increase in the mEPSP frequency in a hypertonic solution. The frequency was found to depend exponentially on the tonicity of the bathing solution. Two different mechanisms were considered: One was instituted by a charge-screening process that would account for the effect of hypertonicity in terms of a reduction of the electronic repulsion between synaptic vesicles and presynaptic membrane as a result of an increased ionic strength (Van der Kloot and Kita, 1973); the other mechanism involved $[Ca^{2+}]_i$ increase generated by increased tonicity (Shimoni *et al.*, 1977). While neither of these mechanisms or models is conclusive, it is of interest that the effect of a hypertonic solution on the bursting of mEPSPs resembles that due to increased $[Ca^{2+}]_o$.

B. Evoked Release

1. Poisson or Binomial Statistics

The Ca^{2+} influx caused by an action potential at the presynaptic terminal includes a synchronous release of many quanta of ACh and generates the fast EPSP at the subsynaptic membrane. The amplitude of a fast EPSP usually exceeds the threshold membrane potential for the generation of an action potential, thereby serving as a main pathway for ganglionic synaptic transmission (see Chapters 5 and 6). The quantal fluctuation of the fast EPSP amplitude in the frog sympathetic ganglion cell was first analyzed by Blackman *et al.* (1963b). The frequency distribution of the amplitude of the fast EPSP recorded in a low-Ca^{2+}/high-Mg^{2+} solution was reasonably fitted by a Poisson distribution, and the amplitude of a single quantum, which was estimated by dividing the mean amplitude of the fast EPSP by the mean quantal content (m) obtained from the number of failures of the EPSPs, was roughly equal to the mean amplitude of the mEPSP. Furthermore, the quantal content estimated from the variance of the fast EPSPs recorded in a low-Ca^{2+}/high-Mg^{2+} solution was in good agreement with that obtained under similar conditions from the number of failures of the fast EPSP in a bullfrog sympathetic ganglion cell (Figure 2). However, this is not the case in Ringer's solution, in which the probability of release is high. McLachlan (1975a) has shown that the binomial statistics describe better than Poisson distribution the fluctuation of the amplitude of the fast EPSPs recorded in Ringer's solution by a selective stimulation of a single preganglionic fiber, and she used the binomial distribution to calculate the probability of release (p) and the available transmitter store (n) during repetitive stimulation of sympathetic ganglia.

2. Effects of Ca^{2+} and Mg^{2+}

Bennett *et al.* (1976) observed the effect of changes in extracellular Ca^{2+} and Mg^{2+} concentrations on the statistical parameters, m, n, and p, of the release of ACh in rabbit superior cervical ganglia. They found that m depends on the 1.5th power of the $[Ca^{2+}]_o$ only in the case of a limited range of $[Ca^{2+}]_o$ (between 0.2 and 0.5 mM). The reason for the deviation from 1.5th-power dependence at concentrations of Ca^{2+} of more than 0.5 mM is the "saturation" of p ($p = 1$) that occurs when $p = \alpha$, α being the probability of quantal release at preganglionic-nerve-terminal varicosities. The available store was dependent on the 0.5th power of $[Ca^{2+}]_o$, while the probability of release was related to the 1st power of $[Ca^{2+}]_o$. On the other hand, these values were inversely proportional to similar powers of the extracellular Mg^{2+} concentration. These results indicate the bimodal role of Ca^{2+} in the mechanism of transmitter release. The

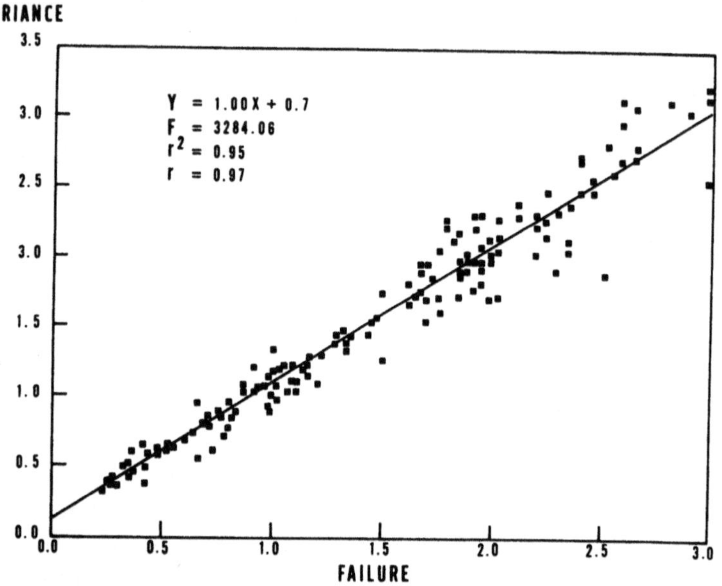

Figure 2. Comparison of quantal contents estimated from the variance of the fast EPSP amplitude with those estimated in terms of the number of failures. Bullfrog sympathetic ganglia. Data were obtained from 164 cells maintained in a low-Ca²/high-Mg^{2+} solution. From E. Kumamoto and K. Kuba (unpublished observations).

dependence of p of the preganglionic terminal on $[Ca^{2+}]_o$ resembles that of the motor-nerve terminal. In contrast, the cooperativity of $[Ca^{2+}]_o$ and n is different from their relationship at the motor-nerve terminal, where it depends on the 3rd power of $[Ca^{2+}]_o$ (Bennett and Florin, 1974). However, it is not known whether this difference in the cooperativity of $[Ca^{2+}]_o$ and n indicates a difference in the cooperativity of $[Ca^{2+}]_i$ and n, since the real relationship between $[Ca^{2+}]_o$ and an increase in $[Ca^{2+}]_i$ generated by the action potential at the presynaptic-nerve terminals is not known. In this context, it is worthwhile to note that there was no cooperative relationship between the magnitude of Ca^{2+} current at the presynaptic terminal of the squid giant synapse and the amplitude of the EPSP (Llinas *et al.*, 1981). Furthermore, it should be kept in mind that the estimation of n and p by means of binomial statistics is based on the assumption that the probabilities of release at each release site and at each release time are homogeneous for a given period of measurement (T. H. Brown *et al.*, 1976; Barton and Cohen, 1977). It was suggested that the number of quanta available for release corresponds to the number of varicosities or the number of synaptic vesicles attached to the presynaptic membrane (Bennett *et al.*, 1976), so that these constitute basic release units. If the conditions of each release site vary in a random fashion, so does the probability of quantal release; the binomial analysis of the release parameters would then not be meaningful. In a qualitative sense, it is clear that the release of ACh takes place in a quantal manner and that the number of quanta released is a function of the probability of release and the number of quanta that are available in the nerve terminals.

C. Regulation of Transmitter Liberation

The efficacy of transmitter release per single presynaptic impulse alters as a function of the previous activity and as a result of the regulating action of the extrinsic transmitter or hormone.

1. Intrinsic Mechanisms

Repetitive stimulation of the preganglionic nerve increases or decreases the amount of ACh released per single presynaptic impulse, depending on the experimental conditions (G. L. Brown and Feldberg, 1936; Birks and McIntosch, 1961). These phenomena, which lead to facilitation, potentiation, or depression of transmitter release, were described for many types of synapses; they are reflected as changes in the amplitude of the fast EPSP or the amount of ACh released in a perfusate of the ganglion.

Reduction of the fast EPSP amplitude or of the amount of ACh released in a perfusate that occurs during the repetitive stimulation was already discussed in the previous section. This depression of transmitter release is apparently due to the decrease in the amount of ACh stored in the presynaptic nerve terminals, since this depression may be related to a decrease in the total amount of ACh content in the ganglion (Birks and McIntosch, 1961) or to the diminution in the value of n (McLachlan, 1975b) or to both. These phenomena are especially marked when the synthesis of ACh is inhibited by HC-3 (Birks and McIntosch, 1961; Bennett and McLachlan, 1972a).

The mode of the facilitation of ACh release has generally been classified as either facilitation, in a narrower sense of this term, or potentiation. The former has a relatively short time–course, decaying with a time constant of tens or hundreds of milliseconds, while the latter has a time constant on the order of seconds or minutes and is conspicuous after tetanic stimulation applied at a high frequency; accordingly, this potentiation is also known as posttetanic potentiation (PTP). PTP of even longer duration—extending for hours—has also been described (see below). The mechanisms for these types of facilitation can be accounted for in terms of phenomena concerning the amount of ACh available for release, on one hand, and $[Ca^{2+}]_i$ in the nerve terminal, on the other.

As already discussed in the previous section, repetitive stimulation of presynaptic fibers significantly increases the ACh store in the nerve terminal. The time–course of the increase in the ACh store paralleled that of the enhancement of ACh released in a perfusate of the ganglion (Birks, 1977). Statistical analysis of the fast EPSP revealed an increase in the readily available store of quanta (n) during the facilitation of the fast EPSP (Bennett *et al.*, 1976). This result is in line with the hypothesis of the role of increased ACh store; it was also interpreted as due to the existence of the residual Ca-complex that is involved in the activation of ACh release (Bennett *et al.*, 1976).

The hypothesis of the residual Ca-complex was first suggested by Katz and Miledi (1968) for motor-nerve terminals; it was later applied by other investigators to many synapses, including sympathetic ganglionic synapses (Tashiro *et al.*, 1976). Bennett *et al.* (1976) found that a rise in quantal release by a test impulse following a conditioning impulse (corresponding to facilitation in the narrower sense) was primarily due to an increased n, while the enhanced quantal content of the successive EPSPs in an EPSP train was due to increases in n and p. These increases in n or p or both were accounted for by the binding of residual Ca^{2+} (entered during the preceding impulses) to Ca-receptors regulating the number of available quanta or the release probability. However, it should be kept in mind that this analysis is based on several assumptions, including the existence of two types of receptors for Ca^{2+} and the homogeneity of the release probabilities at each release site.

Recording the compound fast EPSP with a sucrose-gap method, Zengel *et al.* (1980) described four types of facilitation for the rabbit superior cervical ganglion. The facilitation in the narrower sense of the term was described as having an early component (with a decay time constant of 60 msec) and a late component (with a decay time constant of 400 msec), while the potentiation was described as having, again, two components, augmentation and potentiation, with decay time constants of 7.2 and 88 sec, respectively. These two latter components of potentiation were similar to those distinguished at the motor-nerve terminal (Zengel and Magleby, 1980). The interesting findings of Zengel and Magleby (1980) with regard to both the motor-nerve terminal and the ganglion are that the addition of Ba^{2+}, 0.1–0.2 mM, to a perfusing solution selectively increased the magnitude of the augmentation, but did not affect the other components, while Sr^{2+} enhanced the late component of facilitation only. Whether or not residual Ca^{2+} is present in the form of Ca-receptor complex or free Ca^{2+} in the nerve terminal, it may be inferred on the basis of the residual Ca hypothesis that the activities of several Ca^{2+} systems in the nerve terminal, which include the sequestration of Ca^{2+} in a Ca^{2+} reservoir and the extrusion of Ca^{2+} from the cell, either by a Ca-pump that utilizes ATP or by an Na–Ca exchange mechanism, would determine the rates of various components of facilitation.

It was recently found that in both mammalian and amphibian sympathetic ganglia, repetitive stimulation of the preganglionic nerve potentiates the fast EPSP for more than several tens of minutes or even for hours (T. H. Brown and McAffee, 1982; Koyano *et al.*, 1983, 1985; Kuba *et al.*, 1985; Briggs *et al.*, 1985). This potentiation of sympathetic ganglionic transmission resembles the long-term potentiation of the synaptic efficacy in hippocampal neurons, which was suggested to be a basis for learning and memory (cf. Bliss, 1979); (see also Chapter 21). Although the mechanism of the long-term potentiation of the fast EPSP of mammalian ganglia is not clear because extracellular recordings were employed in the pertinent experiments, the long-term potentiation found in bullfrog sympathetic ganglia was shown to be due to the increase in both the quantal content and the size of the fast EPSP (see Kumamoto and Kuba, 1983b; Koyano *et al.*, 1985). Furthermore, the long-term enhancement in the quantal content depended on the extracellular Ca^{2+}; contrary to PTPs of shorter duration, long-term potentiation was enhanced by lowering the temperature (Koyano *et al.*, 1985; Briggs *et al.*, 1985). Although the residual Ca^{2+} hypothesis might also explain the long-term potentiation of transmitter release, an additional process is needed to keep the intracellular Ca^{2+} at a high level. This mechanism is likely to require little metabolic energy, since lowering the temperature did not depress the enhancement of the quantal content (Koyano *et al.*, 1985). It is of interest that the long-term potentiation, while dependent on Ca^{2+}, was not sensitive to atropine, D-tubourarine, or α- and β-adrenergic blockers (Briggs *et al.*, 1985). Whatever mechanism is involved, these results suggest that the release of transmitter is regulated over long time periods by the preceding presynaptic activities, and both this use-dependent long-term potentiation and the long-term potentiation induced by epinephrine (see below) may contribute to the plastic regulation of the peripheral autonomic nervous system.

Still another presynaptic mechanism may be involved in the regulation of ganglionic responses. This mechanism involves appearance of repetitive presynaptic firing on presynaptic stimulation ("stimulus-bound repetitive firing"), which may be facilitated by drugs or intrinsic factors including ions (Riker and Kosay, 1970; Montoya and Riker, 1982). This subject is discussed in Chapter 13.

2. Extrinsic Mechanisms

The amount of ACh released in response to a presynaptic impulse is regulated by extrinsic mechanisms that involve the action of chemical substances that include neurotransmitters and hormones. A number of regulatory agents act on the presynaptic terminals, as described in Chapter 11. In this chapter, we will deal only with some of the modulatory effects of GABA and catecholamines exerted at the presynaptic terminals of the bullfrog sympathetic ganglion, which were discovered and described recently in our laboratories and those of Kurume University.

a. GABA. Following the discovery in the 1960s and early 1970s of GABA-induced inhibition of transmission in mammalian sympathetic ganglia (DeGroat, 1970), Adams and Brown (1975) suggested that this action is due to a shunting effect of GABA on the postsynaptic membrane in the case of mammalian ganglia, while Koketsu *et al.* (1974) related it to a presynaptic mechanism in that of amphibian ganglia (see Chapter 11). Subsequent studies (Kato *et al.*, 1978; Kato and Kuba, 1980) revealed that GABA causes a significant inhibition of transmitter release in the bullfrog sympathetic ganglion (Figure 3A, B); indeed, the quantal content of the fast EPSP was significantly decreased in the presence of GABA (1.5 μM to 1 mM) without a change in the quantal size. Also, GABA reduced the amplitude of the synaptic current underlying the fast EPSP. On the other

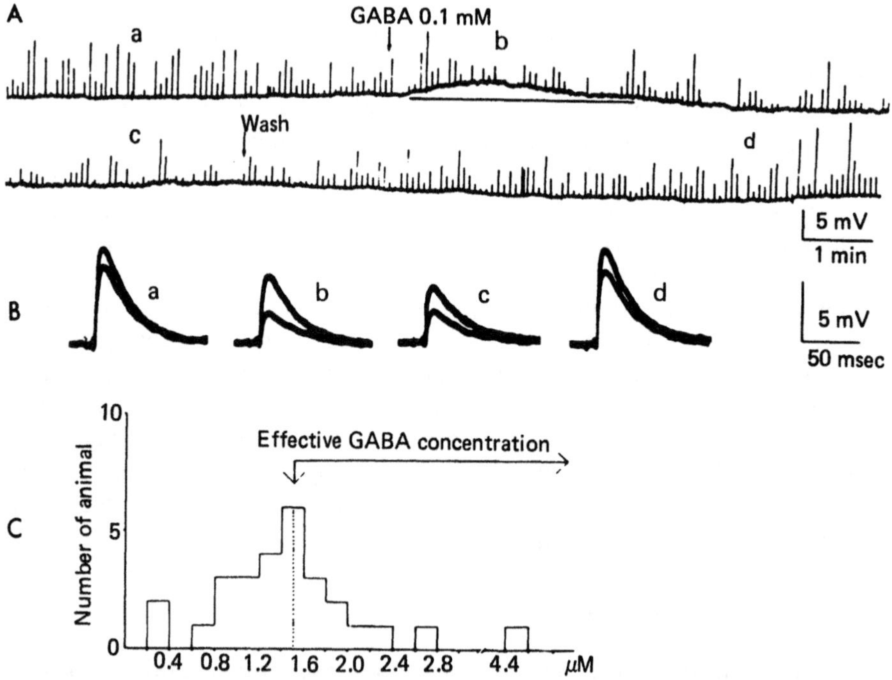

Figure 3. Effects of GABA (100 μM) on the fast EPSP and resting membrane potential in a low-Ca^{2+}/high-Mg^{2+} solution. (A) Records obtained via an ink-writing recorder. (B) Tracings represent photographs of cathode ray oscilloscope recordings. Each record in (B) corresponds to the time marked by the corresponding letter in (A). (C) Distribution of GABA concentration in sera of 28 different bullfrogs. *Ordinate:* number of animals; *abscissa:* GABA concentration. From Kato *et al.* (1980).

hand, GABA decreased only transiently the membrane potential and resistance, and did not affect the amplitude of ACh potentials. All these effects of GABA indicate its presynaptic site of action in the case of the bullfrog ganglion, in contrast to mammalian ganglia, where GABA exerts a major postsynaptic effect (Adams and Brown, 1975), although the existence of a presynaptic effect was also indicated more recently (D. A. Brown and Higgins, 1979).

Using a sucrose-gap technique, Koketsu *et al.* (1974) observed a depolarization of the presynaptic axon or some part of the presynaptic terminals, or both, in the presence of GABA (see also Chapter 11). However, this depolarization may not be responsible for the inhibitory action of GABA and, indeed, may not take place at the presynaptic terminal membrane, since the frequency of mEPSPs was unaffected by GABA. Another pertinent finding is that picrotoxin did not antagonize the action of GABA on the fast EPSP, while it antagonized GABA depolarization of what appeared to be, in the sucrose-gap recording, the nerve terminal; thus, GABA depolarization observed with the sucrose-gap technique may actually not have occurred at the nerve-terminal ending (Koketsu *et al.*, 1974). While it is possible that a membrane depolarization that does not affect the mEPSP frequency (see the previous section) may occur at the membrane of the active release site, lack of effect of an increase in the $[K^+]_o$ on the inhibitory action of GABA seems to further rule out the role of membrane depolarization in the mechanism of presynaptic action of GABA (Kato and Kuba, 1980).

The inhibition of the fast EPSP by GABA almost disappeared in Cl-free solution, indicating the involvement of the Cl conductance (G_{Cl}) of the presynaptic membrane in the GABA action. Thus, it may be suggested that GABA decreases the amplitude of an action potential at the presynaptic membrane by shunting the inward Na current as a result of an increased G_{Cl}; then, the reduced peak level of an action potential would result in decreased Ca^{2+} influx and eventually in a reduction of transmitter output. The observation that the amplitude of an action potential recorded with an extracellular electrode was reduced only slightly by GABA (Kato and Kuba, 1980) may not necessarily be inconsistent with this hypothesis, since the extracellularly recorded event that occurs during the course of the action potential represents membrane current flowing during an action potential, but not the peak of the latter; this explanation hinges on the assumption that there is no change in G_{Na} when G_{Cl} is affected by GABA and that GABA should therefore not affect extracellularly recorded spike amplitude (Kato and Kuba, 1980).

There appears to be no GABA neuron in the bullfrog sympathetic ganglion (see, however, Chapter 11). The concentration of GABA in the serum of the bullfrog was around 1.5 μM (Figure 3C) (Kato *et al.*, 1980), although this figure varied markedly from animal to animal. Since the least effective concentration of GABA amounts to 1.5 μM, it is possible that GABA present in the circulating blood may, under physiological conditions, regulate ganglionic transmission. The primary source of GABA in the blood is not known. However, the observation by Taniguchi *et al.* (1977) that high concentrations of GABA are present in the B cells of the rat pancreas is noteworthy in this context. It must be added that in contrast to that in mammalian ganglia, the uptake of GABA at glial cells (see Chapter 2) does not seem to control the local concentration of GABA in the bullfrog sympathetic ganglion (Kato *et al.*, 1980).

b. Epinephrine. Inhibitory action of epinephrine on the presynaptic terminals of mammalian sympathetic ganglia was evaluated in detail by Christ and Nishi (1971) and is described at length in Chapter 11 (see also Kato *et al.*, 1985). In this chapter, the plastic modulation in bullfrog sympathetic ganglia of ACh release by epinephrine is addressed specifically.

In the course of experiments concerning the possible involvement of cyclic adenosine 3′,5′-monophosphate (cAMP) in the inhibitory action of epinephrine on the presynaptic nerve terminal (Kato *et al.*, 1985), it was found that dibutyryl cAMP (db-cAMP) (0.1–4 mM), which is known to penetrate the membrane more easily than cAMP, exerted diphasic effects on the fast EPSP and its quantal content. First, it reduced slightly the amplitude of the fast EPSP, but not its quantal content. Then, 10–15 min after the removal of db-cAMP, the amplitude and quantal content of the fast EPSP were markedly increased (Figure 4B). The enhancement of the fast EPSP by db-cAMP lasted for more than 2–5 hr. A similar but smaller effect was also observed with a high (4 mM) concentration of cAMP. Adenosine and other nucleotides, such as ATP and ADP, did not exhibit such potentiating actions. On the other hand, a phosphodiesterase inhibitor, isobutylmethylxanthine (10 μM), caused a significant and long-lasting potentiation of the fast EPSP (Kuba *et al.*, 1978; Kato *et al.*, 1985). These findings should be related to the effects of epinephrine on ACh release in bullfrog sympathetic ganglia (Kuba *et al.*, 1981). Epinephrine (10 μM to 1 mM) reduced in most cells the amplitude of the fast EPSP recorded in a low-Ca^{2+}/high-Mg^{2+} solution, and its quantal content (Figure 4A). At 10–15 min after the removal of epinephrine, the values for the fast EPSP amplitude and for the quantal content recovered and, in fact, increased markedly over the control values, this potentiation lasting for as long as 2–5 hr. In the case of some cells, the fast EPSP amplitude remained unchanged or increased even in the presence of epinephrine. Throughout the action of epinephrine (1–10 μM), there was no significant change in the quantal size, the amplitude of ACh potential induced by iontophoretic application of ACh, the resting membrane potential, or the resistance. Thus, it appears that both the inhibitory and the long-lasting facilitatory actions of epinephrine were caused by its presynaptic actions. When these

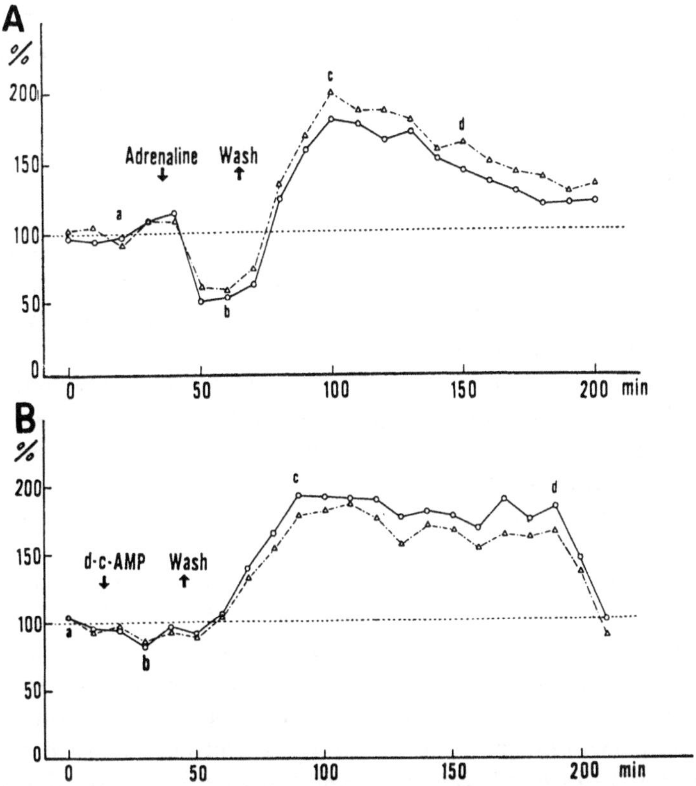

Figure 4. Effects of epinephrine (100 μM); (A) dibutyryl cAMP (1 mM); (B) on the amplitude and quantal content of the fast EPSP recorded from bullfrog sympathetic ganglion cells in a low-Ca^{2+}/high-Mg^{2+} solution. The points represent the mean amplitude (○) and quantal content (△) from 200 fast EPSPs induced at 0.3 Hz; they are expressed (ordinates) in terms of relative magnitude (%) as compared to control values. The abscissae represent time. From Kuba *et al.* (1981).

effects of epinephrine are considered jointly with those of cAMP and db-cAMP, it seems to be highly possible that endogenous cAMP is involved in a mechanism of long-lasting potentiating action of epinephrine, while it does not mediate the inhibitory action seen in the presence of epinephrine.

Spontaneous release of ACh is also enhanced after treatment with epinephrine or db-cAMP, the frequency of mEPSPs being markedly increased by epinephrine and db-cAMP. The adrenoreceptor responsible for the facilitatory action of epinephrine does not appear to be of either α-type or β-type. Both an α-agonist, dopamine (10 μM), and a β-agonist, isoprenaline (10 μM), significantly augmented the amplitude of the fast EPSP for more than 2 hr; furthermore, the potentiating action of epinephrine was suppressed in the presence of either an α-antagonist, phentolamine (10 μM), or a β-antagonist, phenoxybenzamine (10 μM). It is to be noted that the long-lasting potentiation of the fast EPSP by epinephrine was markedly suppressed at low temperature, indicating the participation of a metabolic process in the mechanism of its action (Kumamoto and Kuba, 1983a).

It may be hypothesized that the mechanism of plastic modulatory actions of epinephrine is as follows: When impulses from the central nervous system activate the postganglionic neuron or the internuncial cells [small intensely fluorescent (SIF) cells] or both, these may liberate catecholamines (cf. Libet and Owman, 1974; Suetake *et al.*, 1981) (see below and Chapters 2, 3, 11, and 13) and increase their concentration in the interstitial fluid bathing the presynaptic boutons. Alternatively, the catecholamine concentration in the ganglion may be raised by an increase in its concentration in the circulating blood due to the liberation of catecholamines from the adrenal medulla. This presence of catecholamines would inhibit ACh release, either by reducing Ca^{2+} current normally generated by the action potential (Minota and Koketsu, 1977) (but see Kato *et al.*, 1985) or by affecting the excitation–secretion coupling mechanism; at the same time, the catecholamines would raise the cAMP level in the nerve terminals, as occurs in the case of brain synaptosomes (Kobayashi *et al.*, 1981). Following a decrease in the catecholamine concentration around the nerve terminal, the inhibitory action may subside, while the increased level of cAMP triggers a long-lasting mechanism that leads to the potentiation of ACh release. This mechanism may involve a protein kinase and phosphorylation of a protein important for the excitation–secretion coupling or ACh release or both (Greengard, 1978); this protein may be the so-called protein I (Nestler and Greengard, 1982) or synapsin I (Navone *et al.*, 1984) (see also Chapter 13).

Long-lasting mechanisms for the potentiation of transmitter release induced by epinephrine and presynaptic activity (see above) underlie the concept of neuronal plasticity. A functional alteration of synaptic transmission has been seen in the case of many synapses, both central and peripheral, of various species (cf. Kupferman, 1979; Tsukahara, 1981), including, as shown here, the sympathetic ganglia. For instance, potentiation of the EPSP, lasting for many hours or even several days, has been observed in the case of the granule cells of the dentate gyrus (Bliss and Gardner-Medwin, 1973). The analysis in terms of the presynaptic mechanisms of this plastic potentiation that occurs in the central nervous system (CNS) may be difficult, the application to the CNS of quantal analysis being limited due to the complicated geometric structure of the central neurons. Thus, the potentiation of release induced by the action of epinephrine and/or presynaptic activity in the bullfrog sympathetic ganglion represents a useful model for the analysis of the synaptic plasticity such as that which may occur in the case of processes of learning and memory (Libet, 1970) (see also Chapters 12 and 13).

c. Antidromic Inhibition of ACh Release. It has been suggested that interactions between the postganglionic neurons and the presynaptic nerve terminals may occur in bullfrog sympathetic ganglia (Koketsu *et al.*, 1972; Minota and Koketsu, 1978). Furthermore, it is known that the postganglionic neurons contain epinephrine in their soma; this epinephrine may then act on the presynaptic terminals. This seems to be the case (Miyagawa *et al.*, 1981), since the amplitude of the fast EPSP was significantly decreased for several minutes after the tetanic stimulation of the postganglionic neuron, while the amplitude of the ACh potential induced by nicotinic action of ACh was increased under the same conditions (Figure 5). Therefore, it appears that the reduction of the fast EPSP amplitude after tetanic stimulation of postganglionic neurons may be due to a de-

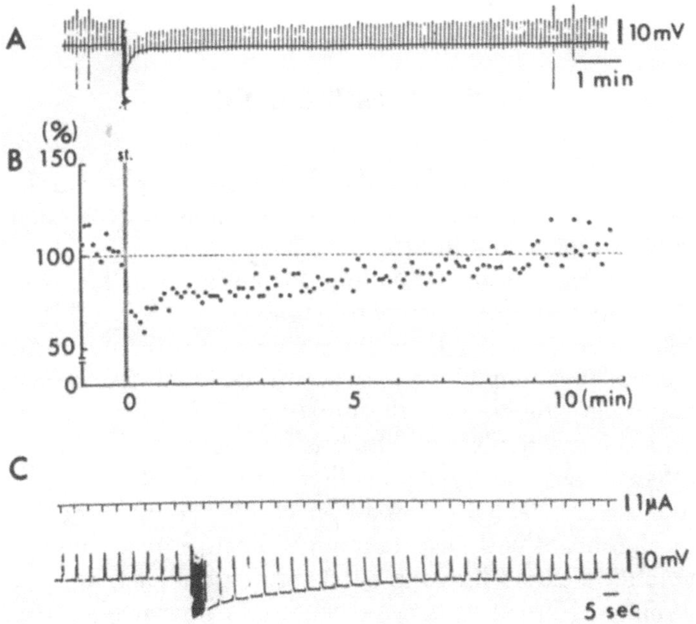

Figure 5. Effects of antidromic activation of the postganglionic neurons on the fast EPSPs recorded in the presence of D-tubocurarine (30 μM) (A, B) and on ACh potentials (C). The records were obtained in Ringer's solution. Fast EPSPs (A) and ACh potentials (C) (induced by iontophoresis of ACh) were recorded by means of a pen-writing oscillograph. Horizontal bars indicate time periods in the course of which tetanic stimulation (20 Hz, 5 sec) was applied to the postganglionic fibers. The relative amplitudes of the fast EPSPs are plotted against time in (B). From Miyagawa *et al.* (1981).

crease in the amount of ACh released caused by epinephrine liberated by the soma.

It may be that the activation of the postganglionic neuron causes a rise in the $[K^+]_o$ around the presynaptic nerve terminals; this might reduce the amplitude of the action potential of the presynaptic terminals, thereby decreasing the quantal release of ACh. However, this possibility seems to be ruled out by the finding that the frequency of mEPSPs decreases after tetanic stimulation of the postganglionic neuron; thus, spontaneous no less than evoked (via the action potential) release of ACh is diminished by the prior activity of the soma. Accordingly, it may be suggested that the tetanus releases from the postganglionic neurons a substance that depresses the presynaptic terminals; such a substance is likely to be epinephrine. This possibility is further substantiated by the finding that fast EPSP amplitude is not decreased, in the presence of epinephrine, after tetanic stimulation of the postganglionic neuron (Miyagawa *et al.*, 1981). Finally, the content of epinephrine in the ganglion cells as evaluated by a histochemical method is markedly reduced by antidromic stimulation (Suetake *et al.*, 1981) (Figure 6).

IV. CONCLUSIONS

This chapter focuses on a number of unresolved problems concerning the mechanism of ganglionic release of ACh, as well as on its special characteristics and potentialities.

The unresolved problems include the controversy between the concepts of vesicular and nonvesicular cytoplasmic release of ACh; it must be stressed, however, that whichever mechanism is involved, ACh release seems to occur in a quantal manner, according to either binomial or Poisson statistics. In this contest, the bursting of mEPSPs is an interesting phenomena the mechanism of which is not clear at present.

Another unresolved problem is that of the role of Ca^{2+} in the control of ACh release as affected by repetitive stimulation; obviously, this role of Ca^{2+} is significant with respect to physiological phenomena of release, since the latter occurs following bursts of neuronal activity (see Chapter 18). In a related context, it is not clear at this time whether or not such

→

Figure 6. Formaldehyde-induced fluorescence photomicrographs of a pair of bullfrog paravertebral sympathetic ganglia. Ganglia were previously incubated for 5 hr in Ringer's containing α-methyl-p-tyrosine (80 μg/ml), an inhibitor of catecholamine synthesis. The specimens were exposed for 3 hr to formaldehyde vapor at 75% relative humidity and 80°C. (A) Unstimulated control ganglion. (B) Ganglion stimulated at 30 Hz for 60 min. From Suetake *et al.* (1981). Photographs kindly provided by Dr. H. Kojima.

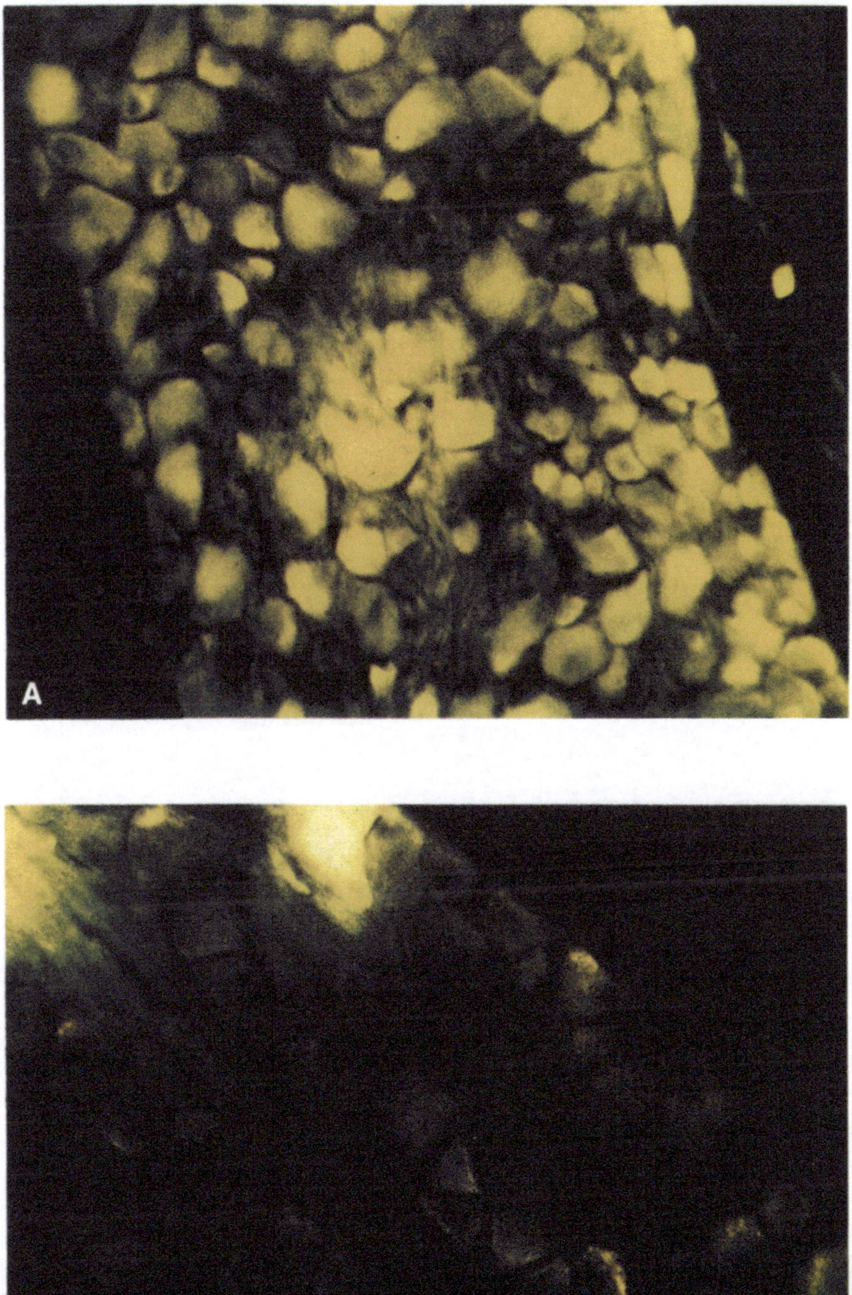
A
B

endogenous substances as GABA, epinephrine, and others (see Chapter 11) that obviously are effective pharmacologically in the control of ganglionic release of ACh are also involved in physiological regulation of this release.

An interesting characteristic of ganglionic release of ACh is its plasticity and capacity for feedback regulation. Previous neuronal activity including antidromic effects, several Ca^{2+} compartments, endogenous substances as well as drugs, cooperativity between release sites, as well as ACh synthesis and storage—all contribute to this plasticity. In this chapter, the involvement of epinephrine and cyclic nucleotides in neuronal plasticity processes was particularly reviewed. These presynaptic mechanisms that modulate ganglionic release of ACh (see Chapter 3) obviously play an important role in the physiological control and regulation of ganglionic transmission and its moment-to-moment control.

Note added in proof. The type of adrenoreceptor for the epinephrine-induced potentiation appears to be α-type, since phenoxybenzamine (1 μM) suppressed the potentiation, but phentolamine (1 μM) did not (K. Kuba and E. Kumamoto, unpublished observation).

ACKNOWLEDGMENTS. Our published and unpublished work was generously supported in part by Grants-in-Aid for Special Project Research from the Ministry of Education, Science and Culture of Japan. The assistance of Mrs. Margaret Gray is gratefully acknowledged.

REFERENCES

Adams, P. R., and Brown, D. A.: Actions of γ-aminobutyric acid on sympathetic ganglion cells. *J. Physiol. (London)* **250**:85–120 (1975).

Barton, S. B., and Cohen, I. S.: Are transmitter release statistics meaningful? *Nature (London)* **268**:267–268 (1977).

Bennett, M. R., and Florin, T.: A statistical analysis of the release of acetylcholine at newly formed synapses in striated muscle. *J. Physiol. (London)* **238**:93–107 (1974).

Bennett, M. R., and McLachlan, E. M.: An electrophysiological analysis of the storage of acetylcholine in preganglionic nerve terminals. *J. Physiol. (London)* **221**:657–668 (1972a).

Bennett, M. R., and McLachlan, E. M.: An electrophysiological analysis of the synthesis of acetylcholine preganglionic nerve terminals. *J. Physiol. (London)* **221**:669–682 (1972b).

Bennett, M. R., and Pettigrew, A. G.: The formation of synapses in amphibian striated muscle during development. *J. Physiol. (London)* **25**:203–239 (1975).

Bennett, M. R., Florin, T., and Pettigrew, A. G.: The effect of calcium ions on the binomial statistics parameters that control acetylcholine release at preganglionic nerve terminals. *J. Physiol. (London)* **257**:597–620 (1976).

Beume, R., and Pott, L.: Dual effect of external calcium on the frequency of miniature synaptic potentials in frog sympathetic ganglion cells. *Pfluegers Arch.* **376**:21–26 (1978).

Birks, R. I.: The relationship of transmitter release and storage to structure in a sympathetic ganglion. *J. Neurocytol.* **3**:133–160 (1974).

Birks, R. I.: A long-lasting potentiation of transmitter release related to an increase in transmitter store in a sympathetic ganglion. *J. Physiol. (London)* **271**:847–862 (1977).

Birks, R. I.: Regulation by paterned preganglionic neural activity of transmitter stores in a sympathetic ganglion. *J. Physiol. (London)* **280**:559–572 (1978).

Birks, R. I., and Fitch, J. C.: Storage and release of acetylcholine in sympathetic ganglion. *J. Physiol. (London)* **240**:125–134 (1974).

Birks, R. I., and McIntosch, F. C.: Acetylcholine metabolism in a sympathetic ganglion. *Can. J. Biochem. Physiol.* **39**:787–827 (1961).

Blackman, J. G., and Purves, R. D.: Intracellular recordings from ganglia of the thoracic sympathetic chain of the guinea-pig. *J. Physiol. (London)* **203**:173–198 (1969).

Blackman, J. G., Ginsborg, B. L., and Ray, C.: Spontaneous synaptic activity in sympathetic ganglion cells of the frog. *J. Physiol. (London)* **167**:389–401 (1963a).

Blackman, J. G., Ginsborg, B. L., and Ray, C.: On the quantal release of the transmitter at a sympathetic synapse. *J. Physiol. (London)* **167**:402–415 (1963b).

Blackman, J. G., Growcroft, P. J., Devine, C. E., Holman, M. E., and Yonemura, K.: Transmission from preganglionic fibers in the hypogastric nerve to peripheral ganglia of male guinea-pigs. *J. Physiol. (London)* **201**:723–743 (1969).

Bliss, T. V. P.: Synaptic plasticity in the hippocampus. *Trends Neurosci.* **2**:42–45 (1979).

Bliss, T. V. P., and Gardner-Medwin, A. R.: Long-lasting potentiation of synaptic transmission in the dentate area of the unanesthetized rabbit following stimulation of the perforant path. *J. Physiol. (London)* **232**:357–374 (1973).

Bornstein, J. C.: Spontaneous multiquantal release of synapses in guinea-pig hypogastric ganglia: Evidence that release can occur in bursts. *J. Physiol. (London)* **282**:375–398 (1978).

Bornstein, J. C.: Some effects of hypertonic solution in the properties of spontaneous transmitter release in the hypogastric ganglia of guinea-pigs. *J. Physiol. (London)* **290**:11–22 (1979).

Briggs, C. A., Brown, T. A., and McAffee, D. A.: Neurophysiology and pharmacology of long-term potentiation in the rat sympathetic ganglion. *J. Physiol. (London)* **359**:503–521 (1985).

Brown, D. A., and Higgins, A. J.: Presynaptic effects of γ-aminobutyric acid in isolated rat superior cervical ganglia. *Br. J. Pharmacol.* **66**:108P–109P (1979).

Brown, G. L., and Feldberg, W.: The acetylcholine metabolism of "sympathetic ganglion." *J. Physiol. (London)* **88**:265–283 (1936).

Brown, T. H., and McAffee, D. A.: Long-term synaptic potentiation in the superior cervical ganglion. *Science* **215**:1411–1413 (1982).

Brown, T. H., Perkel, D. H., and Feldman, M. W.: Evoked neurotransmitter release: Statistical effects of nonuniformity and nonstationarity. *Proc. Natl. Acad. Sci. U.S.A.* **73**:2913–2917 (1976).

Ceccarelli, B., and Hurlbut, W. P.: Vesicle hypothesis of the release of quanta or acetylcholine. *Physiol. Rev.* **60**:396–441 (1980).

Christ, D. D., and Nishi, S.: Site of adrenaline blockade in the superior cervical ganglion of the rabbit. *J. Physiol. (London)* **213**:107–117 (1971).

Cohen, I., Kita, H., and Van der Kloot, W.: The stochastic properties of spontaneous quantal release of transmitter at the frog neuromuscular junction. *J. Physiol. (London)* **236**:341–361 (1974).

Collier, B.: The preferential release of newly synthesized transmitter by a sympathetic ganglion. *J. Physiol. (London)* **205**:341–352 (1969).

DeGroat, W. C.: The actions of γ-aminobutyric acid and related amino acids on mammalian autonomic ganglia. *J. Pharmacol. Exp. Ther.* **172**:384–396 (1970).

DelCastillo, J., and Katz, B.: Quantal component of the end-plate potential. *J. Physiol. (London)* **124**:560–573 (1954).

Dunant, Y., and Israel, M.: When the vesicular hypothesis is no longer the vesicular hypothesis. *Trends Neurosci.* **2**:130–132 (1979).

Dunant, Y., Israel, M., Lesbats, B., and Manaranche, R.: Oscillation of acetylcholine during nerve activity in the *Torpedo* electric organ. *Brain Res.* **125**:123–140 (1977).

Dunant, Y., Corthay, J., Eder, L., and Loctin, F.: Acetylcholine changes during transmission of a single nerve impulse, in: *Cholinergic Mechanisms* (G. Pepeu and H. Ladinsky, eds.), pp. 287–299, Plenum Press, New York (1981).

Eccles, R. M.: Intracellular potentials recorded from mammalian sympathetic ganglion. *J. Physiol. (London)* **130**:572–584 (1955).

Feldberg, W., and Gaddum, J. H.: The chemical transmitter at synapses in a sympathetic ganglion. *J. Physiol. (London)* **81**:305–319 (1934).

Fonnum, F.: Choline acetyltransferase binding to and release from membrane. *Biochem. J.* **109**:389–398 (1968).

Gage, P. W., and Quastel, D. M. J.: Competition between sodium and calcium ions in transmitter release at mammalian neuromuscular junction. *J. Physiol. (London)* **185**:95–123 (1966).

Greengard, P.: Phosphorylated proteins as physiological effectors. *Science* **199**:146–152 (1978).

Hubbard, J. I.: Microphysiology of vertebrate neuromuscular transmission. *Physiol. Rev.* **53**:674–723 (1973).

Kato, E., and Kuba, K.: Inhibition of transmitter release in bullfrog sympathetic ganglia induced by γ-aminobutyric acid. *J. Physiol. (London)* **298**:271–283 (1980).

Kato, E., Koketsu, K., Kuba, K., and Kumamoto, E.: The mechanism of the inhibitory action of adrenaline on transmitter release in bullfrog sympathetic ganglia: Independence of cyclic AMP and calcium ions. *Br. J. Pharmacol.* **84**:435–443 (1985).

Kato, E., Kuba, K., and Koketsu, K.: Presynaptic inhibition by γ-aminobutyric acid in bullfrog sympathetic ganglion cells. *Brain Res.* **153**:389–402 (1978).

Kato, E., Morita, K., Kuba, K., Yamada, S., Kuhara, T., Shinka, T., and Matsumoto, I.: Does γ-aminobutyric acid in blood control transmitter release in bullfrog sympathetic ganglia? *Brain Res.* **195**:208–214 (1980).

Katz, B.: *The Release of Neural Transmitter Substances.* Sherrington Lecture No. 10, Liverpool University Press (1969).

Katz, B., and Miledi, R.: The role of calcium in neuromuscular facilitation. *J. Physiol. (London)* **195**:481–492 (1968).

Katz, B., and Miledi, R.: Transmitter leakage from motor nerve endings. *Proc. R. Soc. London Ser. B* **196**:59–72 (1977).

Kibjakow, A. W.: Über humorale Übertragung der Erregung von einem Neuron auf das andere. *Pfluegers Arch. Gesaute Physiol.* **232**:432–443 (1933).

Kobayashi, K., Kuroda, Y., and Yoshioka, M.: Change of cyclic AMP level in synaptosomes from cerebral cortex: Increase by adenosine derivatives. *J. Neurochem.* **36**:86–91 (1981).

Koketsu, K., Minota, S., and Nakamura, M.: The presynaptic potential produced by postsynaptic activation in bullfrog sympathetic ganglia. *Life Sci.* **11**:263–268 (1972).

Koketsu, K., Shoji, T., and Yamamoto, K.: Effects of GABA on presynaptic nerve terminals in bullfrog (*Rana catesbeiana*) sympathetic ganglia. *Experientia* **30**:382–383 (1974).

Koyano, K., Kuba, K., Kumamoto, E., Minota, S., and Nohmi, M.: Synaptic plasticity in sympathetic ganglia. I. Long-term potentiation induced by presynaptic activities. *J. Physiol. Soc. Jpn.* **45**:535 (1983).

Koyano, K., Kuba, K., and Minota, S.: Long-term potentiation of transmitter release induced by repetitive presynaptic activities in bullfrog sympathetic ganglia. *J. Physiol. (London)* **359**:219–333 (1985).

Kuba, K.: Release of calcium ions linked to the activation of potassium conductance in a caffeine-treated sympathetic neuron. *J. Physiol. (London)* **298**:251–269 (1980).

Kuba, K., and Koketsu, K.: Synaptic events in sympathetic ganglia. *Prog. Neurobiol.* **11**:77–170 (1978).

Kuba, K., Kato, E., Akasu, T., and Koketsu, K.: The role of cyclic AMP in synaptic transmission. *J. Physiol. Soc. Jpn.* **40**:232 (1978).

Kuba, K., Kato, E., Kumamoto, E., Koketsu, K., and Hirai, K.: Sustained potentiation of transmitter release by adrenaline and dibutyryl cyclic AMP in sympathetic ganglia. *Nature (London)* **291**:654–656 (1981).

Kuba, K., Kumamoto, E., Minota, S., and Koyano, K.: Three types of long-term potentiation in bullfrog sympathetic ganglia. Proceedings of Ninth International Symposium on Neurosecretion (E. Kobayashi, H. A. Bern, and A. Urano, eds.), pp. 308–314, Japan Scientific Society Press, Tokyo (1985).

Kumamoto, E., and Kuba, K.: Independence of presynaptic bimodal actions of adrenaline in sympathetic ganglia. *Brain Res.* **265**:344–347 (1983a).

Kumamoto, E., and Kuba, K.: Sustained rise in ACh sensitivity of a sympathetic ganglion cell induced by postsynaptic electrical activities. *Nature (London)* **305**:145–146 (1983b).

Kupferman, I.: Modulatory actions of neurotransmitters. *Annu. Rev. Neurosci.* **2**:447–465 (1979).

Libet, B.: Generation of slow inhibitory and excitatory postsynaptic potentials. *Fed. Proc. Fed. Am. Soc. Exp. Biol.* **29**:1945–1956 (1970).

Libet, B., and Owman, C.: Concomitant changes in formaldehyde-induced fluorescence of dopamine interneurons and in slow inhibitory postsynaptic potentials of the superior cervical ganglion, induced by stimulation of the preganglionic nerve or by a muscarinic agent. *J. Physiol. (London)* **237**:635–662 (1974).

Llinas, R., Steinberg, I. Z., and Walton, K.: Relationship between presynaptic calcium current and postsynaptic potential in squid giant synapse. *Biophys. J.* **33**:323–352 (1981).

Madden, K. S., and Van der Kloot, W.: Surface changes and the effects of calcium on the frequency of miniature end-plate potentials at the frog neuromuscular junction. *J. Physiol. (London)* **276**:227–232 (1978).

Marchbanks, R. M.: The vesicular hypothesis questioned. *Trends Neurosci.* **1**:83–84 (1978).

Marchbanks, R. M.: Interaction of choline transport with acetylcholine release and synthesis, in: *Cholinergic Mechanisms* (G. Pepeu and H. Ladinsky, eds.), pp. 489–496, Plenum Press, New York (1981).

Martin, A. R., and Pilar, A.: Quantal components of the synaptic potential in the ciliary ganglion of the chick. *J. Physiol. (London)* **175**:1–16 (1964).

Matthews, G., and Wickelgren, W. O.: On the effect of calcium on the frequency of miniature end-plate potentials at the frog neuromuscular junction. *J. Physiol. (London)* **266**:91–101 (1977).

McLachlan, E. M.: Analysis of the release of acetylcholine from preganglionic nerve terminals. *J. Physiol. (London)* **245**:447–466 (1975a).

McLachlan, E. M.: Changes in statistical release parameters during prolonged stimulation of preganglionic nerve terminals. *J. Physiol. (London)* **253**:447–491 (1975b).

McLachlan, E. M.: The effects of strontium and barium ions at synapses in sympathetic ganglia. *J. Physiol. (London)* **267**:497–518 (1977).

Minota, S., and Koketsu, K.: Effects of adrenaline on the action potential of sympathetic ganglion cells in bullfrogs. *Jpn. J. Physiol.* **27**:353–366 (1977).

Minota, S., and Koketsu, K.: Recurrent synaptic activation of the bullfrog sympathetic ganglion cells by direct intracellular stimulation. *Jpn. J. Physiol.* **28**:799–806 (1978).

Miyagawa, M., Minota, S., and Koketsu, K.: Antidromic inhibition of acetylcholine release from presynaptic nerve terminals in bullfrog's sympathetic ganglia. *Brain Res.* **224**:305–313 (1981).

Montoya, G. A., and Riker, W. R.: A study of the actions of bromide on frog sympathetic ganglion. *Neuropharmacology* **21**:581–585 (1982).

Navone, F., Greengard, P., and De Camilli, P.: Synapsin I in nerve terminals: Selective association with small synaptic vesicles. *Science* **226**:1209–1211 (1984).

Nestler, E. J., and Greengard, P.: Distribution of protein I and regulation of its state of phosphorylation in the rabbit superior cervical ganglion. *J. Neurosci.* **2**:1011–1023 (1982).

Nishi, S., and Koketsu, K.: Electrical properties and activities of single sympathetic neurons of frogs. *J. Cell. Comp. Physiol.* **55**:15–30 (1960).

Nishi, S., Soeda, H., and Koketsu, K.: Release of acetylcholine from sympathetic preganglionic nerve terminals. *J. Neurophysiol.* **30**:114–134 (1967).

Ohta, Y., and Kuba, K.: Inhibitory action of Ca^{2+} on spontaneous transmitter release at motor nerve terminals in a high K^{+} solution. *Pfluegers Arch.* **386**:29–34 (1980).

Riker, W. K., and Kosay, S.: Drug induction and suppression of stimulus-bound repetition in sympathetic ganglia. *J. Pharmacol. Exp. Ther.* **173**:284–292 (1970).

Sacchi, O., and Perri, V.: Quantal release of acetylcholine from the nerve endings of the guinea-pig superior cervical ganglion. *Pfluegers Arch.* **329**:207–219 (1971).

Shimoni, Y., Alnaes, E., and Rahamimoff, R.: Is hyperosmotic neurosecretion from motor nerve endings a calcium-dependent process? *Nature (London)* **267**:170–172 (1977).

Suetake, K., Kojima, H., Inanaga, K., and Koketsu, K.: Catecholamine is released from nonsynaptic cell-soma membrane: Histochemical evidence in bullfrog sympathetic ganglion cells. *Brain Res.* **205**:436–440 (1981).

Taniguchi, H., Okada, Y., Shimada, C., and Baba, S.: GABA in pancreatic islets. *Arch. Histol. Jpn.* **40**(Suppl.):87–97 (1977).

Tashiro, N., Gallagher, J. P., and Nishi, S.: Facilitation and depression of synaptic transmission in amphibian sympathetic ganglia. *Brain Res.* **118**:45–62 (1976).

Tauc, L.: Are vesicles necessary for release of acetylcholine at cholinergic synapses? *Biochem. Pharmacol.* **27**:3493–3498 (1979).

Tauc, L.: The vesicular hypothesis of acetylcholine release from neurons: Are there alternatives?, in, *Trends in Autonomic Pharmacology*, Vol. 2 (S. Kalsner, ed.), pp. 43–60, Urban and Schwarzenberg, Baltimore (1982).

Taxi, J.: Etude de l'ultrastructure des zones synaptiques dans les ganglions sympathiques de la grenouille. *C. R. Acad. Sci.* **252**:174–176 (1961).

Tsukahara, N.: Synaptic plasticity in the mammalian central nervous system. *Annu. Rev. Neurosci.* **4**:351–379 (1981).

Van der Kloot, W., and Kita, H.: The possible role of fixed membrane surface charges in acetylcholine release at the frog neuromuscular junction. *J. Membr. Biol.* **14**:365–382 (1973).

Whittaker, V. P.: The vesicle hypothesis. *Trends Neurosci.* **2**:55–56 (1979).

Yamamoto, T.: Some observations on the fine structure of sympathetic ganglion in bullfrog. *J. Cell Biol.* **16**:159–170 (1963).

Zengel, J. E., and Magleby, K. L.: Differential effects of Ba^{2+}, Sr^{2+}, and Ca^{2+} on stimulation-induced changes in transmitter release at the frog neuromuscular junction. *J. Gen. Physiol.* **76**:175–211 (1980).

Zengel, J. E., Magleby, K. L., Horn, J. P., McAffee, D. A., and Yarowsky, P. J.: Facilitation, augmentation, and potentiation of synaptic transmission at the superior cervical ganglion of the rabbit. *J. Gen. Physiol.* **76**:213–231 (1980).

Zimmermann, H.: Vesicle recycling and transmitter release. *Neuroscience* **4**:1773–1804 (1979).

Zimmermann, H., and Denston, C. F.: Separation of synaptic vesicles of different functional states from the cholinergic synapses of the *Torpedo* electric organ. *Neuroscience* **2**:715–730 (1977).

11

Presynaptic Modulation

Endogenous Substances with Ganglionic Depressant Actions

DARYL D. CHRIST and NAE J. DUN

I. PRESYNAPTIC MODULATION

Efferent autonomic activity is transmitted from the central nervous system through the autonomic ganglia to target cells. The frequency of action potentials reaching the neuroeffector junctions depends on the efficacy of ganglionic transmission. Many endogenous substances can alter synaptic transmission in sympathetic ganglia (see Chapter 3) and thereby alter target-organ activity. Theoretically, they could act by altering the release of neurotransmitter from the ganglionic neuron, by altering the sensitivity of the postganglionic neuron to the neurotransmitter, or by modulating the postsynaptic action potential configuration. This chapter reviews some of the research relating to the actions of endogenous substances on the release of transmitter from preganglionic neurons. Substances with modulatory postsynaptic actions will be discussed in Chapter 12.

DARYL D. CHRIST • South Bend Center for Medical Education, Indiana University School of Medicine, Notre Dame, Indiana 46556. NAE J. DUN • Department of Pharmacology and Experimental Therapeutics, Loyola University Stritch School of Medicine, Maywood, Illinois 60153.

II. ENDOGENOUS SUBSTANCES WITH PRESYNAPTIC DEPRESSANT ACTIONS

A. Catecholamines

1. Ganglionic Blockade by Catecholamines

The initial report of catecholamine actions on ganglia was provided by Marrazzi (1939a), who observed that epinephrine, administered intraarterially, blocked ganglionic transmission. The blocking action was confirmed by many investigators, and facilitation by catecholamines was also observed (Bulbring, 1944; Lundberg, 1952; Pardo *et al.*, 1963). DeGroat and Volle (1966) concluded that depression and facilitation of ganglionic transmission induced by the catecholamines were mediated by two types of adrenoceptors: The depressant actions of norepinephrine and epinephrine were antagonized by α-adrenoceptor-blocking drugs, whereas the facilitatory actions of epinephrine and isoproterenol were antagonized by β-adrenoceptor-blocking drugs.

The mechanisms of the ganglionic actions of catecholamines were not clearly elucidated in the early studies. The ganglionic depression was accompanied by a ganglionic hyperpolarization; however, the magnitudes of the hyperpolarization and the depression did not correlate well (Lundberg, 1952; McIsaac, 1966; Haefely, 1969). The results of experiments carried out with intracellular electrodes indicated that a presynaptic site was involved in the ganglionic blocking action of the catecholamines (Christ and Nishi, 1971a; Dun and Nishi, 1974; Dun and Karczmar, 1977). First, there was an apparent lack of postsynaptic action of catecholamines: Catecholamines reduced the amplitude of the excitatory postsynaptic potential (EPSP) at concentrations that did not appreciably affect the resting membrane potential of the postganglionic cell. Second, there was no significant change in the electrical properties of the postganglionic cell. Finally, there was no appreciable effect on the sensitivity of the postganglionic cell to iontophoretically applied acetylcholine (ACh) (Figure 1). On the other hand, there was ample evidence for a presynaptic action of catecholamines: The catecholamines significantly depressed the release of ACh from the preganglionic nerve endings, as determined by measuring the quantal content of the EPSP in low $[Ca^{2+}]/[Mg^{2+}]$. The depressant effect of the catecholamines on the EPSP was prevented by α-adrenoceptor antagonists (Christ and Nishi, 1971a; Dun and Nishi, 1974; Dun and Karczmar, 1977).

The results of several other experiments implicate a presynaptic α-adrenergic action for the catecholamines. First, [^{3}H]dihydroergocryptine, an α-adrenoceptor antagonist, was found to bind to membranes of the rat

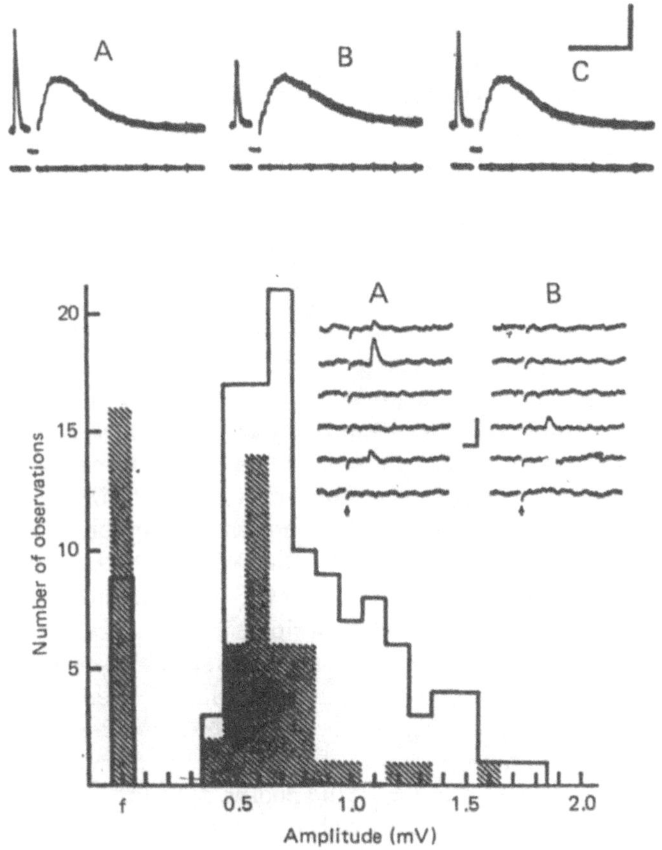

Figure 1. Actions of epinephrine on the EPSPs from rabbit superior cervical ganglia. The upper part of this figure shows EPSPs and ACh potentials. The current pulses (8×10^{-8} A) for iontophoretic applications of ACh are shown in the lower traces. (A, C) Controls. (B) Recorded in the presence of 10^{-5} M epinephrine. Note that the EPSP was reduced by epinephrine, but the ACh potential was unchanged. Calibration: 20 mV and 300 msec. The lower part of this figure contains histograms of EPSP amplitudes in 0.5 mM $CaCl_2$ and 5.5 mM $MgCl_2$. The failures (*f*) are plotted at 10 times the ordinate values. Epinephrine (10^{-5} M) is indicated by the cross-hatched area. *Inset:* Records of EPSPs before (A) and during (B) epinephrine superfusion. Indirect stimulation occurs at the arrow. Calibrations: 2 mV and 10 msec. Note that epinephrine increased the number of failures, but did not change the amplitude of the unit response. From Christ and Nishi (1971a).

superior cervical ganglion (Kafka and Thoa, 1979). The binding was reduced by α-adrenoceptor agonists and antagonists, but not by β-adrenoceptor agonists and antagonists. Sectioning of the preganglionic nerve reduced the binding of [^{3}H]dihydroergocryptine by approximately 50% after 7 days, indicating that many of the binding sites were located on the preganglionic terminals. Second, the efflux of [^{3}H]-ACh induced by stimulation of the preganglionic nerve of the isolated superior cervical ganglion of the cat was significantly reduced by norepinephrine, whereas the spontaneous efflux of [^{3}H]-ACh from the unstimulated ganglia was not depressed by the catecholamines (Martinez and Adler-Graschinsky, 1980). The depressant effect was reduced by an α-adrenoceptor antagonist (Dawes and Vizi, 1973).

Together, these results indicate that catecholamines block ganglionic transmission by acting on α-adrenoceptors on the preganglionic nerve endings. Interestingly, presynaptic ganglionic blockade by dopamine also appeared to be mediated through α-adrenoceptors, rather than through dopamine receptors, since the depressant effect of dopamine was antagonized by α-adrenoceptor antagonists (Dun and Nishi, 1974).

The cellular mechanisms of the presynaptic depressant effect of the catecholamines have been difficult to elucidate. Epinephrine did not change the threshold of presynaptic nerve endings to depolarizing current, indicating that epinephrine did not alter the resting membrane potential of the nerve-terminal membrane (Christ and Nishi, 1971b). On the other hand, catecholamines suppressed several Ca^{2+}-dependent phenomena in postganglionic neurons of the rat superior cervical ganglion, and it has been suggested that the presynaptic depressant action of the catecholamines may also be due to a suppression of voltage-dependent Ca^{2+} conductance changes in the preganglionic nerve terminal (Horn and McAfee, 1980). This "plastic" modulation of ACh release by catecholamines may involve cyclic nucleotides and certain specific proteins that serve as substrates for nucleotide-mediated phosphorylations (Nestler and Greengard, 1982). It should be pointed out that these "plastic" effects may involve facilitation rather than inhibition of ACh release; this possibility was discussed at some length in Chapter 10 and will be returned to later (see Chapter 13).

2. Role of Catecholamines as Inhibitory Modulators

Since there are stores of catecholamines in ganglia (Eranko and Harkonen, 1963), and since preganglionic stimulation can release catecholamines (Lissak, 1939; Bulbring, 1944; Noon *et al.*, 1975; Martinez and Adler-Graschinsky, 1980), it appears that the catecholamines may act as neuromodulators of ganglionic transmission. The site of release of catecholamines has not been determined; however, several sites have been suggested.

Martinez and Adler-Graschinsky (1980) have hypothesized that preganglionic nerve stimulation causes the release of norepinephrine from the dendrites and cell body of cat superior cervical ganglion cells. In this regard, varicosities and storage granules have been observed along the dendrites of sympathetic ganglion cells (Elfvin, 1971; Kondo *et al.*, 1980);

thus, released norepinephrine could act retrogradely to reduce the release of ACh from preganglionic fibers. Interestingly, intense antidromic stimulation of bullfrog sympathetic ganglia reduced the catecholamine fluorescence of postganglionic neurons (Suetake *et al.*, 1981). This indicates that antidromic stimulation may release catecholamines that may then act on the preganglionic nerve endings. In fact, either antidromic or direct stimulation of the ganglion cells of bullfrog sympathetic ganglia was reported to depress the EPSP (Miyagawa *et al.*, 1981).

Second, the catecholamines may be released from specialized cells in the ganglion, the small intensely fluorescent (SIF) cells (Eranko and Harkonen, 1965; Norberg *et al.*, 1966) (see Chapters 3, 10, and 13). The fluorescence correlated well with the presence of dense-core granules that may be the storage sites for the catecholamines (Grillo *et al.*, 1974). The fluorescence was characteristic for both norepinephrine and dopamine; the relative quantities of each amine varied from one ganglion to another and from one species to another (Chiba and Williams, 1975). Catecholamines released from SIF cells could either generate a postsynaptic inhibitory potential (see Chapters 9 and 13) or act on the nerve terminals (Dun and Karczmar, 1981b).

Third, it has been suggested that norepinephrine is released from adrenergic axons that project into the rabbit superior cervical ganglion (Noon *et al.*, 1975). Recently, Bowers and Zigmond (1981) used horseradish peroxidase to demonstrate that neurons from the middle or inferior cervical ganglia of the rat projected axons through the superior cervical ganglion. It is not known whether these fibers terminate on ganglionic neurons or whether they release catecholamines into the superior cervical ganglion when activated.

Finally, concentrations of catecholamines sufficiently high to affect sympathetic ganglia may be present in the blood. Stimulation of sympathetic fibers innervating the adrenal glands suppressed ganglionic transmission (Marrazzi, 1939b). This was probably due to the release from the adrenal medulla of catecholamines that could then exert a systemic action on the ganglia. The depressant action of norepinephrine on the isolated stellate ganglion was markedly potentiated by catecholamine uptake inhibitors; thus, it is possible that in the presence of the uptake inhibitors, circulating catecholamines can suppress transmission in sympathetic ganglia (Christ *et al.*, 1982; Christ and Zitaglio, 1982). Together, circulating catecholamines may exert important pharmacological actions on the ganglia.

Although the results of these experiments support the concept of a modulatory effect of catecholamines on the ganglion, there are two aspects of this matter that must be clarified before this hypothesis can be fully accepted. While catecholamines are very effective in blocking synaptic transmission in mammalian sympathetic ganglia, repetitive conditioning stimulation does not appear to suppress mammalian ganglionic transmission (Libet, 1964), although such inhibition was observed in bullfrog sympathetic ganglia (Koketsu and Nishi, 1967; Nishi and Koketsu, 1967). An inhibitory process must be observed before the catecholamines can be accepted as inhibitory neuromodulators in mammalian ganglia. Also, if the catecholamines are functioning as inhibitory neuromodulators, adrenergic blocking drugs should increase the efficacy of ganglionic transmission. Pertinent indirect evidence was provided for the isolated stellate ganglion by Christ and Zitaglio (1984). In their experiments, tyramine reduced the ganglionic afterdischarges, presumably by releasing endogenous catecholamines from the ganglia, as this effect of tyramine was much weaker on ganglia isolated from animals that were pretreated with reserpine. This was the first demonstration of an effect of endogenous catecholamines in a mammalian ganglion, and further experiments must be performed to determine whether similar effects occur under physiological conditions.

B. Acetylcholine

After being released from preganglionic nerve fibers, ACh interacts with the postsynaptic nicotinic and muscarinic receptors of the autonomic ganglia and initiates the fast and slow synaptic potentials, respectively (see Chapters 5, 6, and 7). A number of studies indicated that ACh may interact as well with nicotinic and muscarinic receptors situated on the preganglionic nerve fibers. That ganglionic transmission may be modified by phenomena involving presynaptic nerve terminals has long been suspected. Riker and Szreniawski (1959) demonstrated that close-arterial injection of ACh to cat superior cervical ganglia initiated antidromic firing in the preganglionic nerve fibers. Dempsher and Riker (1957) reported that rat superior cervical ganglia infected with pseudorabies virus show periodic discharges of impulses along pre- and postganglionic nerve fibers. They suggested that ACh is released spontaneously under these circumstances from preganglionic nerve terminals and is involved in the initiation of discharges. A more recent study utilizing intracellular techniques confirmed that the virus induces presynaptically the release of ACh (Kiraly and Dolivo, 1982). Moreover, close-arterial injection of anticholinesterases evoked a prolonged postganglionic discharge that was not present after chronic preganglionic denervation (Takeshige and Volle, 1962). These results collectively demonstrate that the nerve terminal is pharmacologically reactive.

In fact, Riker felt that this nerve terminal reactivity related primarily to ganglionic transmission, since it initiated such transmission and, when unduly intense, could block it; thus, Riker (1968) attributed the ganglionic blocking effect of ACh in high concentration to its depolarizing action on the unmyelinated nerve terminals, since he showed that in a small number of frog sympathetic ganglion cells, transmission blockade occurred without depolarization of the postsynaptic membrane. Furthermore, when transmission blockade was accompanied by postsynaptic depolarization and when an inward current was applied to cause the return of the membrane potential to the resting level, the ganglionic transmission was not restored. Riker did not define the nature of this nerve-terminal activity or the mechanism of its induction of transmission facilitation or block (see Riker and Kosay, 1970).

Another view of the nerve terminal as the modulatory site was offered by Koelle (1962, 1963). He demonstrated that cholinomimetics release ACh from the ganglia (McKinstry *et al.*, 1963). He also demonstrated the presence of acetylcholinesterase in preganglionic nerve-terminal membranes. He suggested that this cholinesterase relates to the "percussive" role of ACh that is released from the nerve terminal: During its initial release, ACh depolarizes the terminal and causes further, massive release of the transmitter.

Koelle's and Riker's hypotheses require the demonstration of an effect of ACh on the nerve terminal. Using the sucrose-gap method, Koketsu and Nishi (1968) demonstrated that ACh or nicotine induced a slow and transient depolarization of ganglionic nerve terminals that could be blocked by curare or nicotine and was not affected by atropine (Koketsu and Nishi, 1968). Furthermore, ACh and carbachol reduced the threshold and the amplitude of nerve-terminal action potentials (Nishi, 1970; Ginsborg, 1971); this effect was also blocked by D-tubocurarine. These results indicate that the nerve terminals are endowed with nicotinic cholinoceptors; however, activation of these receptors with iontophoretic application of ACh did not appreciably change the frequency of miniature EPSPs (mEPSPs). Nishi (1970) concluded that the presynaptic action of ACh did not alter ganglionic transmission. Also, ACh did not induce the release of [^{3}H]-ACh from the ganglion, except at high concentrations or in the presence of a cholinesterase inhibitor (Brown *et al.*, 1970; Collier and Katz, 1970), and superfusion with ACh did not alter the evoked release of [^{3}H]-ACh. In conclusion, the hypothesis of "percussive" (Koelle, 1962, 1963) ACh release is not substantiated at this time, and the functional significance of presynaptic nicotinic cholinoceptors is uncertain. It is possible, however, that these presynaptic receptors modulate the release of ACh during high-frequency activity of the preganglionic neuron.

More recent studies indicated that muscarinic receptors are also present on the preganglionic nerve terminals. Employed at a concentration of 10^{-5} M, the muscarinic agonist, bethanechol, did not change the surface potential recorded from preganglionic terminals of bullfrog sympathetic ganglia (Koketsu and Yamada, 1982); however, bethanechol did reduce the EPSP in Ringer's solution with low-[Ca^{2+}]/[Mg^{2+}]- or D-tubocurarine. This effect appeared to be due to a reduction in the quantal content of the EPSP. At this concentration, bethanechol did not change the resting membrane potential of the postsynaptic cell, did not change the frequency or amplitude of mEPSPs, and did not change the amplitude of depolarizations evoked by iontophoretic application of ACh. Oxotremorine, another muscarinic agonist, produced similar effects on the isolated myenteric plexus of the guinea pig ileum (Morita *et al.*, 1982). The effects of both agonists were blocked by atropine, and it was concluded that activation of muscarinic cholinoceptors on the presynaptic nerve terminals reduces the release of ACh.

The EPSPs were increased in amplitude by atropine in the myenteric plexus (Morita *et al.*, 1982). While atropine did not increase the amplitude of the EPSPs of the bullfrog ganglion, it did increase their quantal content; this did not result in an increase of the EPSP amplitude because there was a simultaneous reduction in postsynaptic sensitivity to ACh (Koketsu and Yamada, 1982). These results support the conclusion that ACh that is physiologically released from the nerve terminal can inhibit neurotransmitter release by acting at muscarinic cholinoceptors.

Studies of [^{3}H]-ACh release from the myenteric plexus support this conclusion, since scopolamine increased the output of [^{3}H]-ACh (Kilbinger and Wessler, 1983). Release of [^{3}H]-ACh from the cat superior ganglion, however, was not affected by ACh (Collier and Katz, 1970). In fact, atropine reduced the release of [^{3}H]-ACh from this ganglion (Martinez and Adler-Graschinsky, 1980). These data indicate that the role of muscarinic receptors on the nerve terminals may differ from one site to another. It should be added that presynaptic muscarinic receptors are also present in the central cholinergic synapses (see Chapter 21).

C. γ-Aminobutyric Acid

The effects of γ-aminobutyric acid (GABA) on sympathetic neurons and on ganglionic transmission have been studied extensively (see also Chapters 12 and 13). GABA blocked transmission in mammalian ganglia (Matthews and Roberts, 1961). Since the blockade occurred at concentrations of GABA that depolarized the ganglion cells, it was hypothesized that this blockade was postsynaptic and due to membrane depolarization (DeGroat, 1970). Adams and Brown (1975) reported that the membrane depolarization induced by GABA in mammalian ganglia was accompanied by an increase in membrane conductance, had a reversal potential near −42 mV, and was markedly attenuated by substitution of external chloride with a larger impermeable anion. On the basis of these results, it was concluded that GABA induced a large increase in the membrane conductance to chloride and that this shunting effect resulted in a reduction of the EPSP.

The conclusions based on the results of the experiments on bullfrog sympathetic ganglion cells were quite different; in their case, GABA appeared to inhibit synaptic transmission by a presynaptic mechanism (Kato *et al.*, 1978; Kato and Kuba, 1980). Indeed, while GABA reduced the amplitude of the EPSP, the postsynaptic depolarizing action of GABA was small and accompanied by little or no change in membrane conductance. Moreover, tolerance developed rapidly to the depolarizing action of GABA, but there was little tolerance to the blocking action of GABA. Furthermore, GABA did not affect the depolarization induced by iontophoretically applied ACh. This suggests that the sensitivity of the postsynaptic membrane was not appreciably changed by GABA. Finally, GABA reduced the quantal content of the EPSP in a low-[Ca^{2+}]/[Mg^{2+}] solution, but did not significantly change the frequency or amplitude of the miniature EPSPs. Taken together, these results support the hypothesis that ganglionic blockade by GABA is mediated through a presynaptic mechanism.

The exact mechanism by which GABA reduced the transmitter output is not yet clear. Experiments carried out with extracellular recording methods indicated that GABA depolarized the preganglionic nerve endings (Koketsu *et al.*, 1974). This depolarization of the nerve terminal appeared to be mediated by an increase in chloride conductance; thus, this effect of GABA was similar to its postsynaptic action (Koketsu *et al.*, 1974). However, as already mentioned, GABA did not significantly change the frequency of the mEPSPs, which indicates that terminal-membrane depolarization was not the primary cause of the reduction of neurotransmitter release.

A small, detectable amount of GABA is present in sympathetic ganglia (Bertilsson and Costa, 1976) (see also Chapter 3). The ganglion cells did not appear to take up [^{3}H]-GABA; however, nonneuronal cells of the ganglion exhibited an uptake mechanism and accumulated [^{3}H]-GABA (Young *et al.*, 1973; Bowery *et al.*, 1979b). GABA that was taken up could be released by high [K^+] through a calcium-dependent mechanism (Bowery *et al.*, 1979a).

The physiological significance of the effect of GABA on ganglionic transmission has not been elucidated (see also Chapters 12 and 13). The concentration of GABA that is normally present in the blood appears to be near the concentration that is capable of depressing ganglionic transmitter release in the bullfrog (Kato *et al.*, 1980), and it is possible that the circulating GABA may exert a modulatory effect on presynaptic and postsynaptic membranes of the ganglia (see Chapter 12).

D. Prostaglandin E_1

Interest in prostaglandin E_1 (PGE_1) was generated by the report that it was synthesized in and released from rat superior cervical ganglia (Davis *et al.*, 1971; Webb *et al.*, 1978). At low concentrations (0.01–0.5 μ M), PGE_1 reduced the amplitude of the EPSP without changing the resting membrane potential or input resistance of the ganglion cell (Dun, 1980; Bellazzi *et al.*, 1982). Also, the amplitude of the depolarization evoked by iontophoretically applied ACh was not affected. At high concentrations (>1 μM), PGE_1 depolarized the membrane and reduced the membrane resistance. These results indicate that at low concentrations, PGE_1 acted presynaptically to reduce the release of neurotransmitter; at high concentrations, however, it exerted a postsynaptic action (Dun, 1980). The physiological role of PGE_1 in sympathetic ganglia is not known.

E. Enkephalins

The most conspicuous effect of enkephalins on the sympathetic ganglia is the depression of synaptic transmission (Konishi *et al.*, 1979; Dun and Karczmar, 1981b). At concentrations of 0.1–2 μM, met-enkephalin or leu-enkephalin consistently and reversibly depressed the EPSP and noncholinergic synaptic potential in most cells of the guinea pig inferior mesenteric ganglion. There was very little change in the resting membrane potential or input resistance (Konishi *et al.*, 1979; Dun and Karczmar, 1981b; Jiang *et al.*, 1982). A small hyperpolarization (<5 mV) and a reduction in input resistance by the enkephalins were observed in a few neurons. Enkephalins did not appreciably affect the depolarization due to iontophoretically applied ACh. Since substance P may be the neurotransmitter for the noncholinergic synaptic potential (Dun and Jiang, 1982) (see Chapter 8), the effect of enkephalins on the ganglionic response to superfusion with substance P was also analyzed; enkephalins did not affect this response (Dun and Jiang, 1982). Thus, in the case of the guinea pig inferior mesenteric ganglion, enkephalins do not alter the sensitivity of the ganglion cell membrane to the transmitter substances and depress ganglionic transmission via a presynaptic action, contrary to what may occur in the case of enteric ganglia (see Chapter 16). On the other hand, enkephalins appeared to inhibit ganglion transmission of the cat ciliary ganglion by both pre- and postsynaptic actions, since they reduced ACh output and hyperpolarized and increased the conductance of the membrane (Katayama and Nishi, 1984). In all these studies, the depressant effect of the enkephalins was prevented by the narcotic antagonists naloxone and naltrexone, which suggests that enkephalins reduce the release of neurotransmitters from preganglionic nerve endings by acting on specific presynaptic opioid receptors.

Enkephalins have been found in sympathetic ganglia (Hughes *et al.*, 1977; DiGiulio *et al.*, 1978; Schultzberg *et al.*, 1978, 1979). The distribution and localization of enkephalinlike immunoreactivity appeared to vary among species and among ganglia from a single species (see also Chapter 3). Dense networks of enkephalin-positive fibers were observed in inferior mesenteric ganglia of the guinea pig, yet only a few enkephalin-positive fibers could be identified in the superior cervical ganglion of the same species (Schultzberg *et al.*, 1979). Also, some SIF cells in sympathetic ganglia of the guinea pig and bullfrog exhibited enkephalinlike immunoreactivity (Schultzberg *et al.*, 1979; Kondo and Yui, 1981) (see Chapter 2).

Do endogenously released enkephalins have a physiological role in modulation of cholinergic or noncholinergic transmission? This problem was explored in the inferior mesenteric ganglion of the guinea pig by analyzing the effects of opioid antagonists on ganglionic potentials. Naloxone had no appreciable effect on the amplitude of the EPSPs evoked by single stimuli. However, a depression of EPSPs generated by repetitive nerve stimulation was significantly reduced by naloxone (Konishi *et al.*, 1981; Dun and Karczmar, 1981b; Jiang *et al.*, 1982); in fact, the amplitude of noncholinergic potentials evoked by repetitive stimulation was markedly enhanced by an opioid antagonist (Figure 2). These results suggested that enkephalins may be important in the modulation of transmitter release during high-frequency stimulation of the ganglia.

F. Serotonin (5-Hydroxytryptamine)

Depending on the experimental conditions, 5-hydroxytryptamine (5-HT) either depolarized or hyperpolarized sympathetic neurons and either depressed or facilitated ganglionic transmission (Trendelenburg, 1956; Machova and Boska, 1969; DeGroat and Lally, 1973; Haefely, 1974; Wallis and Woodward, 1974). Hirai and Koketsu (1980) observed that low concentrations of 5-HT (1–30 μM) produced an increase in the amplitude and the quantal content of the EPSP of the bullfrog sympathetic ganglia in a low-$[Ca^{2+}]/[Mg^{2+}]$ solution, while at higher concentrations (0.1–1 mM), 5-HT reduced the amplitude and quantal content of the EPSP.

Superfusion of the rabbit superior cervical ganglion with 5-HT (1–100 μM) reversibly depressed the amplitude of the EPSP without affecting the potential generated by iontophoretically applied ACh (Dun and Karczmar, 1981a) (Figure 3). 5-HT did not change noticeably the resting membrane potential or input resistance in the majority of the ganglion cells (Figure 4A), although in some of the cells, it did produce a transient

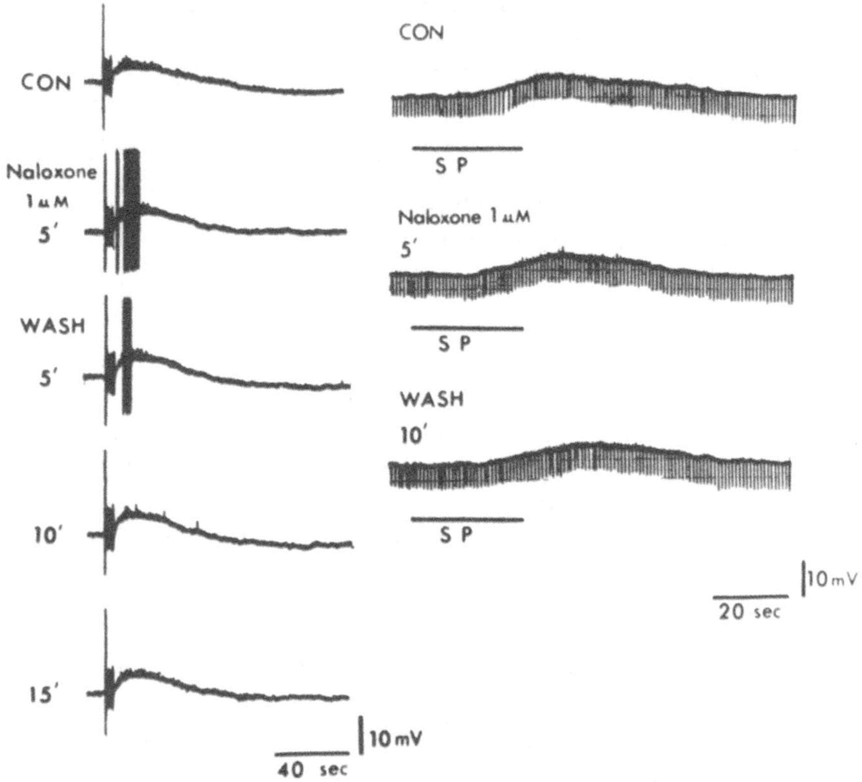

Figure 2. Effects of naloxone on nerve-evoked noncholinergic depolarizations (left column) and membrane depolarizations induced by bath applications of substance P (right column) on an inferior mesenteric ganglion cell of the guinea pig. The ganglion was continuously superfused with a Krebs solution containing D-tubocurarine (0.05 mM). Left column: The noncholinergic potentials were elicited by stimulation of hypogastric nerves (30 Hz for 5 sec). Note that the nicotinic transmission was only partially suppressed by curare, and as a result, orthodromic spike potentials and subthreshold potentials can be seen at the beginning of each trace. The noncholinergic depolarization was markedly enhanced 5 min after naloxone (1 µM), leading to intense neuronal discharges at the peak of the depolarization. Note that the intense discharges during the depolarization represented neuronal action potentials and the intensity of discharge was directly related to the magnitude of noncholinergic depolarization. The noncholinergic response returned to near control levels 15 min after the washout of naloxone. Right column: Membrane depolarization induced by substance P (SP) (1 µM) as indicated by the bars. The SP-induced depolarization was not noticeably changed during naloxone superfusion or 10 min after the superfusion was stopped.

depolarization that was accompanied by a fall in membrane resistance. When the membrane potential and input resistance returned to normal in the presence of 5-HT, the synaptic potentials were still reduced in amplitude (Figure 4B). The mean quantal content, but not the quantal size, of the EPSP was significantly reduced by 5-HT in a low-[Ca^{2+}]/[Mg^{2+}] solution. All these results indicate that 5-HT reduces the EPSP by an action

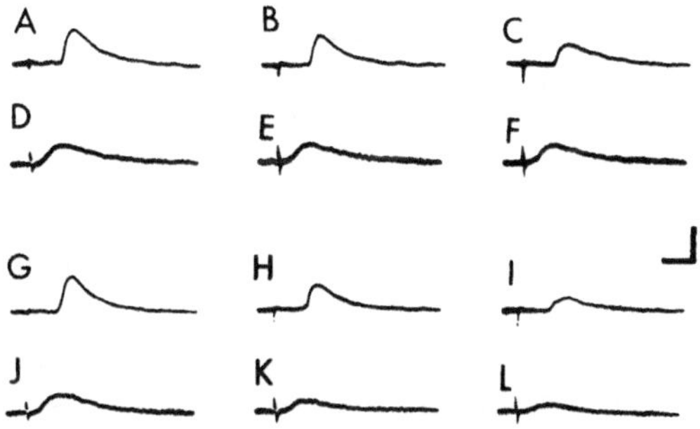

Figure 3. Effects of 5-HT (50 μM) and D-tubocurarine (10 μM) on the synaptic potentials and ACh potentials of a rabbit superior cervical ganglion cell. ACh potentials were induced by iontophoretic application of ACh (35 nA; 8-msec pulse duration). (A–C, G–I) Synaptic potentials; (D–F, J–L) ACh potentials. (A, D) Control; (B, E) 1 min (C, F) 2 min after 5-HT superfusion. Note that there is a reduction of the amplitude of the EPSP, but no detectable change of the ACh potential. (G, J) Control; (H, K) 45 sec (I, L) 75 sec after curare superfusion. Note that both the EPSP and the ACh potential were depressed by curare. Calibration: (A–C, G–I) 20 mV and 20 msec; (D–F, J–L) 20 mV and 100 msec.

on the presynaptic nerve endings. However, the inhibitory action of 5-HT on the EPSP was not affected by the 5-HT antagonists lysergic acid diethylamide and methysergide; in fact, either of these drugs reversibly diminished the EPSP (Dun and Karczmar, 1981a).

5-HT has been detected by means of biochemical and immunofluorescent methods in rabbit (Dun *et al.*, 1980) and rat (Verhofstad *et al.*, 1981) superior cervical ganglia, respectively. The ganglionic content of 5-HT in the rabbit superior cervical ganglion was not appreciably altered by preganglionic denervation; however, reserpine reduced the content to 50% of the control values. These findings indicate that 5-HT is not localized in the preganglionic neuron (Dun *et al.*, 1980). It must be added

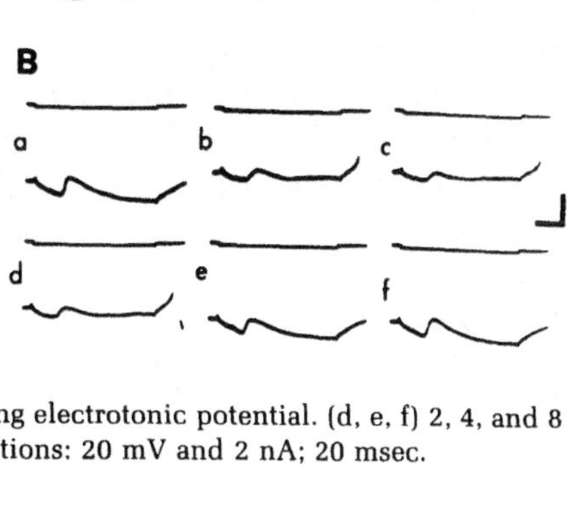

Figure 4. Effects of 5-HT on the membrane potential, membrane resistance, and synaptic potential of two ganglion cells. The upper trace of each pair of tracings represents the current pulse used to induce the hyperpolarizing electrotonic potential (lower trace). The current trace also represents the zero potential level. The synaptic potential elicited by orthodromic stimulation was superimposed on the electrotonic potential. In cell (A): (a) control; (b, c) 65 and 120 sec after 5-HT (100 μM) superfusion, respectively; (d, e, f) 2, 4, and 8 min after washing with Krebs solution, respectively. Note that the membrane potential and input resistance were not detectably changed, whereas the synaptic potential was depressed by 5-HT. In cell (B): (a) control; (b, c) 45 and 90 sec after 5-HT (50 μM) superfusion. Note that there was a reduction of the membrane potential and a decrease in the amplitude of the hyperpolarizing electrotonic potential. (d, e, f) 2, 4, and 8 min after washing with Krebs solution. Calibrations: 20 mV and 2 nA; 20 msec.

that 5-HT immunofluorescence was present in the SIF cells of the rat superior cervical ganglion (Verhofstad *et al.*, 1981). In fact, Hadjiconstantinou *et al.* (1982) provided evidence indicating that 5-HT release may be modulated by preganglionic impulses to the SIF cells.

In view of its presence in the sympathetic ganglia and its effectiveness in depressing the release of neurotransmitter from the preganglionic nerve fibers, 5-HT may play a role as a presynaptic modulator of ganglionic transmission. The physiological stimuli capable of releasing 5-HT and the exact site of this release must still be elucidated.

G. Histamine

Acting on H_1 and H_2 preganglionic receptors, histamine facilitates and depresses, respectively, ACh release (see Chapter 10).

III. CONCLUSIONS

For many years, autonomic ganglia were considered to be simple neuronal synapses, their transmission involving primarily the actions of ACh on postsynaptic nicotinic receptors. As it became known that catecholamines, enkephalins, 5-HT, and prostaglandins are endogenous to ganglia (see Chapters 2 and 3), studies of the effects of exogenous application of these substances were initiated. The early experiments utilized extracellular methods; this tended to emphasize postsynaptic phenomena, such as changes of postsynaptic membrane potential and of postsynaptic membrane excitability, whereas it was difficult to observe presynaptic phenomena with these methods. With the advent of intracellular methods, it became apparent that many endogenous substances also exerted presynaptic actions. Thus, catecholamines, GABA, enkephalins, 5-HT, and PGE_1 produce blockade without altering the chemosensitivity or electrical properties of the postsynaptic membranes of the ganglion. Accordingly, the role of presynaptic mechanisms in modulating ganglionic transmission is receiving more emphasis.

In fact, evidence presented in this chapter indicates that any or a combination of catecholamines, prostaglandins, 5-HT, and enkephalins could function as presynaptic inhibitory modulators. Indeed, in view of the potent presynaptic blocking action of catecholamines, it may be speculated that physiologically, this presynaptic action is functionally more important than the postsynaptic inhibitory action of catecholamines (see Chapters 9 and 13). This speculation is even more tempting in view of the universality of the blocking action of catecholamines on ganglionic transmission, whereas inhibitory postsynaptic potentials and the SIF cells are not present in all ganglia of all species (see Chapters 9 and 13). Future studies should reveal the physiological conditions under which the presynaptic modulatory actions of endogenous substances can be manifested in the ganglion.

REFERENCES

Adams, P. R., and Brown, D. A.: Actions of γ-aminobutyric acid on sympathetic ganglion cells. *J. Physiol. (London)* **250**:85–120 (1975).

Bellazzi, O., Biondi, C., Pieu, G. B., Capuzzo, A., Ferretti, M. E., Trevisani, A., and Peri, V.: Electrophysiological evidence for a PGE-mediated presynaptic control of acetylcholine output in the guinea-pig superior cervical ganglion. *Brain Res.* **236**:383–391 (1982).

Bertilsson, L., and Costa, E.: Mass fragmentographic quantitation of glutamic acid and γ-aminobutyric acid in cerebellar nuclei and sympathetic ganglia of rats. *J. Chromatogr.* **118**:395–402 (1976).

Bowers, C. W., and Zigmond, R. E.: Sympathetic neurons in lower cervical ganglia send axons through the superior cervical ganglion. *Neuroscience* **6**:1783–1791 (1981).

Bowery, N. G., Brown, D. A., and Marsh, S.: γ-Aminobutyric acid efflux from sympathetic glial cells: Effect of "depolarizing" agents. *J. Physiol. (London)* **293**:75–101 (1979a).

Bowery, N. G., Brown, D. A., White, R. D., and Yamini, G.: [^{3}H] γ-aminobutyric acid uptake into neuroglial cells of rat superior cervical sympathetic ganglia. *J. Physiol. (London)* **293**:51–74 (1979b).

Brown, D. A., Jones, K. B., Halliwell, J. V., and Quilliam, J. P.: Evidence against a presynaptic action of acetylcholine during ganglionic transmission. *Nature (London)* **226**:958–959 (1970).

Bulbring, E.: The action of adrenaline on transmission in the superior cervical ganglion. *J. Physiol. (London)* **103**:55–67 (1944).

Chiba, T., and Williams, T. H.: Histofluorescence characteristics and quantification of small intensely fluorescent (SIF) cells in sympathetic ganglia of several species. *Cell Tissue Res.* **162**:331–342 (1975).

Christ, D., and Nishi, S.: Site of adrenaline blockade in the superior cervical ganglion of the rabbit. *J. Physiol. (London)* **213**:107–117 (1971a).

Christ, D., and Nishi, S.: Effects of adrenaline on nerve terminals in the superior cervical ganglion of the rabbit. *Br. J. Pharmacol.* **41**:331–338 (1971b).

Christ, D., and Zitaglio, T.: Interactions of catecholamine uptake inhibitors and norepinephrine on autonomic ganglia. *Eur. J. Pharmacol.* **81**:509–513 (1982).

Christ, D., and Zitaglio, T.: Blockade of ganglionic afterdischarges by tyramine may be mediated by endogenous catecholamines. *J. Pharmacol. Exp. Ther.* **228**:376–379 (1984).

Christ, D., Curry, J., and Zitaglio, T.: Potentiation of the ganglionic blocking action of norepinephrine by cocaine. *J. Pharmacol. Exp. Ther.* **220**:97–101 (1982).

Collier, B., and Katz, H. S.: The release of acetylcholine by acetylcholine in the cat's superior cervical ganglion. *Br. J. Pharmacol.* **39**:428–438 (1970).

Davis, H. A., Horton, E. W., Jones, K. B., and Quilliam, J. P.: Identification of prostaglandins in prevertebral venous blood after preganglionic stimulation of the cat superior cervical ganglion. *Br. J. Pharmacol.* **42**:569–583 (1971).

Dawes, P. M., and Vizi, E. S.: Acetylcholine release from the rabbit isolated superior cervical ganglion preparation. *Br. J. Pharmacol.* **48**:225–232 (1973).

DeGroat, W. C.: The actions of α-aminobutyric acid and related amino acids on mammalian autonomic ganglia. *J. Pharmacol. Exp. Ther.* **172**:384–396 (1970).
DeGroat, W. C., and Lalley, P. M.: Interaction between picrotoxin and 5-hydroxytryptamine in the superior cervical ganglion of the cat. *Br. J. Pharmacol.* **48**:233–244 (1973).
DeGroat, W. C., and Volle, R. L.: The actions of the catecholamines on transmission in the superior cervical ganglion of the cat. *J. Pharmacol. Exp. Ther.* **154**:1–13 (1966).
Dempsher, J., and Riker, W. K.: The role of acetylcholine in virus-infected sympathetic ganglia. *J. Physiol. (London)* **139**:145–156 (1957).
DiGiulio, A. M., Yang, H.-Y. T., Lutold, B., Fratta, W., Hong, J., and Costa, E.: Characterization of enkephalin-like material extracted from sympathetic ganglia. *Neuropharmacology* **17**:989–992 (1978).
Dun, N. J.: Inhibition of ACh release by prostaglandin E_1 in the rabbit superior cervical ganglion. *Neuropharmacology* **19**:1137–1140 (1980).
Dun, N. J., and Jiang, Z. G.: Non-cholinergic excitatory transmission in inferior mesenteric ganglia of the guinea-pig: Possible mediation by substance P. *J. Physiol. (London)* **325**:145–159 (1982).
Dun, N. J., and Karczmar, A. G.: The presynaptic site of action of norepinephrine in the superior cervical ganglion of guinea pig. *J. Pharmacol. Exp. Ther.* **200**:328–335 (1977).
Dun, N. J., and Karczmar, A. G.: Evidence for a presynaptic inhibitory action of 5-hydroxytryptamine in a mammalian sympathetic ganglion. *J. Pharmacol. Exp. Ther.* **217**:714–718 (1981a).
Dun, N. J., and Karczmar, A. G.: Multiple mechanisms in ganglionic transmission, in: *Cholinergic Mechanisms* (G. Pepeu and H. Ladinsky, eds.), pp. 109–118, Plenum Press, New York (1981b).
Dun, N. J., and Nishi, S.: Effects of dopamine on the superior cervical ganglion of the rabbit. *J. Physiol. (London)* **239**:155–164 (1974).
Dun, N. J., Ingerson, A., and Karczmar, A. G.: A neurochemical and neurophysiological study of serotonin in the superior cervical ganglion of the rabbit. *Soc. Neurosci. Abstr.* **6**:216 (1980).
Elfvin, L.-G.: Ultrastructural studies on the synaptology of the inferior mesenteric ganglion of the cat. III. The structure and distribution of the axodendritic and dendrodendritic contacts. *J. Ultrastruct. Res.* **37**:432–448 (1971).
Eranko, O., and Harkonen, M.: Histochemical demonstration of fluorogenic amines in the cytoplasm of sympathetic ganglion cells of the rat. *Acta Physiol. Scand.* **58**:285–286 (1963).
Eranko, O., and Harkonen, M.: Monoamine-containing small cells in the superior cervical ganglion of the rat and an organ composed of them. *Acta Physiol. Scand.* **63**:511–512 (1965).
Ginsborg, B. L.: On the presynaptic acetylcholine receptors in sympathetic ganglia of the frog. *J. Physiol. (London)* **216**:237–246 (1971).
Grillo, M. A., Jacobs, L., and Comroe, J. H.: A combined fluorescence histochemical and electron microscopic method for studying special monoamine-containing cells (SIF cells). *J. Comp. Neurol.* **153**:1–14 (1974).
Hadjiconstantinou, M., Potter, P. E., and Neff, N. H.: Trans-synaptic modulation via muscarinic receptors of serotonin-containing small intensely fluorescent cells of superior cervical ganglion. *J. Neurosci.* **2**:1836–1839 (1982).
Haefely, W. E.: Effects of catecholamines in the cat superior cervical ganglion and their postulated role as physiological modulators of ganglionic transmission. *Prog. Brain Res.* **31**:61–72 (1969).
Haefely, W. E.: The effects of 5-hydroxytryptamine and some related compounds on the cat superior cervical ganglion *in situ*. *Naunyn-Schmiedeberg's Arch. Pharmacol.* **281**:145–165 (1974).

Hirai, K., and Koketsu, K.: Presynaptic regulation of the release of acetylcholine by 5-hydroxytryptamine. *Br. J. Pharmacol.* **70**:499–500 (1980).

Horn, J. P., and McAfee, D. A.: Alpha-adrenergic inhibition of calcium-dependent potentials in rat sympathetic neurones. *J. Physiol. (London)* **301**:191–204 (1980).

Hughes, J., Kosterlitz, H. W., and Smith, T. W.: The distribution of methionine-enkephalin and leucine-enkephalin in the brain and peripheral tissues. *Br. J. Pharmacol.* **61**:639–647 (1977).

Jiang, Z. G., Simmons, M. A., and Dun, N. J.: Enkephalinergic modulation of non-cholinergic transmission in mammalian prevertebral ganglia. *Brain Res.* **235**:185–191 (1982).

Kafka, M. S., and Thoa, N. B.: α-Adrenergic receptors in the rat superior cervical ganglion. *Biochem. Pharmacol.* **28**:2485–2489 (1979).

Katayama, Y., and Nishi, S.: Sites and mechanisms of actions of enkephalin in the feline parasympathetic ganglion. *J. Physiol. (London)* **351**:111–121 (1984).

Kato, E., and Kuba, K.: Inhibition of transmitter release in bullfrog sympathetic ganglia induced by γ-aminobutyric acid. *J. Physiol. (London)* **298**:271–283 (1980).

Kato, E., Kuba, K., and Koketsu, K.: Presynaptic inhibition by γ-aminobutyric acid in bullfrog sympathetic ganglion cells. *Brain Res.* **153**:398–402 (1978).

Kato, E., Morita, K., Kuba, K., Yamada, S., Kuhara, T., Shinka, T., and Matsumoto, I.: Does γ-aminobutyric acid in blood control transmitter release in bullfrog sympathetic ganglia? *Brain Res.* **195**:208–214 (1980).

Kilbinger, H., and Wessler, I.: The variation of acetylcholine release from myenteric neurones with stimulation frequency and train length. *Naunyn-Schmiedeberg's Arch. Pharmacol.* **324**:130–133 (1983).

Kiraly, M., and Dolivo, M.: Alteration of the electrophysiological activity in sympathetic ganglia infected with a neurotropic virus. I. Presynaptic origin of the spontaneous bioelectric activity. *Brain Res.* **240**:43–54 (1982).

Koelle, G. B.: A new general concept of neurohumoral functions of acetylcholine and acetylcholinesterase. *J. Pharm. Pharmacol.* **14**:65–90 (1962).

Koelle, G. B.: Cytological distributions and physiological functions of cholinesterases, in: *Cholinesterases and Anticholinesterase Agents* (G. B. Koelle, ed.), Handbuch der Experimentellen Pharmakolologie, Ergwk., Vol. 15, pp. 187–298, Springer-Verlag, Berlin (1963).

Koketsu, K., and Nishi, S.: Characteristics of the slow inhibitory postsynaptic potential of bullfrog sympathetic ganglion cells. *Life Sci.* **6**:1827–1836 (1967).

Koketsu, K., and Nishi, S.: Cholinergic receptors at sympathetic preganglionic nerve terminals. *J. Physiol. (London)* **196**:293–310 (1968).

Koketsu, K., and Yamada, M.: Presynaptic muscarinic receptors inhibiting active acetylcholine release in the bullfrog sympathetic ganglion. *Br. J. Pharmacol.* **77**:75–82 (1982).

Koketsu, K., Shoji, T., and Yamamoto, K.: Effects of GABA on presynaptic nerve terminals in bullfrog (*Rana catesbiana*) sympathetic ganglia. *Experientia* **30**:382–383 (1974).

Kondo, H., and Yui, R.: Enkephalin-like immunoreactivity in the SIF cells of sympathetic ganglia of frogs. *Biomed. Res.* **2**:338–340 (1981).

Kondo, H., Dun, N. J., and Pappas, G. D.: A light and electron microscopic study of the rat superior cervical ganglion cells by intracellular HRP-labeling. *Brain Res.* **197**:193–199 (1980).

Konishi, S., Tsunoo, A., and Otsuka, M.: Enkephalins presynaptically inhibit cholinergic transmission in sympathetic ganglia. *Nature (London)* **282**:515–516 (1979).

Konishi, S., Tsunoo, A., and Otsuka, M.: Enkephalin as a transmitter for presynaptic inhibition in sympathetic ganglia. *Nature (London)* **294**:80–82 (1981).

Libet, B.: Slow synaptic responses and excitatory changes in sympathetic ganglia. *J. Physiol. (London)* **174**:1–25 (1964).

Lissak, K.: Liberation of acetylcholine and adrenaline by stimulating isolated nerves. *Am. J. Physiol.* **127**:263–271 (1939).

Lundberg, A.: Adrenaline and transmission in the sympathetic ganglion of the cat. *Acta Physiol. Scand.* **26**:252–263 (1952).

Machova, J., and Boska, D.: The effect of 5-hydroxytryptamine, dimethylphenylpiperazinium and acetylcholine on transmission and surface potential in the cat sympathetic ganglion. *Eur. J. Pharmacol.* **7**:152–158 (1969).

Marrazzi, A. S.: Electrical studies on the pharmacology of autonomic synapses. II. The action of a sympathomimetic drug (epinephrine) on sympathetic ganglia. *J. Pharmacol. Exp. Ther.* **65**:395–404 (1939a).

Marrazzi, A. S.: Adrenergic inhibition at sympathetic synapses. *Am. J. Physiol.* **127**:738–744 (1939b).

Martinez, A. E., and Adler-Graschinsky, E.: Release of norepinephrine induced by preganglionic stimulation of the isolated superior cervical ganglion of the cat. *J. Pharmacol. Exp. Ther.* **212**:527–532 (1980).

Matthews, R. J., and Roberts, B. J.: The effect of gamma-aminobutyric acid on synaptic transmission in autonomic ganglia. *J. Pharmacol. Exp. Ther.* **132**:19–22 (1961).

McIsaac, R. J.: Ganglion blocking properties of epinephrine and related amines. *Neuropharmacology* **5**:15–26 (1966).

McKinstry, D. N., Koenig, E., Koelle, W. A., and Koelle, G. B.: The release of acetylcholine from a sympathetic ganglion by carbachol: Relationship to the functional significance of the localization of acetylcholinesterase. *Can. J. Biochem. Physiol.* **41**:2599–2609 (1963).

Miyagawa, M., Minota, S., and Koketsu, K.: Antidromic inhibition of acetylcholine release from presynaptic nerve terminals in bullfrog's sympathetic ganglia. *Brain Res.* **224**:305–313 (1981).

Morita, K., North, R. A., and Tokimasa, T.: Muscarinic presynaptic inhibition of synaptic transmission in myenteric plexus of guinea-pig ileum. *J. Physiol. (London)* **333**:141–149 (1982).

Nestler, E. J., and Greengard, P.: Distribution of protein I and regulation of its state of phosphorylation in the rabbit superior cervical ganglion. *J. Neurosci.* **2**:1011–1023 (1982).

Nishi, S.: Cholinergic and adrenergic receptors at sympathetic preganglionic nerve terminals. *Fed. Proc.* **29**:1957–1965 (1970).

Nishi, S., and Koketsu, K.: Origin of ganglionic inhibitory postsynaptic potential. *Life Sci.* **6**:2049–2055 (1967).

Noon, J. P., McAfee, D. A., and Roth, R. H.: Norepinephrine release from nerve terminals within the rabbit superior cervical ganglion. *Naunyn-Schmiedeberg's Arch. Pharmacol.* **291**:139–162 (1975).

Norberg, K.-A., Ritzen, M., and Ungerstedt, U.: Histochemical studies on a special catecholamine-containing cell type in sympathetic ganglia. *Acta Physiol. Scand.* **67**:260–270 (1966).

Pardo, E. G., Cato, J., Gijon, E., and Alonso DeFlorida, F.: Influence of several adrenergic drugs on synaptic transmission through the superior cervical and the ciliary ganglia of the cat. *J. Pharmacol. Exp. Ther.* **139**:296–303 (1963).

Riker, W. K.: Ganglion cell depolarization and transmission block by ACh: Independent events. *J. Pharmacol. Exp. Ther.* **159**:345–352 (1968).

Riker, W. K., and Kosay, S.: Drug induction and suppression of stimulus-bound repetition in sympathetic ganglia. *J. Pharmacol. Exp. Ther.* **173**:284–292 (1970).

Riker, W. K., and Szreniawski, Z.: The pharmacological reactivity of presynaptic nerve terminals in a sympathetic ganglion. *J. Pharmacol. Exp. Ther.* **126**:233–238 (1959).

Schultzberg, M., Hökfelt, T., Lundberg, J. M., Terenius, L., Elfvin, L.-G., and Elde, R.: Enkephalin-like immunoreactivity in nerve terminals in sympathetic ganglia and adrenal medulla and in adrenal medullary gland cells. *Acta Physiol. Scand.* **103**:475–477 (1978).

Schultzberg, M., Hökfelt, T., Terenius, L., Elfvin, L.-G., Lundberg, J. M., Brandt, J., Elde, R. P., and Goldstein, M.: Enkephalin immunoreactive nerve fibres and cell bodies in sympathetic ganglia of the guinea-pig and rat. *Neuroscience* **4**:249–270 (1979).

Suetake, K., Kojima, H., Inanaga, K., and Koketsu, K.: Catecholamine is released from nonsynaptic cell-soma membrane: Histochemical evidence in bullfrog sympathetic ganglion cells. *Brain Res.* **205**:436–440 (1981).

Takeshige, C., and Volle, R. L.: Bimodal response of sympathetic ganglia to acetylcholine following eserine or repetitive preganglionic stimulation. *J. Pharmacol. Exp. Ther.* **138**:66–73 (1962).

Trendelenburg, U.: The action of 5-hydroxytryptamine on the nictitating membrane and on the superior cervical ganglion of the cat. *Br. J. Pharmacol.* **11**:74–80 (1956).

Verhofstad, A. A. J., Steinbusch, H. W. M., Penke, B., Varga, J., and Joosten, H. W. J.: Serotonin-immunoreactive cells in the superior cervical ganglion of the rat: Evidence for the existence of separate serotonin and catecholamine-containing small ganglionic cells. *Brain Res.* **212**:39–49 (1981).

Wallis, D. I., and Woodward, B.: The facilitatory actions of 5-hydroxytryptamine and bradykinin in the superior cervical ganglion of the rabbit. *Br. J. Pharmacol.* **51**:521–531 (1974).

Webb, J. G., Saelens, D. A., and Halushka, P. V.: Biosynthesis of prostaglandin E by rat superior cervical ganglia. *J. Neurochem.* **31**:13–19 (1978).

Young, J. A. C., Brown, D. A., Kelly, J. S., and Schon, F.: Autoradiographic localization of sites of [^{3}H] γ-aminobutyric acid accumulation in peripheral autonomic ganglia. *Brain Res.* **63**:479–486 (1973).

12

Postsynaptic Modulation

K. KOKETSU and T. AKASU

I. INTRODUCTION

In the sympathetic ganglia, transmission of nerve impulses from preganglionic neurons to ganglion cells is modulated both presynaptically and postsynaptically. As has aleady been described in Chapter 3, transmission in sympathetic ganglia results in the generation of the fast excitatory postsynaptic potential (EPSP), slow EPSP, late slow EPSP, and slow inhibitory postsynaptic potential (IPSP) of ganglion cells (cf. Koketsu, 1969; Kuba and Koketsu, 1978). The sizes of these postsynaptic potentials are regulated in a very complicated manner by both presynaptic and postsynaptic modulations (see Chapter 3).

The postsynaptic modulation takes place at the subsynaptic membrane, where the subsynaptic receptors are located, and also at the postsynaptic membrane, where the action potential is generated. A most common mode of postsynaptic modulation is via the modulation of the resting membrane potential of postsynaptic neurons. In this case, the ganglionic postsynaptic potential mediated by a particular neurotransmitter is modulated by the changes in the resting membrane potential or conductance of ganglion cells. In addition, the postsynaptic modulation often operates via changes in the action potentials of the ganglion cells. In this case, a biogenic substance (synaptic modulator) directly modulates the excitation of ganglion cells by changing the configuration of action potentials generated in these neurons (Kuba and Koketsu, 1978; Kupfermann, 1979; Siegelbaum and Tsien, 1983; Koketsu, 1984). Still another type of post-

K. KOKETSU and T. AKASU • Department of Physiology, Kurume University School of Medicine, Kurume, Japan.

synaptic modulation involves a change in the sensitivity of the receptor located at the subsynaptic membrane, and it was reported recently that some biogenic substances modulate the sensitivity of the nicotinic receptor that mediates the fast EPSP in sympathetic ganglion cells (Koketsu, 1984).

II. MODULATION OF THE RESTING POTENTIAL

According to the ionic theory (Hodgkin, 1951, 1958), the membrane potential of neurons is expressed by Goldman's equation

$$E = -\frac{RT}{F}\ln\frac{P_K\,[K]_i + P_{Na}\,[Na]_i + P_{Cl}\,[Cl]_o}{P_K\,[K]_o + P_{Na}\,[Na]_o + P_{Cl}\,[Cl]_i} \quad (1)$$

where E is the intracellular membrane potential; R, T, and F are the gas constant, absolute temperature, and Faraday's constant, respectively; $[K]_i$ and $[K]_o$ are the concentrations of potassium ions in the intra- and extracellular fluids, respectively; and $[Na]_i$ and $[Na]_o$ and $[Cl]_i$ and $[Cl]_o$ reflect intra- and extracellular concentrations of sodium and chloride ions, respectively. Finally, P_K, P_{Na}, and P_{Cl} are the permeability constants for the respective ions.

As expressed by equation (1), the membrane potential is assumed to be the diffusion potential, which is determined by the intra- and extracellular concentrations of Na, K, and Cl ions and by the ratio of $P_{Na} : P_K : P_{Cl}$. Thus, according to this equation, the membrane potential can be expected to be modulated by changes in the intracellular ionic concentrations and also by changes in the membrane permeabilities. Both the intracellular ionic concentrations and the membrane permeabilities would be changed by the postsynaptic modulatory actions of many biogenic substances (modulators). These modulatory actions induce changes in the intracellular ionic concentrations that occur very slowly, whereas changes in the membrane permeabilities take place relatively rapidly.

Equation (1) is based on the assumption that active ionic transport does not contribute directly to the potential difference across the membrane. However, active ionic transport is not always electrically neutral; it often seems to be electrogenic (Koketsu, 1971; Thomas, 1972). Thus, some fraction of the membrane potential must be directly associated with the active transport of some ions. The electrogenic sodium pump seems to constitute the most important active ionic transport that must be taken into consideration in the analysis of the modulatory action of biogenic substances on the membrane potential (Koketsu, 1971).

Indeed, much experimental evidence indicates that the Na^+-K^+ pump of many kinds of excitable cells is modulated by the action of neurotransmitters or neurohormones (Koketsu, 1971; Kuba and Koketsu; 1978; Phillis and Wu, 1981; Kaibara *et al.*, 1982).

A. Catecholamines

Catecholamines produce membrane hyperpolarization in sympathetic ganglion cells (Lundberg, 1952; DeGroat and Volle, 1966) (cf. also Chapters 9 and 13). In the case of amphibian sympathetic ganglion cells, hyperpolarizing responses produced by the direct action of epinephrine were recorded by the sucrose-gap method (Nakamura and Koketsu, 1972; Koketsu and Nakamura, 1976). Epinephrine-induced hyperpolarization of bullfrog ganglion cells is markedly depressed by ouabain or by lowering the temperature; it was therefore suggested that epinephrine-induced hyperpolarization may be associated with the electrogenic Na^+ pump (Nakamura and Koketsu, 1972; Koketsu and Nakamura, 1976; Kaibara *et al.*, 1982).

Epinephrine-induced hyperpolarization of mammalian sympathetic ganglion cells was recorded by means of intracellular microelectrodes (Libet and Kobayashi, 1968; Kobayashi and Libet, 1970; Christ and Nishi, 1971). In the past, no changes in the membrane conductance of mammalian sympathetic ganglia appeared to occur during catecholamine-induced hyperpolarization (Kobayashi and Libet, 1970; Ivanov and Skok, 1980). More recently, however, Nakamura and Nishi (1982), using the voltage-clamp method, found that the membrane resistance was decreased in rabbit superior cervical ganglia during catecholamine-induced hyperpolarization; furthermore, in their hands, catecholamine-induced hyperpolarization reversed its polarity at a membrane potential level that was close to the K^+ equilibrium potential. On the basis of these observations, they suggested that catecholamine-induced hyperpolarization resulted from an increase in the K^+ conductance. Similarly, Smith (1984) found that epinephrine-induced hyperpolarization of frog sympathetic ganglia was also produced by an activation of K^+ conductance.

On the other hand, the facilitatory action of catecholamine on ganglionic transmission may also occur, as first reported by Bulbring (1944). Furthermore, DeGroat and Volle (1966a) found that catecholamines produce a depolarization of mammalian ganglia, and membrane depolarization produced by epinephrine was demonstrated by means of intracellular microelectrodes in mammalian (Christ and Nishi, 1971) and bullfrog (Nakamura and Koketsu, 1972) sympathetic ganglion cells.

Ivanov and Skok (1980) reported that norepinephrine-produced hyperpolarization was followed by depolarization in 55% of neurons of the rabbit superior cervical ganglia; they found that norepinephrine-induced depolarization was usually associated with a decrease in the input membrane resistance (Ivanov and Skok, 1980).

B. 5-Hydroxytryptamine

It is well known that 5-hydroxytryptamine (5-HT) exerts excitatory actions on sympathetic ganglia (Trendelenburg, 1957). Robertson (1954) first demonstrated that 5-HT stimulated mammalian sympathetic ganglion cells, and this stimulant action of 5-HT was confirmed subsequently (Trendelenburg, 1957; Bindler and Gyermek, 1961; Hertzler, 1961; Gyermek and Bindler, 1962a,b; DeGroat and Volle, 1966b; Jaramillo and Volle, 1968; Wallis and Woodward, 1973) (see also Chapter 13). It was suggested in these reports that 5-HT facilitated synaptic transmission by its depolarizing action on the ganglion cells. Indeed, 5-HT induced depolarization was observed in bullfrog sympathetic ganglion cells (Watanabe and Koketsu, 1973). Wallis and North (1978) analyzed the nature of depolarization induced by 5-HT in rabbit superior cervical ganglion cells; it appeared that the 5-HT-induced depolarization is associated with a decrease in the membrane resistance, and Wallis and North suggested that the increase of G_{Na} and G_{Ca} was involved in the 5-HT-induced depolarization. In the case of the guinea pig coeliac ganglion, endogenous 5-HT acted as a depolarizer and was responsible for the slow noncholinergic EPSP (Dun *et al.*, 1984).

On the other hand, Eccles and Libet (1961) reported that 5-HT exerted an inhibitory action on the synaptic transmission of sympathetic ganglia. Biphasic responses to 5-HT, namely, a depolarization followed by a hyperpolarization, were observed subsequently (DeGroat and Lalley, 1973). Wallis and Woodward (1973) found that 5-HT-induced hyperpolarization was blocked by ouabain and suggested that an electrogenic Na^+ pump mechanism may be involved. Interestingly, 5-HT-induced depolarization of bullfrog sympathetic ganglion cells shifted to a hyperpolarization in the presence of nicotine, after the latter depolarized the cell membrane and completely blocked the fast EPSP (Watanabe and Koketsu, 1973). The amplitude of 5-HT-induced hyperpolarization recorded in the presence of nicotine was markedly augmented by moderate conditioning hyperpolarization and depressed during intense hyperpolarization. Thus, it seemed that 5-HT produced hyperpolarization of the ganglion cells when the intracellular Na^+ concentration was increased by a depolarizer, such as acetylcholine (ACh) or nicotine (Shirasawa and Koketsu, 1978). Furthermore, 5-HT-induced hyperpolarization was very sensitive to ouabain, as reported earlier by Wallis and Woodward (1973) (see above), and it was eliminated in Na^+-free Li^+ solution (Shirasawa and Koketsu, 1978).

Thus, 5-HT-induced hyperpolarization may be associated with an activation of the electrogenic Na^+ pump (Watanabe and Koketsu, 1973).

It appears, then, that 5-HT exerts potent pharmacological effects on the ganglia; in addition to the postsynaptic effects described herein, it exerts presynaptic actions that were already described in Chapter 11. While 5-HT is present in the ganglia or their small intensely fluorescent (SIF) cells (see Chapters 2 and 3), it is not clear at this time whether its pharmacological effects reflect its physiological, ganglionic actions.

C. Adenosine Triphosphate

Since 1972, when the concept of purinergic transmission was proposed by Burnstock (1972), there have appeared many studies of possible physiological actions of adenosine and its derivatives in the central and peripheral nervous systems (Phillis, 1977; Burnstock, 1981; Stone, 1981). With regard to the latter, it has been reported that in bullfrog sympathetic ganglia, ATP and related compounds depolarized the postganglionic neurons (Nakamura *et al.*, 1974; Siggins *et al.*, 1977).

ATP-induced depolarization was associated with an increase of membrane resistance (Figure 1) (Siggins *et al.*, 1977; Akasu *et al.*, 1981c, 1983a,b). The voltage-current relationship of ganglion cells obtained in the presence of ATP was steeper than the control relationship at hyperpolarized potential levels, and the two curves crossed at approximately -90 mV, which is the K^+ equilibrium potential (Akasu *et al.*, 1981c, 1983a,b). These results suggest that the ATP-induced depolarization is produced by the inactivation of the K^+ conductance. The potency sequence of purines with respect to producing depolarization was: ATP > ADP > AMP; adenosine exerted no depolarizing action, suggesting that the depolarization was mediated through the P_2 purinoceptor (Burnstock, 1981). ATP-induced depolarization was not sensitive to the effect of tetraethylammonium (TEA), a potent inhibitor of the delayed rectifier K^+ current. Thus, it was suggested that the K^+ channel, which is closed by the action of ATP and responsible for production of the ATP-induced depolarization, is actually the M channel (Figure 1) (Akasu *et al.*, 1981c, 1983a,b) (see also Chapters 4 and 7).

Synaptic vesicles isolated from *Torpedo* electroplaques contained both ATP and ACh (Whittaker *et al.*, 1972). It was demonstrated that ACh is capable of releasing ATP from presynaptic nerve terminals in the rat diaphragm muscle (Silinsky and Hubbard, 1973; Silinsky, 1975) and in the frog sympathetic ganglion (Silinsky and Ginsborg, 1983). Thus, it is possible that ATP may be released from preganglionic nerve terminals and depolarize the postganglionic sympathetic neurons.

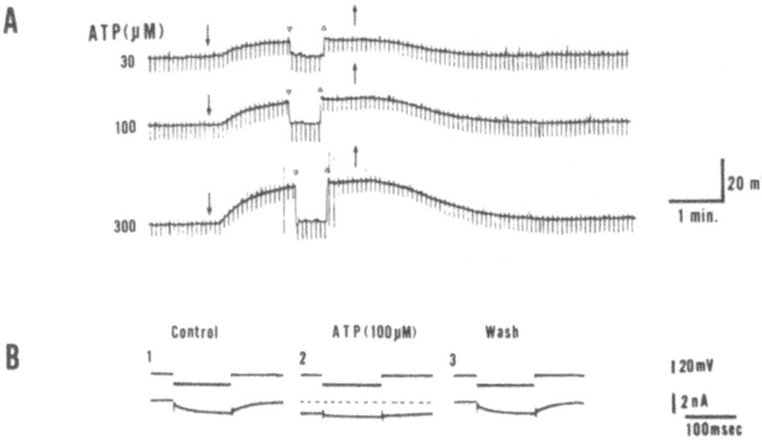

Figure 1. Effects of ATP on resting membrane potential and resistance (A) and slow inward relaxation [M current (I_M)] recorded under voltage-clamp conditions (B). (A) Downward and upward arrows represent the beginning and end of treatment of a ganglion cell with ATP (30–100 μM), respectively. Electrotonic potentials were produced by hyperpolarizing square current pulses constantly applied through a recording electrode. ATP depolarizations were nullified by anodal current between (▽) and (△) to measure the membrane resistance. (B) Slow inward relaxations were produced by an application of hyperpolarizing step-command (200 msec) to the ganglion cell clamped at −30 mV. (1, 2) Obtained before 10 min after application of ATP (100 μM) to Ringer's solution containing tetrodotoxin (1.4 μM), respectively. (3) Obtained 10 min after withdrawal of ATP. The upper traces indicate the membrane potential, while the lower traces indicate the membrane currents. The horizontal broken lines indicate zero current. Note the shift in inward direction (inward current downward) of the membrane current and the marked depression of slow inward relaxation by ATP. From Akasu *et al.* (1983b).

D. γ-Aminobutyric Acid

DeGroat (1970) was the first to describe a depolarizing action of γ-aminobutyric acid (GABA) on sympathetic ganglion cells of the cat. Adams and Brown (1973, 1975) used intracellular recording to demonstrate GABA-induced depolarization of rat superior cervical ganglion cells. The depolarization produced by GABA was associated with a decrease in the membrane resistance of the ganglion cell, and it was reversed at around −42 mV by depolarization of the resting membrane potential. Replacement of Cl^- with a larger anion resulted in suppression of the GABA response. Finally, GABA markedly depressed the action potential of the ganglion cell. On the basis of these experimental results, Adams and Brown (1975) concluded that GABA causes a large increase in the membrane conductance for Cl^- (G_{Cl}) of the ganglion cell.

GABA-induced depolarization was also demonstrated in the case of bullfrog sympathetic ganglion cells; both the sucrose-gap technique (Koketsu *et al.*, 1974) and intracellular microelectrode methods were employed (Koketsu *et al.*, 1974; Kato *et al.*, 1978a; Kato and Kuba, 1980). This GABA-induced depolarization was very small; with GABA concentrations of 0.1–1 mM, it amounted to only a few millivolts (Kato and Kuba, 1980). In view of the smallness of this effect, the presynaptic mechanism of GABA-generated synaptic depression should be seriously considered (see below and Chapter 11).

It was reported that neuroglial cells in mammalian sympathetic ganglia are able to take up GABA from external solution containing as low a concentrations of GABA as 1 μM (Bowery and Brown, 1972; Young *et al.*, 1973). Therefore, neuroglial cells may regulate the extracellular GABA concentration, affecting the membrane excitability of postganglionic neurons of the sympathetic ganglia (Brown and Galvan, 1977). In this connection, the concentration of GABA in plasma was estimated at 2–4 μM in the case of cats (Crowshaw *et al.*, 1967) and at 0.37–4.4 μM in that of bullfrogs (Kato *et al.*, 1980). Thus, it is possible that GABA in the plasma exerts a regulating action on the excitability of postganglionic membranes as well as on transmitter release from preganglionic nerve terminals (see Chapter 11).

III. MODULATION OF THE ACTION POTENTIAL

During their studies of the ionic mechanism of the slow EPSP and slow IPSP, it was found by Koketsu and his associates (Koketsu, 1974, 1977; Kuba and Koketsu, 1975, 1976a; Koketsu and Minota, 1975; Minota and Koketsu, 1977) that ACh and catecholamines modulate the voltage-dependent currents during generation of action potentials in the bullfrog sympathetic ganglion cell. Both ACh and catecholamines appear to depress the increases in G_{Ca} and G_K during the initiation of action potentials.

The mechanism underlying the modulatory action of biogenic substance on voltage-dependent ionic currents is unknown. The density of an ionic current induced at a voltage-clamped membrane potential can be expressed by the product of (1) the total number of functionable ion channels, (2) the probability for opening of these channels at a fixed membrane potential level, and (3) the ionic current passing through a single opened channel, or its conductance. Thus, the mechanism of the modulatory effect of biogenic substances on the action potential may be clarified by quantification of these three parameters, as well as in terms of the equilibrium potential of the pertinent ions (Koketsu, 1984).

A. Acetylcholine

The afterhyperpolarization of an action potential produced by direct or antidromic stimulation was found to be significantly and reversibly depressed during the course of the slow EPSP or ACh-induced slow depolarization of curarized sympathetic ganglion cells of bullfrogs (Kuba and Koketsu, 1975, 1976a). This suppression of the amplitude of afterhyperpolarization was clearly seen even when the depolarization was canceled out by application of an appropriate amount of anodal current. Furthermore, the maximum rates of rise and fall of the spike were both significantly depressed by the muscarinic effect of ACh (Figure 2). These actions of ACh are not due to its shunting effect on the ganglionic cell membrane, since the membrane resistance was generally either increased or unchanged at the resting or depolarized levels of membrane potential during the course of the muscarinic effect of ACh (Kuba and Koketsu, 1974, 1976b).

Bullfrog sympathetic neurons can generate action potentials in isotonic Ca^{2+} solution in which all the NaCl has been replaced with equimolar $CaCl_2$ (Koketsu and Nishi, 1969). Both the Ca^{2+} and the K^+ currently underlying the generation of Ca^{2+} spikes induced in this isotonic Ca^{2+} solution were depressed by ACh (Kuba and Koketsu, 1976a). ACh decreased the peak amplitude and maximum rates of rise and fall of the Ca^{2+} spike and increased the afterdepolarization following the spike (Figure 2). The membrane resistance was markedly increased even at the depolarized or resting membrane potential level (Figure 2). The Ca^{2+}-dependent K^+ current of the action potential of bullfrog sympathetic ganglion cells was also directly depressed by the muscarinic action of ACh (Tokimasa, 1984).

These effects of ACh on the action potentials were blocked by atropine, suggesting that they are mediated by the muscarinic receptor located somewhere in the neuronal membrane (Kuba and Koketsu, 1975, 1976a; Tokimasa, 1984). Presumably, the muscarinic receptor modulating the action potentials of sympathetic ganglion cells is linked with the ionic channel, the voltage-dependent channels being responsible for the initiation of action potentials. This muscarinic receptor-channel complex may serve as a functional unit concerned with the modulation of the voltage-dependent current. In keeping with this concept, Akasu and Koketsu (1980) reported that the muscarinic depressant effect of ACh on the action potentials was subject to desensitization during a sustained application of ACh to the ganglion cells.

Voltage-clamp experiments were recently carried out to evaluate the concept that the Ca^{2+} and K^+ currents that participate in the initiation of the action potentials of bullfrog sympathetic ganglion cells are de-

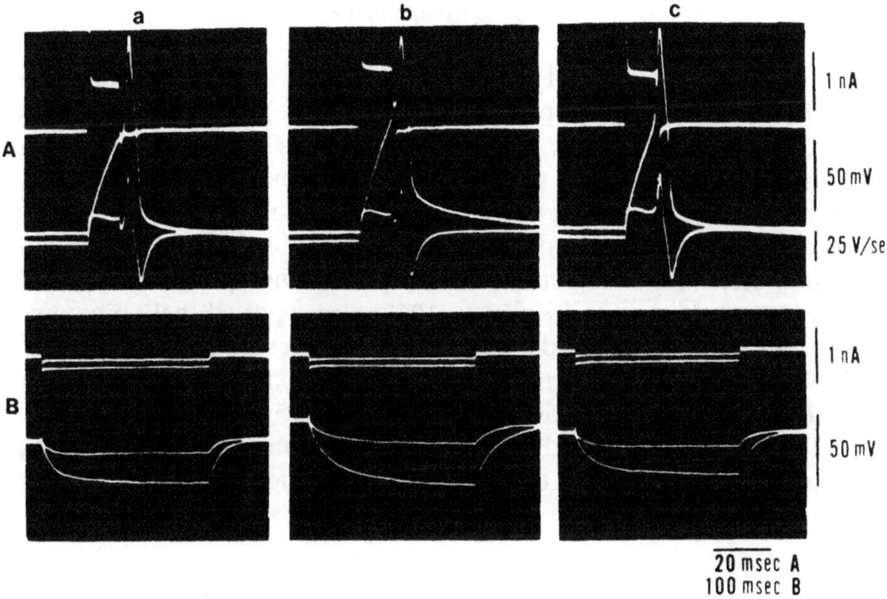

Figure 2. Muscarinic effect of ACh on the Ca^{2+} spikes (A) and on the membrane resistance (B) in isotonic Ca^{2+} solution containing 2 mM K^+. (a) Before the application of ACh (0.27 mM); (b) during the application of ACh, during which the membrane was depolarized by 5 mV in (A) (not shown; see below) and by 10 mV in (B) [see the level of the lower recording in B (b); (c) after removal of ACh. (A, B) Recordings were obtained from two different cells that showed resting potentials of −60 mV (A) and −55 mV (B). In all the records shown in (A), the membrane potential was held at −85 mV to eliminate inactivation of Ca^{2+} conductance by ACh-induced depolarization. The upper traces in (A) and (B) indicate the current passed to the cell through a recording electrode. The baselines of the upper trace in (A) also indicate a level about 5 mV negative with respect to the zero level. The baselines of the upper traces in (B) represent the zero levels for membrane potential. The middle recordings in (A) and lower recordings in (B) indicate membrane potentials, while the lower tracings in (A), indicate first derivatives of the membrane potential. Bullfrog sympathetic ganglion cell. From Kuba and Koketsu (1976b).

pressed by the muscarinic action of ACh (Akasu, 1981; Akasu and Koketsu, 1981a, 1982). Indeed, Akasu and Koketsu (Akasu, 1981; Akasu and Koketsu, 1981a, 1982) showed that both the slow inward current carried by Ca^{2+} (Akasu and Koketsu, 1981b) and an outward current were depressed by the muscarinic action of ACh; these investigators suggested that the outward current was composed of the TEA-sensitive delayed rectifier K^+ current (I_{k_1}) and the TEA-insensitive K^+ current (I_{k_2}). The TEA-insensitive K^+ current may correspond to the M current (Brown and Adams, 1980). The M current is depressed by the muscarinic effect of ACh during the generation of action potentials (Brown and Adams, 1980; Adams *et al.*, 1982).

B. Epinephrine

The action potential of bullfrog sympathetic ganglia was modulated as to its configuration by epinephrine (adrenaline) (Koketsu and Minota, 1975; Minota and Koketsu, 1977). The amplitude of the afterhyperpolarization was depressed and the duration of the spike was prolonged after application of epinephrine; the maximum rates of rise and fall of the spike were significantly decreased. Similar effects on the action potential of the ganglion cell were observed in the presence of norepinephrine, but not isoproterenol. Minota and Koketsu (1977) also reported that epinephrine depressed the Ca^{2+}-dependent action potential evoked in isotonic Ca^{2+} solution (Figure 3), since the amplitude and maximum rates of rise and fall of this potential were depressed by epinephrine.

These results suggested that epinephrine depresses the Ca^{2+} currents and possibly also the K^+ currents responsible for generation of action

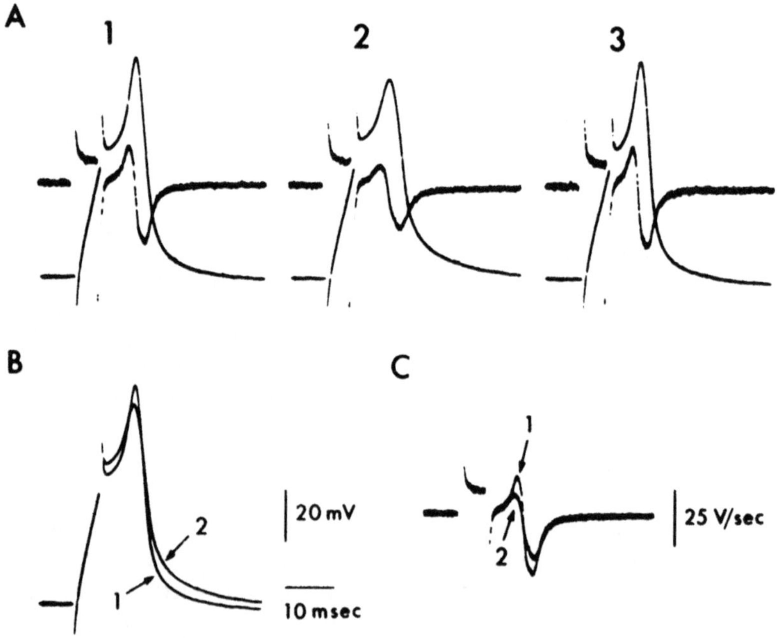

Figure 3. Effects of epinephrine (adrenaline) (3×10^{-4} M) on Ca-potential recorded in isotonic $CaCl_2$ solution. Cathodal pulses of 5-msec duration were used for direct stimulations. (A) Records were taken before (1) and 4 min after (2) an application of epinephrine, and 5 min after its withdrawal (3). To demonstrate suppression of the positive afterpotential and the maximum rate of fall by epinephrine, the lower and upper traces of records (1) and (2) in (A) are superimposed in (B) and (C), respectively, and numbered as in (A). The resting membrane was fixed at -70 mV by applying an anodal current through an intracellular recording microelectrode. Bullfrog sympathetic ganglion cell. From Minota and Koketsu (1977).

potentials. Both the amplitude and the duration of the action potentials, which exhibit a characteristic plateau phase in a Ringer's solution containing TEA and tetrodotoxin (TTX) and are therefore generated by increases in G_{Ca} and G_K, were also decreased by epinephrine; their maximum rates of rise and fall were also decreased (Minota and Koketsu, 1977). In the case of mammalian sympathetic ganglia, Horn and McAfee (1979, 1980) reported that norepinephrine depressed the amplitude of the Ca^{2+}-dependent action potential that was recorded in a Ringer's solution containing TTX and TEA.

The results obtained recently in voltage-clamp experiments confirmed that catecholamines inhibit the G_{Ca} and G_K during generation of action potentials of bullfrog sympathetic ganglion cells (Akasu and Koketsu, 1981a; Koketsu and Akasu, 1982; Galvan and Adams, 1982), since the slow inward Ca^{2+} current was depressed in these experiments by epinephrine or norepinephrine.

C. Adenosine Triphosphate

ATP is still another endogenous compound that exerts characteristic effects on the action potential. ATP (100 μM) markedly reduced the peak amplitude of the afterhyperpolarization and the maximum rate of fall of the action potential in a Ringer's solution containing TEA (20 mM); TEA should have depressed the delayed rectifier K^+ conductance. These results suggest that ATP depressed a G_K system that is TEA-insensitive and is therefore probably different from the delayed rectifier G_K system (Akasu *et al.*, 1981c, 1983a,b).

Voltage-clamp experiments were carried out to analyze the effect of ATP on the voltage-dependent K^+ conductance of ganglion cells (Akasu *et al.*, 1981c, 1983b). Again, ATP appeared to depress a G_K system that is different from the delayed rectifier K^+ current and that may correspond to the M current (I_M). This characteristic effect of ATP on the G_K system may depolarize the resting membrane and inhibit the afterhyperpolarization of action potentials in bullfrog sympathetic ganglion cells. Such an effect resembles that of ACh (Akasu, 1981; Akasu and Koketsu, 1981a,b 1982). A structurally related compound, UTP, was also found to suppress the M current (Adams *et al.*, 1982). Akasu *et al.* (1981c) noted, however, that ATP may depress to some extent the delayed rectifier K^+ current during initiation of action potentials. Recently, Henon and McAfee (1983) reported that the Ca^{2+}-dependent action potential of the rat superior cervical ganglion cell was depressed by adenosine. However, the Ca^{2+} spike and slow inward Ca^{2+} current of the bullfrog sympathetic ganglion cell were not significantly blocked by ATP or related compounds (Akasu *et al.*, 1983a).

D. Polypeptides

Endogenous polypeptides, such as luteinizing-hormone-releasing hormone (LH-RH) and substance P, were recently found to depress the voltage-dependent ion currents during the generation of action potentials in bullfrog sympathetic neurons. Adams and Brown (1980) (see also Adams *et al.*, 1982) demonstrated that the M current was strongly depressed by LH-RH. Subsequently, it was reported that the K^+ current, including both the delayed rectifier K^+ current (I_{k_1}) and the M current (I_{k_2}), seem to be depressed by LH-RH (Akasu *et al.*, 1983e) and substance P (Akasu *et al.*, 1983f; Adams *et al.*, 1983). It should be noted that the afterhyperpolarization of the action potential was also suppressed during generation of the late slow EPSP (Akasu *et al.*, 1983e). The inward Na^+ current also seems to be inhibited by LH-RH (Akasu *et al.*, 1983e) or substance P (Akasu *et al.*, 1983f).

IV. MODULATION OF RECEPTOR SENSITIVITY

ACh released from presynaptic nerve terminals binds to the nicotinic receptor located in the subsynaptic membrane; hence, it produces a fast EPSP. When the postsynaptic potential becomes large enough to initiate the action potential, signal information supplied by the presynaptic nerves is transmitted from the presynaptic to the postsynaptic neurons. In general, the efficiency of monosynaptic transmission is represented by the amplitude of the postsynaptic potential generated at the postsynaptic membrane. The amplitude of the postsynaptic potential is determined not only by the amplitude of postsynaptic current but also by the resting potential or conductance of the postsynaptic membrane. Therefore, actual synaptic events produced by transmitters such as ACh should be defined by agonist-induced postsynaptic currents rather than postsynaptic potentials. The amplitude of the postsynaptic current can be measured by means of the voltage-clamp method. The amplitude of the postsynaptic current is determined primarily by the amount of neurotransmitter (A) released from the presynaptic membrane and the sensitivity of the receptor (R) located at the subsynaptic membrane, as shown in equation (2):

$$A + R \underset{}{\overset{K_m}{\rightleftharpoons}} AR \underset{\alpha}{\overset{\beta}{\rightleftharpoons}} AR^* \qquad (2)$$

In this equation, A and R are the amount of ACh and number of nicotinic receptors, respectively; AR and AR* represent the closed and open states of the receptor-channel complex; and K_m, α, and β are rate constants.

The sensitivity of the receptor to a neurotransmitter can therefore be estimated by measuring the amplitude of postsynaptic current induced by direct application of a constant amount of the transmitter to the receptor.

That the sensitivity of the nicotinic ACh receptor is altered by the actions of many exogenous substances is well known. Thus, neuromuscular or ganglionic cholinergic transmission is inhibited by several pharmacological agents and toxins (Witkop and Grossinger, 1983). These exogenous substances are nicotinic antagonists (or blockers), and the mechanism of their blockade of the nicotinic transmission has been extensively explored. According to their mode of blocking action, nicotinic antagonists have been generally divided into two groups, competitive and noncompetitive antagonists, although more extensive subdividing of these drugs is also possible as discussed previously (Chapter 6). There are also pharmacological drugs and toxins, that facilitate nicotinic transmission, and it is of interest that in the case of anticholinesterases, some of their facilitatory as well as blocking actions may not be due to ACh accumulation, but may depend on direct sensitizing and receptor actions (Koppanyi and Karczmar, 1951) (see also Chapter 13).

Much experimental evidence accumulated recently indicates that the sensitivity of the nicotinic ACh receptor of the sympathetic ganglion cells as well as skeletal muscle end-plates is modulated not only by exogenous substances but also by endogenous substances. These endogenous substances (modulators) either augment or depress the sensitivity of the nicotinic ACh receptor. Interestingly, they block these receptors via mechanisms similar to those underlying the blocking action of exogenous nicotinic antagonists, since receptor sensitivity is depressed by endogenous substances either competitively or noncompetitively.

In the case of the nicotinic ACh receptor-ion channel complex, combination of ACh with the specific site of the receptor molecule leads to the opening of the ion channels and an increase in conductance [see equation (2) and Chapter 6]. If this is the case, the density of the ACh current can be expressed by the product of (1) the number of functional channels, (2) the probability of their opening, and (3) the unit ion current passing through a single channel. If a combination of ACh with the receptor site activates the associated ion channels, the number of functional ion channels is determined by the number of receptor-ion channel complexes and the affinity of ACh for receptor sites. Accordingly, ACh current may be changed with the alteration of (1) the total number of receptor-ion channel complexes, (2) the affinity of ACh for the receptor sites, (3) the probability of opening of ion channels, (4) the conductance of a single ion channel, and (5) the equilibrium potential of the ion in question.

Endogenous substances seem to modulate the nicotinic ACh receptor sensitivity either directly or via noncholinergic ganglionic receptors; finally, they may modulate the sensitivity of other than cholinergic receptors. To quote a few examples, the amplitude of the slow EPSP in sympathetic ganglion cells was found to be potentiated for an extended period of time by catecholamines (Libet *et al.*, 1975) (see also Chapter 13), suggesting that the sensitivity of the muscarinic ACh receptors may be increased by the activation of catecholaminergic receptors; the slow EPSC, mediated through muscarinic receptors, is blocked by LH-RH, the putative transmitter for the late slow EPSP (Jan *et al.*, 1979; Kuffler and Sejnowski, 1983), and ATP modulates the GABA-induced response at the soma membrane of bullfrog primary afferent neurons (Morita *et al.*, 1984).

A. Catecholamines

Catecholamines such as norepinephrine, epinephrine, and isoproterenol depress the sensitivity of the nicotinic receptor of sympathetic ganglion cells as well as skeletal muscle end-plates (Koketsu, 1981; Koketsu *et al.*, 1982a–c). The amplitude of ACh potentials evoked by ion-

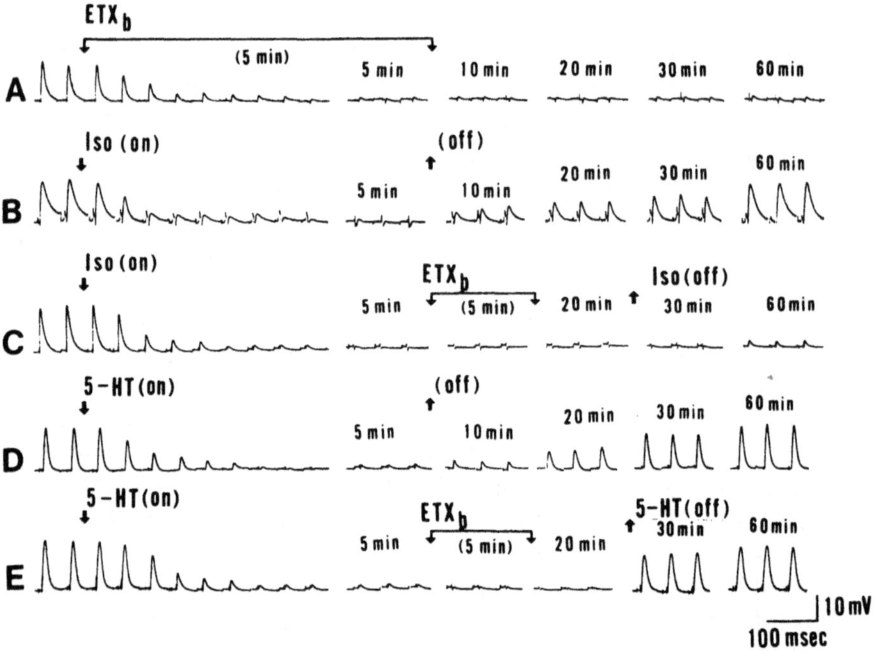

Figure 4. Effects of 5-hydroxytryptamine (5-HT) and isoproterenol (Iso) on irreversible blocking action of erabutoxin (ETX_b)-b on ACh potentials produced by iontophoretic applications of ACh to the skeletal muscle end-plate. (A) Effect of ETX_b (1 μM) on ACh potentials. The horizontal line indicates the application of ETX_b; the time of application is shown in parentheses. (B, D): Effects of Iso (1 mM) and 5-HT (1 mM) on ACh potentials. Downward and upward arrows indicate the beginning of application and withdrawal of these agents, respectively. (C, E) Effects of previous application of Iso and 5-HT, respectively, on the irreversible blocking action of ETX_b. Iso and 5-HT were applied between the downward and upward arrows. The horizontal lines indicate the time for application of ETX_b (5 min). Note that 5-HT protected the inhibitory effect of ETX_b while Iso did not.

tophoretic application of ACh to ganglion cells and employed for the evaluation of receptor sensitivity was reversibly reduced by epinephrine without causing changes of either membrane conductance or resting membrane potential; of course, catecholamines affect ACh response by their hyperpolarizing actions as well (see Chapter 9). Voltage-clamp analysis revealed that ACh current was also reduced by epinephrine. The order of the potency of catecholamines with respect to reducing the sensitivity of the bullfrog neurons to ACh was: isoproterenol > epinephrine > norepinephrine; this suggests that α-action of catecholamines was involved.

The mechanism of the inhibitory effect of isoproterenol on the nicotinic ACh receptor sensitivity of bullfrog ganglion cells was analyzed by Koketsu (1981) and Koketsu *et al.* (1982a–c). An S-shaped curve of the dose-response relationship of the ACh current was obtained by plotting the amplitude of ACh current against the logarithm of relative ACh quantity applied iontophoretically; isoproterenol shifted this curve downward. The Lineweaver-Burk plot revealed that isoproterenol decreased V_{max} without significantly changing K_m. On the other hand, the reversal potential of the fast EPSP and the falling phase of the fast EPSP was not changed by isoproterenol. On the basis of these results, Koketsu and his associates (Koketsu, 1981; Koketsu *et al.*, 1982a–c) suggested that catecholamines may act on the allosteric (or catalytic) site of the receptor-ionic channel complex without affecting the receptor site (Koketsu *et al.*, 1982b) (see also Figure 4) and that they may inhibit the receptor sensitivity by decreasing the conductance of open channels without changing their lifetime (Koketsu *et al.*, 1982a).

B. 5-Hydroxytryptamine

The sensitivity of the nicotinic ACh receptor of bullfrog sympathetic ganglion cells as well as muscle end-plate was found to be depressed by the action of 5-HT (Akasu *et al.*, 1981a; Hirai *et al.*, 1981), since the amplitude of ACh potentials of the ganglion cells was reversibly reduced by 5-HT. That the depression of ACh sensitivity by 5-HT was involved in this effect was clearly demonstrated by means of the voltage-clamp technique, since the ACh current was depressed by 5-HT. The S-shaped log dose-response curve of nicotinic ACh potentials recorded from ganglion cells was shifted to the right by 5-HT, and the Lineweaver-Burk plot of the dose-response curve showed that 5-HT increases K_m without changing V_{max} significantly. These results suggest that 5-HT decreases the affinity of the ACh molecule for the recognition site of the receptor-ionic channel complex (Akasu *et al.*, 1981a; Hirai *et al.*, 1981).

5-HT therefore seems to inhibit the sensitivity of the nicotinic receptors of bullfrog sympathetic ganglion cells in a manner that differs from that of catecholamines; namely, 5-HT seems to interact with a specific ACh-binding site, whereas catecholamines interact with a catalytic site of the ACh receptor-ionic channel complex. This hypothesis was further evaluated at the end-plate by studying the interaction between catecholamines or 5-HT and erabutoxin-b (ETX-b), which is known to bind irreversibly, similarly to α-bungarotoxin, with the specific nicotinic ACh receptor site (Kato *et al.*, 1978b). Whereas ETX-b block of the response of the end-plate to iontophoretic application of ACh could not be antagonized by isoproterenol, 5-HT was effective in protecting the endplate from ETX-b; i.e., following the application of the reversible blocker, 5-HT, and ETX-b, the ACh response reappeared after the wash (Figure 4) (see also Koketsu *et al.*, 1982b). These results are consistent with the notion that in the ganglia, 5-HT interacts directly with the specific ACh-binding sites, whereas isoproterenol acts on other sites of the ACh receptor-ionic channel complex (Koketsu *et al.*, 1982a,b). It should be pointed out, however, that the action of α-bungarotoxin (and presumably of ETX-b as well) on the ganglionic nicotinic receptor differs from its action on the end-plate receptor (Dun and Karczmar, 1980) (cf. Chapter 13).

C. Histamine

Histamine reduced the amplitude of the ACh potential at the frog skeletal muscle end-plate, suggesting that it may depress the sensitivity of the nicotinic ACh receptor to ACh (Scuka, 1973). Such a modulatory action of histamine on the sensitivity of the nicotinic receptor was confirmed recently with regard to both frog skeletal muscle end-plates and bullfrog sympathetic ganglion cells (Ohta *et al.*, 1984; Ariyoshi *et al.*, 1984). Histamine appears to shift the dose-response relationship between ACh concentration and ACh current to the right, suggesting that histamine competitively depresses the ACh current. In agreement with this notion, the ETX-b block of neuromuscular transmission was prevented by a previous application of histamine (Ohta *et al.*, 1984).

D. Polypeptides

It was found recently that at low concentrations, LH-RH, an endogenous polypeptide, markedly depresses the sensitivity of the nicotinic ACh receptors of bullfrog sympathetic ganglion cells as well as frog skeletal muscle end-plates (Akasu *et al.*, 1983c). The ACh current of the end-plate was depressed in a dose-dependent manner, and its dose-response curve was shifted downward in the presence of LH-RH; thus, the effect of LH-RH was similar to that of catecholamines (see above). This suggests that LH-RH inhibits the nicotinic ACh receptor sensitivity in a noncompetitive manner (Akasu *et al.*, 1983c).

That substance P may block the nicotinic receptor has been known since the studies by Ryall and Belcher (1977) of its actions on the Renshaw cells. This property of substance P was recently expanded to include peripheral synapses, since Akasu *et al.* (1983d) showed by means of voltage-clamp experiments that substance P decreased the sensitivity of the nicotinic ACh receptor of both bullfrog sympathetic ganglion cells and frog skeletal muscle end-plates. Further analysis of the effects of substance P indicated that its blocking is noncompetitive, and thus its action is similar to that of LH-RH and of catecholamines. Substance P exerts a similar inhibitory action on the ACh response of isolated chromaffin cells (Clapham and Neher, 1984). Thus, besides exerting transmitter effects on the ganglion (Jan *et al.*, 1979) (see Chapters 8 and 13) as well, substance P acts as a postsynaptic modulator of the nicotinic receptors.

E. Adenosine Triphosphate

It has been suggested that the sensitivity of the cholinergic receptor of the rat diaphragm end-plate may be augmented by a direct action of high-energy phosphate adenine nucleotides (Ewald, 1976). More recently, Akasu *et al.* (1981b) analyzed the facilitatory effect of ATP and its derivatives on the nicotinic end-plate and ganglionic receptor sensitivity by using the voltage-clamp technique and Lineweaver-Burk analysis of the effect of ATP on the current-dose relationship of a nicotinic agonist; they also measured the effect of ATP on the reversal of the ACh current. While ATP increased the V_{max}, it did not affect the K_m or the falling phase and the reversal potential of the postsynaptic current. These results suggested that ATP increased the sensitivity of the nicotinic receptor of both the ganglion and the end-plate by increasing either the conductance of the channels or the total number of available channels (Akasu *et al.*, 1981b). The sequence of the potencies of the facilitatory actions of ATP and its derivatives was: ATP > ADP > AMP; adenosine did not exhibit any appreciable effect on receptor sensitivity (Akasu *et al.*, 1981b).

F. Cyclic Adenosine 3′,5′-Monophosphate

Libet and Tosaka (1970) (see also Libet *et al.*, 1975) reported not only that cyclic AMP (cAMP) and dopamine augmented the slow EPSP, but also that this augmentation persisted for a long time after their application.

Libet (Libet *et al.*, 1975) (see also Libet, 1979) felt that this phenomenon was mediated by the muscarinic effect of presynaptically released ACh or SIF cells and the concomitant liberation of dopamine from SIF cells; dopamine activation on the dopaminergic receptor of postganglionic neurons increased the level of cAMP in these neurons, AMP augmenting the slow EPSP. Similarly, applied to bullfrog sympathetic ganglion cells, cAMP caused long-term potentiation of the late slow EPSP (Nishi and Katayama, 1981). To Libet (1979), this cyclic-nucleotide-mediated long-term potentiation of the slow EPSP constituted a model for memory processes (see also Chapter 21). Incidentally, Libet's proposal concerning the augmenting action of cAMP on the slow EPSP is distinct from the notion that cyclic nucleotides are involved in the generation of the slow EPSP and of the slow IPSP (Greengard, 1976) (see Chapters 9 and 13).

V. CONCLUSIONS

This chapter concerns regulation of ganglionic transmission via postsynaptic modulation, i.e., via mechanisms other than neurotransmitter-generated ganglionic potentials. This modulation of particularly nicotinic cholinergic transmission is exerted in sympathetic ganglia by postsynaptic modulatory actions of many biogenic (endogenous) substances. These postsynaptic actions involve (1) the modulation of the resting membrane potential or conductance of ganglion cells, (2) the modulation of the action potential of ganglion cells, and (3) the modulation of the sensitivity of subsynaptic receptors of the membrane of ganglion cells. The experimental evidence supporting these modulatory actions of endogenous substances was reviewed, and the mechanisms underlying these modulatory actions were discussed. The multiplicity of substances and mechanisms that appear to be available for postsynaptic modulation underlies the capacity of the ganglion and its neurons for subtle regulation of transmission.

REFERENCES

Adams, P. R., and Brown, D. A.: Action of γ-aminobutyric acid (GABA) on rat sympathetic ganglion cells. *Br. J. Pharmacol.* **47:**639P–640P (1973).

Adams, P. R., and Brown, D. A.: Actions of γ-aminobutyric acid on sympathetic ganglion cells. *J. Physiol. (London)* **250:**85–120 (1975).

Adams, P. R., and Brown, D. A.: Luteinizing hormone-releasing factor and muscarinic agonists act on the same voltage-sensitive K^+-current in bullfrog sympathetic neurones. *Br. J. Pharmacol.* **68:**353–355 (1980).

Adams, P. R., Brown, D. A., and Constanti, A.: Pharmacological inhibition of the M-current. *J. Physiol. (London)* **332**:223–262 (1982).

Adams, P. R., Brown, D. A., and Jones, S. W.: Substance P inhibits the M-current in bullfrog sympathetic neurones. *Br. J. Pharmacol.* **79**:330–333 (1983).

Akasu, T.: Voltage clamp analysis of muscarinic effects on action potentials in sympathetic ganglion cells of bullfrogs. *Neurosci. Lett.* **6**(Suppl.):59 (1981).

Akasu, T., and Koketsu, K.: Desensitization of the muscarinic receptor controlling action potentials of sympathetic ganglion cells in bullfrogs. *Life Sci.* **27**:2261–2267 (1980).

Akasu, T., and Koketsu, K.: Modulatory actions of neurotransmitters on voltage-dependent membrane currents in bullfrog sympathetic neurones. *Kurume Med. J.* **28**:345–348 (1981a).

Akasu, T., and Koketsu, K.: Voltage-clamp studies of a slow inward current in bullfrog sympathetic ganglion cells. *Neurosci. Lett.* **26**:259–262 (1981b).

Akasu, T., and Koketsu, K.: Modulation of voltage-dependent currents by muscarinic receptor in sympathetic neurones of bullfrog. *Neurosci. Lett.* **29**:41–45 (1982).

Akasu, T., Hirai, K., and Koketsu, K.: 5-Hydroxytryptamine controls ACh-receptor sensitivity of bullfrog sympathetic ganglion cells. *Brain Res.* **211**:217–220 (1981a).

Akasu, T., Hirai, K., and Koketsu, K.: Increase of acetylcholine-receptor sensitivity by adenosine triphosphate: A novel action of ATP on ACh-sensitivity. *Br. J. Pharmacol.* **74**:505–507 (1981b).

Akasu, T., Hirai, K., and Koketsu, K.: Action of ATP as an inhibitor of TEA-insensitive K^+ current in bullfrog sympathetic ganglion cell. 8th International Congress of Pharmacology, Abstract 396 (1981c).

Akasu, T., Hirai, K., and Koketsu, K.: Modulatory actions of ATP on membrane potentials of bullfrog sympathetic ganglion cells. *Brain Res.* **258**:313–317 (1983a).

Akasu, T., Hirai, K., and Koketsu, K.: Modulatory actions of ATP on nicotinic transmission in bullfrog sympathetic ganglia, *Physiology and Pharmacology of Adenosine Derivatives*, (J. W. Daly, Y. Kuroda, J. W. Phillips, H. Shimizu, and M. Ui, eds.) pp. 165–171 Raven Press, New York (1983b).

Akasu, T., Kojima, M., and Koketsu, K.: Luteinizing hormone-releasing hormone modulates nicotinic ACh-receptor sensitivity in amphibian cholinergic transmission. *Brain Res.* **279**:347–351 (1983c).

Akasu, T., Kojima, M., and Koketsu, K.: Substance P modulates the sensitivity of the nicotinic receptor in amphibian cholinergic transmission. *Br. J. Pharmacol.* **80**:123–131 (1983d).

Akasu, T., Nishimura, T., and Koketsu, K.: Modulation of action potential during the late slow excitatory postsynaptic potential in bullfrog sympathetic ganglia. *Brain Res.* **280**:349–354 (1983e).

Akasu, T., Nishimura, T., and Koketsu, K.: Substance P inhibits the action potentials in bullfrog sympathetic ganglion cells. *Neurosci. Lett.* **41**:161–166 (1983f).

Ariyoshi, M., Tokimasa, T., Ohta, Y., and Koketsu, K.: Effects of histamine on the sensitivity of frog nicotinic receptor-ionic channel complex. *Neurosci. Lett.* **17**(*Suppl.*):127 (1984).

Bindler, E. H., and Gyermek, L.: Influence of 5-HT antagonists on the ganglionic stimulant action of 5-HT and DMPP. *Fed. Proc. Fed. Am. Soc. Exp. Biol.* **20**:319 (1961).

Bowery, N. G., and Brown, D. A.: γ-Aminobutyric acid uptake by sympathetic ganglia. *Nature (London)* **238**:89–91 (1972).

Brown, D. A., and Adams, P. R.: Muscarinic suppression of a novel voltage-sensitive K^+ current in a vertebrate neurone. *Nature (London)* **283**:673–676 (1980).

Brown, D. A., and Galvan, M.: Influence of neuroglial transport on the action of γ-amino-butyric acid on mammalian ganglion cells. *Br. J. Pharmacol.* **59**:373–378 (1977).

Bulbring, E.: The action of adrenaline on transmission in the superior cervical ganglion. *J. Physiol. (London)* **103**:55–67 (1944).

Burnstock, G.: Purinergic nerves. *Pharmacol. Rev.* **24**:509–581 (1972).

Burnstock, G.: Neurotransmitters and trophic factors in the autonomic nervous system. *J. Physiol. (London)* **313**:1–35 (1981).

Christ, D. D., and Nishi, S.: Site of adrenaline blockade in the superior cervical ganglion of the rabbit. *J. Physiol. (London)* **213**:107–117 (1971).

Clapham, D. E., and Neher, E.: Substance P reduces acetylcholine-induced currents in isolated bovine chromaffin cells. *J. Physiol. (London)* **347**:255–277 (1984).

Crowshaw, K., Jessup, S. J., and Ramwell, P. W.: Thin-layer chromatography of 1-dimethylaminonaphthalene-5-sulphonyl derivatives of amino acids present in superfusates of cat cerebral cortex. *Biochem J.* **103**:79–85 (1967).

DeGroat, W. C.: The actions of γ-aminobutyric acid and related amino acids on mammalian autonomic ganglia. *J. Pharmacol. Exp. Ther.* **172**:384–396 (1970).

DeGroat, W. C., and Lalley, P. M.: Interaction between picrotoxin and 5-hydroxytryptamine in the superior cervical ganglion of the cat. *Br. J. Pharmacol.* **48**:233–244 (1973).

DeGroat, W. C., and Volle, R. L.: The actions of the catecholamines on transmission in the superior cervical ganglion of the cat. *J. Pharmacol. Exp. Ther.* **154**:1–13 (1966a).

DeGroat, W. C., and Volle, R. L.: Interactions between the catecholamines and ganglionic stimulating agents in sympathetic ganglia. *J. Pharmacol. Exp. Ther.* **154**:200–215 (1966b).

Dun, N. J., and Karczmar, A. G.: Blockade of ACh potentials by α-bungarotoxin in rat superior cervical ganglion cells. *Brain Res.* **196**:536–540 (1980).

Dun, N. J., Kiraly, M., and Ma, R. C.: Evidence for a serotonin-mediated slow excitatory potential in the guinea-pig coeliac ganglia. *J. Physiol. (London)* **351**:61–76 (1984).

Eccles, R. M., and Libet, B.: Origin and blockade of the synaptic responses of curarized sympathetic ganglia. *J. Physiol. (London)* **157**:484–503 (1961).

Ewald, D. A.: Potentiation of postjunctional cholinergic sensitivity of rat diaphragm muscle by high-energy-phosphate adenine nucleotides. *J. Membr. Biol.* **29**:47–65 (1976).

Galvan, M., and Adams, P. R.: Control of calcium current in rat sympathetic neurons by norepinephrine. *Brain Res.* **244**:135–144 (1982).

Greengard, P.: Possible role for cyclic nucleodites and phosphorylated membrane proteins in postsynaptic actions of neurotransmitters. *Nature (London)* **260**:101–108 (1976).

Gyermek, L., and Bindler, E.: Blockade of the ganglionic stimulant action of 5-hydroxytryptamine. *J. Pharmacol. Exp. Ther.* **135**:344–348 (1962a).

Gyermek, L., and Bindler, E.: Action of indole alkylamines and amidines on the inferior mesenteric ganglion of the cat. *J. Pharmacol. Exp. Ther.* **138**:159–164 (1962b).

Henon, B. K., and McAfee, D. A.: The ionic basis of adenosine receptor actions on postganglionic neurones in the rat. *J. Physiol. (London)* **336**:607–620 (1983).

Hertzler, E. C.: 5-Hydroxytryptamine and transmission in sympathetic ganglia. *Br. J. Pharmacol.* **17**:406–413 (1961).

Hirai, K., Akasu, T., and Koketsu, K.: Effect of 5-hydroxytryptamine on the nicotinic acetylcholine receptor-channel complex. *Neurosci. Lett.* **6**:(Suppl.):68 (1981).

Hodgkin, A. L.: The ionic basis of electrical activity in nerve and muscle. *Biol. Rev.* **26**:339–409 (1951).

Hodgkin, A. L.: Ionic movements and electrical activity in giant nerve fibres. *Proc. R. Soc. London Ser. B* **148**:1–37 (1958).

Horn, J. P., and McAfee, D. A.: Norepinephrine inhibits calcium-dependent potentials in rat sympathetic neurons. *Science* **204**:1233–1235 (1979).

Horn, J. P., and McAfee, D. A.: Alpha-adrenergic inhibition of calcium-dependent potentials in rat sympathetic neurones. *J. Physiol. (London)* **301**:191–204 (1980).

Ivanov, A. Ya., and Skok, V. I.: Slow inhibitory postsynaptic potentials and hyperpolarization evoked by noradrenaline in the neurones of mammalian sympathetic ganglion. *J. Auton. Nerv. Syst.* **1**:255–263 (1980).

Jan, Y. N., Jan, L. Y., and Kuffler, S. W.: A peptide as a possible transmitter in sympathetic ganglia of the frog. *Proc. Natl. Acad. Sci. U.S.A.* **76**:1501–1505 (1979).

Jaramillo, J., and Volle, R. L.: A comparison of the ganglionic stimulating and blocking properties of some nicotinic drugs. *Arch. Int. Pharmacol.* **174**:88–97 (1968).

Kaibara, K., Koketsu, K., Akasu, T., and Miyagawa, M.: A kinetic analysis of the facilitatory action of adrenaline. *Pflugers Arch.* **392**:304–306 (1982).

Kato, E., and Kuba, K.: Inhibition of transmitter release in bullfrog sympathetic ganglia induced by γ-aminobutyric acid. *J. Physiol. (London)* **298**:271–283 (1980).

Kato, E., Kuba, K., and Koketsu, K.: Presynaptic inhibition by γ-aminobutyric acid in bullfrog sympathetic ganglion cells. *Brain Res.* **153**:398–402 (1978a).

Kato, E., Kuba, K., and Koketsu, K.: Effects of erabutoxins on neuromuscular transmission in frog skeletal muscles. *J. Pharmacol. Exp. Ther.* **204**:446–453 (1978b).

Kato, E., Morita, K., Kuba, K., Yamada, S., Kuhara, T., Shinka, T., and Matsumoto, I.: Does γ-aminobutyric acid in blood control transmitter release in bullfrog sympathetic ganglia? *Brain Res.* **195**:208–214 (1980).

Kobayashi, H., and Libet, B.: Actions of noradrenaline and acetylcholine on sympathetic ganglion cells. *J. Physiol. (London)* **208**:353–372 (1970).

Koketsu, K.: Cholinergic synaptic potentials and the underlying ionic mechanisms. *Fed. Proc. Fed. Am. Soc. Exp. Biol.* **28**:101–112 (1969).

Koketsu, K.: The electrogenic sodium pump. *Adv. Biophys.* **2**:77–112 (1971).

Koketsu, K.: Synaptic transmission in the autonomic nervous system. Proceedings of the International Union of Physiological Sciences Vol. 10, New Delhi, 1974.

Koketsu, K.: Neurohumoral controls of neurone activities, in: *Neurohumoral Correlates of Behaviour* (S. Subrahmanyam, ed.), pp. 21–34, Thomson Press, Faridabad, Haryana, (1977).

Koketsu, K.: Electropharmacological actions of catecholamine in sympathetic ganglia: Multiple modes of actions to modulate the nicotinic transmission. *Jpn. J. Pharmacol.* **31**(Suppl.):27P–28P (1981).

Koketsu, K.: Modulations of receptor sensitivity and action potentials by transmitters in vertebrate neurones, *Jpn. J. Physiol.* **34**:945–960 (1984).

Koketsu, K., and Akasu, T.: Modulation of the slow inward Ca^{2+} current by adrenaline in bullfrog sympathetic ganglion cells. *Jpn. J. Physiol.* **32**:137–140 (1982).

Koketsu, K., and Minota, S.: The direct action of adrenaline on the action potentials of bullfrog's (*Rana catesbeiana*) sympathetic ganglion cells. *Experientia* **31**:822–823 (1975).

Koketsu, K., and Nakamura, M.: The electrogenesis of adrenaline-hyperpolarization of sympathetic ganglion cells in bullfrogs. *Jpn. J. Physiol.* **26**:63–77 (1976).

Koketsu, K., and Nishi, S.: Calcium and action potentials of bullfrog sympathetic ganglion cells. *J. Gen. Physiol.* **53**:608–623 (1969).

Koketsu, K., Shoji, T., and Yamamoto, K.: Effects of GABA on presynaptic nerve terminals in bullfrog (*Rana catesbiana*) sympathetic ganglia. *Experientia* **30**:382–383 (1974).

Koketsu, K., Akasu, T., Miyagawa, M., and Hirai, K.: Modulation of nicotinic transmission by biogenic amines in bullfrog sympathetic ganglia. *J. Auton. Nerv. Syst.* **6**:47–53 (1982a).

Koketsu, K., Akasu, T., Miyagawa, M., and Hirai, K.: Biogenic antagonists of the nicotinic receptor: Their interactions with erabutoxin. *Brain Res.* **250**:391–393 (1982b).

Koketsu, K., Miyagawa, M., and Akasu, T.: Catecholamine modulates nicotinic ACh-receptor sensitivity. *Brain Res.* **236**:487–491 (1982c).

Koppanyi, T., and Karczmar, A. G.: Contribution to the study of the mechanism of action of cholinesterase inhibitors. *J. Pharmacol. Exp. Ther.* **101**:327–343 (1951).

Kuba, K., and Koketsu, K.: Ionic mechanism of the slow excitatory postsynaptic potential in bullfrog sympathetic ganglion cells. *Brain Res.* **81**:338–342 (1974).

Kuba, K., and Koketsu, K.: Direct control of action potentials by acetylcholine in bullfrog sympathetic ganglion cells. *Brain Res.* **89**:166–169 (1975).

Kuba, K., and Koketsu, K.: The muscarinic effects of acetylcholine on the action potential of bullfrog sympathetic ganglion cells. *Jpn. J. Physiol.* **26**:703–716 (1976a).

Kuba, K., and Koketsu, K.: Analysis of the slow excitatory postsynaptic potential in bullfrog sympathetic ganglion cells. *Jpn J. Physiol.* **26**:651–669 (1976b).

Kuba, K., and Koketsu, K.: Synaptic events in sympathetic ganglia. *Progr. Neurobiol.* **11**:77–169 (1978).

Kuffler, S. W., and Sejnowski, T. J.: Peptidergic and muscarinic excitation at amphibian sympathetic synapses. *J. Physiol. (London)* **341**:257–278 (1983).

Kupfermann, I.: Modulatory actions of neurotransmitters. *Annu. Rev. Neurosci.* **2**:447–465 (1979).

Libet, B.: Which postsynaptic action of dopamine is mediated by cyclic AMP? *Life Sci.* **24**:1043–1058 (1979).

Libet, B., and Kobayashi, H.: Electrogenesis of slow postsynaptic potentials in sympathetic ganglion cells. *Fed. Proc. Fed. Am. Soc. Exp. Biol.* **27**:750 (1968).

Libet, B., and Tosaka, T.: Dopamine as a synaptic transmitter and modulator in sympathetic ganglia: A different mode of synaptic action. *Proc. Natl. Acad. Sci. U.S.A.* **67**:667–673 (1970).

Libet, B., Kobayashi, H., and Tanaka, T.: Synaptic coupling into the production and storage of a neuronal memory trace. *Nature (London)* **258**:155–157 (1975).

Lundberg, A.: Adrenaline and transmission in the sympathetic ganglion of the cat. *Acta Physiol. Scand.* **26**:252–263 (1952).

Minota, S., and Koketsu, K.: Effects of adrenaline on the action potential of sympathetic ganglion cells in bullfrogs. *Jpn. J. Physiol.* **27**:353–366 (1977).

Morita, K., Katayama, Y., Koketsu, K., and Akasu, T.: Actions of ATP on the soma of bullfrog primary afferent neurons and its modulating action on the GABA-induced response. *Brain Res.* **293**:360–363 (1984).

Nakamura, M., and Koketsu, K.: The effect of adrenaline on sympathetic ganglion cells of bullfrogs. *Life Sci.* **11**:1165–1173 (1972).

Nakamura, T., and Nishi, S.: Analysis of the slow hyperpolarizations induced by catecholamines, acetylcholine and preganglionic volleys in rabbit superior cervical ganglion cells. *Neurosci. Lett.* **9**(Suppl.):78 (1982).

Nakamura, M., Hayashi, H., Hirai, K., and Koketsu, K.: Effects of ATP on sympathetic ganglia from bullfrogs. *Jpn. J. Pharmacol.* **24**(Suppl.):134 (1974).

Nishi, S., and Katayama, Y.: The non-cholinergic excitatory transmission in sympathetic ganglia, in: *Advances in Physiological Sciences*, Vol. 4, Physiology and Excitable Membranes (J. Salanki, ed.), pp. 323–327, Pergamon Press, New York (1981).

Ohta, Y., Ariyoshi, M., and Koketsu, K.: Histamine as an endogenous antagonist of nicotinic ACh-receptor. *Brain Res.* **306**:370–373 (1984).

Phillis, J. W.: The role of cyclic nucleotides in the CNS. *Can. J. Physiol. Sci.* **4**:151–195 (1977).

Phillis, J. W., and Wu, P. H.: Catecholamines and the sodium pump in excitable cells. *Prog. Neurobiol.* **17**:141–184 (1981).

Robertson, P. A.: Potentiation of 5-hydroxytryptamine by the true-cholinesterase inhibitor 284C51. *J. Physiol. (London)* **125**:37P–38P (1954).

Ryall, R. W., and Belcher, G.: Substance P selectively blocks nicotinic receptors on Renshaw cells: A possible synaptic inhibitory mechanism. *Brain Res.* **137**:376–380 (1977).

Scuka, M.: Analysis of the effects of histamine on the end-plate potentials. *Neuropharmacology* **12**:441–450 (1973).

Shirasawa, Y., and Koketsu, K.: An analysis of 5-HT hyperpolarization of sympathetic ganglion cells. *Jpn. J. Pharmacol.* **28**:57–60 (1978).

Siegelbaum, S. A., and Tsien, R. W.: Modulation of gated ion channels as a mode of transmitter action. *Trends Neurosci.* **6**:307–313 (1983).

Siggins, G. R., Gruol, D. L., Padjen, A. L., and Forman, D. S.: Purine and pyrimidine mononucleotides depolarise neurones of explanted amphibian sympathetic ganglia. *Nature (London)* **270**:263–265 (1977).

Silinsky, E. M.: On the association between transmitter secretion and the release of adenine nucleotides from mammalian motor nerve terminals. *J. Physiol. (London)* **247**:145–162 (1975).

Silinsky, E. M., and Ginsborg, B. L.: Inhibition of acetylcholine release from preganglionic frog nerves by ATP but not adenosine. *Nature (London)* **305**:327–328 (1983).

Silinsky, E. M., and Hubbard, J. I.: Release of ATP from rat motor nerve terminals. *Nature (London)* **243**:404–405 (1973).

Smith, P. A.: Examination of the role of the electrogenic sodium pump in the adrenaline-induced hyperpolarization of amphibian neurons. *J. Physiol. (London)* **347**:377–395 (1984).

Stone, T. W.: Physiological roles for adenosine and adenosine 5′-triphosphate in the nervous system. *Neuroscience* **6**:523–555 (1981).

Thomas, R. C.: Electrogenic sodium pump in nerve and muscle cells. *Physiol. Rev.* **52**:563–594 (1972).

Tokimasa, T.: Muscarinic agonists depress calcium-dependent g_K in bullfrog sympathetic neurones. *J. Auton. Nerv. Syst.* **10**:107–116 (1984).

Trendelenburg, U.: The action of histamine, pilocarpine and 5-hydroxytryptamine on transmission through the superior cervical ganglion. *J. Physiol. (London)* **135**:66–72 (1957).

Wallis, D. I., and North, R. A.: The action of 5-hydroxytryptamine on single neurones of the rabbit superior cervical ganglion. *Neuropharmacology* **17**:1023–1028 (1978).

Wallis, D. I., and Woodward, B.: The depolarizing action of 5-hydroxytryptamine on sympathetic ganglion cells. *Br. J. Pharmacol.* **49**:168P (1973).

Watanabe, S., and Koketsu, K.: 5-HT hyperpolarization of bullfrog sympathetic ganglion cell membrane. *Experientia* **29**:1370–1372 (1973).

Whittaker, V. P., Dowdall, M. J., and Boyne, A. F.: The storage and release of acetylcholine by cholinergic nerve terminals: Recent results with non-mammalian preparations. *Biochem. Soc. Symp.* **36**:49–68 (1972).

Witkop, B., and Grossinger, E.: Amphibian alkaloids. *Alkaloids* **21**:139–253 (1983).

Young, J. A. C., Brown, D. A., Kelly, J. S., and Schon, F.: Autoradiographic localization of sites of [^{3}H] γ-aminobutyric acid accumulation in peripheral autonomic ganglia. *Brain Res.* **63**:479–486 (1973).

13

Pharmacology of Synaptic Ganglionic Transmission and Second Messengers

ALEXANDER G. KARCZMAR and NAE J. DUN

I. INTRODUCTION

Intensive investigations in the past ten years have clearly demonstrated that multiple mechanisms interact in the physiological and pharmacological control of synaptic transmission at autonomic ganglia. Indeed, this junction comprises multiple pre- and postsynaptic receptors that are involved in the regulation and modulation of transmission (see Chapter 3); additionally, there is an interplay between potentials generated at some of these receptors, leading to facilitation or inhibition of transmission. Furthermore, it was shown recently that several endogenous peptides, amines, and cyclic nucleotides that are present in the ganglia or released in the vicinity of the junctional sites may affect ganglionic transmission and thus may participate in physiological and pharmacological control of transmission. In all, the autonomic ganglia cannot be considered today as simple relay stations; they should rather be looked on as centers for integration of diverse afferent inputs arising from the preganglionic neurons as well as from peripheral effector sites.

To account for these phenomena, we will emphasize in this chapter pharmacological effects of substances that are present in the ganglia, such as γ-aminobutyric acid (GABA), 5-hydroxytryptamine (5-HT) (serotonin),

ALEXANDER G. KARCZMAR and NAE J. DUN • Department of Pharmacology and Experimental Therapeutics, Stritch School of Medicine, Loyola University Medical Center, Maywood, Illinois 60153.

and catecholamines; possible participation in the modulation of ganglionic transmission of some of these endogenous compounds via their release from ganglionic interneurons, small intensely fluorescent (SIF) cells, will also be discussed. We will also stress actions of anticholinesterases, drugs that act via the cholinergic system of the ganglia and that may, as well, exert noncholinergic ganglionic effects; finally, actions of nonendogenous compounds such as drugs that block ganglionic transmission will be described.

II. PHARMACOLOGICAL EFFECTS OF COMPOUNDS THAT ARE PRESENT IN THE GANGLION

A. γ-Aminobutyric Acid

A measurable amount of GABA is present in sympathetic ganglia, where GABA appears to be concentrated in glial components; furthermore, it was found that glial cells in mammalian sympathetic ganglia can take up GABA at an external concentration as low as 1 μM (Bowery and Brown, 1972; Young *et al.*, 1973). Since the concentration of GABA in the plasma of cats may be as high as 2–4 μM (Crowshaw *et al.*, 1967), GABA present in the plasma may exert a regulatory effect on the excitability of the pre- and postsynaptic membrane via the GABA transport system of the ganglionic glial cells. In this context, GABA was found to exert a depolarizing effect postsynaptically (Adams and Brown, 1973, 1975) and, presynaptically, a depressant effect on transmitter release (Kato *et al.*, 1978). These actions are discussed in detail in Chapters 10, 11, and 12.

B. Serotonin (5-Hydroxytryptamine)

The effects of 5-HT on sympathetic ganglia and on ganglionic transmission were studied in the early 1950s by Trendelenburg, who used the nictitating membrane contraction as an index. It was shown that 5-HT excites the cat superior cervical ganglia directly, that the stimulatory action is not affected by ganglionic blocking agents, and that ganglia exhibit marked tachyphylaxis after a single injection of 5-HT (Trendelenburg, 1956). Since these early studies of Trendelenburg, a number of investigators found that the effects of 5-HT on sympathetic ganglia and on their transmission are variable and complex, both depolarization and hyperpolarization of the ganglionic membrane and both facilitation and depression of ganglionic transmission having been observed following 5-HT administration or application (DeGroat and Lally, 1973; Machova and

Boska, 1969; Haefely, 1974) (see also Chapter 12).

The direct depolarizing effect of 5-HT on autonomic neurons was confirmed by means of intracellular recording techniques (Wallis and North, 1978; Skok and Selyanko, 1979). Membrane depolarization induced by 5-HT is always associated with a decrease in membrane resistance; thus, it was suggested that it is caused by an increase of G_{Na}, G_K, and G_{Ca}, as is the depolarization of the ganglion cells evoked by the nicotinic action of acetylcholine (ACh) (Wallis and North, 1978; Skok and Selyanko, 1979).

A measurable amount of 5-HT is present in mammalian paravertebral ganglia (Gertner *et al.*, 1959; Dun *et al.*, 1980). The exact localization of 5-HT was not carried out until an immunohistofluorescent technique for 5-HT was developed and used to study the distribution of 5-HT in rat superior cervical ganglia. In this ganglion, 5-HT immunofluorescence was found to be localized in certain SIF cells that differed from the catecholamine-containing SIF cells (Verhofstad *et al.*, 1981). The 5-HT-positive SIF cells were located near blood vessels; they exhibited varicose processes (Verhofstad *et al.*, 1981). Recently, it was found that muscarinic agonists increase, whereas muscarinic antagonists and decentralization lower, the 5-HT content of the ganglia; this suggests that 5-HT-containing SIF cells may be controlled by preganglionic cholinergic inputs acting on muscarinic receptors (Hadjiconstantinou *et al.*, 1982) (see also Chapter 7).

In addition to the postsynaptic depolarizing action, 5-HT depresses ACh release from the presynaptic fibers. The relevant information can be found in Chapter 11.

Although 5-HT is present in, and active on, paravertebral ganglia, 5-HT does not appear to serve as a ganglionic transmitter in generating a transmitter-specific membrane potential. However, in the case of myenteric and prevertebral ganglia of the guinea pig, 5-HT may act as a transmitter mediating a slow excitatory postsynaptic potential (EPSP). Indeed, when applied to the myenteric neurons, 5-HT caused a slow depolarization associated with an increase of membrane excitability (Wood and Mayer, 1979), and this depolarization as well as the slow EPSP elicited in these neurons were blocked by methysergide; accordingly, it was suggested that 5-HT is the mediator of the slow EPSP (Wood and Mayer, 1979). However, the results of Wood and Mayer (1979) were at variance with the observations of North and associates (Katayama and North, 1978; Morita *et al.*, 1980), since these investigators found that 5-HT caused, in the myenteric neurons that exhibit a slow EPSP, hyperpolarization rather than depolarization of the membrane (S. M. Johnson *et al.*, 1981). These findings and the earlier observation by Johnson and his associates (see S. M. Johnson *et al.*, 1981) that substance P (SP), when applied to the myenteric neurons, mimics the slow EPSP suggest that SP rather than 5-HT may be the mediator of the slow EPSP in the myenteric plexus. It is conceivable that there may be two populations of myenteric neurons and that one of these populations is subserved by 5-HT and the other by SP. In fact, the heterogeneity of transmitters responsible for the generation of the slow EPSP was clearly demonstrated with regard to the prevertebral ganglia, since SP and 5-HT generated a slow EPSP in inferior mesenteric and coeliac ganglia, respectively (Dun and Jiang, 1982; Dun and Kiraly, 1983; Kiraly *et al.*, 1983; Dun and Ma, 1984; Dun *et al.*, 1984a; see also Simmons, 1985).

The evidence that 5-HT may serve as a transmitter in peripheral autonomic ganglia appears to be most complete in the case of guinea pig coeliac ganglia (Kiraly *et al.*, 1983; Dun *et al.*, 1984a). Repetitive stimulation of splanchnic nerves elicited a slow noncholinergic EPSP in the majority of coeliac neurons (Figure 1); the slow EPSP was suppressed by cyproheptadine, a 5-HT-receptor antagonist (Haigler and Aghajanian, 1977) and was enhanced by the 5-HT-reuptake inhibitor fluoxetine (Fuller and Wong, 1977) and tryptophan, a 5-HT precursor. Application of 5-HT to coeliac neurons that exhibit a slow EPSP mimicked the latter (Figure 1); furthermore, the slow EPSP was reversibly suppressed by excess 5-HT. Last, networks of nerve fibers exhibiting immunoreactivity to 5-HT were observed in the coeliac ganglia (Kiraly *et al.*, 1983; Dun *et al.*, 1984a). Collectively, these results strongly suggest that the mediator of the slow EPSP in the majority of coeliac neurons is 5-HT (Dun *et al.*, 1984a). The origin of the 5-HT-containing fibers in the coeliac ganglia is at present not known; it remains to be studied whether they are of central or peripheral origin or both.

C. Histamine

Earlier investigations showed that the effects of histamine on autonomic ganglia are weak and inconsistent (Konzett, 1952; Trendelenburg, 1954). However, a weak facilitatory effect of histamine on autonomic ganglionic response to submaximal preganglionic stimulation was noted by several investigators (Trendelenburg, 1956, 1957; Iorio and McIsaac, 1966). Brimble and Wallis (1973), on the other hand, described biphasic effects of histamine in the case of rabbit superior cervical ganglia; using the sucrose-gap method, they reported that the H_1-receptor blocker mepyramine enhanced the histamine-induced synaptic depression, whereas the H_2-receptor blocker burimamide attenuated the facilitatory action of histamine. Thus, it appears that both H_1 and H_2 receptors are present in the sympathetic ganglia and that activation of these receptors may regulate the efficacy of synaptic transmission. The site and mechanism of the facilitatory and inhibitory effect of histamine in sympathetic ganglia were not pursued further until recently. By means of intracellular recording

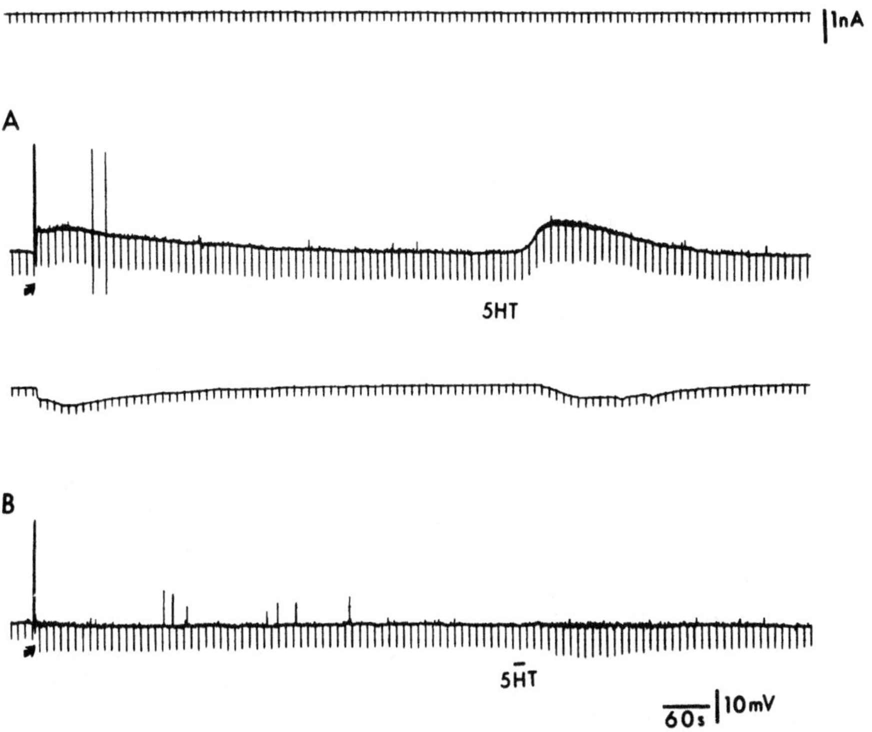

Figure 1. Similar membrane resistance increase during the noncholinergic EPSP and 5-HT depolarization in a guinea pig coeliac ganglion cell. (A) Repetitive stimulation of the left splanchnic nerves (20 Hz, 2 sec; indicated by the curved arrow) elicited a burst of action potentials (dark vertical tracing) followed by a noncholinergic EPSP. The spikes in this and subsequent recordings are attenuated because of the limitation of the frequency response of the pen recorder. The preparation was superfused continuously with a Krebs solution containing atropine (1 μM). The top tracings in (A) and (B) represent the amount of current (downward deflections of upper tracings), and the lower tracings represent membrane potential change. The amplitude of hyperpolarizing electrotonic potentials was used to monitor the cell input resistance change. 5-HT (10 μM) was applied to the ganglion for 15 sec as indicated. The slow depolarization elicited by nerve stimulation and by 5-HT was accompanied by an increase in membrane resistance (as shown by an increase of the amplitude of the electrotonic potentials) of 57 and 50%, respectively. (B) Noncholinergic EPSP and 5-HT depolarization were elicited in the same manner as in (A). However, membrane depolarization was prevented by passage of hyperpolarizing currents to manually clamp the membrane potential at the resting level. Under these conditions, the membrane resistance increase was 75 and 66%, respectively. From Kiraly *et al.* (1983).

techniques, Yamada *et al.* (1982) presented evidence indicating that there are indeed two types of histamine receptors in bullfrog sympathetic ganglia. They also found that histamine exerts a facilitatory or an inhibitory action depending on the concentrations used. Thus, low (a few micromolar) and high ($\geq$ 100 μM) concentrations of histamine increased and depressed, respectively, the amplitude of the fast EPSP and its quantal content. Furthermore, the facilitatory effect could be blocked by an H_1-receptor blocker, whereas the inhibitory effect was suppressed by the H_2 blocker cimetidine. On the other hand, the postsynaptic membrane sensitivity to iontophoretically applied ACh and the amplitude of miniature EPSPs (mEPSPs) were not appreciably affected by either low or high concentrations of histamine. These results collectively suggest that the effects of histamine on sympathetic ganglionic transmission are presynaptic in nature, since histamine may both facilitate and depress the release of ACh. The mechanism underlying these effects appears to be related to the presence of both H_1 and H_2 receptors in the sympathetic ganglia, and the activation of these receptors may regulate the efficacy of synaptic transmission. The molecular causation of the specific facilitatory and inhibitory actions of histamine on the ganglia is not clear. Histamine has been found to increase cyclic AMP (cAMP) and cGMP levels in sympathetic ganglia by differentially activating H_2 and H_1 receptors, respectively (Study and Greengard, 1978). Whether or not the cyclic nucleotide system is causally related to the dualistic ganglionic effect of histamine remains to be studied.

In the myenteric ganglia, histamine causes a slow depolarization the characteristics of which are similar to these of the depolarization produced by either substance P or 5-HT (Nemeth *et al.*, 1984). However, histamine is ruled out as a mediator of the noncholinergic EPSP, since the latter is not blocked by histamine antagonists that prevent the histamine-induced depolarization (Nemeth *et al.*, 1984). Nevertheless, as the authors pointed out, histamine released from mast cells could affect neuronal excitability.

D. Peptides

Intensive studies of the biological role of peptides were triggered in the past few years by the availability of synthetic compounds appropriate for electrophysiological studies and of immunohistofluorescent techniques applicable for the localization of endogenous peptides. A number of peptides, notably angiotensin II, SP, somatostatin, enkephalins and endorphins, thyrotropic-releasing hormone, luteinizing-hormone-releasing hormone (LH-RH), melanocyte-stimulating hormone, and vasopressin, are widely distributed in the central nervous system and affect its function (see Chapters 2 and 3) (Hökfelt *et al.*, 1984; Bloom, 1972; Nicoll, 1972, 1976) as well as certain specific behaviors (Severs and Daniels-Severs, 1973; Moss and McCann, 1975; Barker, 1975). In the present context, the peptides of particular interest are those that are present in the autonomic ganglia or active at the autonomic sites or both (see Sejnowski, 1982; Dun, 1983) (see also Chapters 2 and 3). Histofluorescence study so far has revealed the presence of SP, vasoactive intestinal polypeptide (VIP), enkephalins, cholecystokinin, dynorphin, bombesin, and gastrin-releasing peptide in nerve fibers of various sympathetic ganglia; particularly rich networks of fibers exhibiting immunohistofluorescence indicating the presence of these peptides are found in guinea pig prevertebral ganglia (see Hökfelt *et al.*, 1977a, 1980; Dun, 1983) (see also Table 1 in Chapter 3). The peptides somatostatin and avian pancreatic peptide (Lundberg *et al.*, 1980), on the other hand, appear to be localized in the noradrenergic sympathetic neurons (see also Hökfelt *et al.*, 1980). It should be stressed that there are species differences with respect to the localization of peptides in sympathetic ganglia as well as among various ganglia in the same species (Hökfelt *et al.*, 1980). For example, SP is found to be localized principally in nerve fibers in the coeliac–superior mesenteric and inferior mesenteric ganglia (Hökfelt *et al.*, 1977b), whereas it appears to be concentrated in postganglionic neurons of rat superior cervical ganglia (Robinson *et al.*, 1980; Kessler *et al.*, 1981). SP, then, may exert different physiological functions in these two groups of sympathetic ganglia.

The effects of peptides on sympathetic neurons and on ganglionic transmission are the subject of detailed description in Chapter 8. Suffice it to say at this time that evidence is particularly strong with respect to SP and LH-RH as transmitters mediating the slow noncholinergic excitatory potential in guinea pig inferior mesenteric (Dun and Jiang, 1982; Tsunoo *et al.*, 1982; Dun and Kiraly, 1983; Jiang *et al.*, 1983) and bullfrog sympathetic ganglia (Y. N. Jan *et al.*, 1979, 1980; L. Y. Jan and Y. N. Jan, 1982), respectively. Applied as drugs to either ganglion, both SP and LH-RH exert potent depolarizing effects on the postsynaptic membranes of the ganglia in question (see Chapter 8). On the other hand, enkephalins appear to act as inhibitory transmitters in reducing the output of transmitters from preganglionic fibers, both cholinergic and peptidergic release being affected (Konishi *et al.*, 1979, 1980; Jiang *et al.*, 1982). Other peptides, such as angiotensin II, exert potent pre- and postsynaptic effects. Angiotensin II is one of the most potent ganglionic stimulants tested in this laboratory (Dun *et al.*, 1978b) (see also Chapter 8). These presynaptic effects were described in detail in Chapter 11.

Another interesting peptide present in prevertebral ganglia, VIP (Hökfelt *et al.*, 1977a), caused slow depolarization in a portion of prevertebral ganglion cells and markedly increased the amplitude and duration of the muscarinic slow EPSP evoked synaptically and of the membrane depolarization induced by methacholine. These results suggest that the peptide may be a neuromodulator effective in enhancing the sensitivity of postsynaptic muscarinic receptors to ACh (Mo and Dun, 1984).

In addition to the aforementioned peptides, the pharmacological actions of a number of peptides that are present in sympathetic ganglia (see above) remain to be investigated.

III. SMALL INTENSELY FLUORESCENT CELLS AND GANGLIONIC TRANSMISSION

A. Location and Morphology

Small granular cells that exhibited marked yellow fluorescence following exposure to formaldehyde vapor were first demonstrated in rat superior cervical ganglia by Eränkö and Harkonen (1963, 1965); these SIF cells differed from principal ganglion cells with respect to their size (SIF cells being smaller than the principal ganglionic neurons) and their histofluorescent characteristics (Eränkö and Harkonen, 1965). SIF cells occur in sympathetic ganglia of a variety of mammals (Bjorklund *et al.*, 1970; Williams *et al.*, 1977) and amphibians (Weight and Weitsen, 1977); they are usually present in clusters located around blood vessels (Matthews and Raisman, 1969; Elfvin *et al.*, 1975). The chemical nature of the amine of the SIF cells has been investigated extensively (Bjorklund *et al.*, 1970; Libet and Owman, 1974; Elfvin *et al.*, 1975). This nature appears to depend on the species; for example, guinea pig and rabbit SIF cells contain primarily norepinephrine (Elfvin *et al.*, 1975) and dopamine (Libet and Owman, 1974), respectively. In addition, SIF cells of rat superior cervical ganglia may contain 5-HT (see above) (Verhofstad *et al.*, 1981), and an enkephalinlike peptide may be present in SIF cells of amphibian paravertebral ganglia (Kondo and Yui, 1981).

In the case of the rat superior cervical ganglion, the sympathetic preganglionic fibers appear to make synaptic contacts with SIF cells; the processes of these cells in turn make synaptic contacts with cell processes of the sympathetic neurons (Siegrist *et al.*, 1968; Matthews and Raisman, 1969; Williams and Palay, 1969). Thus, SIF cells of the rat superior cervical ganglion appear to exhibit the nature of an interneuron. Similarly, SIF cells may also act as interneurons in the superior cervical ganglion of the rabbit, although their number is small (Libet and Owman, 1974; Williams *et al.*, 1977; Dail and Evans, 1978) and their efferent synapses are rare (Dail and Evan, 1978). Finally, efferent synapses of SIF cells were not observed in sympathetic ganglia of the bullfrog (Weight and Weitsen, 1977).

In terms of their morphology, fine structure, chemical composition, and function, two or more types of SIF cells could be distinguished in superior cervical ganglia of several species (Williams *et al.*, 1975, 1976, 1977). Type I SIF cells were situated among principal ganglion cells; they were generally solitary and possessed one or more long, ramifying cell processes. Type II SIF cells were characteristically located next to blood vessels in the stroma or fibrous capsule of the ganglion; they were generally arranged in clusters, and their processes were sparse, short, and oriented toward blood vessels (Williams *et al.*, 1977). The distribution of Type I and Type II SIF cells varied widely among several species; 80% of SIF cells of the rabbit belonged to Type I, whereas 99% of cat SIF cells exhibited the characteristics of Type II cells (Williams *et al.*, 1977). It was suggested on the basis of the distribution and morphology of the two types of SIF cells that Type I cells may be interneurons that are responsible for the generation of the slow inhibitory postsynaptic potential (IPSP) in sympathetic ganglia, whereas Type II cells may conceivably subserve the function of enhancement of the slow EPSP (Williams *et al.*, 1977). Thus, the proposed functions of these two types of SIF cells were diametrically opposite, since Type I cells may provide an inhibitory input onto the postsynaptic membrane, while Type II cells may augment the facilitatory postsynaptic function of the slow EPSP (Libet, 1976; Williams *et al.*, 1977).

Although the contention that at least Type I SIF cells function as interneurons has received some experimental support (Libet and Owman, 1974; Williams *et al.*, 1977), discordant findings were obtained in a morphological study of rat superior cervical ganglia in which the ultrathin serial section method was utilized (Kondo, 1977). In this study, the morphological characteristics of SIF cells were found to be similar to those of the chief cells of the carotid body; furthermore, most of the fibers abutting on the SIF cells of the ganglia appeared to be sensory in nature, similar to the innervation of the chief cells of the carotid body; Kondo (1977) therefore suggested that SIF cells may subserve a chemoreception function, as do the chief cells of the carotid body. Alternatively, the results obtained in this laboratory suggest that if SIF cells are capable of releasing catecholamines, whether dopamine or norepinephrine, their primary effect would be to depress transmitter release from preganglionic fibers (see below and Chapter 11).

The discovery of SIF cells and the recent description of several types of SIF cells have added an interesting dimension to the organization of autonomic ganglia; the clarification of their physiological function may provide insight into the regulatory mechanisms of the autonomic nervous system.

B. Catecholamines and Cyclic Nucleotides, Small Intensely Fluorescent Cells, and Ganglionic Transmission

SIF cells may be involved in transmission via mechanisms mediated by catecholamines or second messengers such as cyclic nucleotides or both. These mechanisms and the effects of the substances in question are described below.

1. Catecholamines

Interest in the effects of catecholamines on ganglionic transmission dates back to the 1930s, when Marrazzi (1939) reported that epinephrine depresses ganglionic transmission. Since that time, a number of reports dealing with catecholamines and ganglionic transmission have appeared (DeGroat and Volle, 1966; McIsaac, 1966; Christ and Nishi, 1971; Dun and Nishi, 1974; Dun and Karczmar, 1977a) (see also Chapters 9 and 12).

According to the hypothesis of Libet (1970, 1979), catecholamines, specifically dopamine in the case of rabbit superior cervical ganglia, are released from SIF cells and then interact with the adrenergic receptors of the postganglionic neurons to generate the slow IPSP. This hypothesis, the role of dopamine in the generation of the IPSP, and the pertinent experiments are discussed in detail in Chapter 9; on the whole, it may appear today that the IPSP, at least in some species, is generated by a special type of muscarinic action of synaptically released ACh, rather than by a catecholamine that is released from SIF cells. Since the response is constituted by a hyperpolarizing potential, catecholamines, whether released from SIF cells or present in the circulation, may conceivably elevate the threshold for the action potential. Additionally, catecholamines released from SIF cells may also exert a very different effect, since they may enhance the slow muscarinic EPSP via a cAMP-dependent mechanism (Libet, 1979; Kobayashi and Tosaka, 1983). It must be stressed that according to Libet and his collaborators, these two actions of catecholamines on the sympathetic ganglia are independent and generated by different mechanisms (see below). Finally, while postsynaptic mechanisms are stressed in this chapter, it should not be forgotten that catecholamines exert presynaptic actions (see Nishi, 1970 and Chapter 11) that are so potent that these actions must be considered when the primary site of the physiological function of catecholamines is being discussed.

2. Cyclic Nucleotides and Ganglionic Transmission

Cyclic adenosine 3′,5′-monophosphate (cAMP) was discovered by Sutherland and Rall (1958) as a heat-stable factor accumulated in liver homogenates exposed to epinephrine or glucagon. Cyclic guanosine 3′,5′-monophosphate (cGMP) was discovered subsequently (Ashman *et al.*, 1963). Further investigations, particularly of Greengard (cf. Greengard, 1976; Nestler and Greengard, 1984; Nestler *et al.*, 1984), stressed the role of cyclic nucleotides as second messengers. This concept underwent several stages of experimental development [compare Greengard (1976) with Nestler *et al.* (1984)]; in its present stage, it can be presented as follows: First messengers, i.e., neurotransmitters and hormones, may, in some instances, cause physiological effects by "directly inducing . . . allosteric changes in the receptor–ion channel complex." However, "a common and important property of many ion channels" . . . is that they are regulated . . . "by phosphorylation of proteins . . . through some second messenger" (Nestler *et al.*, 1984), this phosphorylation constituting the specific response of the target cell to the hormone or transmitter (Sutherland *et al.*, 1968; Greengard, 1976). Cyclic nucleotides serve as second messengers as their intracellular levels are augmented by neurotransmitters and hormones leading to activation of specific protein kinases [sometimes neurotransmitters can cause this activation without an assist of second messengers (Sutherland *et al.*, 1968; Greengard, 1976)]. Prominent second messengers present in the nervous system include cAMP and cGMP, as well as calcium. Over the past years, several proteins were identified as specific substrates for the phosphorylating action of cAMP, cGMP, or calcium/calmodulin-dependent protein kinases (Greengard, 1982). The function of cAMP was investigated in detail with regard to several tissues including the cerebellum, the hypothalamus, and the cortex, and it is of interest that much evidence resulted from the studies of this function in the ganglia by Greengard and his associates (see below); similarly, its activation by several transmitters including dopamine (see below), ACh, and norepinephrine was also evaluated in the ganglia. On the other hand, the function of cGMP in biological tissues has not been studied as intensively. It was proposed that cGMP may act as a second messenger for the actions of various neurotransmitters including ACh (Greengard, 1976) and histamine (see above) (Study and Greengard, 1978).

Insofar as the autonomic ganglia are concerned, the evidence supporting the hypothesis of the second-messenger role of cyclic nucleotides first came from biochemical studies demonstrating that the electrical stimulation of preganglionic nerve fibers produces an increase in the ganglionic contents of cAMP or cGMP (McAfee *et al.*, 1971; Greengard and Kebabian, 1974; Weight *et al.*, 1974) and that the exogenous application of appropriate agonists increases the levels of cyclic nucleotides in slices of sympathetic ganglia (Kebabian and Greengard, 1971; Greengard and Kebabian, 1974; Kebabian *et al.*, 1975). Second, these data were followed by pharmacological evaluation. Using the sucrose-gap method, McAfee and Greengard (1972) found that in rabbit superior cervical ganglia, exogenous mono- or dibutyryl cAMP and cGMP hyperpolarize and depolarize the ganglia, respectively. Furthermore, theophylline, a phosphodiesterase inhibitor, potentiated the hyperpolarizing effect of dopamine and augmented the amplitude of the slow IPSP (McAfee and Greengard, 1972). Cyclic AMP and cGMP thus appeared to fulfill the requirements of the intracellular messengers of slow synaptic potentials in sympathetic ganglia (Greengard, 1976).

These results appear, however, to be controversial. Also using the sucrose-gap method, Dun and Karczmar (1977b), Akasu and Koketsu (1977), and Busis *et al.* (1978) found that cAMP or its derivatives exerted no effect on the membrane potential or the action potential of rabbit and bullfrog sympathetic ganglia. While theophylline was found to increase the amplitude of the P potential (the extracellular counterpart of the slow IPSP) of rabbit and bullfrog sympathetic ganglia (Dun and Karczmar, 1977b; Busis *et al.*, 1978), this effect was not augmented by the application of cAMP. Actually, the ganglionic action of theophylline is complex and species-dependent (Dun and Karczmar, 1977b), and theophylline depressed the slow IPSP of bullfrog sympathetic ganglia (Akasu and Koketsu, 1977). Furthermore, intracellular studies carried out in these laboratories showed that dibutyryl cAMP, applied iontophoretically, elicited no hyperpolarizing response in rabbit ganglion cells that, however, could be clearly hyperpolarized by iontophoretically applied dopamine (Figure 2) (Dun *et al.*, 1977). Similarly, negative results were obtained on injection of cAMP into single neurons of the rabbit superior cervical ganglion (Kobayashi *et al.*, 1978). In the case of rat superior cervical ganglion cells, intracellular injection of cAMP depolarized the cell membrane and blocked

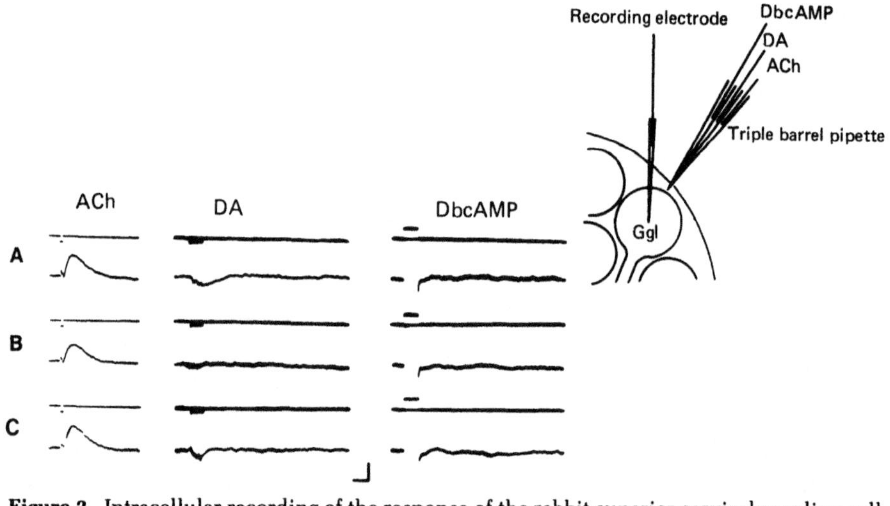

Figure 2. Intracellular recording of the response of the rabbit superior cervical ganglion cell to iontophoretic applications of ACh, dopamine (DA), and dibutyryl cAMP (DbcAMP). (A) Control responses. (B) Responses after 10 min of haloperidol (0.1 μM) superfusion. (C) Responses after 30 min washing with Krebs solution. The upper tracing of each recording represents the current pulse. ACh potentials were elicited by a current pulse of 7-msec duration. The DA and DbcAMP responses were elicited by tetanic (30-Hz) current pulses of 30-msec duration. Note that DbcAMP application did not mimic the membrane hyperpolarization induced by iontophoretic application of DA. Records were taken from the same ganglion cell. Vertical bar of the calibration mark: 10 mV and 1×10^{-7} A for ACh potentials, 4 mV and 1×10^{-6} A for DA and DbcAMP responses. Horizontal bar of the calibration mark: 40 msec for ACh, 4 sec for DA and DbcAMP responses. The diagram of the experimental arrangement is illustrated at the upper right corner. From Dun *et al.* (1977).

action potentials (Gallagher and Shinnick-Gallagher, 1977); these effects were opposite to those that would be expected if cAMP were the intracellular mediator of the slow IPSP (see also Chapter 9).

Although cAMP does not appear to be the intracellular second messenger for the generation of the slow IPSP in sympathetic ganglia, some evidence has been obtained suggesting that it is involved in the cellular process leading to an enhancement of slow muscarinic depolarization induced by dopamine in rabbit superior cervical ganglia (Libet *et al.*, 1975; Kobayashi *et al.*, 1978). Thus, cAMP applied extracellularly (Libet *et al.*, 1975) or intracellularly (Kobayashi *et al.*, 1978) markedly enhanced the muscarinic slow EPSP. Furthermore, recent evidence suggests that activation of the D_1 type of dopamine receptors of rabbit superior cervical ganglia by dopamine released from SIF cells is responsible for the postsynaptic increase of cAMP; this increase leads, in turn, via an as yet unidentified cellular process, to the enhancement of the slow muscarinic EPSP (Mochida *et al.*, 1981). In bullfrog and guinea pig sympathetic neurons, a similar enhancement of the peptide-mediated late slow EPSP by cAMP has been reported (Dun *et al.*, 1984b) (see Chapter 8). The concept that cAMP is involved in the cellular process leading to long-term enhancement of a heterosynaptic reponse is interesting and may constitute the basis for neuronal plasticity (see Chapter 21).

Similarly perplexing is the question of the postsynaptic effect of cGMP, which was alluded to in Chapter 8. It was proposed by Greengard (1976) that cGMP that may be generated by the muscarinic action of ACh on sympathetic ganglia is involved in the generation of the muscarinic slow EPSP. Indeed, cGMP or its derivatives depolarized sympathetic ganglion cells of the rabbit and rat whether applied by perfusion (Dun *et al.*, 1978a; Hashiguchi *et al.*, 1978), iontophoresis (Dun *et al.*, 1977), or intracellular injection (Gallagher and Shinnick-Gallagher, 1977); bullfrog sympathetic cells showed no consistent response to cGMP (Busis *et al.*, 1978; Weight *et al.*, 1978). However, membrane depolarization induced by cGMP was accompanied by an increase in membrane conductance (Dun *et al.*, 1978a), contrary to what occurs during the slow EPSP. Similarly, intracellular injection of cGMP into ganglion cells was found to cause a membrane depolarization of the rat superior cervical ganglion with an increase of membrane resistance (Gallagher and Shinnick-Gallagher, 1977). Furthermore, the cGMP-induced membrane depolarization was frequently followed by a long-lasting membrane hyperpolarization; a decrease in membrane conductance was observed in the course of the latter (Dun *et al.*, 1977, 1978a).

More recently, however, Hashiguchi *et al.* (1982) reported that in rabbit superior cervical ganglia, the muscarinic depolarization—the equivalent of the slow EPSP—is composed of two distinct components; at membrane potentials close to or more positive than -60 mV, the suppression of M channels can account for the generation of the muscarinic EPSP, whereas at membrane potentials more negative than −60 mV, muscarinic depolarization was not accompanied by any change in membrane conductance. Since cGMP was found to cause a membrane depolarization with no change in membrane conductance, it was hypothesized that cGMP may be responsible for the initiation of the muscarinic depolarization at membrane potentials more negative than −60 mV (Hashiguchi *et al.*, 1982).

It is of interest in this context that muscarinic agonists may mediate membrane permeability changes and receptor effects also by inhibition of cAMP-induced phosphorylations, attenuation of adenylate cyclase, and diminution of cAMP levels (J. H. Brown, 1979; Watanabe *et al.*, 1984). These effects of muscarinic agonists were described particularly for cardiac muscle, where they may be involved in muscarinic inhibition of cardiac β-adrenergic stimulation (J. H. Brown, 1979; Watanabe *et al.*, 1984). The possibility that these effects of muscarinic agonists—that is, inhibition of the cAMP system—may be involved in ganglionic phenomena has not been studied as yet.

At this time, it must be concluded that the electrophysiological characteristics of cGMP-induced depolarization and its time pattern do not resemble those of the slow EPSP and that cGMP is not involved in the generation of the slow EPSP or of the muscarinic effects of ACh. Thus, despite considerable efforts that have been expended in the past few years, the role of cyclic nucleotides in postsynaptic ganglionic function remains to be understood (see Volle, 1975; Volle *et al.*, 1982). It is conceivable that cyclic nucleotides may participate in a variety of cellular processes, some of which may not be detectable with our present technology.

While the involvement of cAMP in postsynaptic processes is controversial or even untenable, cAMP appears to regulate the processes of transmitter release by presynaptic nerve terminals. Indeed, Kuba *et al.* (1981) reported that in bullfrog sympathetic ganglia, catecholamines cause a long-lasting (for hours) enhancement of the amplitude of the fast EPSP and its quantal content; this effect can be mimicked by cAMP and by phosphodiesterase inhibitors (for details, see Chapter 11).

C. Phosphatidate Metabolism

The role of phosphatidylinositol metabolism and phosphatidate turnover in transmission resembles that of cyclic nucleotides, since the activation of the phosphatidylinositol cycle depends on phosphokinases such as phosphokinase C and since this activation is induced by a number of neurotransmitters including ACh ("PI response") (Hokin and Hokin, 1953; Nishizuka, 1984; Fisher *et al.*, 1984); finally, phosphatidylinositol processes and phosphokinase C appear to be involved in the regulation of such transmission-related parameters as membrane conductance and permeability and transmitter release. Thus, these processes serve, similarly to those involving cyclic nucleotides, as second messengers and should be mentioned here as such, although their role in ganglionic transmission has not yet been explored. In fact, there is evidence indicating that the PI, cyclic nucleotides, and Ca^{2+} mobilization systems are interconnected (see Nishizuka, 1984) or that they underlie the membrane permeability changes and transport processes in muscarinically innervated exocrine glands (see also Koelle, 1963). First, Hokin and Hokin (1953) described the role of this system, in connection with Ca^{2+} and Na^+, K^+-ATPase activity, as underlying the membrane permeability changes and transport processes in muscarinically innervated exocrine glands (see also Koelle, 1963). Subsequently, it was shown that muscarinic agonists serve as specific stimulators of phosphatidate turnover (cf. Michell, 1975) that includes the formation of Ca^{2+} carriers such as phosphatidic acid (Oosako and Deguchi, 1981). Furthermore, cholinomimetics stimulate, via cardiac muscarinic receptors, phosphatidylinositol turnover (the "PI response"), and this may underlie muscarinically induced permeability changes of the membrane including possibly such effects as muscarinic atrial hyperpolarization (S. L. Brown and Heller-Brown, 1983; Heller-Brown and Brown-Masters, 1984).

It is not known at this time whether this phenomenon is related to any specific ganglionic response such as the slow EPSP, nor is it clear whether the primary ganglionic transmission, i.e., the nicotinic pathway, engenders the PI response. Indeed, this question is unresolved at this time (see, for example, Larrabee and Leicht, 1965); on the whole (see DeRobertis, 1971; Michell, 1975), it appears probable that not only muscarinic but also nicotinic stimulation may markedly augment phosphatidic metabolism, particularly because of the high affinity of nicotinic agonists for pertinent components of the phosphatidate cycle (Cho *et al.*, 1978). In so stimulating phosphatidic metabolism, nicotinic agonists may also engender Ca^{2+} fluxes and affect postsynaptic (and presynaptic) membrane permeability (see also Ohta *et al.*, 1985). These interesting phenomena and their relationship to transmission as well as the modulation of transmission deserve further attention.

IV. EFFECTS OF ANTICHOLINESTERASES ON THE GANGLIA

Anticholinesterases (antiChEs) were first employed in R. M. Eccles's classic studies of ganglionic transmission in the 1930s and in Feldberg's studies of the release of ACh from the ganglia. Their investigations and the subsequent work in the 1950s and 1960s of Eccles, Volle and Koelle, Koppanyi and Karczmar, and Koketsu and Nishi (for references, see Zaimis, 1963; Karczmar, 1970; Haefely, 1980; Volle, 1980) established several important aspects of the action of antiChEs including organophosphorus agents.

First, primary nicotinic transmission is facilitated by antiChEs of both carbamate and organophosphorus types over considerable ranges of doses (Koppanyi and Karczmar, 1951; Holaday *et al.*, 1954; Takeshige and Volle, 1962; Mo *et al.*, 1985). Second, antiChEs, including organophosphorus agents, produce changes in ganglionic excitability, since they markedly enhance muscarinic transmission; this action results in postsynaptic repetitive discharges that can be blocked by atropine (Volle, 1962; Koketsu *et al.*, 1968).

Third, antiChEs, including organophosphorus drugs, seem to exert presynaptic actions that result in repetitive firing (Riker and Szreniawski, 1959). It is not quite clear whether or not this effect is related to that described by Volle and Koelle (cf. Koelle, 1963; Volle, 1962, 1980). Nor is it clear whether or not this action of antiChEs is due to accumulation of ACh at the postsynaptic membrane and its retrograde ["percussive" (Koelle, 1963)] action on the nerve terminal (see Chapter 11) or to their direct nerve-terminal action as proposed by Riker (see Riker and Szreniawski, 1959) [for counterargument, see Zaimis (1963) and Chapter 11]. finally, it is not clear whether the presynaptic action of antiChEs is facilitatory or inhibitory with respect to release of ACh (Koelle, 1963; Nishi, 1970, 1974) (see also Chapter 11).

Fourth, antiChEs exert effects on the ganglionic membrane that appear to be independent of the inhibition of ChEs and of accumulation of ACh (Koppanyi *et al.*, 1947; Koppanyi and Karczmar, 1951).

Fifth, organophosphorus agents and carbamates, when utilized at high doses, block ganglionic transmission. This has been known since the 1950s (Koppanyi and Karczmar, 1951), and while pertinent studies are relatively few (cf. Volle, 1980), these early results were convincingly confirmed recently for several organophosphorus agents including soman (Kirsch *et al.*, 1981; Yarowsky *et al.*, 1984) and for physostigmine (eserine) (Mo *et al.*, 1985). However, the data available so far are inconsistent, since the block seemed to occur, in the case of diisopropylfluorophosphate (DFP) at least, at widely different doses, the blocking dose depending on the investigator and the paradigm employed [compare, for example, Koppanyi and Karczmar (1951) with Holaday *et al.* (1954), Kamijo and Koelle (1952), and Yarowsky *et al.* (1984)]. Also, these studies revealed a lack of parallelism between inhibition of ChE (or AChE) and the occurrence of the block (see also Zaimis, 1963; Haefely, 1980). Nor is it quite clear whether the block is due solely to the effect on the nicotinic receptor or also depends on the status of muscarinic receptors [as stressed recently by

Kirsch *et al.* (1981)]. It should be noted that the majority of these studies were carried out by extracellular or surface recording techniques; this was the case in the recent study of Kirsch *et al.* (1981) concerning the action of soman on postsynaptic action potentials. Thus, the effects of antiChEs on the resting membrane potential and other membrane properties as well as the site and mechanism of the potentiating and blocking effects of antiChEs on ganglionic transmission have not yet been investigated in depth. On the other hand, intracellular methods were used by Mo *et al.* (1985) and Yarowsky *et al.* (1984) in their studies of the action of eserine on, respectively, the rabbit and rat superior cervical ganglion. While Yarowsky *et al.* (1984) implicated the block of release of ACh as a possible site of the blocking action of soman, Mo *et al.* (1985) showed that ganglionic block due to eserine, neostigmine, and DFP (Dun and Karczmar, unpublished data) is postsynaptic in nature. Also, in their hands, blocking concentrations of eserine did not alter the membrane properties of ganglionic neurons, while Yarowsky *et al.* (1984) found that soman caused a loss of membrane resistance and membrane resting potential.

Finally, activators of phosphorylated ChEs, such as NaF and several oximes, antagonize the ganglionic effects of organophosphorus agents (Koketsu, 1966).

The direct effects of antiChEs, including organophosphorus agents, warrant emphasis. The classic explanation of the mechanism by which organophosphorus antiChEs cause pharmacological effects, toxicity, and/or lethality concerns the phosphorylation of the active site of ChE (see, for example, Usdin, 1970) and the resulting accumulation of ACh. While this mechanism is clearly involved in many aspects of the action of antiChEs generally, and organophosphorus agents especially, many of the old results, already alluded to, as well as the recent data suggest that toxic and pharmacological effects of antiChEs at various sites may not be limited to inhibition of AChE (Jovic, 1974; Van Meter *et al.*, 1978; cf. Holmstedt, 1959, 1963; Karczmar, 1963, 1970, 1984). In fact, the aforementioned ganglionic blockade caused by antiChEs may well be due to their direct actions independent of ChE inhibition and ACh accumulation (Koppanyi and Karczmar, 1951; Mo *et al.*, 1985). Thus, eserine and neostigmine, used at relatively high concentrations, blocked not only the EPSP but also responses induced by iontophoretic application of ACh and of carbachol, a cholinomimetic that is not hydrolyzable by ChEs; similarly, several antiChEs, including carbamates and organophosphorus compounds, blocked ganglionic responses to both ACh and nicotine (Koppanyi *et al.*, 1947). Also, at the neuromyal junction and in the CNS, antiChEs seem to exert direct actions (cf. Karczmar, 1984).

Several mechanisms may be involved in the direct action of antiChEs. These may involve channels or channel–receptor macromolecules (Mo *et al.*, 1985; Kuba *et al.*, 1974). Furthermore, ACh-receptor sensitivity appeared to be reduced following antiChE treatment (Milosevic, 1970). In addition, morphopathological actions of organophosphorus agents that involve neuromyal as well as central sites were described by several investigators (Karczmar, 1963; Engle *et al.*, 1973; Laskowski *et al.*, 1977). Finally, some of the organophosphorus agents exert neurotoxic effects that may depend on several mechanisms (Abou-Donia *et al.*, 1979; Petras, 1981; M. K. Johnson and Richardson, 1983).

V. GANGLIONIC STIMULATING AND BLOCKING AGENTS

Langley and Dickinson (1889) were the first to demonstrate the stimulating and paralyzing action of nicotine on cat superior cervical ganglia (see also Chapter 6). The considerable interest in the action of ganglionic blocking agents during the 1950s and 1960s stemmed from the usefulness of these agents as tools in the study of ganglionic transmission and in antihypertensive therapy (see, however, Chapter 20).

Paton and Perry (1953) classified ganglionic blocking agents "according to whether they act like acetylcholine or by preventing its action." They termed the former "depolarizing" and the latter "competitive" blocking agents. More complex categorizing of ganglionic blockers is described in detail in Chapter 6.

Nicotine is the prototype of depolarizing blocking agents. When nicotine and nicotinic agonists such as acetylcholine, 1,1-dimethyl-4-phenylpiperazinium, and nicotinic cholinomimetics are applied to isolated ganglia by iontophoresis or superfusion, or when they are administered systemically in effective doses to intact animals, they exert, at least initially, a depolarizing action that, under certain circumstances, is fast in nature (fast ACh potential) (Koketsu, 1969; Nishi, 1974). This effect depends on the cholinergic receptor as the recognition site and on the resulting channel effect, as described in detail in Chapters 5 and 6. Quantitatively, the effect varies from compound to compound (see Chapter 6) and, to a lesser extent, from ganglion to ganglion (Gyermek, 1980; D. A. Brown, 1980). Thus, stimulated *in situ*, ganglia exert appropriate effects on their effector organs, such as the nictitating membrane, heart, and blood vessels (evoking complex ino- and chronotropic as well as depressor or pressor actions) in the case of the sympathetic ganglia, and the bladder and heart in that of the parasympathetic ganglia (for detailed descriptions, see Koelle, 1975; Karczmar, 1967; Levy and Vassale, 1982).

In the continuous presence of nicotinic agonists, the cell membrane gradually repolarizes to its initial level, but synaptic transmission is blocked, as demonstrated by both surface and intracellular recording (Eccles, 1956; Pascoe, 1956; Ginsborg and Guerrero, 1964). The question that is often raised is whether the block of transmission produced by nicotinic agents results from the depolarization itself or from the interaction of the compounds with the receptors, or a combination of both; the possibility that these compounds block transmission by interfering with presynaptic nerve-terminal conduction has also been suggested (Riker, 1968) (see also Chapters 11 and 21). Although the question has not been resolved unequivocally, the available evidence appears to indicate that the depolarizing action of nicotinic agents renders the cell inexcitable by (1) inactivating the inward Na^+ current responsible for spike generations and (2) increasing the outward K^+ current responsible for the repolarization. Furthermore, these agents reduce the amplitude of the EPSP by short-circuiting the membrane. This effect may relate to effects on channel half-life that also occur with competitive blockers such as hexamethonium (see below and Chapter 6). These two combined effects appear to be sufficient to account for the block of transmission during the initial phase of depolarization (see also D. A. Brown, 1980).

However, during prolonged application of nicotinic agonists, the membrane depolarization and conductance increase produced by the agonist may diminish and the direct membrane excitability, as measured by the antidromic spike, may be restored, while ganglionic transmission remains blocked (Ginsborg and Guerrero, 1964). This type of nondepolarizing block appears to resemble the process observed in skeletal muscle and termed "desensitization" or "receptor inactivation." The phenomenon of desensitization block has been investigated extensively in the case of the neuromuscular junction (Thesleff, 1955; cf. Nastuk and Wolfson, 1976; Karczmar and Ohta, 1981). On the other hand, since the initial reports (Krivoy and Wills, 1956; Ginsborg and Guerrero, 1964) that a second application of nicotinic agonist produces less depolarization in frog sympathetic ganglia than the first one and that continuous presence of ACh also diminishes postsynaptic responsiveness, a systematic investigation of ganglionic desensitization has not been carried out.

It is entirely possible that mechanisms involved in the agonist-induced desensitization of the neuromyal junction obtain as well as the ganglia; these include conformational change in the receptor that renders it inactive with respect to the channel (Gage, 1976) and the related phenomena described recently by Albuquerque and his associates, consisting of changed (increased) receptor affinity and special actions on the channel in its open conformation (Akaike *et al.*, 1984). It is of interest, however, that *in situ*, even with large doses of antiChEs, repeated administrations of ACh or nicotine continue to exert potent and, in fact, exaggerated stimulatory ganglionic actions (Koppanyi and Karczmar, 1951) (see above). Thus, whether or not desensitization constitutes a functionally important step in agonist-induced late phase block remains to be investigated.

The actions of nondepolarizing competitive ganglionic blockers have been investigated more extensively than those of the depolarizers, because of their potential application in clinical medicine. A large number of compounds were screened for their ganglionic blocking activities in the 1950s and 1960s (Mason, 1967; Gyermek, 1980), among which tetramethylammonium and tetraethylammonium (Acheson and Moe, 1946), hexamethonium, pentolinium, chlorisondamine, trimethydinium, and pentacynium were most potent or important or both (Trcka, 1980). More recently, additional drugs such as mecamylamine and trimethaphan were developed (Erina, 1980; Aviado, 1980) (see also Chapter 20). A detailed description of the structure–activity relationship of the ganglionic blocking activity of mono-, bis-, and tris-quarternary as well as secondary amines was provided by Trcka (1980).

Earlier investigations of ganglionic blocking agents in which extracellular recording and end-organ responses were employed to monitor ganglionic activity did not reveal any major differences in the sites and mechanisms of action of various ganglionic blockers. These studies provided information indicating that the compounds in question act by competing with ACh (and nicotine) for nicotinic postsynaptic sites as originally suggested by Paton and Perry (Paton and Perry, 1953; Mason, 1967). Recent studies employing more refined electrophysiological techniques provided useful information with regard to the site and mechanism of action and selectivity of the action of several ganglionic blockers; the conclusions of these studies are not necessarily in agreement with the early concepts.

Lees and Nishi (1972) analyzed by means of intracellular recording techniques the ganglionic blocking actions of mecamylamine, hexamethonium, and D-tubocurarine on the rabbit superior cervical ganglion. Employed in concentrations that depressed ganglionic transmission, these compounds exerted no appreciable effect on the resting membrane potential and excitability of neurons to direct or antidromic stimulation. However, these investigators found that mecamylamine depressed the quantal content in a train of EPSPs, whereas hexamethonium and D-tubocurarine had no appreciable effect on this parameter. It was therefore concluded that hexamethonium and D-tubocurarine reduce transmission solely by their postsynaptic actions, whereas mecamylamine exerts both pre- and postsynaptic inhibitory action (Lees and Nishi, 1972). It should be stressed, however, that the presynaptic inhibitory action is not an essential component of the blocking action of mecamylamine (Lees and Nishi, 1972).

The recent application of voltage-clamp and noise-analysis techniques provides even more sensitive analysis of the action of ganglionic blocking agents. For example, Selyanko *et al.* (1981) reported that hexamethonium appears to act mainly noncompetitively, since it shortens the lifetime of the channel opened by ACh in the case of rabbit superior cervical ganglion cells; on the other hand, D-tubocurarine was found to be a competitive blocker at the resting membrane potential levels (−50 to −60 mV) and noncompetitive blocker at more negative potential levels. It was further reported that the selectivity of short-chain bis-ammonium compounds in blocking the ganglionic fast excitatory current and ACh current is correlated well with their channel-blocking activities (Skok *et al.*, 1983; see also Blackman, 1970). Additional molecular mechanisms of the action of nondepolarizing blockers are described in Chapters 12 and particularly 6. It appears that while the term "nondepolarizing" still applies to the substances in question, they should no longer be regarded as noncompetitive blockers. Further investigations of the ACh receptor–channel mechanism and of their sensitivity to various blocking agents should provide information useful in the design of blocking agents specific for sympathetic and parasympathetic ganglia.

VI. MUSCARINIC DRUGS

Investigations carried out with microelectrode methods indicated that muscarinic drugs and muscarinic cholinomimetics such as acetyl-β-methylcholine (methacholine) exert effects similar or identical to effects of ACh on the same receptors. Thus, they evoke the slow muscarinic depolarization and the corresponding extracellularly recorded potential, the late negative. These effects, including the molecular and ionic mechanisms—such as generation of the M current—involved in the ganglionic muscarinic response, were reviewed in detail in Chapter 7. Several muscarinic drugs were studied to a lesser extent with intracellular methods; among them are the prototype drug muscarine and its derivatives (Waser, 1961), pilocarpine and oxotremorine.

Certain other effects arise as a consequence of the muscarinic stimulation, namely, postsynaptic repetitive firing, particularly in the presence of antiChEs (see Volle, 1966) (see also Section IV of this chapter). Furthermore, muscarinic cholinomimetics are capable of inducing, in ganglia capable of generating the slow IPSP on repetitive presynaptic stimulation, the corresponding inhibitory potential; again, this phenomenon was described in detail in Section III.B, of this chapter, as well as in Chapters 9 and 12.

There is an additional, interesting phenomenon that can be elicited by muscarinic drugs and muscarinic cholinomimetics: the effects of these drugs that can be obtained *in situ* or *in vivo* with respect to end-organs that usually respond to nicotinic drugs. The first demonstration of this kind is probably due to Cushny (1910), who obtained, upon intravenous administration of pilocarpine and physostigmine, contractile responses of the pregnant cat uterus that were similar to those of hypogastric-nerve stimulation; this effect was antagonized by atropine. These results were confirmed and expanded by Dale and Laidlaw (1912), who showed that nicotine and pilocarpine both stimulate (given intravenously or applied to the ganglion) the cat superior cervical ganglion, thus evoking nictitating membrane response; furthermore, in their hands, application of pilocarpine to appropriate sites caused suprarenal secretory activity and the uterine response. A similar demonstration of atropine-sensitive response of an end-organ was offered by Koppanyi (1932), who found that pilocarpine contracts the cat nictitating membrane when injected into the circulation of the superior cervical ganglion. The findings of Cushny, Dale and Laidlaw, and Koppanyi were subsequently confirmed by others with respect to pilocarpine and such compounds as muscarine (cf. Ambache *et al.*, 1956; DeGroat and Volle, 1963; Volle, 1966; Haefely, 1980). It is of interest that Ambache *et al.* (1956) also reported that the response evoked by muscarine was blocked by hexamethonium in the case of the normal but not of the denervated cat's superior cervical ganglion. Subsequently, several additional muscarinic compounds were shown to be similarly effective, including a butyltrimethyl ammonium derivative (McN-A-343) (Roszkowski, 1961) and a pyrrolidyl acetate methobromide derivative (AHR-602) (Franko *et al.*, 1963). Furthermore, Trendelenburg (1954) clearly established that the nictitating membrane effect of pilocarpine was due to ganglionic stimulation, and the dose–effect relationship of this action was subsequently described by Iorio and McIsaac (1966). These compounds could also induce, besides an effect on the nictitating membrane, pressor actions (Roszkowski, 1961; Levy and Ahlquist, 1962), and it is of interest that several antiChE agents such as neostigmine and physostigmine could also evoke pressor actions (Hilton, 1961; Long and Eckstein, 1961; Varagic, 1955); among these agents were also organophosphorus antiChEs such as sarin (Dirnhuber and Cullumbine, 1955; Van Meter *et al.*, 1978), although many other organophosphorus drugs do not affect the blood pressure. It appears that the effect in question is due both to vasoconstriction and to increased cardiac output (Long and Eckstein, 1961). It should be stressed in this context that while many antiChEs evoke this pressor action and, in particular, potentiate pressor action evoked by intravenous or intraarterial administration (into the circulation of sympathetic ganglia) of ACh in the presence of atropine (Koppanyi and Karczmar, 1951), the antiAChEs listed above produce this action even in the absence of atropine. Furthermore, these pressor effects seem not to depend on the release of catecholamines from adrenal medulla and are at least partially due to their ganglionic effects (Koppanyi *et al.*, 1940; see also Volle, 1966), although various forms of the catecholamine–cholinergic link (Burn and Rand, 1965; for further references, see Karczmar, 1963) may contribute to this phenomenon.

What should be stressed in the present context is that just as the muscarinic uterine response, surprisingly enough, was blocked under certain circumstances by hexamethonium (Ambache *et al.*, 1956) (see above), the pressor responses to muscarinic agonists and antiChEs can also be antagonized by nicotinic anticholinergics such as hexamethonium (Dirnhuber and Cullumbine, 1955; Varagic, 1955). This point is controversial, however, and it appears generally that the effects in question can be blocked with atropinics rather than with hexamethonium (Long and Eckstein, 1961; see also Karczmar, 1967).

With the advent of the isolated ganglionic preparations and particularly of microelectrode methods, *in situ* studies of end-organ response or studies of preparations that include end-organs are not being pursued actively at this time. This is regrettable, particularly in view of the following aspect of this matter. The physiological significance of the slow muscarinic response, the slow EPSP, and the related responses is generally understood to concern the modulation of the postsynaptic membrane and the excitability of the latter (see Chapters 3 and 7), while the fast EPSP constitutes the transmissive or effective postsynaptic response. Yet the findings described in this section strongly suggest that the muscarinic postsynaptic response may be productive and effective on its own, and result in the response of end-organs. This point deserves both attention and further studies.

Another question, unresolved at this time, is whether muscarinic ganglionic receptors constitute a homogeneous group of receptors (see also Chapter 7). In fact, while a large number of ganglionic muscarinic agonists and antagonists have been studied since the early investigations of Koppanyi (1932) and Ambache *et al.* (1956) (for reviews, see Gyermek, 1967, 1980; Karczmar, 1967), no particular differences between the actions of these drugs were found until very recently. That is, similar effects were obtained *in vitro* and *in situ* (see above) with pilocarpine, methacholine, carbachol, and bethanechol (Nishi, 1974; Mo *et al.*, 1985) and such synthetics as McN-A-343 and AHR 602 (see above). However, as already noted, it appeared quite early that under certain conditions, the effects of muscarinics were antagonizable with nicotinic blockers such as hexamethonium (see above). This finding is of interest in view of the recent proposition that muscarinic receptors, similarly to adrenergic receptors, may be quite heterogeneous. Thus, it was suggested that there are at least two subclasses of muscarinic receptors, M_1 and M_2. M_1 receptors may use the phosphatidylinositol system as its second messenger, while M_2 receptors may link with the cyclic nucleotides (Vickroy *et al.*, 1984). Also, and of particular interest in this context, these two subclasses of receptors may differ in their distribution; the M_1 receptors may be present in sympathetic ganglia (D. A. Brown *et al.*, 1980), while the M_2 receptors may predominate in smooth muscle, in the heart, and in the CNS (cf. Vickroy *et al.*, 1984). Furthermore, these two receptor subclasses may be differentiated pharmacologically, since at the ganglia, the M_1 receptors may be specifically excited and blocked by McN-A-343 and pirenzepine, respectively (D. A. Brown *et al.*, 1980), while both M_1 and M_2 receptors are excited by ACh and conventional muscarinic agonists such as oxotremorine and blocked by atropine and the irreversible ligand quinuclidinyl benzilate. More recently, Ashe and Yarosh (1984) found that pirenzepine blocked the slow EPSP but not the slow IPSP or fast EPSP; they do not seem to have compared the effects of pirenzepine with those of atropine or to have studied the action of pirenzepine on the slow ACh potential. On the other hand, they found that at concentrations that did not affect the slow EPSP, the fast EPSP, or the noncholinergic (late slow) EPSP, gallamine blocked the slow IPSP; thus, Ashe and Yarosh (1984) felt that gallamine and pirenzepine differentiate between muscarinic receptors underlying the muscarinic IPSP and the slow EPSP.

It must be pointed out, however, that studies of the possible heterogeneity of the muscarinic ganglionic receptors have barely been initiated, and definitive conclusions cannot be presented as yet.

VII. TOXINS

First, let it be emphasized that many toxins, whether of amphibian, marine, insect, or microbial origins, are available at present. These toxins may exert effects at both muscarinic and nicotinic receptors and serve effectively in the analysis of receptor structure and function (Witkop and Brossi, 1984; Witkop and Gossinger, 1983; Daly, 1982). Yet, while these toxins have been used extensively in this manner with respect to the neuromyal junction, they have been studied much less extensively in the case of the ganglion. α-Bungarotoxin (αBT) is exceptional in this regard, and its ganglionic effects will be reviewed (see also Chapter 6).

One of the more intriguing and often controversial aspects of autonomic pharmacology concerns the action of αBT on ganglionic nicotinic receptors. Since its isolation from the venom of *Bungarus multicinctus*, αBT has been repeatedly shown to bind specifically and irreversibly to nicotinic receptors of the skeletal muscle membrane, thereby blocking the physiological response of the muscle to ACh (Lee, 1970). As a result, this toxin has been extensively and successfully used as a probe of the nicotinic receptors of the muscle membrane.

The status of αBT with respect to the nicotinic sites on the neuronal membrane is less clearly established. In a number of investigations, αBT was found to be ineffective in blocking nicotinic transmission of cat and rat superior cervical ganglia (Chou and Lee, 1969; D. A. Brown and Fumagalli, 1977; Dun and Karczmar, 1980), chick sympathetic ganglia (Carbonetto *et al.*, 1978), and ganglia of the guinea pig myenteric plexus (Bursztajn and Gershon, 1977). On the other hand, applied at relatively high concentrations for long periods of time, αBT was shown to block cholinergic transmission of bullfrog sympathetic ganglia (Marshall, 1981) and chick ciliary ganglia (Chiappinelli and Zigmond, 1978). The reason for the differential blocking action of αBT on some ganglionic nicotinic receptors but not on others is not clear. It should be mentioned, however, that saturable binding sites for αBT were shown to be present in rat superior cervical ganglia despite the inability of the toxin to block nicotinic transmission (D. A. Brown and Fumagalli, 1977). This suggests that the αBT-binding sites on the postsynaptic membrane of these ganglia are unrelated to the sites that initiate the fast EPSP. The problem that thus arises may be resolved if there are at least two distinct subclasses of nicotinic receptors in rat sympathetic ganglia. In the pertinent experiments (Dun and Karczmar, 1980), αBT applied by superfusion to rat superior cervical ganglia depressed but did not completely block the depolarization induced by iontophoretic application of carabachol or ACh. The depressant effect of αBT was reversible after a prolonged period of wash, in contrast to the irreversibility of the blocking action of αBT at the neuromuscular junction. On the other hand, the fast EPSP elicited in the same cell by nerve stimulation was not at all affected by αBT. These findings were interpreted as indicating that there are at least two subtypes of nicotinic receptors: receptors that mediate the junctional synaptic potential and are not affected by αBT and receptors that mediate extrajunctional effects that can be blocked by αBT (Dun and Karczmar, 1980). Interestingly, in a recent study, a fraction of αBT designated as Toxin 11-S1, different from the fraction of αBT (Toxin 11-S2) used by Dun and Karczmar (1980), exhibited reverse characteristics; i.e., it blocked the synaptic but not the extrasynaptic nicotinic receptor (Quik and Lamarca, 1982). In light of these results obtained with different fractions of the toxin, it is imperative that in future research the homogeneity of the toxin used be ascertained. Interestingly, it was reported in a recent study that toxins obtained from *Dendroaspis viridis* venom blocked cholinergic transmission of skeletal muscle, but not that of frog sympathetic ganglia (Quik *et al.*, 1981). Again, this result should be interpreted with caution, since the homogeneity of the toxins in question has not been established. Altogether, the results obtained with the BT-like toxins suggest that nicotinic receptors on autonomic ganglia may be pharmacologically and physiologically different from those of the skeletal muscle. Further work in this area is needed, particularly since still other toxins apparently capable of exerting both ganglionic and neuromyal actions, such as batrachotoxin (Kayaalp *et al.*, 1970), have been studied only to a limited degree.

VIII. GENERAL ANESTHETICS AND BARBITURATES

The first convincing evidence that general anesthetics and barbiturates affect ganglionic transmission by a mechanism other than the blockade of axonal conduction or lowering of the oxygen consumption of the ganglia was provided by Larrabee and his associates (Larrabee and Holaday, 1952; Larrabee and Posternak, 1952; Larrabee *et al.*, 1952). These investigators showed that ether, chloroform, or pentobarbital blocked synaptic transmission of cat stellate ganglia at concentrations lower than were necessary to block axonal conduction. Interestingly, for reasons yet to be determined, urethane blocked axonal conduction prior to impairment of ganglionic transmission. Accordingly, urethane is widely employed as the general anesthetic of choice in experimental studies of ganglionic transmission.

The possible sites and mechanisms of the blocking action of general anesthetics and barbiturates on autonomic ganglia were examined by a number of investigators (see Nistri and Quilliam, 1980). Christ (1977) observed that halothane, used in concentrations similar to those used in clinical anesthesia, blocked in hamster stellate ganglia nicotinic compound action potentials elicited by a low frequency of nerve stimulation as well as discharges induced by nicotinic agonists. Muscarinic discharges induced by a high frequency of stimulation, but not those evoked by bath application of muscarinic agonists, were also suppressed by comparable concentrations of halothane. These findings are compatible with the interpretation that halothane affects two ganglionic sites, since it may act postsynaptically in depressing the chemosensitivity to ACh of the postsynaptic nicotinic receptor and may depress ACh release from presynaptic nerve endings during repetitive nerve stimulation (Christ, 1977). The reasons for this differential action of halothane with respect to low and high frequency of stimulation are not known.

That general anesthetic agents selectively affect nicotinic transmission was further illustrated by the action of ketamine, a dissociative anesthetic agent, on bullfrog sympathetic ganglia. At lower concentrations (<1 mM), ketamine reversibly blocked the nicotinic fast EPSP as well as the fast ACh depolarization induced by iontophoretic application of ACh, without appreciably affecting the muscarinic slow EPSP and the peptidergic late slow EPSP; the passive membrane properties as well as the slow muscarinic depolarization induced by iontophoretic application of ACh were likewise not significantly changed (Gallagher *et al.*, 1976). These findings clearly suggested that ketamine at low concentrations preferentially blocks nicotinic ACh receptors. On the other hand, employed at higher concentrations, ketamine caused a membrane depolarization, probably by decreasing membrane conductance to K^+, as well as nonselectively blocked nicotinic, muscarinic, and peptidergic transmission (Gallagher *et al.*, 1976).

The nature and mechanism of the blockade of nicotinic receptors by general anesthetics and barbiturates are not entirely clear. Nicoll (1978) suggested that pentobarbital blocks nicotinic receptors of bullfrog sympathetic ganglia by interacting with postsynaptic Na^+ channels as suggested by Barker (1975) with regard to invertebrate neurons and by Adams (1976) for the frog neuromuscular junction.

IX. COMMENTS AND CONCLUSIONS

As stated in the Introduction, the autonomic ganglia constitute complex systems so that their transmission function is subject to modulations (see Chapter 3) of many types. The physiological mechanisms involved are described elsewhere in this book; thus, presynaptic regulation or modulation is discussed in Chapters 10 and 11, while postsynaptic modulations are discussed in detail in Chapter 12. In addition, the interplay between postsynaptic potentials that presumably occurs under physiological circumstances depends particularly on the slow potentials as described in Chapters 7, 8, and 9 with respect to sympathetic ganglia and in Chapters 15 and 16 with respect to parasympathetic and enteric ganglia, respectively.

It should be emphasized that these regulations depend on endogenous substances present in the ganglia (see Chapter 3). These substances may, in fact, contribute to the regulation of transmission in their role as transmitters, i.e., substances released from preganglionic terminals that, in this particular case, do not initiate transmission but act as modulators; in other instances, these substances may act via their presence in the glia or blood. These considerations explain the use in the preceding paragraph of the term "physiological" in the description of the regulations in question. Yet pharmacological employment and pharmacological effects of these and additional, exogenous substances also reflect on the same concept of the modulation and regulation of transmission and of the multifactorial nature of this regulation.

Thus, used as drugs or exogenous substances, compounds such as GABA, serotonin, histamine, several peptides, and catecholamines frequently exert both pre- and postsynaptic effects, although sometimes one or the other effect predominates, as, for example, the presynaptic action in the case of histamine and enkephalins (see also Chapter 11). Sometimes, when both sites of action are involved, as in the case of catecholamines, it may not be easy to decide which site is primarily involved in the putative physiological function of the substance in question. Indeed, while the postsynaptic actions of catecholamines were stressed in this chapter, their presynaptic actions are so potent (see Chapter 11) that the early description by Marrazzi (1939) of the block of ganglionic transmission by epinephrine and the subsequent demonstration of similar effects of other catecholamines may well reflect the primary involvement of the presynaptic site in catecholamine-induced depression of ganglionic junction.

In this context of post- and presynaptic receptors, the coupling of these receptors with the channel protein and the mechanism by means of which they generate permeability changes are of great importance. Nucleotide-mediated phosphorylation was proposed by Greengard (1976) on the basis, initially, of his ganglionic investigations as a general mechanism underlying effects of transmitters on neuronal membranes and their permeability ("second-messenger" mechanism). While this mechanism may hold elsewhere, it appears to be untenable with respect to the ganglia, whether in the case of muscarinic or nicotinic stimulation. What, then, is the role of activation of nucleotides that has been dependably demonstrated to occur in ganglionic neurons? This important question apparently cannot be answered at this time. Nor can there be assigned, at this time, a specific role to the activation of phosphatidate turnover that apparently can be generated in the ganglia and that may engender permeability changes, interact with the cyclic nucleotides (Michell, 1975; Nishizuka, 1984), and/or affect Ca^{2+} fluxes, thus constituting, similarly to nucleotides, a second-messenger mechanism; this particular story appears to be, at the ganglia, indeed incomplete.

The actions of purely pharmacological substances such as certain cholinomimetics deserve some comment. Perhaps special attention should be focused on the pharmacological effects of muscarinic cholinomimetics that include choline esters as well as such alkaloids as the prototypical substance muscarine with its analogues, and pilocarpine, or synthetics such as oxotremorine, pirezenpine, and McN-A-343. The particular interest here should not center solely on the molecular postsynaptic actions of these compounds, since these actions, whenever studied by microelectrode means, do not seem to differ from the muscarinic effects of ACh itself; what is noteworthy is that these compounds, via their muscarinic ganglionic agonist action, exert effects on end-organs including the heart, the blood vessels, and the nictitating membrane. Many of these effects—such as those on the nictitating membrane—should be expected only from the productive, ganglionic transmission, i.e., primarily the nicotinic transmission. Thus, the end-organ effectiveness of the muscarinic compounds in question suggests that the muscarinic ganglionic site is capable of an overt functional effect and that such a muscarinic effect may occur under physiological conditions.

Still another point deserving further study concerns the possible heterogeneity of ganglionic muscarinic receptors. Do they differ from muscarinic brain and smooth muscle receptors? Are they affected specifically by certain antagonists such as pirenzepine? Does the muscarinic IPSP response differ pharmacologically from the slow EPSP?

AntiChEs may serve as examples of pharmacological agents, presumably never endogenous in nature. In this case, their action may be of toxicological interest, particularly in the case of the organophosphorus agents; however, they also serve to clarify the function of the transmitter ACh, since they reveal certain possible consequences of augmentation of muscarinic ganglionic function, such as the increase in ganglionic excitability resulting in repetitive postsynaptic discharges. Unexpectedly, antiChEs appear to exert ganglionic actions that do not seem to depend on accumulation of ACh; thus, the ganglion may be added to other structures including the neuromyal junction and the CNS, in which direct actions of antiChEs were demonstrated. It is of particular interest to find out whether these direct effects of antiChEs that are both pharmacological (Karczmar and Ohta, 1981; Akaike *et al.*, 1984) and morphological (Karczmar, 1984) in nature correspond to some hitherto unknown receptor sites or physiological function or both.

Somewhat similarly to antiChEs, toxins such as αBT and batrachotoxin may serve as tools for exploring ganglionic receptors, particularly since the ganglionic effects of αBT seem to underline the differences between nicotinic ganglionic and end-plate receptors. Indeed, present findings obtained with αBT suggest that there may be subtypes of nicotinic receptors; in fact, some of these may be extrajunctional in nature.

Still another category of drugs that affect ganglionic transmission is the ganglionic blockers. Among the drugs that may be included in this category are depolarizing and nondepolarizing blockers, anesthetics and barbiturates, all of which were discussed in this chapter, as well as many miscellaneous neurotropic drugs such as neuroleptic phenothiazines, butyrophenones and rauwolfia alkaloids, anticonvulsants and stimulants such as strychnine, and minor tranquilizers. This list could be extended just about endlessly (Nistri and Quillam, 1980; Mason, 1967); indeed, the ganglia are depressed by many agents, reflecting, in part, the fact that synaptic transmission is more readily affected by drugs than axonal conduction, a concept made prominent in the 1950s by Larrabee and Posternak (1952); also, the presynaptic site involved in the release of ACh appears to be vulnerable to many drugs (see Chapters 10 and 11). Furthermore, in many instances, the effect of these drugs cannot be considered specific, since their ganglionic effects are achievable only at relatively high concentrations compared with the concentrations capable of exerting effects at such target sites as the spinal presynaptic inhibitory site or GABA receptors in the case of barbiturates (Nicoll, 1978); in fact, controversy frequently surrounds the interpretation of these effects, as in the case of monoamine oxidase inhibitors (see Nistri and Quillam, 1980).

The ganglionic blockade exerted by opioids and enkephalins, on the other hand, is of considerable interest (see this chapter and Chapters 6 and 11), as is the blockade evoked by depolarizing and nondepolarizing blockers. Certain novel aspects of this matter should be emphasized. First, the phenomenon of desensitization may be a component of the blocking action of the depolarizers; this is a basic event that throws light on the functioning of the receptor–channel macromolecule and that is characteristic for the action not only of ACh but also of other neurotransmitters, but that has been studied to a limited extent in the case of the ganglia. Second, it appears that nondepolarizers such as hexamethonium exert effects that extend beyond their competitive interaction with ACh, the latter action being previously considered as the primary mechanism of their action (see also Chapter 6).

Still another point concerning the nondepolarizing blockers should be emphasized. Many of these drugs were used in the past clinically, and while this use is on the wane (see Chapter 20), some of them—such as trimethaphan and mecamylamine—are still being employed therapeutically. It must be stressed, then, that the reason for this diminution in use relates to the fact that the drugs in question do not seem to exhibit differential specificity with regard to their action on sympathetic vs. parasympathetic ganglia; to the contrary, they seem to depress both. Thus, while at one time or another these drugs were employed in the treatment of hypertension—an effect requiring sympathetic blockade—a general ganglionic depression ensued, rendering their use difficult or undesirable (see Chapter 20). It is to be hoped, however, that further advances in the knowledge of the structure–activity relationships of the pertinent agents may lead to a fortunate breakthrough and to development of ganglionic blockers that affect one type of ganglia or another specifically and differentially.

REFERENCES

Abou-Donia, M. B., Graham, D. G., and Komeil, A. A.: Delayed neurotoxicity of O-ethyl O-2,4-dichlorophenylphosphonothioate: Effects of single oral dose on hens. *Toxicol. Appl. Pharmacol.* **49:**293–303 (1979).

Acheson, G. H., and Moe, G. K.: The action of tetraethylammonium ion on the mammalian circulation. *J. Pharmacol. Exp. Ther.* **87:**220–236 (1946).

Adams, P. R.: Drug blockade of open end plate channels. *J. Physiol. (London)* **260:**531–552 (1976).

Adams, P. R., and Brown, D. A.: Action of γ-aminobutyric acid (GABA) on rat sympathetic ganglion cells. *Br. J. Pharmacol.* **47:**639–640 (1973).

Adams, P. R., and Brown, D. A.: Actions of γ-aminobutyric acid on sympathetic ganglion cells. *J. Physiol.* **250:**85–120 (1975).

Akaike, A., Ikeda, S. R., Brookes, N., Pascuzzo, G. J., Rickett, D. L., and Albuquerque, E. X.: The nature of interaction of pyridestigmine with the nicotinic receptor–ionic channel complex. II. Patch clamp studies. *Mol. Pharmacol.* **25**:102–112 (1984).

Akasu, T., and Koketsu, K.: Effects of dibutyryl cyclic adenosine 3′,5′-monophosphate and theophylline on the bullfrog sympathetic ganglion cells. *Br. J. Pharmacol.* **60**:331–336 (1977).

Ambache, N., Perry, W. L. M., and Robertson, P. A.: The effect of muscarine on perfused superior cervical ganglia of cats. *Br. J. Pharmacol.* **11**:442–448 (1956).

Ashe, J. H., and Yarosh, C. A.: Differential and selective antagonism of the slow-IPSP and slow-EPSP by gallamine and pirenzepine in rabbit superior cervical ganglion. *Neuropharmacology* **23**:1321–1329 (1984).

Ashman, D. F., Lipton, R., Melicow, M. M., and Price, T. D.: Isolation of adenosine 3′,5′-monophosphate and guanosine 3′,5′-monophosphate from rat urine. *Biochem. Biophys. Res. Commun.* **11**:3303–334 (1963).

Aviado, D. M.: Action of ganglion-blocking agents on the cardiovascular system, in: *Pharmacology of Ganglionic Transmission* (D. A. Kharkevich, ed.), pp. 237–266, Springer-Verlag, Berlin (1980).

Barker, J. L.: CNS depressants: Effects on postsynaptic pharmacology. *Brain Res.* **92**:35–55 (1975).

Bjorklund, A., Cegrell, L., Falck, B., Ritzen, M., and Rosengren, E.: Dopamine-containing cells in sympathetic ganglia. *Acta Physiol. Scand.* **78**:334–338 (1970).

Blackman, J. G.: Dependence on membrane potential of the blocking action of hexamethonium at a sympathetic ganglionic synapse. *Proc. Univ. Otago. Med. Sch.* **48**:4–5 (1970).

Bloom, F. E.: Amino acids and polypeptides in neuronal function. *Neurosci. Res. Prog. Bull.* **10**:121–251 (1972).

Bowery, N. G., and Brown, D. A.: γ-aminobutyric acid uptake by sympathetic ganglia. *Nature (London)* **238**:89–91 (1972).

Brimble, M. J., and Wallis, D. I.: Histamine H_1 and H_2-receptors at a ganglionic synapse. *Nature (London)* **246**:156–158 (1973).

Brown, D. A.: Locus and mechanism of action of ganglion-blocking agents, in: *Pharmacology of Ganglionic Transmission* (D. A. Kharkevich, ed.), pp. 185–235, Springer-Verlag, Berlin (1980).

Brown, D. A., and Fumagalli, L.: Dissociation of α-bungarotoxin binding and receptor block in the rat superior cervical ganglion. *Brain Res.* **129**:165–168 (1977).

Brown, D. A., Forward, A., and Marsh, S.: Antagonist discrimination between ganglionic and ideal muscarinic receptors. *Br. J. Pharmacol.* **71**:362–364 (1980).

Brown, J. H.: Cholinergic inhibition of catecholamine-stimulable cyclic AMP accumulation in murine atria. *J. Cyclic Nucleotide Res.* **5**:423–433 (1979).

Brown, S. L., and Heller-Brown, J.: Muscarinic-regulation of phosphatidyl metabolism in atria. *Mol. Pharmacol.* **24**:351–356 (1984).

Burn, J. H., and Rand, M. J.: Acetylcholine in adrenergic transmission. *Annu. Rev. Pharmacol.* **5**:163–182 (1965).

Bursztajn, S., and Gershon, M. D.: Discrimination between nicotinic receptors in vertebrate ganglia and skeletal muscle by alpha-bungarotoxin and cobra venoms. *J. Physiol. (London)* **269**:17–31 (1977).

Busis, N. A., Weight, F. F., and Smith, P. A.: Synaptic potentials in sympathetic ganglia: Are they mediated by cyclic nucleotides? *Science* **200**:1079–1081 (1978).

Carbonetto, S. T., Fambrough, D. M., and Muller, K. J.: Non-equivalence of α-bungarotoxin receptors and acetylcholine receptors in chick sympathetic neurons. *Proc. Natl. Acad. Sci. U.S.A.* **75**:1016–1020 (1978).

Chiappinelli, V. A., and Zigmond, R. E.: Alpha-bungarotoxin blocks nicotinic transmission in the avian ciliary ganglion. *Proc. Natl. Acad. Sci. U.S.A.* **75**:2999–3003 (1978).

Cho, T. M., Cho, J. S., and Loh, H. H.: Alteration of the physiochemical properties of triphosphoinositide by nicotine ligands. *Proc. Natl. Acad. Sci. U.S.A.* **75**:784–788 (1978).

Chou, T. C., and Lee, C. Y.: Effect of whole and fractionated cobra venom on sympathetic ganglionic transmission. *Eur. J. Pharmacol.* **8**:326–330 (1969).

Christ, D. D.: Effects of halothane on ganglionic discharges. *J. Pharmacol. Exp. Ther.* **200**:336–342 (1977).

Christ, D. D., and Nishi, S.: Site of adrenaline blockade in the superior cervical ganglion of the rabbit. *J. Physiol. (London)* **213**:107–117 (1971).

Crowshaw, K., Jessup, S. J., and Ramwell, P.: Thin-layer chromatography of 1-dimethylaminoaphthalene-5-sulphonyl derivatives of amino acids present in superfusates of cat cerebral cortex. *Biochem. J.* **103**:79–85 (1967).

Cushny, A. R.: The action of atropine, pilocarpine and physostigmine. *J. Physiol. (London)* **41**:233–245 (1910).

Dail, W. G., Jr., and Evans, A. P.: Ultrastructure of adrenergic terminals and SIF cells in the superior cervical ganglion of the rabbit. *Brain Res.* **148**:469–477 (1978).

Dale, H. H., and Laidlaw, P. P.: The significance of the suprarenal capsules in the action of certain alkaloids. *J. Physiol. (London)* **45**:1–26 (1912).

Daly, J. W.: Alkaloids of neotropical poison frogs (Dendrobatidae). *Fortschr. Chem. Org. Naturst.* **41**:205–340 (1982).

DeGroat, W. C., and Lally, P. M.: Interaction between picrotoxin and 5-hydroxytryptamine in the superior ganglion of the cat. *Br. J. Pharmacol.* **48**:233–244 (1973).

DeGroat, W. C., and Volle, R. L.: Ganglionic actions of oxotremorine. *Life Sci.* **2**:618–623 (1963).

DeGroat, W. C., and Volle, R. L.: The action of the catecholamines on the transmission in the superior cervical ganglion of the cat. *J. Pharmacol. Exp. Ther.* **154**:1–13 (1966).

DeRobertis, E.: Molecular biology of synaptic receptors. *Science* **171**:963–971 (1971).

Dirnhuber, P., and Cullumbine, H.: The effect of anticholinesterase agents on the rat's blood pressure. *Br. J. Pharmacol.* **10**:12–15 (1955).

Dun, N. J.: Peptide hormones and transmission in sympathetic ganglia, in: *Autonomic Ganglia* (L. G. Elfvin, ed.), pp. 345–366, John Wiley, Chichester (1983).

Dun, N. J., and Jiang, Z. G.: Non-cholinergic excitatory transmission in inferior mesenteric ganglia of the guinea pig: Possible mediation by substance P. *J. Physiol. (London)* **325**:145–159 (1982).

Dun, N., and Karczmar, A. G.: The presynaptic site of action of norepinephrine in the superior cervical ganglion of guinea pig. *J. Pharmacol. Exp. Ther.* **200**:328–335 (1977a).

Dun, N. J., and Karczmar, A. G.: A comparison of the effect of theophylline and cyclic adenosine 3′:5′-monophosphate on the superior cervical ganglion of the rabbit by means of the sucrose-gap method. *J. Pharmacol. Exp. Ther.* **202**:89–96 (1977b).

Dun, N. J., and Karczmar, A. G.: Blockade of ACh potentials by alpha-bungarotoxin in rat superior cervical ganglion cells. *Brain Res.* **196**:536–540 (1980).

Dun, N. J., and Kiraly, M.: Capsaicin causes release of a substance P-like peptide in guinea-pig inferior mesenteric ganglia. *J. Physiol. (London)* **340**:107–120 (1983).

Dun, N. J., and Ma, R. C.: Slow non-cholinergic excitatory potentials in neurones of the guinea-pig coeliac ganglia. *J. Physiol. (London)* **351**:47–60 (1984).

Dun, N., and Nishi, S.: Effects of dopamine on the superior cervical ganglion of the rabbit. *J. Physiol. (London)* **239**:155–164 (1974).

Dun, N. J., Kaibara, K., and Karczmar, A. G.: Dopamine and adenosine 3′,5′-monophosphate responses of single mammalian sympathetic neurons. *Science* **197**:778 (1977).

Dun, N. J., Kaibara, K., and Karczmar, A. G.: Muscarinic and cGMP induced membrane potential changes: Differences in electrogenic mechanisms. *Brain Res.* **150**:658 (1978a).

Dun, N. J., Nishi, S., and Karczmar, A. G.: An analysis of the effect of angiotensin II on mammalian ganglion cells. *J. Pharmacol. Exp. Ther.* **204**:669–675 (1978b).

Dun, N. J., Karczmar, A. G., and Ingerson, A. A.: A neurochemical and neurophysiological study of serotonin in the superior cervical ganglia. *Soc. Neurosci. Abstr.* **6**:216 (1980).

Dun, N. J., Kiraly, M., and Ma, R. C.: Evidence for a serotonin mediated slow excitatory potential in the guinea-pig coeliac ganglia. *J. Physiol. (London)* **351**:61–76 (1984a).

Dun, N. J., Jiang, Z. G., and Mo, N.: Long-term facilitation of peptidergic transmission by catecholamines in guinea-pig inferior mesenteric ganglia. *J. Physiol. (London)* **357**:37–50 (1984b).

Eccles, R. M.: The effects of nicotine on synaptic transmission in the sympathetic ganglion. *J. Pharmacol. Exp. Ther.* **118**:26–38 (1956).

Elfvin, J. G., Hökfelt, T., and Goldstein, M.: Fluorescence microscopical, immunohistochemical and ultrastructural studies on sympathetic ganglia of guinea pig, with special reference to the SIF cells and their catecholamine content. *J. Ultrastruct. Res.* **51**:377–396 (1975).

Engle, A. G., Lambert, F. H., and Santa, T.: Study of long-term anticholinesterase therapy: Effects on neuromuscular transmission and on motor end-plate structure. *Neurology* **23**:1273–1281 (1973).

Eränkö, O., and Harkonen, M.: Histochemical demonstration of fluorogenic amines in the cytoplasm of sympathetic ganglion cells of the rat. *Acta Physiol. Scand.* **58**:285–286 (1963).

Eränkö, O., and Harkonen, M.: Monoamine-containing small cells in the superior cervical ganglion of the rat and an organ composed of them. *Acta Physiol. Scand.* **63**:511–512 (1965).

Erina, E. V.: Ganglion-blocking agents in internal medicine, in: *Pharmacology of Ganglionic Transmission* (D. A. Kharkevich, ed.), pp. 417–438, Springer-Verlag, New York (1980).

Fisher, S. K., Van Rooigen, L. A. A., and Agranoff, B. W.: Renewed interest in the polyphosphoinositides. *Trends Biochem. Sci.* **9**:53–56 (1984).

Franko, B. V., Ward, J. W., and Alphin, R. S.: Pharmacologic studies of N-benzyl-3-pyrrolidylacetate methobromide (AHR-602), a ganglion stimulating agent. *J. Pharmacol. Exp. Ther.* **139**:25–30 (1963).

Fuller, R. W., and Wong, D. T.: Inhibition of serotonin reuptake. *Fed. Proc. Fed. Am. Soc. Exp. Biol.* **36**:2154–2158 (1977).

Gage, P. W.: Generation of end-plate potentials. *Physiol. Rev.* **56**:177–247 (1976).

Gallagher, J. P., and Shinnick-Gallagher, P.: Cyclic nucleotides injected intracellularly into rat superior cervical ganglion cells. *Science* **198**:851–852 (1977).

Gallagher, J. P., Dun, N., Higashi, H., and Nishi, S.: Actions of ketamine on synaptic transmission in frog sympathetic ganglia. *Neuropharmacology* **15**:139–143 (1976).

Gertner, S. B., Passonen, M. K., and Giarman, N. J.: Studies concerning the presence of 5-hydroxytryptamine (serotonin) in the perfusate from the superior cervical ganglion. *J. Pharmacol. Exp. Ther.* **127**:268–275 (1959).

Ginsborg, B., and Guerrero, S.: On the action of depolarizing drugs on sympathetic ganglion cell of the frog. *J. Physiol. (London)* **172**:189–206 (1964).

Greengard, P.: Possible role for cyclic nucleotides and phosphorylated membrane proteins in postsynaptic actions of neurotransmitters. *Nature (London)* **260**:101–108 (1976).

Greengard, P.: Phosphorylated proteins as physiological effectors. *Harvey Lect.* **75**:277–331 (1982).

Greengard, P., and Kebabian, J. W.: Role of cyclic AMP in synaptic transmission in the mammalian peripheral nervous system. *Fed. Proc. Fed. Am. Soc. Exp. Biol.* **33**:1059–1067 (1974).

Gyermek, L.: Ganglionic stimulant and depressant agents, in: *Drugs Affecting the Peripheral Nervous System*, Vol. 1 (A. Burger, ed.), pp. 149–326, Marcel Dekker, New York (1967).

Gyermek, L.: Methods for examination of ganglionic blocking agents, in: *Pharmacology of*

Ganglionic Transmission (D. A. Kharkevich, ed.), pp. 63–121, Springer-Verlag, Berlin (1980).
Hadjiconstantinou, M., Potter, P. E., and Neff, N. H.: Trans-synaptic modulation via muscarinic receptors of serotonin-containing small intensely fluorescent cells of superior cervical ganglion. *J. Neurosci.* **2**:1836–1939 (1982).
Haefely, W.: The effects of 5-hydroxytryptamine and some related compounds on the cat superior cervical ganglion *in situ*. *Naunyn-Schmiedeberg's Arch. Pharmacol.* **281**:145–165 (1974).
Haefely, W.: Non-nicotinic chemical stimulation of autonomic ganglia, in: *Pharmacology of Ganglionic Transmission* (D. A. Kharkevich, ed.), pp. 313–357, Springer-Verlag, Berlin (1980).
Haigler, H. J., and Aghaganian, G. K.: Serotonin receptors in the brain. *Fed. Proc. Fed. Am. Soc. Exp. Biol.* **36**:2159–2164 (1977).
Heller-Brown, J., and Brown-Masters, S.: Muscarinic regulation of phosphatidylinositol turnover and cyclic nucleotide metabolism in the heart. *Fed. Proc. Fed. Am. Soc. Exp. Biol.* **43**:2613–2617 (1984).
Hashiguchi, T., Ushiyama, N. S., Kobayashi, H., and Libet, B.: Does cyclic GMP mediate the slow excitatory synaptic potential in sympathetic ganglia? *Nature (London)* **271**:267–268 (1978).
Hashiguchi, T., Kobayashi, H., Tosaka, T., and Libet, B.: Two muscarinic depolarizing mechanisms in mammalian sympathetic neurons. *Brain Res.* **242**:378–382 (1982).
Hilton, J. G.: The pressor response to neostigmine after ganglionic blockade. *J. Pharmacol. Exp. Ther.* **132**:23–28 (1961).
Hökfelt, T., Elfvin, L.-G., Schultzberg, M., Fuxe, K., Said, S. E., Mutt, V., and Goldstein, M.: Immunohistochemical evidence of vasoactive intestinal polypeptide-containing neurons and nerve fibres in sympathetic ganglia. *Neuroscience* **2**:885–896 (1977a).
Hökfelt, T., Elfvin, L.-G., Schultzberg, M., Goldstein, M., and Nilsson, G.: On the occurrence of substance P containing fibers in sympathetic ganglia: Immunohistochemical evidence. *Brain Res.* **132**:29–41 (1977b).
Hökfelt, T., Johansson, O., Ljungdabe, A., Lundberg, J. M., and Schultzberg, M.: Peptidergic neurons. *Nature (London)* **284**:515–521 (1980).
Hökfelt, T., Johansson, O., and Goldstein, M.: Chemical anatomy of the brain. *Science* **225**:1326–1334 (1984).
Hokin, M. R., and Hokin, L. E.: Enzyme secretion and the incorporation of P^{32} into phospholipids of pancreas slices. *J. Biol. Chem.* **203**:967–977 (1953).
Holaday, D. A., Kamijo, K., and Koelle, G. B.: Facilitation of ganglionic transmission following inhibition of cholinesterase by DFP. *J. Pharmacol. Exp. Ther.* **111**:241–254 (1954).
Holmstedt, B.: Pharmacology of organophosphorus inhibitors. *Pharmacol. Rev.* **11**:567–688 (1959).
Holmstedt, B.: Structure–activity relationships of the organophosphorus anticholinesterase agents, in: *Handbuch der experimentellen Pharmakologie, Ergänzungswerk*, Vol. 15, *Cholinesterases and Anticholinesterase Agents* (G. B. Koelle, ed.), pp. 428–485, Springer-Verlag, Berlin (1963).
Iorio, L. C., and McIsaac, R. J.: Comparison of the stimulating effects of nicotine, pilocarpine and histamine on the superior cervical ganglion of the cat. *J. Pharmacol. Exp. Ther.* **151**:430–437 (1966).
Jan, L. Y., and Jan, Y. N.: Peptidergic transmission in sympathetic ganglia of the frog. *J. Physiol.* **327**:219–246 (1982).
Jan, Y. N., Jan, L. Y., and Kuffler, S. W.: A peptide as a possible transmitter in sympathetic ganglia of the frog. *Proc. Natl. Acad. Sci. U.S.A.* **76**:1501–1505 (1979).
Jan, Y. N., Jan, L. Y., and Kuffler, S. W.: Further evidence for peptidergic transmission in sympathetic ganglia. *Proc. Natl. Acad. Sci. U.S.A.* **77**:5008–5012 (1980).

Jiang, Z. G., Simmons, M. A., and Dun, N. J.: Enkephalinergic modulation of non-cholinergic transmission in mammalian prevertebral ganglia. *Brain Res.* **235**:185–191 (1982).

Jiang, Z. G., Dun, N. J., and Karczmar, A. G.: Substance P: A putative sensory transmitter in mammalian autonomic ganglia. *Science* **217**:739–741 (1983).

Johnson, M. K., and Richardson, R. J.: Biochemical endpoints: Neurotoxic esterase. *Neurotoxicology* **4**:311–320 (1983).

Johnson, S. M., Katayama, Y., Morita, K., and North, R. A.: Mediators of slow synaptic potentials in the myenteric plexus of the guinea-pig ileum. *J. Physiol. (London)* **320**:175–186 (1981).

Jovic, R. C.: Correlation between signs of toxicity and some biochemical changes in rats poisoned by soman. *Eur. J. Pharmacol.* **25**:159–164 (1974).

Kamijo, K., and Koelle, G. B.: The relationship between cholinesterase inhibition and ganglionic transmission. *J. Pharmacol. Exp. Ther.* **105**:349–357 (1952).

Karczmar, A. G.: Ontogenetic effects of anticholinesterase agents, in: *Handbuch der Experimentellen Pharmakologie, Ergänzungswerk*, Vol. 15, *Cholinesterases and Anticholinesterase Agents* (G. B. Koelle, ed.), pp. 179–186, Springer-Verlag, Berlin (1963).

Karczmar, A. G.: Pharmacologic, toxicologic and therapeutic properties of anti-cholinesterase agents, in: *Physiological Pharmacology*, Vol. 3 (W. S. Root and F. G. Hofman, eds.), pp. 163–322, Academic Press, New York (1967).

Karczmar, A. G.: History of the research with anticholinesterase agents, in: *Anticholinesterase Agents, Internatl. Encyclopedia Pharmacol. Therap.*, Vol. 1, Section 13 (A. G. Karczmar, ed.), pp. 1–44, Pergamon Press, Oxford (1970).

Karczmar, A. G.: Acute and long-lasting central actions of organophosphorus agents. *Fundam. Appl. Toxicol.* **2**:S1–S17 (1984).

Karczmar, A. G., and Ohta, Y.: Neuromyo-pharmacology as related to anticholinesterase action. *Fundam. Appl. Toxicol.* **1**:135–142 (1981).

Katayama, Y., and North, R. A.: Does substance P mediate slow synaptic excitation within the myenteric plexus? *Nature (London)* **274**:387–388 (1978).

Kato, E., Kuba, K., and Koketsu, K.: Presynaptic inhibition by γ-aminobutyric acid in bullfrog sympathetic ganglion cells. *Brain Res.* **153**:398–402 (1978).

Kayaalp, S. O., Albuquerque, E. X., and Warnick, J. E.: Ganglionic and cardiac actions of batrachotoxin. *Eur. J. Pharmacol.* **12**:10–18 (1970).

Kebabian, J. W., and Greengard, P.: Dopamine-sensitive adenyl cyclase: Possible role in synaptic transmission. *Science* **174**:1346–1349 (1971).

Kebabian, J. W., Steiner, A. L., and Greengard, P.: Muscarinic cholinergic regulation of cyclic guanosine 3′,5′-monophosphate in autonomic ganglia: Possible role in synaptic transmission. *J. Pharmacol. Exp. Ther.* **193**:474–488 (1975).

Kessler, J. A., Adler, J. E., Bohn, M. C., and Black, I. R.: Substance P in principal sympathetic neurons: Regulation by impulse activity. *Science* **214**:335–336 (1981).

Kiraly, M., Ma, R. C., and Dun, N. J.: Serotonin mediates a slow excitatory potential in mammalian celiac ganglia. *Brain Res.* **275**:378–383 (1983).

Kirsch, D., Hauser, W., and Wegar, N.: Effect of bispyridinium oximes HGG12 and HGG42 and ganglion blocking agents on synaptic transmission and NAD(P)H fluorescence in the superior cervical ganglion of the rat after soman poisoning *in vitro*. *Fundam. Appl. Toxicol.* **1**:169–176 (1981).

Kobayashi, H., and Tosaka, T.: Slow synaptic actions in mammalian sympathetic ganglia, with special reference to the possible roles played by cyclic nucleotides, in: *Autonomic Ganglia* (L.-G. Elfvin, ed.), pp. 281–307, John Wiley, (1983).

Kobayashi, H., Hashiguchi, T., and Ushiyama, N. S.: Postsynaptic modulation of excitatory process in sympathetic ganglia by cyclic AMP. *Nature (London)* **271**:268–270 (1978).

Koelle, G. B.: Cytological distributions and physiological functions of cholinesterases, in: *Handbuch der Experimentellen Pharmakologie, Ergänzungswerk*, Vol. 15, *Cholinester-*

ases and Anticholinesterase Agents (G. B. Koelle, ed.), pp. 187–298, Springer-Verlag, Berlin (1963).

Koelle, G. B.: Neurohumoral transmission and the autonomic nervous system, in: *The Pharmacological Basis of Therapeutics*, 5th ed. (L. S. Goodman and A. Gilman, eds.), pp. 404–444, Macmillan, New York (1975).

Koketsu, K.: Restorative action of fluoride on synaptic transmission blocked by organophosphorus anticholinesterase. *Int. J. Neuropharmacol.* **5**:247–254 (1966).

Koketsu, K.: Cholinergic synaptic potentials and the underlying ionic mechanisms. *Fed. Proc. Fed. Am. Soc. Exp. Biol.* **28**:101–112 (1969).

Koketsu, K., Nishi, S., and Noda, Y.: Effects of physostigmine on the discharge and slow postsynaptic potentials of bullfrog sympathetic ganglia. *Br. J. Pharmacol.* **34**:177–188 (1968).

Kondo, H.: Innervation of SIF cells in the superior cervical and nodose ganglia: An ultrastructural study with serial sections. *Biol. Cell.* **30**:253–264 (1977).

Kondo, H., and Yui, R.: Enkephalin-like immunoreactivity in the SIF cells of sympathetic ganglia of frogs. *Biomed. Res.* **2**:338–340 (1981).

Konishi, S., Tsunoo, A., and Otsuka, M.: Enkephalins presynaptically inhibit cholinergic transmission in sympathetic ganglia. *Nature (London)* **282**:515–516 (1979).

Konishi, S., Tsunoo, A., Yanaihara, N., and Otsuka, M.: Peptidergic excitatory and inhibitory synapses in mammalian sympathetic ganglia: Roles of substance P and enkephalin. *Biomed. Res.* **1**:528–536 (1980).

Konzett, H.: The effect of histamine on isolated sympathetic ganglia. *J. Mt. Sinai Hosp.* **19**:149 (1952).

Koppanyi, T., and Karczmar, A. G.: Contribution to the study of mechanism of action of cholinesterase inhibitors. *J. Pharmacol. Exp. Ther.* **101**:327–343 (1951).

Koppanyi, T. Studies on the synergism and antagonism of drugs. I. The non-parasympathetic antagonism between atropine and the miotic alkaloids. *J. Pharmacol. Exp. Ther.* **46**:395–405 (1932).

Koppanyi, T., Linegar, C. R., and Herwick, R. P.: Analysis of the vasopressor and other "nicotinic" actions of acetylcholine. *Am. J. Physiol.* **130**:346–357 (1940).

Koppanyi, T., Karczmar, A. G., and King, T. O.: The effects of tetraethyl pyrophosphate on sympathetic ganglionic activity. *Science* **106**:492–493 (1947).

Krivoy, W. A., and Wills, J. H.: Adaptation to constant concentrations of acetylcholine. *J. Pharmacol.* **116**:220–226 (1956).

Kuba, K., Albuquerque, E. X., Daly, J., and Barnard, E. A.: A study of the irreversible cholinesterase inhibitor, diisopropylfluorophosphate on time course of endplate currents in frog sartorius muscle. *J. Pharmacol. Exp. Ther.* **189**:499–512 (1974).

Kuba, K., Kato, E., Kumamoto, E., Koketsu, K., and Hirai, K.: Sustained potentiation of transmitter release by adrenaline and dibutyryl cyclic AMP in sympathetic ganglia. *Nature (London)* **291**:654–656 (1981).

Langley, J. H., and Dickinson, W. C.: On the local paralysis of the peripheral ganglia and on the connection of different classes of nerves with them. *Proc. R. Soc. London* **46**:423–431 (1889).

Larrabee, M. G., and Holaday, D. A.: Depression of transmission through sympathetic ganglia during general anesthesia. *J. Pharmacol. Exp. Ther.* **105**:400–408 (1952).

Larrabee, M. G., and Leicht, W. S.: Metabolism of phosphatidyl inositol and other lipids in active neurones of sympathetic ganglia and other peripheral nervous tissues: The site of the inositide effect. *J. Neurochem.* **12**:1–13 (1965).

Larrabee, M. G., and Posternak, J. M.: Selective action of anesthetics on synapses and axons in mammalian sympathetic ganglia. *J. Neurophysiol.* **15**:91–114 (1952).

Larrabee, M. G., Ramos, J. G., and Bulbring, E.: Effect of anesthetics on oxygen consumption and on synaptic transmission in sympathetic ganglia. *J. Cell. Comp. Physiol.* **40**:461–494 (1952).

Laskowski, W. B., Olson, W. H., and Dettbarn, W. D.: Initial ultrastructural abnormalities at the motor endplate produced by a cholinesterase inhibitor. *Exp. Neurol.* **57**:13–33 (1977).

Lee, C. Y.: Elapid neurotoxins and their mode of action. *Clin. Toxicol.* **3**:457–472 (1970).

Lees, G. M., and Nishi, S.: Analysis of the mechanism of action of some ganglion-blocking drugs in the rabbit superior cervical ganglion. *Br. J. Pharmacol.* **46**:78–88 (1972).

Levy, M. N., and Vassale, M. (eds.): Excitation and Neural Control of The Heart. American Physiological Society, Bethesda, Maryland (1982).

Levy, B., and Ahlquist, R. P.: A study of sympathetic ganglion stimulants. *J. Pharmacol. Exp. Ther.* **137**:219–228 (1962).

Levy, M. N., and Vassale, M.: Excitation and Neural Control of The Heart. American Physiological Society, Bethesda, Maryland (1982).

Libet, B.: Generation of slow inhibitory and excitatory postsynaptic potentials. *Fed. Proc. Fed. Am. Soc. Exp. Biol.* **29**:1945–1956 (1970).

Libet, B.: The SIF cells as a functional dopamine releasing interneuron in the rabbit superior cervical ganglion, in: *SIF Cells: Structure and Function of the Small Intensely Fluorescent Sympathetic Cells* (O. Eranko, ed.), pp. 163–177, Fogarty International Center Proceedings No. 30, U. S. Government Printing Office, Washington, D. C. (1976).

Libet, B.: Which postsynaptic action of dopamine is mediated by cyclic AMP? *Life Sci.* **24**:1043–1058 (1979).

Libet, B., and Owman, C.: Concomitant changes in formaldehyde-induced fluorescence of dopamine interneurons and in slow inhibitory postsynaptic potentials of the rabbit superior cervical ganglion, induced by stimulation of the preganglionic nerve or by a muscarinic agent. *J. Physiol. (London)* **237**:635–662 (1974).

Libet, B., Kobayashi, H., and Tanaka, T.: Synaptic coupling into the production and storage of a neuronal memory trace. *Nature (London)* **258**:155–157 (1975).

Long, J. P., and Eckstein, J. W.: Ganglionic actions of neostigmine methyl sulphate. *J. Pharmacol. Exp. Ther.* **133**:216–222 (1961).

Lundberg, J. H., Hökfelt, T., Anggard, A., Kimmel, J., Goldstein, M., and Markey, K.: Coexistence of an avian pancreatic polypeptide (APP) immunoreactive substance and catecholamine in some peripheral and central neurons. *Acta. Physiol. Scand.* **110**:107–109 (1980).

Machova, H., and Boska, D.: The effect of 5-hydroxytryptamines, dimethylphenylpiperazinium and acetylcholine on transmission and surface potential in the cat sympathetic ganglion. *Eur. J. Pharmacol.* **7**:152–158 (1969).

Marrazzi, A. S.: Electrical studies on the pharmacology of autonomic synapses. The action of a sympathetomimetic drug (epinephrine) on sympathetic ganglia. *J. Pharmacol. Exp. Ther.* **65**:395–404 (1939).

Marshall, L. M.: Synaptic localization of α-bungarotoxin binding which blocks nicotinic transmission at frog sympathetic neurons. *Proc. Natl. Acad. Sci. U.S.A.* **78**:1948–1952 (1981).

Mason, D. F. J.: Ganglion-blocking drugs, in: *Physiological Pharmacology*, Vol. 3 (W. S. Root and F. G. Hoffman, eds.), pp. 363–388, Academic Press, New York (1967).

Matthews, M. R., and Raisman, G.: The ultrastructure and somatic efferent synapses of small granule-containing cells in the superior cervical ganglion. *J. Anat.* **105**:255–282 (1969).

McAfee, D. A., and Greengard, P.: Adenosine 3′,5′-monophosphate: Electrophysiological evidence for a role in synaptic transmission. *Science* **178**:310–312 (1972).

McAfee, D. A., Schorderet, M., and Greengard, P.: Adenosine 3′,5′-monophosphate in nervous tissue: Increase associated with synaptic transmission. *Science* **171**:1156–1158 (1971).

McIsaac, R. J.: Ganglionic blocking properties of epinephrine and related amines. *Int. J. Neuropharmacol.* **5**:14–26 (1966).
Michell, R. H.: Inositol phospholipids and cell surface receptor function. *Biochim. Biophys. Acta* **415**:81–147 (1975).
Milošević, M. P.: Acetylcholine content in the brain of rats treated with paraoxon and obicloxime. *Br. J. Pharmacol.* **39**:732–737 (1970).
Mo, N., and Dun, N. J.: Vasoactive intestinal polypeptide facilitates muscarinic transmission in mammalian sympathetic ganglia. *Neurosci. Lett.* **52**:19–23 (1984).
Mo, N., Dun, N. J., and Karczmar, A. G.: Facilitation and inhibition of ganglionic transmission by eserine. *Neuropharmacology* (in press) (1985).
Mochida, S., Kobayashi, H., Tosaka, T., Ito, J., and Libet, B.: Specific dopamine receptor mediates the production of cyclic AMP in the rabbit sympathetic ganglia and thereby modulates the muscarinic postsynaptic responses. *Adv. Cyclic Nucleotides Res.* **14**:685 (1981).
Morita, K., North, R. A., and Katayama, Y.: Evidence that substance P is a neurotransmitter in myenteric plexus. *Nature (London)* **287**:151–152 (1980).
Moss, R. L., and McCann, S. M.: Action of luteinizing hormone-releasing factor (LRF) in the initiation of lordosis behavior in the estrone-primed ovariectomized female rat. *Neuroendocrinology* **17**:309–318 (1975).
Nastuk, W. L., and Wolfson, C. H.: Cholinergic receptor desensitization. *Ann. N. Y. Acad. Sci.* **274**:1639–1646 (1967).
Nemeth, P. R., Ort, C. A., and Wood, J. D.: Intracellular study of effects of histamine on electrical behavior of myenteric neurones in guinea-pig small intestine. *J. Physiol. (London)* **355**:411–426 (1984).
Nestler, E. J., and Greengard, P.: *Protein Phosphorylation in the Neurons System.* John Wiley, New York (1984).
Nestler, E. J., Walaas, S. I., and Greengard, P.: Neuronal phosphoroproteins: Physiological and clinical implications. *Science* **225**:1357–1364 (1984).
Nicoll, R. A.: Peptide receptors in CNS, in: *Handbook of Psychopharmacology*, Vol. 4 (L. L. Iverson, S. D. Iverson, and S. H. Snyder, eds.), pp. 229–263, Plenum Press, New York (1972).
Nicoll, R. A.: Promising peptides. *Neurosci. Symp.* **1**:99–122 (1976).
Nicoll, R. A.: Pentobarbital: Differential postsynaptic actions on sympathetic ganglion cells. *Science* **199**:451 (1978).
Nishi, S.: Cholinergic and adrenergic receptors at sympathetic preganglionic nerve terminals. *Fed. Proc. Fed. Am. Soc. Exp. Biol.* **29**:1957–1965 (1970).
Nishi, S.: Ganglionic transmission, in: *The Peripheral Nervous System* (J. I. Hubbard, ed.), pp. 225–256, Plenum Press, New York (1974).
Nishizuka, Y.: Turnover of inositol phospholipids and signal transduction. *Science* **225**:1365–1370 (1984).
Nistri, A., and Quilliam, J. P.: Ganglion activity of centrally acting neurotropic agents, in: *Pharmacology of Ganglionic Transmission* (D. A. Kharkevich, ed.), pp. 359–384, Springer-Verlag, Berlin (1980).
Ohta, Y., Karczmar, A. G., and Karczmar, G. S.: Effects of phosphatidic acid on amphibian neuromyal transmission. *Neuropharmacology* (in press) (1985).
Oosako, S., and Deguchi, T.: Stimulation by phosphatidic acid of calcium influx and cyclic GMP synthesis in neuroblastoma cells. *J. Biol. Chem.* **256**:10,945–10,948 (1981).
Pascoe, J. E.: The effects of acetylcholine and other drugs in the isolated superior cervical ganglion. *J. Physiol. (London)* **132**:242–255 (1956).
Paton, W. D. M., and Perry, W. L. M.: The relationship between depolarization and block in the cat's superior cervical ganglion. *J. Physiol. (London)* **119**:43–57 (1953).
Petras, J. M.: Soman neurotoxicity. *Fundam. Appl. Toxicol.* **1**:242 (1981).

Quik, M., and Lamarca, M. V.: Blockade of transmission in rat sympathetic ganglia by a toxin which co-purifies with α-bungarotoxin. *Brain Res.* **238**:385–399 (1982).

Quik, M., Smith, P. A., Padjen, A. L., and Collier, B.: A critical evaluation of the use of toxins from *Dendroaspis viridis* to block nicotinic responses at central and ganglionic synapses. *Brain Res.* **209**:129–143 (1981).

Riker, W. K.: Ganglion cell depolarization and transmission block by ACh: Independent events. *J. Pharmacol. Exp. Ther.* **159**:345–352 (1968).

Riker, W. K., and Szreniawski, Z.: The pharmacological reactivity of presynaptic nerve terminals in a sympathetic ganglion. *J. Pharmacol. Exp. Ther.* **126**:233–238 (1959).

Robinson, S. E., Schwartz, J. P., and Costa, E.: Substance P in the superior cervical ganglion and the submaxillary gland of the rat. *Brain Res.* **182**:11–17 (1980).

Roszkowski, A. P.: An unusual type of sympathetic ganglion stimulant. *J. Pharmacol. Exp. Ther.* **132**:156–170 (1961).

Sejnowski, T. J.: Peptidergic transmission in sympathetic ganglia. *Fed. Proc. Fed. Am. Soc. Exp. Biol.* **41**:2923–2928 (1982).

Selyanko, A. A., Derkach, V. A., and Skok, V. I.: Effects of some ganglion-blocking agents on fast excitatory postsynaptic currents in mammalian sympathetic ganglion neurons, in: *Advances in Physiological Sciences*, Vol. 4, *Physiology of Excitable Membranes* (J. Salanki, ed.), pp. 329–342, Akademiai Kiado, Budapest (1981).

Severs, W. B., and Daniels-Severs, A. E.: Effects of angiotensin on the central nervous system. *Pharmacol. Rev.* **25**:415–449 (1973).

Siegrist, G., Dolivo, M., Dunant, Y., Foroglou-Kerameus, C., De Ribaupierre, F., and Rouiller, C.: Ultrastructure and function of chromaffin cells in the superior cervical ganglion of the rat. *J. Ultrastruct. Res.* **25**:381–407 (1968).

Simmons, M. A.: The complexity and diversity of synaptic transmission in the prevertebral sympathetic ganglia. *Prog. Neurobiol.* **24**:43–93 (1985).

Skok, V. I., and Selyanko, A. A.: Acetylcholine and serotonin receptors in mammalian sympathetic ganglion neurons, in: *Integrative Functions of the Autonomic Nervous System* (C. McC. Brooks, K. Koizumi, and A. Sato, eds.), pp. 248–253, University of Tokyo Press, Tokyo (1979).

Skok, V. I., Selyanko, A. A., and Derkach, V. A.: Channel-blocking activity is a possible mechanism for a selective ganglionic blockade. *Pfluegers Arch.* **398**:169–171 (1983).

Study, R. Z., and Greengard, P.: Regulation by histamine of cyclic nucleotide levels in sympathetic ganglia. *J. Pharmacol. Exp. Ther.* **207**:767–778 (1978).

Sutherland, E. W., and Rall, T. W.: Fractionation and characterization of a cyclic adenine ribonucleotide formed by tissue particles. *J. Biol. Chem.* **232**:1077–1091 (1958).

Sutherland, E. W., Robison, G. A., and Butcher, R. W.: Some aspects of the biological role of adenosine 3′,5′-monophosphate (cyclic AMP). *Circulation* **37**:279–306 (1968).

Takeshige, C., and Volle, R. L.: Bimodal response of sympathetic ganglia to acetylcholine following eserine or repetitive preganglionic stimulation. *J. Pharmacol. Exp. Ther.* **138**:66–73 (1962).

Thesleff, S. The mode of neuromuscular block caused by acetylcholine, nicotine, decamethonium and succinylcholine. *Acta Physiol. Scand.* **34**:218–231 (1955).

Trcka, V.: Relationship between chemical structure and ganglion-blocking activity. b) Tertiary and secondary amines, in: *Pharmacology of Ganglionic Transmission* (D. A. Kharkevich, ed.), pp. 155–184, Springer-Verlag, New York (1980).

Trendelenburg, U.: The action of histamine and pilocarpine on the superior cervical ganglion and adrenal glands of the cat. *Br. J. Pharmacol.* **9**:481–487 (1954).

Trendelenburg, U.: The action of 5-hydroxytryptamine on the nictitating membrane and on the superior cervical ganglion of the cat. *Br. J. Pharmacol. Chemother.* **11**:74–80 (1956).

Trendelenburg, U.: The action of histamine, pilocarpine and 5-hydroxytryptamine on transmission through the superior cervical ganglion. *J. Physiol. (London)* **35**:66–72 (1957).

Tsunoo, A., Konishi, S., and Otsuka, M.: Substance P as an excitatory transmitter of primary afferent neurons in guinea-pig sympathetic ganglia. *Neurosci.* **7**:2025–2037 (1982).

Usdin, E.: Reactions of cholinesterases with substrate inhibitors and reactivators, in: *Anticholinesterase Agents* (A. G. Karczmar, ed.), Internatl. Encyclop. Pharmacol. Ther., Vol. 1, Section 13, pp. 45–354, Pergamon Press, Oxford (1970).

Van Meter, W. G., Karczmar, A. G., and Fiscus, R. R.: CNS effects of anticholinesterases in the presence of inhibited cholinesterases. *Arch. Int. Pharmacodyn.* **23**:249–260 (1978).

Varagic, V.: The action of eserine on the blood pressure of the rat. *Br. J. Pharmacol.* **10**:349–353 (1955).

Verhofstad, A. A. J., Steinbusch, H. W. M., Penke, B., Varga, J., and Joosten, H. W. J.: Serotonin-immunoreactive cells in the superior cervical ganglion of the rat: Evidence for the existence of separate serotonin- and catecholamine-containing small ganglionic cells. *Brain Res.* **212**:39–49 (1981).

Vickroy, T. W., Watson, M., Yamamura, H. I., and Roeske, W. R.: Agonist binding to multiple muscarinic receptors. *Fed. Proc. Fed. Am. Soc. Exp. Biol.* **43**:2785–2790 (1984).

Volle, R. L.: The actions of several ganglion blocking agents on the postganglionic discharge induced by diisopropyl phosphorofluoridate (DFP) in sympathetic ganglia. *J. Pharmacol. Exp. Ther.* **135**:45–53 (1962).

Volle, R. L.: Muscarinic and nicotinic stimulant actions at autonomic ganglia, in: *Ganglionic Blocking and Stimulating Agents* (A. G. Karczmar, ed.), pp. 1–10, Intern. Encyclopedia Pharmacol. Therap., Section 12, Vol. 1, Pergamon Press, Oxford (1966).

Volle, R. L.: Cellular pharmacology of autonomic ganglia, in: *Cellular Pharmacology of Excitable Tissues* (T. Narahashi, ed.), pp. 89–140, Charles C Thomas, Springfield, Illinois (1975).

Volle, R. L.: Ganglionic actions of anticholinesterase agents, catecholamines, neuromuscular blocking agents and local anesthetics, in: *Pharmacology of Ganglionic Transmission* (D. A. Kharkevich, ed.), pp. 385–410, Springer-Verlag, Berlin (1980).

Volle, R. L., Quenzer, L. F., and Patterson, B. A.: The regulation of cyclic nucleotides in a sympathetic ganglia. *J. Auton. Nerv. Syst.* **6**:65–72 (1982).

Wallis, D. I., and North, R. A.: The action of 5-hydroxytryptamine on single neurones of the rabbit superior cervical ganglion. *Neuropharmacology* **17**:1023–1028 (1978).

Waser, P. G.: Chemistry and pharmacology of muscarine, muscarone and some related compounds. *Pharmacol. Rev.* **13**:465–515 (1961).

Watanabe, A. M., Lindemann, J. P., and Fleming, J. W.: Mechanisms of muscarinic modulation of protein phosphorylation in intact ventricles. *Fed. Proc. Fed. Am. Soc. Exp. Biol.* **43**:2618–2623 (1984).

Weight, F. F., and Weitsen, H. A.: Identification of small intensely fluorescent (SIF) cells as chromaffin cells in bullfrog sympathetic ganglia. *Brain Res.* **128**:213–226 (1977).

Weight, F. F., Petzold, G., and Greengard, P.: Guanosine 3′,5′-monophosphate in sympathetic ganglia: Increase associated with synaptic transmission. *Science* **186**:942–944 (1974).

Weight, F. F., Smith, A. A., and Schulman, J. A.: Postsynaptic potential generation appears independent of synaptic elevation of cyclic nucleotides in sympathetic neurons. *Brain Res.* **158**:197–202 (1978).

Williams, T. H., and Palay, S. L.: Ultrastructure of the small neurons in the superior cervical ganglion. *Brain Res.* **15**:17–34 (1969).

Williams, T. H., Block, A. C., Chiba, T., and Bhalla, R. C.: Morphology and biochemistry of small, intensely fluorescent cells of sympathetic ganglia. *Nature (London)* **256**:315–317 (1975).

Williams, T. H., Chiba, T., Black, A. C., Jr., Bhalla, R. C., and Jew, J.: Species variation in SIF cells of superior cervical ganglia: Are there two functional types?, in: *SIF Cells: Structure and Function of the Small Intensely Fluorescent Sympathetic Cells* (O. Eranko, ed.), pp. 143–162, Fogarty International Center Proceedings No. 30, DHEW Publ. (NIH) (1976).

Williams, T. H., Black, A. C., Chiba, T., and Jew, J. Y.: Species differences in mammalian SIF cells, in: *Advances in Biochemical Psychopharmacology* (E. Costa and G. L. Gessa, eds.), pp. 505–511, Raven Press, New York (1977).

Witkop, B., and Brossi, A.: Natural toxins and drug development, in: *Natural Products and Drug Development* (P. Krogsgaard, S. Brogger Christensen, and H. Kofod, eds.), pp. 283–300, Munksgaard, Copenhagen (1984).

Witkop, B., and Gossinger, E.: Amphibian alkaloids. *Alkaloids* **21**:139–259 (1983).

Wood, J. D., and Mayer, C. J.: Serotonergic activation of tonic-type enteric neurons in guinea-pig small intestine. *J. Neurophysiol.* **42**:582–593 (1979).

Yamada, M., Tokimasa, T., and Koketsu, K.: Effects of histamine on acetylcholine release in bullfrog sympathetic ganglia. *Eur. J. Pharmacol.* **82**:15–20 (1982).

Yarowsky, P., Fowler, J., Taylor, G., and Weinreich, D.: Noncholinesterase actions of an irreversible acetylcholinesterase inhibitor on synaptic transmission and membrane properties in autonomic ganglia. *Cell. Mol. Neurobiol.* **4**:351–366 (1984).

Young, J. A. C., Brown, D. A., Kelly, J. S., and Schon, F. S.: Autoradiographic localization of sites of ^{3}H-aminobutyric acid accumulation in peripheral autonomic ganglia. *Brain Res.* **63**:479–486 (1973).

Zaimis, E. J.: Actions at autonomic ganglia, in: *Handbuch der Experimentellen Pharmakologie, Ergänzungswerk, Cholinesterase and Anticholinesterase Agents* (G. B. Koelle, ed.), pp. 530–569, Springer-Verlag, Berlin (1963).

III

Parasympathetic and Enteric Ganglia and Their Neuropharmacology

14

Excitatory Transmission in Parasympathetic Ganglia

JOEL P. GALLAGHER
and PATRICIA SHINNICK-GALLAGHER

I. INTRODUCTION

The five sections that follow will deal with transmission in parasympathetic ganglia excluding the enteric ganglia (myenteric plexus and Auerbach plexus); these latter ganglia and their transmission are discussed separately (see Chapter 16). The distinction between parasympathetic and enteric transmission stems from the observation made by Langley (1900) that there is no evidence to indicate whether all or some of the neurons in the gastrointestinal plexuses of Auerbach and Meissner constitute postganglionic neurons located in the cranial or sacral pathways. As a result, since the classification by Langley (1900) of the autonomic nervous system, such a distinction has persisted (Gershon, 1981), although it is not accepted by all (Burnstock, 1979).

Similarly to sympathetic ganglia, parasympathetic ganglia constitute more than simple relay systems, since many of them show several types of postsynaptic responses as well as presynaptic regulatory mechanisms. Furthermore, in the case of the pelvic ganglia of the guinea pig and of the vesical pelvic ganglia of the cat, there are complex synaptic connections between the adrenergic and cholinergic systems, since cholinergic nerve terminals enclose both adrenergic and cholinergic neurons, while adrenergic terminals enclose, and synapse with, cholinergic neurons [see Chap-

JOEL P. GALLAGHER and PATRICIA SHINNICK-GALLAGHER • Department of Pharmacology, University of Texas Medical Branch, Galveston, Texas.

ter 21, and the recent reviews of Wood (1983) and Owman *et al.* (1983)]. Finally, at least in the case of the pelvic ganglia, besides catecholamines and acetylcholine (ACh), peptides and enkephalins participate in the regulation of transmission (see below and Chapters 2 and 21).

II. EXTRACELLULAR STUDIES

Transmission in parasympathetic ganglia has received little attention as compared to transmission in sympathetic ganglia. Investigations have been infrequent because parasympathetic ganglia, unlike sympathetic ganglia, are usually located within the walls (intramural ganglia) (Skok, 1980) of the organ they innervate, making isolation of the ganglion and especially of its postganglionic fibers difficult. As a result, parasympathetic ganglionic transmission has been investigated with extracellular recording techniques only with respect to the following preparations: cat and monkey ciliary ganglion (Whitteridge, 1937), chick ciliary ganglion (Martin and Pilar, 1963a,b), cat ciliary ganglion (Melnitchenko and Skok, 1969; Schaffner and Haefely, 1974), cat vesical pelvic ganglion (DeGroat, 1970, 1975; DeGroat and Ryall, 1968, 1969; DeGroat and Saum, 1971, 1972, 1976; Griffith *et al.*, 1981a; Saum and DeGroat, 1973), and cat colonic ganglia (DeGroat and Krier, 1976).

III. INTRACELLULAR STUDIES

Similarly, intracellular recording of parasympathetic ganglionic transmission has been achieved with only a few preparations. Two ganglia have been studied with intracellular recording *in vivo*, namely, the cat ciliary ganglion (Melnitchenko and Skok, 1970) and the hamster submandibular ganglion (Suzuki and Kusano, 1978). The *in vitro* preparations, many of which were also used for extracellular recording, include: cat ciliary ganglion (Melnitchenko and Skok, 1969, 1970; Nishi and Christ, 1971), pelvic ganglia of the guinea pig (Crowcroft and Szurszewski, 1971), cardiac ganglia in the frog heart (Dennis *et al.*, 1971; Harris *et al.*, 1971), mudpuppy cardiac ganglia (Hartzell *et al.*, 1977; Roper, 1976a,b), submandibular ganglia of rat (Ascher *et al.*, 1979; Lichtman, 1977) and hamster (Suzuki and Kusano, 1978; Suzuki and Volle, 1979), and cat vesical pelvic ganglia (Booth and DeGroat, 1978, 1979, Gallagher *et al.*, 1982; Griffith, 1980; Griffith *et al.*, 1978, 1979, 1980a,b, 1981b).

IV. TYPES OF EXCITATORY TRANSMISSION IN PARASYMPATHETIC GANGLIA

A. Excitatory Nicotinic Cholinergic Transmission

In 1937, Whitteridge (1937) demonstrated that synaptic transmission in the cat ciliary ganglion can be blocked with nicotine. Since then, all studies of parasympathetic ganglionic transmission led to the proposition that ACh is the main transmitter candidate. In addition to cholinergic excitatory transmission, electrical transmission (Martin and Pilar, 1963a,b; Roper, 1976a) and noncholinergic transmission have been suggested as alternative modes of parasympathetic transmission.

An intense in-depth analysis supporting the assumption that ACh is the transmitter at parasympathetic ganglia was presented by Dennis *et al.* (1971). They demonstrated that normal transmission involved the release of ACh from preganglionic neurons with subsequent activation of a nicotinic cholinergic receptor and production of excitatory postsynaptic potentials (EPSPs). Similar to the skeletal neuromuscular junction, the EPSP recorded intracellularly from the cell body of parasympathetic postganglionic neurons is made up of quantal components, miniature EPSPs, the frequency of which is calcium-dependent (Dennis *et al.*, 1971). The permeability changes initiated at the postsynaptic membrane by exogenously applied ACh and the endogenous synaptic transmitter appeared identical, since the ionic fluxes for both responses had the same equilibrium potential. Figure 1 provides an example of the conductance change and reversal potential of the fast EPSPs and response to ACh as recorded from the cat parasympathetic vesical pelvic ganglia. Furthermore, the receptors that are activated by the synaptic transmitter were desensitized by application of ACh. These and additional experiments (Ascher *et al.*, 1979; DeGroat and Saum, 1976; Gallagher *et al.*, 1982; Griffith, 1980; Harris *et al.*, 1971; Hartzell *et al.*, 1977; Roper, 1976a,b) leave little doubt that ACh is the principal transmitter responsible for transmission at parasympathetic ganglia.

B. Electrical Transmission

It is worth commenting on the early statement of Whitteridge (1937); in describing the transmission process of the ciliary ganglion, he pointed

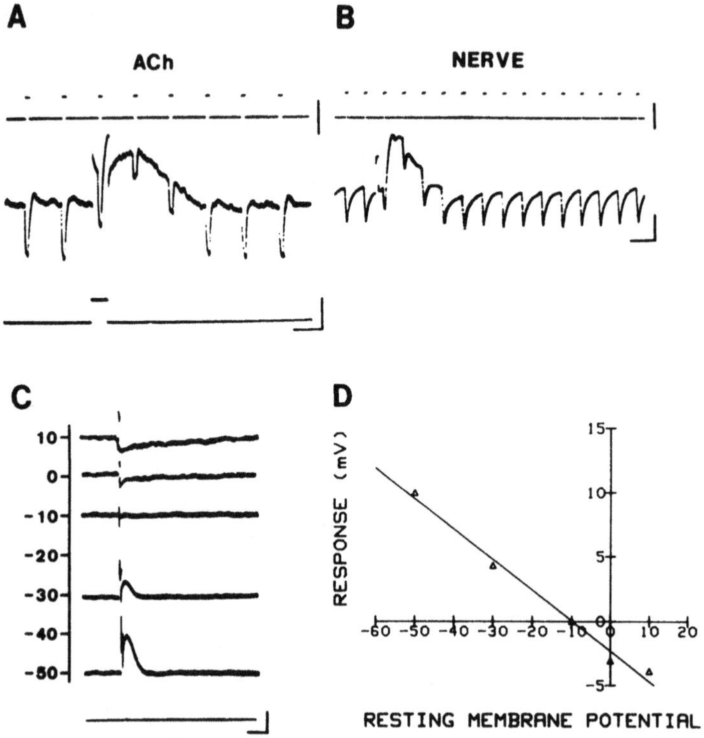

Figure 1. Change in membrane conductance and reversal potentials for the fast EPSP. (A) Change in the size of anodal pulses during an ACh iontophoretic-induced depolarization (10 nA, bottom). (B) Change in anodal pulse size during a fast EPSP (10 mV × 20 msec; 1 nA. (C) Reversal of an ACh iontophoretic depolarization as a function of membrane potential (500 msec × 100 nA). (D) Graph of the responses shown in (C). From Griffith (1980).

out that "the cat ciliary ganglion is not in any way typical of parasympathetic ganglia." What he had in mind is that this ganglion shows poor integrative properties, there being little regulation of the spike, contrary to what happens in other mammalian parasympathetic ganglia. On the other hand, the cat ciliary ganglion is also different from the avian ciliary ganglion, specifically that of the chick, and the mudpuppy cardiac ganglion. In these two cases, electrical coupling is present, constituting another mechanism for excitatory transmission in parasympathetic ganglia. However, the electrical coupling in the chick occurs across the chemical synapse that courses from preganglionic to postganglionic neurons (Martin and Pilar, 1963a,b), whereas in the mudpuppy, electrical junctions are present between adjacent principal neurons (Roper, 1976a). The fact that both electrical coupling and transmission appear to be present in these avian and amphibian parasympathetic ganglia makes them different from sympathetic (and from mammalian parasympathetic) ganglia, but similar to certain neuronal systems of the brain (Sotelo and Palay, 1970; Baker and Llinas, 1971). As was pointed out by Roper (1976a), electrical junctions are well suited for transmitting slow potential changes, which may be similar to the slow synaptic potentials that are a means of modulation and/or integration of synaptic transmission in sympathetic ganglia (Libet, 1970) (see Chapter 21) and the brain (Krnjevic *et al.*, 1971).

C. Slow Excitatory Muscarinic Transmission

Only two parasympathetic ganglia show slow (i.e., noticeable after the initiation of the fast EPSP) potential changes, the mudpuppy cardiac ganglion (Hartzell *et al.*, 1977) and the cat vesical (bladder) pelvic ganglion (Griffith *et al.*, 1980b, 1981a,b). Of these, the former exhibits a slow inhibitory postsynaptic potential (slow IPSP) only, while the latter exhibits both a slow IPSP and a slow EPSP. Recent information indicates that vasoactive intestinal polypeptide (VIP) may facilitate the slow EPSP of these ganglia, since it enhanced postganglionic discharge elicited by muscarinic agonists or organophosphorus anticholinesterases (Kawatani *et al.*, 1985); VIP had no effect on the nicotinic transmission of the bladder ganglion. DeGroat (see Kawatani *et al.*, 1986) suggested that VIP facilitates the muscarinic responses, in both sympathetic and parasympathetic ganglia, as a cotransmitter of ACh; this is consistent with the distribution of VIP in the peripheral neuron system (see Chapter 21). Slow IPSPs in parasympathetic ganglia will be discussed in Chapter 15.

It was suggested on the basis of the early studies that slow EPSPs serve as physiologically significant modulators of sympathetic ganglionic transmission (Libet, 1965, 1970; Hilton and Steinberg, 1966; A. M. Brown, 1967). The slow EPSP recorded in the cat vesical pelvic ganglion results from activation of a muscarinic receptor, since it can be mimicked by application of bethanechol and blocked by atropine (Griffith *et al.*, 1980b, 1981b). The muscarinic depolarization is accompanied by a decreased conductance (probably to potassium) and therefore may be similar to that seen in sympathetic ganglia (Weight and Votava, 1970; D. A. Brown and Adams, 1980) (see Chapter 12). The slow EPSP recorded intracellularly may be responsible for the muscarinic facilitatory mechanism (DeGroat and Booth, 1980) in vesical pelvic ganglia.

A unique feature of cat parasympathetic vesical pelvic ganglia is that their slow responses, the IPSPs and EPSPs, may be recorded following the fast EPSPs or action potentials without the use of any blocking drugs (Figure 2). Together, these results suggest that both slow potentials are

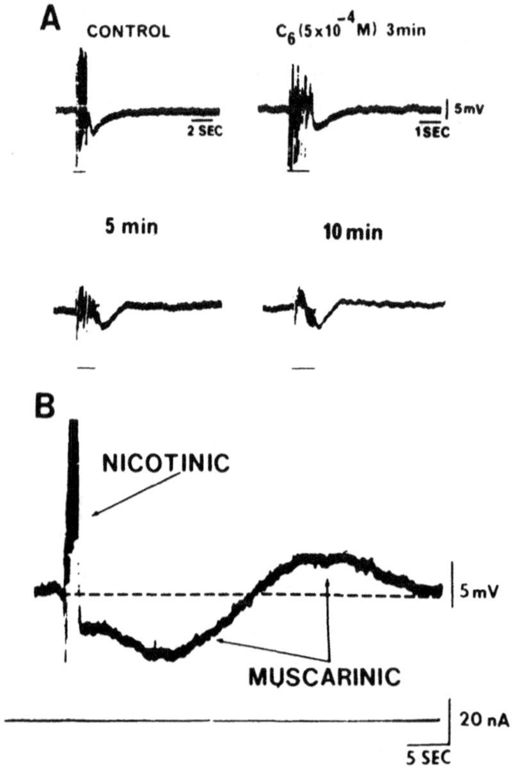

Figure 2. (A) Recordings of both action potentials and a slow IPSP after train stimulation (40 Hz for 1 sec, during solid bar) in the cat vesical pelvic ganglion. In the control tracing, a series of action potentials is followed by a slow IPSP; both the fast and the slow potentials were recorded prior to the application of any antagonist. The addition of hexamethonium (C_6) gradually blocks (3–5 min) the action potentials, and the slow IPSP is more clearly seen. After 10 min of C_6, the action potentials are almost completely blocked, and the size and duration of the slow IPSP recorded after nicotinic receptor antagonism were very similar to those recorded in the control tracing. (B) Nicotinic and muscarinic receptor activation following ACh iontophoresis in the cat vesical pelvic ganglion. A series of iontophoretic pulses (10 Hz for 1 sec) were applied to a cell not previously treated with any antagonists. A sequence of three potentials, the fast depolarization (peaks cut off), a slow hyperpolarization, and finally a slow depolarization could all be recorded. The slow potentials in this cell were not masked by the presence of action potentials. From Griffith (1980).

effective under physiological conditions in the modulation of parasympathetic ganglionic transmission.

V. SPONTANEOUS FIRING

Another unique feature of the cat pelvic parasympathetic ganglion is the presence of spontaneously firing action potentials. These potentials have been recorded in 5% (DeGroat and Booth, 1980) to 10% (Griffith *et al.*, 1980a) of the total population of neurons that have been studied. The spontaneous firing is not attributed to injury, since passive and active membrane properties of these neurons were indistinguishable from those of quiescent cells and remained stable for the duration of the recording. It is possible that spontaneous neuronal activity in the vesical pelvic ganglion arises from endogenously active cells ("pacemaker cells"). Such activity may function to supply efferent "tone" to the smooth muscle of the bladder. Another possibility is that electrical coupling may occur in these ganglia, via specialized junctions, between sensory fibers traveling from the bladder wall through the ganglion and the principal ganglion cells; such junctions have already been observed in other parasympathetic ganglia, e.g., the chick ciliary ganglion (Martin and Pilar, 1963a,b) and the mudpuppy cardiac ganglion (Roper, 1976a). Such electrical coupling could provide a positive feedback mechanism that involves stretch or pain receptor afferents, leading to an increase in the efficacy of micturition.

VI. FACILITATION OF TRANSMISSION

This discussion so far has concerned postsynaptic excitatory cholinergic mechanisms affecting parasympathetic ganglionic transmission. Presynaptic facilitation and inhibition of ACh release were explored with respect to only two parasympathetic ganglia. The first pertinent observation was that of Whitteridge (1937), who simply stated that facilitation of transmission occurs in cat ciliary ganglia. More recently, DeGroat and Booth (1980) and Griffith (1980) concluded that enhanced transmitter release at cat vesical pelvic ganglia accounts for the temporal facilitation seen with this preparation. This temporal facilitation is characterized by a considerable increase in EPSP amplitude and postganglionic firing during repetitive stimluation of the preganglionic nerves at frequencies spanning the normal discharge rates (1–10 Hz) of sacral preganglionic neurons (DeGroat and Booth, 1980).

In addition to this presynaptic cholinergic mechanism that may enhance excitatory cholinergic transmission, other modulatory phenomena involving both pre- and postsynaptic adrenergic mechanisms (DeGroat and Booth, 1980; Griffith *et al.*, 1979) may be present in cat vesical ganglia. Also, inhibitory opioid receptors (of the delta type) appear to be located at the presynaptic nerve endings of this ganglion (Simonds *et al.*, 1983) (see Chapter 15). Thus, of all the parasympathetic ganglia studied thus far, this particular ganglion seems to be the most complex (see also Chapters 12, 15, and 21). It will be interesting to examine the pharmacology of this preparation and explore other mechanisms that may affect excitatory transmission through this ganglion.

There are other parasympathetic ganglia (otic, sublingual, and sphenopalatine) that as yet have not been investigated to any degree (Skok, 1973). It is most likely that excitatory transmission in these ganglia will also be cholinergic, but whether or not these ganglia will exhibit modulatory mechanisms such as those present in cat parasympathetic ganglia is yet to be determined.

VII. CONCLUSIONS

As in the other autonomic ganglia, the fast nicotinic cholinergic response constitutes the primary mechanism of transmission in parasympathetic ganglia; while in isolated cases concerning nonmammalian parasympathetic ganglia electric transmission may be present, it subserves a modulatory rather than a transmissive role.

The transmission of parasympathetic ganglia has been studied to a much lesser extent than that of sympathetic or enteric ganglia, and it may be too early to generalize. At present, it appears that as compared to sympathetic ganglia, parasympathetic ganglia transmit in a manner that is generally relatively simple. Thus, only a few parasympathetic ganglia—mudpuppy cardiac and cat vesicle pelvic—exhibit a slow muscarinic potential, and it is not known at this time whether or not additional slow—inhibitory or excitatory—potentials may be evoked. Similarly, little is known as to modulatory influences—whether of facilitatory or inhibitory nature—that may be exerted at the presynaptic nerve terminals of parasympathetic ganglia. However, initial information that is available suggests that inhibitory and excitatory presynaptic sites may be present in cat parasympathetic ganglia.

It may be shown in the future that parasympathetic ganglia are endowed with special modulatory mechanisms, as reflected, for instance, in the capacity of some of these ganglia to fire spontaneously; indeed, this phenomenon may express a pacemaker or a sensory mechanism, or both, mediated by an electrical coupling.

ACKNOWLEDGMENTS. The authors' research reported herein was supported in part by Grant NS 16228.

REFERENCES

Ascher, P., Large, W. A., and Rang, H. P.: Studies on the mechanism of action of acetylcholine antagonists on rat parasympathetic ganglion cells. *J. Physiol. (London)* **295**:139–170 (1979).

Baker, R., and Llinas, R.: Electronic coupling between neurons in the rat mesencephalic nucleus. *J. Physiol. (London)* **203**:550–570 (1971).

Booth, A. M., and DeGroat, W. C.: A study of recruitment in vesical parasympathetic ganglia of the cat using intracellular recording techniques. *Fed. Proc. Fed. Am. Soc. Ex. Biol.* **37**:526 (1978).

Booth, A. M., and DeGroat, W. C.: A study of facilitation in vesical parasympathetic ganglia of the cat using intracellular recording techniques. *Brain Res.* **169**:388–392 (1979).

Brown, A. M.: Cardiac sympathetic adrenergic pathways in which synaptic transmission is blocked by atropine sulfate. *J. Physiol. (London)* **191**:271–288 (1967).

Brown, D. A., and Adams, P. R.: Muscarinic suppression of a novel voltage-sensitive K-current in a vertebrate neurone. *Nature* **283**:673–676 (1980).

Burnstock, G.: Non-adrenergic, non-cholinergic autonomic neurotransmission mechanisms, in: *Autonomic Neuroeffector Junction* (G. Burnstock, M. D. Gershon, T. Hökfelt, L. H. Iversen, H. W. Kosterlitz, and J. H. Szurszewski, eds.), Neurosciences Research Program Bulletin No. 17, pp. 388–391, MIT Press, Cambridge (1979).

Crowcroft, P. J., and Szurszewski, J. H.: A study of the inferior mesenteric and pelvic ganglia of guinea-pigs with intracellular electrodes. *J. Physiol. (London)* **219**:421–441 (1971).

DeGroat, W. C.: The effects of glycine, GABA and strychnine on sacral parasympathetic preganglionic neurones. *Brain Res.* **18**:542–544 (1970).

DeGroat, W. C.: Excitation and inhibition of sacral parasympathetic neurons by visceral and cutaneous stimuli in the cat. *Brain Res.* **87**:201–211 (1975).

DeGroat, W. C., and Booth, A. M.: Inhibition and facilitation in parasympathetic ganglia of the urinary bladder. *Fed. Proc. Fed Am. Soc. Exp. Biol.* **39**:2990–2996 (1980).

DeGroat, W. C., and Krier, J.: An electrophysiological study of the sacral parasympathetic pathway to the colon of the cat. *J. Physiol. (London)* **260**:425–445 (1976).

DeGroat, W. C., and Ryall, R. W.: The identification and characteristics of sacral preganglionic neurones. *J. Physiol. (London)* **196**:563:577 (1968).

DeGroat, W. C., and Ryall, R. W.: Reflexes to sacral parasympathetic neurones concerned with micturition in the cat. *J. Physiol. (London)* **200**:87–108 (1969).

DeGroat, W. C., and Saum, W. R.: Adrenergic inhibition in mammalian parasympathetic ganglia. *Nature (London) New Biol.* **231**:188–189 (1971).

DeGroat, W. C., and Saum, W. R.: Sympathetic inhibition of the urinary bladder and of pelvic ganglionic transmission in the cat. *J. Physiol. (London)* **220**:297–314 (1972).

DeGroat, W. C., and Saum, W. R.: Synaptic transmission in parasympathetic ganglia in the urinary bladder of the cat. *J. Physiol. (London)* **256**:137–158 (1976).

Dennis, M. J., Harris, A. J., and Kuffler, S. W.: Synaptic transmission and its duplication by focally applied acetylcholine in parasympathetic neurons in the heart of the frog. *Proc. R. Soc. London Ser. B* **177**:509–539 (1971).

Gallagher, J. P., Griffith, W. H., III, and Shinnick-Gallagher, P.: Cholinergic transmission in cat parasympathetic neurones. *J. Physiol. (London)* **332**:473–486 (1982).

Gershon, M. D.: The enteric nervous system. *Annu. Rev. Neurosci.* **4**:227–272 (1981).

Griffith, W. H., III: The physiology and pharmacology of a mammalian parasympathetic ganglion. Ph.D. dissertation, University of Texas, Galveston (1980).

Griffith, W. H., III, Shinnick-Gallagher, P., and Gallagher, J. P.: An intracellular investigation of the cat pelvic parasympathetic ganglion. *Fed. Proc. Fed. Am. Soc. Exp. Biol.* **37**:527 (1978).

Griffith, W. H., III, Gallagher, J. P., and Shinnick-Gallagher, P.: Action of norepinephrine on neuronal activity recorded from cat parasympathetic ganglia. *Fed. Proc. Fed. Am. Soc. Exp. Biol.* **38**:272 (1979).

Griffith, W. H., III, Gallagher, J. P., and Shinnick-Gallagher, P.: An intracellular investigation of cat vesical pelvic ganglia. *J. Neurophysiol.* **43**:343–354 (1980a).

Griffith, W. H., III, Gallagher, J. P., and Shinnick-Gallagher, P.: Cholinergic slow potentials influence parasympathetic ganglionic transmission. *Soc. Neurosci. Abstr.* **6**:68 (1980b).

Griffith, W. H., III, Gallagher, J. P., and Shinnick-Gallagher, P.: Sucrose-gap recordings of nerve-evoked potentials in mammalian parasympathetic ganglia. *Brain Res.* **209**:446–451 (1981a).

Griffith, W. H., III, Gallagher, J. P., and Shinnick-Gallagher, P.: Mammalian parasympathetic ganglia fire spontaneous action potentials and transmit slow potentials, in: *Advances in Physiological Sciences* Vol. 4, *Phyisology of Excitable Membranes* (J. Salánki, ed.), pp. 347–350 Adadémiai Kiadó, Budapest (1981b).

Harris, A. J., Kuffler, S. W., and Dennis, M. J.: Differential chemosensitivity of synaptic and extrasynaptic area of the neuronal surface membrane in parasympathetic neurons of the frog, tested by microapplication of acetylcholine. *Proc. R. Soc. London Ser. B* **177**:541–553 (1971).

Hartzell, H. C., Kuffler, S. W., Stickgold, R., and Yoshikami, D.: Synaptic excitation and inhibition resulting from direct action of acetylcholine on two types of chemoreceptors on individual parasympathetic neurones. *J. Physiol. (London)* **271**:817–846 (1977).

Hilton, J. G., and Steinberg, M.: Effects of ganglion and parasympathetic blocking drugs upon the pressor response elicited by elevation of the intracranial fluid pressure. *J. Pharmacol. Exp. Ther.* **153**(2):285–291 (1966).

Kawatani, M., Rutigliano, M., and DeGroat, W. C.: Selective facilitatory effects of vasoactive intestinal polypeptide on muscarinic mechanisms in sympathetic and parasympathetic ganglia of the cat, in: *Dynamics of Cholinergic Function* (I. Hanin, ed.), Plenum Press, New York (in press) (1986).

Krnjevic, K., Pumain, R., and Renaud, L.: The mechanism of excitation by acetylcholine in the cerebral cortex. *J. Physiol. (London)* **215**:247–268 (1971).

Langley, J. N.: The sympathetic and other related systems of nerves, in: *Schafer Textbook of Physiology*, **2**:616–696, Pentland, Edinburgh (1900).

Libet, B.: Slow synaptic responses and excitatory changes in sympathetic ganglia. *J. Physiol. (London)* **174**:1–25 (1965).

Libet, B.: Generation of slow inhibitory and excitatory synaptic potentials. *Fed. Proc. Fed. Am. Soc. Exp. Biol.* **29**:1945–1956 (1970).

Lichtman, J. W.: The reorganization of synaptic connections in the rat submandibular ganglion during post-natal development. *J. Physiol. (London)* **273**:155–177 (1977).

Martin, A. R., and Pilar, G.: Dual mode of synaptic transmission in the avian ciliary ganglion. *J. Physiol. (London)* **168**:464–475 (1963a).

Martin, A. R., and Pilar, G.: Transmission through the ciliary ganglion of the chick. *J. Physiol. (London)* **168**:464–475 (1963b).

Melnitchenko, L. V., and Skok, V. I.: Electrophysiological study of the ciliary ganglion in the cat (in Russian). *Neurophysiology* **1**:101–108 (1969).

Melnitchenko, L. V., and Skok, V. I.: Natural electrical activity in mammalian parasympathetic ganglion neurons. *Brain Res.* **23**:277–279 (1970).

Nishi, S., and Christ, D. D.: Electrophysiological and anatomical properties of mammalian parasympathetic ganglion cells. *Proceedings of the International Union of Physiological Sciences*, Vol. 9, p. 421, German Physiological Society, Munich 1971.

Owman, C., Alan, P., and Sjoberg, N.-O.: Pelvic autonomic ganglia: Structure, transmitters and steroid influence, in: *Autonomic Ganglia* (L.-G. Elfvin, ed.), pp. 125–143, John Wiley, London (1983).

Roper, S.: Electrophysiological study of chemical and electrical synapses on neurones in the parasympathetic cardiac ganglion in the mudpuppy, *Necturus maculosus*: Evidence for intrinsic ganglionic innervation. *J. Physiol. (London)* **254**:427–454 (1976a).

Roper, S.: The acetylcholine sensitivity of the surface membrane of multiply-innervated parasympathetic ganglion cells in the mudpuppy before and after partial denervation. *J. Physiol. (London)* **254**:455–473 (1976b).

Saum, W. R., and DeGroat, W. C.: Nicotinic and muscarinic mechanisms in pelvic parasympathetic ganglia in the urinary bladder of the cat. *Fed. Proc. Fed. Am. Soc. Exp. Biol.* **32**:800 (1973).

Schaffner, R., and Haefely, W.: The chiliary ganglion of the cat: Pharmacological aspects. *Naunyn-Schmiederberg's Arch. Pharmacol.* **282**(Suppl.):R83 (1974).

Simonds, W. F., Booth, A. M., Thor, K. B., Ostrowski, N. L., Nagel, J. R., and DeGroat, W. C.: Parasympathetic ganglia: Naloxone antagonizes inhibition by leucine-enkephalin and GABA. *Brain Res.* **271**:365–370 (1983).

Skok, V. I.: *Physiology of Autonomic Ganglia*. Igaku Shoin, Tokyo (1973).

Skok, V. I.: Ganglionic transmission: Morphology and physiology, in: *Handbood of Experimental Pharmacology,* Vol. 53, *Pharmacology of Ganglionic Transmission* (D. Kharkevich, ed.), pp. 9–30, Springer-Verlag, New York (1980).

Sotelo, C., and Palay, S. L.: The fine structure of the lateral vestibular nucleus in the rat. II. Synaptic organization. *Brain Res.* **18**:93–115 (1970).

Suzuki, T., and Kusano, K.: Hyperpolarization potentials induced by Ca^{++}-mediated K^{+}-conductance increase in hamster submandibular ganglion cells. *J. Neurobiol.* **9**:367–392 (1978).

Suzuki, T., and Volle, R. L.: Nicotinic, muscarinic and adrenergic receptors in a parasympathetic ganglion. *J. Pharmacol. Exp. Ther.* **211**:252–256 (1979).

Weight, F. F., and Votava, J.: Slow synaptic excitation in sympathetic ganglion cells: Evidence for synaptic inactivation of potassium conductance. *Science* **170**:755–757 (1970).

Whitteridge, D.: The transmission of impulses through the ciliary ganglion. *J. Physiol. (London)* **89**:99–111 (1937).

Wood, J. D.: Neurophysiology of parasympathetic and enteric ganglia, in: *Autonomic Ganglia* (L.-G. Elfvin, ed.), pp. 367–398, John Wiley, London (1983).

15

Inhibition in Parasympathetic Ganglia

PATRICIA SHINNICK-GALLAGHER, JOEL P. GALLAGHER, and WILLIAM H. GRIFFITH III

I. INTRODUCTION

Cholinergic function is relatively simple in some parasympathetic ganglia, in particular the pelvic ganglia of the rat (Blackman *et al.*, 1969), the cat ciliary ganglion (Nishi and Christ, 1971), the parasympathetic ganglion of the frog heart (Dennis *et al.*, 1971), and the rat submandibular ganglion (Lichtman, 1977). These ganglia do not appear to possess integrative capabilities, since there is an absence of slow potentials and/or catecholaminergic interneurons such as small intensely fluorescent (SIF) cells that are thought to be responsible for modulating synaptic transmission (Libet, 1970). In the ganglia that are endowed with SIF cells and/or exhibit such slow potentials as the slow inhibitory postsynaptic potentital (IPSP), this modulation may be due to the release from SIF cells of catecholamines or other substances, and it can occur postsynaptically, via the change in postsynaptic membrane potential or affinity, or presynaptically, via change in the release of transmitters. For full discussion of this subject and alternative explanations of the generation of IPSPs, see Chapters 9 and 13. While apparently devoid of this type of integrative capacity, such parasympathetic ganglia as pelvic ganglia exhibit other types of regulations,

PATRICIA SHINNICK-GALLAGHER and JOEL P. GALLAGHER • Department of Pharmacology, University of Texas Medical Branch, Galveston, Texas. WILLIAM H. GRIFFITH III • Department of Pharmacology, School of Medicine, Texas A and M University, College Station, Texas.

particularly those that result from interactions between several transmitters (see below and Chapter 14).

Recent studies have suggested that other parasympathetic ganglia have integrating or modulatory properties previously demonstrated only in the case of sympathetic neurons (DeGroat and Saum, 1971, 1972; Griffith *et al.*, 1979b, 1980a, 1981a,b; Johnson *et al.*, 1977). The pertinent studies performed on parasympathetic neurons are few, and the studies involving inhibition in parasympathetic ganglia are even fewer. This chapter will examine inhibition in parasympathetic neurons; myenteric neurons will be considered in Chapter 16.

Inhibition can result from a number of different phenomena; for the purpose of this review, we will define it as exhibiting one or more of the following events: (1) decrease in ganglionic firing, (2) hyperpolarization of the ganglionic membrane, and (3) decrease in neurotransmitter release.

II. EXTRACELLULAR STUDIES

A. *In Vivo* Investigations

The most thorough investigation of inhibition in parasympathetic ganglia in which extracellular recording methods were used was carried out on the cat vesical pelvic ganglion by DeGroat and co-workers. These ganglia lie on the surface of the urinary bladder and contain three types of principal ganglion cells, cholinergic, adrenergic, and purinergic, as well as SIF cells; these "parasympathetic ganglia" are unique, since they receive innervation from both divisions of the autonomic nervous system (Burnstock, 1978; Burnstock *et al.*, 1978a,b; Dail, 1976; Elbadawi and Schenk, 1968, 1973; Hamberger and Norberg, 1965; Kuntz and Mosely, 1936; Langley and Anderson, 1895, 1896).

DeGroat and Saum (1971, 1972) demonstrated that stimulation of the hypogastric (sympathetic) nerve inhibited transmission in the vesical ganglia and that application of exogenous catecholamines mimicked the effect of the nerve stimulation. The inhibitory effects of endogenous and exogenous catecholamines were blocked by α-adrenergic blocking agents. In an early study, Tum Suden and Marrazzi (1951) also observed that epinephrine depressed transmission in ciliary ganglia.

Intraarterial injection of cholinomimetic agents produced a biphasic depression of ganglionic transmission; both phases were blocked by atropine, but only the late depression was blocked by α-adrenergic blocking agents (DeGroat and Saum, 1971). DeGroat and Saum (1971) proposed that the late response was due to muscarinic activation of a peripheral adrenergic neuron, since transection of the hypogastric nerves did not affect the response. However, since atropine did not block the inhibitory effect of hypogastric nerve stimulation, they proposed that the peripheral adrenergic neuron, which they thought to be an SIF cell, must also be synaptically activated by nicotinic receptors. The situation is made even more complex by the observation that hypogastric nerve stimulation can excite the vesical neurons directly (DeGroat *et al.*, 1979). It appears that adrenergic inhibition may play an important physiological role, since stimulation of bladder afferents induces reflex firing in the hypogastric nerve and inhibits ganglionic transmission (DeGroat and Lalley, 1972; DeGroat and Theobald, 1976a).

A purinergic mechanism involving ATP has been proposed for postganglionic excitatory transmission to the bladder musculature (Burnstock, 1978; Burnstock *et al.*, 1978a,b). DeGroat and Theobald (1976b) were the first to implicate purinergic inhibition in bladder ganglia. They found that ATP administered intraarterially depressed ganglionic transmission at one tenth the dose required to excite bladder muscle; the depressant effects of ATP were partially postsynaptic and mediated through a P_1 receptor (DeGroat and Booth, 1980).

B. *In Vitro* Investigations

In extracellular studies in which the sucrose-gap technique was used to record inhibition *in vitro* in vesical pelvic ganglia (Griffith *et al.*, 1979b, 1981a), a slow IPSP was observed following blockade of nicotinic transmission by hexamethonium (5×10^{-4} M). The muscarinic antagonist atropine (10^{-7} M) blocked the slow IPSP response and unmasked a long afterhyperpolarization (LAH). The LAH was observed after high-frequency stimulation and persisted in a Ca^{2+}-free high-Mg^{2+} solution (Figure 1). Griffith *et al.* (1979b, 1981a) proposed that LAH is an electrical through-fiber response that results from repetitive nerve stimulation. Through-fiber responses have also been recorded from vesical ganglia *in vivo* (DeGroat and Saum, 1971).

Cholinomimetics acting on muscarinic receptors also hyperpolarized the vesical ganglion, but neither the β-antagonist L-propranolol (10^{-6} M) nor the α-antagonist phentolamine (10^{-6} M) affected the slow IPSP or the methacholine-induced hyperpolarization (Griffith *et al.*, 1979b, 1981a). It appears, therefore, that the slow IPSP recorded *in vitro* may correlate with the early muscarinic depression recorded *in vivo* in the cat vesical ganglion (DeGroat and Saum, 1972), but not with the slow IPSP of the sympathetic ganglia that according to Libet (Libet, 1970; Eccles and Libet, 1961) may be mediated by catecholamines released from the SIF cells (see, however, Chapters 3, 9, and 13).

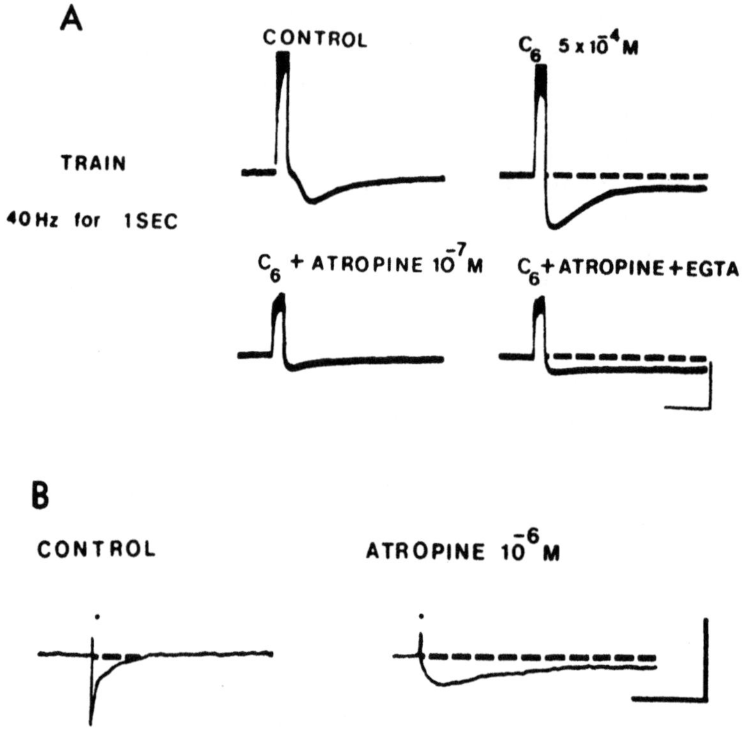

Figure 1. Sucrose-gap recordings of the slow IPSP and LAH. (A) In the control tracing, summation of the spike afterhyperpolarization is shown following train stimulation (40 Hz for 1 sec). After the addition of hexamethonium (C_6), both a slow IPSP and an LAH can be recorded. The slow IPSP was blocked by atropine, while the LAH was still present after atropine, and after block of synaptic transmission by an EGTA (1 mM) solution. Calibration: 1 mV × 4 sec. (B) Pen-chart recording of the slow IPSP and LAH indicates clearly the time-course of the LAH. The dot signifies train stimulation (40 Hz for 1 sec). Calibration: 1 mV × 1 min.

III. INTRACELLULAR STUDIES

A. Presynaptic Inhibition

In sympathetic ganglia, catecholamines inhibit acetylcholine (ACh) release (Christ and Nishi, 1971) (see Chapters 3 and 11), but relatively few intracellular studies have been carried out with regard to the effects of catecholamines on parasympathetic ganglia. It seems possible that presynaptic catecholaminergic inhibition may be important in vesical pelvic ganglia because of the preponderance of adrenergic fibers surrounding cholinergic principal ganglion cells and fibers, as well as adrenergic principal ganglion cells and SIF cells (Elbadawi and Schenk, 1973; Hamberger and Norberg, 1965). In addition, the vesical pelvic ganglion is the only autonomic ganglion in which endogenous adrenergic inhibition of ganglionic firing was demonstrated (DeGroat and Saum, 1971).

We have examined the effect of norepinephrine (NE) on vesical pelvic ganglia. The orthodromically induced action potential was depressed in 11 of 17 neurons superfused with NE (10^{-6}–10^{-4} M), but NE (10^{-5}–10^{-4} M) did not affect iontophoretically induced ACh depolarization (Figure 2). The effect of NE appeared to be independent of membrane potential because NE still depressed orthodromic spikes even when the membrane potential was maintained by anodal or cathodal current injection. Furthermore, DeGroat and Booth (1980) have shown that both NE and dopamine in concentrations ranging from 1×10^{-5} to 5×10^{-3} M depressed excitatory postsynaptic potential (EPSP) amplitude and the number of spikes elicited per stimulus of the pelvic nerve. Since NE did not affect the ACh potential, it is fairly clear that the inhibitory effect of NE on cell firing is due primarily to a presynaptic action. However, preliminary experiments in DeGroat's laboratory have not yet demonstrated an adrenergic presynaptic inhibitory response during stimulation of the hypogastric nerve *in vitro* (DeGroat *et al.*, 1979).

Recently, DeGroat and his associates presented evidence indicating that leucine-enkephalin may be involved in this presynaptic inhibition (Simonds *et al.*, 1983). Since enkephalins are present in preganglionic neurons of the sacral band and are transported peripherally in the sacral ventral roots (Glazer and Basbaum, 1980), DeGroat suggested that leucine-enkephalin may be an inhibitory cotransmitter in the sacral parasympathetic outflow that acts at the cholinergic nerve terminals to reduce ACh release.

B. Postsynaptic Inhibition

Many more data are available regarding postsynaptic than presynaptic inhibition in parasympathetic ganglia. It could be shown that cholinergic, adrenergic, purinergic, and GABAergic agonists directly hyperpolarize the ganglionic membrane potential when applied by superfusion.

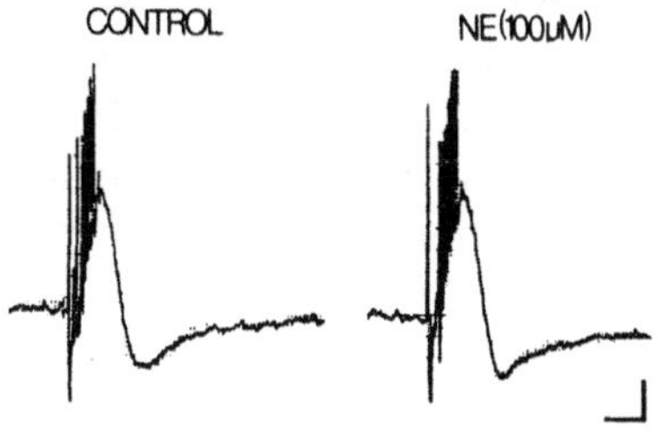

Figure 2. Effect of NE on depolarization induced by iontophoretically applied ACh (50 nA × 100 msec). NE (100 μM), applied by superfusion for 6 min, depolarized the resting membrane potential (3 mV), but had little effect on ACh-induced depolarization. Calibration: 5 mV × 400 msec.

1. Adrenergic Inhibition

Catecholamine effects on parasympathetic ganglia appear to be quite complex, as is the case with sympathetic ganglia (see Chapters 9, 11, 12, and 13). Suzuki and Volle (1979) reported that NE (10^{-4} M) depolarized the membrane and decreased membrane resistance in 11 of 13 cells of the hamster submandibular ganglion; in 4 of 5 cells, a hyperpolarization with an increased membrane resistance was observed. DeGroat and Booth (1980) reported that NE (10^{-5}–5×10^{-3} M) depolarized 70% of sensitive cells in cat vesical ganglia; in some cells, a biphasic effect was observed but in their hands, the postsynaptic responses to NE were not accompanied by a change in membrane resistance. The α-antagonist dihydroergotamine blocked the effects of NE on hamster submandibular (Suzuki and Volle, 1979) and cat vesical pelvic ganglia (DeGroat and Booth, 1980). In our laboratory, we also recorded complex effects of NE on cat vesical ganglia (Griffith *et al.*, 1979a). In 23 neurons, NE (10^{-6}–10^{-3} M) depolarized the membrane; in 19 neurons, NE (10^{-7}–10^{-4} M) induced a hyperpolarization; and in some neurons, no response was observed. Hyperpolarization was induced by lower concentrations of NE than those capable of causing depolarization, and phentolamine (10^{-6} M) antagonized the effects of NE. NE did not appear to cause any clear change in membrane resistance.

Further detailed studies of NE hyperpolarizations and depolarizations in cat vesical parasysmpathetic ganglia (Shinnick-Gallagher *et al.*, 1983; Akasu *et al.*, 1985; Nakamura *et al.*, 1984) indicated that the effects of NE are mediated by two subtypes of α-adrenoceptors; α_2-receptors appeared to underlie membrane hyperpolarization, while α_1-receptors activated membrane depolarization. Potassium channels seemed to be involved in the NE responses (Akasu *et al.*, 1985). The characteristics of these channels were analyzed further by means of pharmacological evaluation of NE responses and the study of the effect on these responses of the membrane voltage.

The NE hyperpolarization was accompanied by an increase in membrane conductance. NE hyperpolarizations became smaller as the membrane was hyperpolarized and reversed polarity beyond -100 mV. The NE hyperpolarization was blocked by yohimbine (1 μM), and α_2-adrenoceptor antagonist, by calcium antagonists, Cd, Mn, and Co, and by intracellular injection of EGTA; it was depressed by intracellular injection and extracellular superfusion of cesium (Cs), but not by superfusion of tetraethylammonium (TEA). These data are consistent with the hypothesis that α_2-adrenoceptor activation produces a membrane hyperpolarization that is mediated through a calcium-dependent potassium conductance (Akasu *et al.*, 1985).

Conversely, an NE deplorarization was associated with a decrease in membrane conductance and was blocked by prazosin (0.1–1 μM), an α_1-

adrenoceptor antagonist. NE deplorarization also became smaller at hyperpolarized membrane potentials, and their polarity was reversed at approximately −90 mV. In contrast to NE hyperpolarization, NE depolarization was not affected by low-Ca solution, calcium antagonists, or intracellular injection of EGTA, nor was it blocked by Cs or TEA, but it was depressed 70% by 10 μM barium. These data suggest that α_1-adrenoceptor activation mediates membrane deplorarization through closure of a voltage-insensitive potassium channel (Akasu *et al.*, 1985).

Thus, the complexity and the inconsistency of the effects of NE obtained previously in various laboratories may be explained if NE acts on two subtypes of α-adrenoceptors that mediate opposing conductance changes. Finally, it is interesting to note that in contrast to their effects on other parasympathetic mammalian ganglia, catecholamines (10^{-4} M) had no effect on the membrane potential of the cardiac parasympathetic ganglia of the mudpuppy (Hartzell *et al.*, 1977) or of cat colonic ganglia (DeGroat and Krier, 1976).

NE was also shown to affect spontaneously active neurons of cat vesical pelvic ganglia (DeGroat and Booth 1980; Griffith 1979a). These spontaneously active neurons exhibited normal resting potentials, membrane resistances, and action potentials, and continued firing for hours (Griffith *et al.*, 1980a). DeGroat and Booth (1980) reported that NE (10^{-5}–10^{-4} M) did not block spontaneous firing in 11 neurons; in 4 neurons, firing increased. In our hands (Griffith *et al.*, 1979a), NE decreased spon-

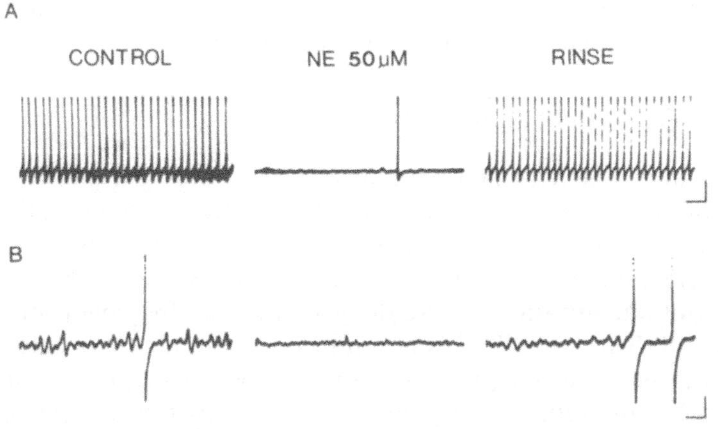

Figure 3. Effects of NE on spontaneously firing cells. (A) Spontaneous action potentials were recorded in the control physiological solution; they were reduced in frequency or completely eliminated in the presence of NE. Calibration: 20 mV × 0.5 sec. (B) In another cell, the small spontaneous membrane fluctuations were also reversibly depressed, as indicated in a recording carried out at higher sensitivity. Calibration: 5 mV × 0.2 sec.

taneous firing in 50% of the cells, increased firing in 25% of the cells, and effect on the remainder (Figure 3). A small change in the membrane potential of a rapidly firing neuron is difficult to measure; thus, in some cells, no correlation could be made between the effect of NE on membrane potential on the one hand and firing rate on the other. In other cells, however, decrease or increase in firing frequency reflected membrane hyperpolarization or depolarization, respectively. Because hyperpolarization of the membrane decreases firing frequency in these neurons while depolarization increases firing (Griffith *et al.*, 1980a, 1981b), it seems likely that the effects of NE on spontaneously firing neurons may correlate with its membrane effects.

Recently, while analyzing NE effects mediated through α_1- and α_2-adrenoceptor subtypes (Shinnick-Gallagher *et al.*, 1983; Nakamura *et al.*, 1984), we found that NE activation of α_2-adrenoceptors inhibited cell firing and that the NE-induced conductance increase, although rather small (0.5–6.5 nS), was probably sufficient to shunt the current responsible for the spontaneous firing (Akasu *et al.*, 1985). In contrast, NE depolarizations were often accompanied by membrane potential fluctuations, increased membrane "noise" in the baseline recording, and cell firing at the peak of the response (Akasu *et al.*, 1985), indicating that neuronal excitability was increased during activation of α_1-adrenoceptors.

Most important, DeGroat *et al.* (1979) have not yet been successful in demonstrating endogenous postsynaptic adrenergic inhibition during stimulation of the hypogastric nerve *in vitro*. On the other hand, cholinergic inhibition (i.e., a slow IPSP) has been demonstrated with respect to parasympathetic ganglia *in vitro* (Griffith *et al.*, 1979b, 1981a,b; Hartzell *et al.*, 1977), and it appears that cholinergic inhibition plays an important role in parasympathetic ganglionic transmission.

2. Cholinergic Inhibition

A slow IPSP was first demonstrated *in vitro* in a parasympathetic ganglion by Hartzell *et al.* (1977). These investigators showed that a slow IPSP lasting several seconds can be recorded from cardiac parasympathetic ganglion cells of the mudpuppy following the fast EPSP and that its magnitude and time–course depended on the frequency and number of stimuli. The slow IPSP was mimicked by iontophoretically applied ACh and bethanechol (BCh) and selectively blocked by atropine (5×10^{-9} M). Slow biphasic responses to BCh or ACh were apparently not present. ACh responses persisted in low-Ca^{2+}/high-Mg^{2+} solution and therefore were not evoked via an interneuron. It was further shown that both the slow IPSP and ACh hyperpolarization resulted from an increased potassium conductance. Hartzell *et al.* (1977) concluded that ACh released from preganglionic terminals acted directly on a muscarinic receptor of the postsynaptic membrane.

When applied to mammalian submandibular ganglia, BCh (2.5×10^{-5} M) depolarized some neurons and hyperpolarized others; biphasic responses were not recorded (Suzuki and Volle, 1979). Each response was associated with an increase in membrane resistance, was blocked by atropine (5×10^{-6} M), persisted in low-Ca^{2+}/high-Mg^{2+} solution, and was not antagonized by dihydroergotamine (1.5×10^{-6} M) (Suzuki and Volle, 1979). It is important to note that in submandibular parasympathetic ganglia of hamsters, responses to the cholinergic agonist applied exogenously were not correlated with an endogenous physiological response.

In cat vesical pelvic ganglia, both a slow IPSP and a slow EPSP were recorded in 31% of the neurons, whereas in 48% of the neurons only a slow IPSP was observed (Griffith *et al.*, 1980b, 1981b). Therefore, an endogenous cholinergic inhibitory response can be recorded from approximately 80% of vesical neurons (Griffith *et al.*, 1980b, 1981b). Both the slow IPSP and the slow EPSP could be mimicked by iontophoretic application of ACh. In cells where a slow IPSP as well as a slow EPSP were present, ACh would hyperpolarize and then depolarize the membrane. On the other hand, in cells in which only a slow IPSP was evoked by presynaptic stimulation, only a hyperpolarizing ACh action could be demonstrated. The cat vesical pelvic ganglion seems to be a unique mammalian parasympathetic ganglion, since it exhibits biphasic muscarinic responses.

In the vesical pelvic ganglia, the α-adrenoceptor antagonist phentolamine (10^{-6} M; $N = 6$) and the β-antagonist L-propranolol (10^{-6} M; $N = 4$) failed to block the slow IPSP (Griffith *et al.*, 1981b). These data further support the notion of a direct action of ACh on the postsynaptic membrane. Furthermore, the reversal potential of ACh hyperpolarization was shifted to 40 mV in high-potassium solution, indicating that ACh hyperpolarization and the slow IPSP are due to an increased potassium conductance (Gallagher *et al.*, 1982); this finding confirms the results of Hartzell *et al.* (1977). Thus, it seems that the muscarinic hyperpolarization of mammalian cat vesical pelvic ganglia (Gallagher *et al.*, 1982) is similar to that of amphibian parasympathetic ganglia (Hartzell *et al.*, 1977), since both seem to be due to decreased membrane resistance; however, it is different from that of hamster submandibular ganglia, in which BCh causes an increase in membrane resistance (Suzuki and Volle, 1979).

An important question to be asked is: What is the physiological role of the slow IPSP in parasympathetic ganglionic transmission? It must be noted that the function of slow potentials in ganglionic transmission has not been defined in most cases (Skok, 1973). This question could be conveniently addressed in the case of cat vesical pelvic ganglia, since the slow inhibitory muscarinic component of transmission is quite large in these ganglia; thus, the effects of the slow IPSP on the firing of spontaneously active neurons could be studied (Griffith *et al.*, 1981b). These neurons fire spontaneously in the presence of nicotinic and muscarinic antagonists, when the ganglion is bathed in low-Ca^{2+}/high-Mg^{2+} solution.

In cells exhibiting only a slow IPSP, spontaneous activity was inhibited by presynaptic stimulation whereas in neurons in which both a slow IPSP and a slow EPSP could be evoked, spontaneous action potential frequency was first decreased and then increased by presynaptic stimulation (Griffith *et al.*, 1981b; Gallagher *et al.*, 1982). Furthermore, exogenously applied ACh and BCh mimicked the effect of the slow potentials on spontaneous firing. Thus, it appears that the slow IPSP and slow EPSP indeed alter ganglionic firing of spontaneously active neurons and that the slow synaptic potentials may alter transmission or excitability (see also Chapters 3 and 13).

It should be mentioned that spontaneous hyperpolarizing potentials have been recorded in amphibian cardiac parasympathetic ganglia (Hartzell *et al.*, 1977) and that these potentials may be due to changes in K^+ conductance, since they disappear with hyperpolarization and increase in size with depolarization. In this context, we have recorded spontaneous subthreshold potentials in cat vesical pelvic ganglia. These potentials likewise disappear with hyperpolarization; however, they appear to be triphasic in nature (Griffith *et al.*, 1980a). Spontaneous hyperpolarizing potentials due to a Ca^{2+}-mediated K^+-conductance increase appear to be present in the hamster submandibular ganglion (Suzuki and Kusano, 1978). These potentials differ from those recorded in the amphibian cardiac ganglion (Hartzell *et al.*, 1977) and cat vesical pelvic ganglion (Griffith *et al.*, 1980a), since in the hamster they were long and slow, occurred at low and irregular frequencies, and had a long duration (3500 msec).

3. GABAergic Inhibition

Cat vesical pelvic ganglia exhibited a biphasic response to γ-aminobutyric acid (GABA) (Mayer *et al.*, 1981). The response consisted of an initial depolarization associated with a conductance increase followed by a delayed hyperpolarization associated with a conductance decrease. The ionic mechanism for the initial depolarization appears to be identical with that shared with many other GABA-receptor-mediated responses, namely, an increase in chloride conductance. On the other hand, a decreased chloride conductance appears mainly responsible for the subsequent, hyperpolarizing response (Mayer *et al.*, 1983). The threshold for directly induced action potential is increased during the hyperpolarization, which appears, then, to play an inhibitory role with respect to transmission.

4. Purinergic Inhibition

Earlier *in situ* experiments with vesical parasympathetic ganglia suggested that a purinergic modulatory mechanism may control urinary bladder function. The presence of a noncholinergic, nonadrenergic nerve supply to the urinary bladder has been postulated (Burnstock *et al.*, 1978a,b) and termed purinergic (Burnstock, 1978), since these nerves have been shown to release ATP (Burnstock *et al.*, 1978b). Furthermore, quinacrine-positive purinergic ganglion cell bodies, nerve fibers, and nerve bundles were observed in the guinea pig urinary bladder, where ATP is thought to mediate atropine-resistant smooth muscle contraction (Burnstock *et al.*, 1978a). Because of the possible neurotransmitter function of the purines, we recently tested the chemosensitivity of vesical parasympathetic neurons to the effects of ATP and related compounds (Shinnick-Gallagher *et al.*, 1984; Akasu *et al.*, 1984).

Bath application (100 nM–1 mM) and pressure application of ATP produced a fast depolarizing (D) response in 52% of the neurons. In some neurons (10%), the fast depolarization was followed by a slow hyperpolarization lasting 1–1.5 min. Bath application (500 nM–1 mM) and pressure application of adenosine produced a hyperpolarizing (H) response in most parasympathetic neurons (92%). The duration of the H response ranged from 1 to 1.5 min.

The relative order of potency for producing a D response was ATP > ADP >> AMP > adenosine, whereas that for producing the H response was 2-chloroadenosine > adenosine > AMP >> ADP > ATP. Caffeine (1 mM) selectively blocked the H response but not the D response, suggesting that the H response is mediated by a P_1 purinoceptor (Burnstock, 1978).

The ATP depolarization was associated with a decrease in membrane resistance, reversed polarity at -7.0 mV, and was dependent on the concentration of Na^+ and K^+ ions. On the other hand, the adenosine hyperpolarization was accompanied by a decrease in membrane resistance; polarity was reversed at -98 mV, and the reversal was dependent on the concentration of extracellular K^+ not Na^+ ions. These results suggested that the ATP depolarization was mediated through activation of Na^+ and K^+ conductance, while the adenosine hyperpolarization was produced by activation of K^+ conductance (Akasu *et al.*, 1984; Shinnick-Gallagher *et al.*, 1984). Furthermore, spontaneous firing of action potentials recorded in cat parasympathetic neurons at their resting membrane potential (-50 to -60 mV) was facilitated during an ATP depolarization and inhibited during an adenosine hyperpolarization, suggesting that the purines may modulate the excitability of cat vesical parasympathetic neurons.

Subsequently, we recorded in the presence of nicotinic and muscarinic antagonists a slow hyperpolarizing synaptic potential (HSP) with high-intensity (8–30 V, 40 Hz, 250 msec) stimulation of the preganglionic nerve trunk (Akasu *et al.*, 1984). The noncholinergic slow HSP had an amplitude of about 5 mV and a duration of 30 sec. Both the adenosine hyperpolarization and the slow HSP were blocked by caffeine (1 mM), a P_1-purinoceptor antagonist (Burnstock, 1978). Adenosine deaminase, the enzyme that catabolizes adenosine to inosine and NH_3, depressed the amplitude of the slow HSP by 80% and the adenosine hyperpolarization by 90%. Conversely, dipyridamole, which blocks the uptake of adenosine, increased the amplitude and duration of the slow HSP and the adenosine hyperpolarization. Both the slow HSP and the adenosine hyperpolarization were associated with a decrease in membrane resistance, became smaller as the membrane was hyperpolarized, and eventually reversed their polarity at a membrane potential of −94 mV. These data satisfied the pharmacological and electrophysiological criteria establishing adenosine as the neurotransmitter mediating the slow HSP in vesical neurons (Akasu *et al.*, 1984; Akasu, Shinnick-Gallagher, and Gallagher, unpublished).

5. Electrogenic Inhibition

Suzuki and Kusano (1978) used the hamster submandibular parasympathetic ganglion to study two electrogenic types of hyperpolarizing potentials, the spike-induced hyperpolarizing afterpotential and the hyperpolarizing phase of the synaptic potential. The reversal potential of either response approximated the potassium equlibrium potential; both responses seemed to be due to a Ca^{2+}-mediated K^+-conductance increase. Suzuki and Kusano (1978) felt that these potentials contribute to the regulation of low-frequency repetitive firing in hamster submandibular ganglion cells.

IV. CONCLUSIONS

These studies indicate that integration and modulation are possible in parasympathetic ganglia and may in fact be at least as complex as the mechanisms observed in sympathetic ganglia. Indeed, both pre- and postsynaptic mechanisms are involved, as well as interaction between sympathetic and parasympathetic neurons (see also Chapter 14). Furthermore, besides ACh, the catecholamines, GABA, and the purines, ATP and adenosine exert modulatory effects on transmission of parasympathetic ganglia. In this context, it is of interest that cholinergic, adrenergic, purinergic, and GABAergic agonists seem to be capable of direct hyperpolarizing action on the postsynaptic membrane. The physiological role of these inhibitory potentials is not completely clear, but evidence as to their effect on neuronal excitability, spontaneous firing, and other processes has been adduced.

ACKNOWLEDGMENTS We thank Drs. Betty Williams and A. G. Karczmar for their careful review of the manuscript. The authors' research reported herein was supported in part by Grant NS16228.

REFERENCES

Akasu, T., Shinnick-Gallagher, P., and Gallagher, J. P.: Adenosine mediates a slow hyperpolarizing synaptic potential in autonomic neurons. *Nature (London)* **311**:62–65 (1984).

Akasu, T., Gallagher, J. P., Nakamura, T., Shinnick-Gallagher, P., and Yoshimura, M.: Noradrenaline hyperpolarization and depolarization in cat vesical parasympathetic neurons. *J. Physiol. (London)* **361**:165–184 (1985).

Blackman, J. G., Crowcroft, P. F., Devine, C. E., Holman, M. E., and Yonemura, K.: Transmission from preganglionic fibers in the hypogastric nerve to peripheral ganglia of male guinea-pigs. *J. Physiol. (London)* **201**:723–743 (1969).

Burnstock, G. A.: Basis for distinguishing two types of purinergic receptor, in: *Cell Membrane Receptors for Drugs and Hormones: A Multidisciplinary Approach* (L. Bolis and R. W. Straub, eds.), pp. 107–108, New York, Raven Press (1978).

Burnstock, G., Cocks, T., Crowe, R., and Kasakov, L.: Purinergic innervation of the guinea-pig urinary bladder. *Br. J. Pharmacol.* **61**:125–138 (1978a).

Burnstock, G., Cocks, T., Kasakov, L., and Wong, N. K.: Direct evidence for ATP release from non-adrenergic, mon-cholinergic ("purinergic") nerves in the guinea pig taenia coli and bladder. Eur. *J. Pharmacol.* **49**:145–149 (1978b).

Christ, D. D., and Nishi, S.: Site of adrenaline blockade in the superior cervical ganglion of the rabbit. *J. Physiol. (London)* **213**:107–117 (1971).

Dail, W. C.: Histochemical and fine structure studies of SIF cells in the major pelvic ganglion of the rat, in: *SIF Cells: Structure and Function of the Small Intensely Fluorescent Sympathetic Cells*, (O. Eranko, ed.), Fogarty International Center Proceedings, No. 30, pp. 8–18, U.S. Government Printing Office, Washington, D.C. (1976).

DeGroat, W. C., and Booth, A. M.: Inhibition and facilitation in parasympathetic ganglia of the urinary bladder. *Fed. Proc. Fed. Am. Soc. Exp. Biol.* **39**:2990–2996 (1980).

DeGroat, W. C., and Krier, J.: An electrophysiological study of the sacral parasympathetic pathway to the colon of the cat. *J. Physiol. (London)* **260**:425–445 (1976).

DeGroat, W. C., and Lalley, P. M.: Reflex firing in the lumbar sympathetic outflow to activation of vesical afferent fibers. *J. Physiol. (London)* **226**:289–309 (1972).

DeGroat, W. C., and Saum, W. R.: Adrenergic inhibition in mammalian parasympathetic ganglia. *Nature (London)* **213**:188–189 (1971).

DeGroat, W. C., and Saum, W. R.: Sympathetic inhibition of the urinary bladder and pelvic transmission in the cat. *J. Physiol. (London)* **220**:297–314 (1972).

DeGroat, W. C., and Theobald, R. J.: Sympathetic inhibitor reflexes to the urinary bladder and bladder ganglia evoked by electrical stimulation of vesical afferents. *J. Physiol. (London)* **259**:223–237 (1976a).

DeGroat, W. C., and Theobald, R. J.: Effects of ATP, cyclic AMP and related nucleotides on transmission in parasympathetic ganglia. *Pharmacologist* **18**:185 (1976b).

DeGroat, W. C., Booth, A. M., and Krier,J.: Interaction between sacral parasympathetic and lumbar sympathetic inputs to pelvic ganglia, in: *Integrative Functions of the Autonomic Nervous Systems* (C. McC Brooks, K. Koizumi, and A. Sato, eds.), pp. 234–247, Elsevier/North-Holland, Amsterdam (1979).

Dennis, M. J., Harris, A. J., and Kuffler, S. W.: Synaptic transmission and its duplication by focally applied acetylcholine in parasympathetic neurons in the heart of the frog. *Proc. R. Soc. London Ser. B.* **171**:509–539 (1971).

Eccles, R. M., and Libet, B.: Origin and blockade of the synaptic responses of curarized sympathetic ganglia. *J. Physiol. (London)* **157**:484–503 (1961).

Elbadawi, A., and Schenk, E. A.: A new theory of the innervation of bladder musculature. Part I. Morphology of the intrinsic vesical innervation apparatus. *J. Urol.* **99**:585–587 (1968).

Elbadawi, A., and Schenk, E. A.: Parasympathetic and sympathetic postganglionic synapses in ureterovesical autonomic pathways. *Z. Zellforsch.* **146**:147–154 (1973).

Gallagher, J. P., Griffith, W. H., III, and Shinnick-Gallagher, P.: Cholinergic transmission in cat parasympathetic neurons. *J. Physiol. (London)* **332**:473–486 (1982).

Glazer, E. J., and Basbaum, A. I.: Leucine enkephalin: Localization in and axoplasmic transport by sacral parasympathetic preganglionic neurons. *Science* **208**:1479–1481 (1980).

Griffith, W. H., III, Gallagher, J. P., and Shinnick-Gallagher, P.: Action of norepinephrine on neuronal activity recorded from cat parasympathetic ganglia. *Fed. Proc. Fed. Am. Soc. Exp. Biol.* **38**:272 (1979a).

Griffith, W. H., III, Gallagher, J. P., and Shinnick-Gallagher, P.: Inhibitory potentials recorded from mammalian parasympathetic ganglia. *Soc. Neurosci. Abstr.* **5**:2498 (1979b).

Griffith, W. H., III, Gallagher, J. P., and Shinnick-Gallagher, P.: An intracellular investigation of cat vesical pelvic ganglia. *J. Neurophysiol.* **43**:343–354 (1980a).

Griffith, W. H., III, Gallagher, J. P., and Shinnick-Gallagher, P.: Cholinergic slow potentials influence parasympathetic ganglionic transmission. *Soc. Neurosci. Abstr.* **6**:68 (1980b).

Griffith, W. H., III, Gallagher, J. P., and Shinnick-Gallagher, P.: Sucrose gap recordings of nerve-evoked potentials in mammalian parasympathetic ganglia. *Brain Res.* **209**:446–451 (1981a).

Griffith, W. H., III, Gallagher, J. P., and Shinnick-Gallagher, P.: Mammalian parasympathetic ganglia fire spontaneous action potentials and transmit slow synaptic potentials, in: Adv. Physiol. Sci., Vol. 4, *Physiology of Excitable Membranes* (J. Salanki, ed.), pp. 347–350, Oxford, Pergamon Press (1981b).

Hamberger, B., and Norberg, K.-A.: Adrenergic synaptic terminals and nerve cells in bladder ganglia of the cat. *Int. J. Neuropharmacol.* **4**:44 (1965).

Hartzell, H. C., Kuffler, S. W., Stickgold, R., and Yoshikami, D.: Synaptic excitation and inhibition resulting from direct action of acetylcholine on two types of chemoreceptors on individual amphibian parasympathetic neurons. *J. Physiol. (London)* **271**:817–846 (1977).

Johnson, D. A., Beach, R., Alanis, J., and Pilar, G.: Presynaptic receptors may regulate acetylcholine release in parasympathetic ganglia. *Soc. for Neurosci. Abstr.* **7**:279 (1977).

Kuntz, A., and Mosely, R. L.: An experimental analysis of the pelvic autonomic ganglia in the cat. *J. Comp. Neurol.* **64**:63–74 (1936).

Langley, J. N., and Anderson, H. K.: The innervation of the pelvic and adjoining viscera. Part II. The bladder. *J. Physiol. (London)* **19**:71–84 (1895).

Langley, J. N., and Anderson, H. K.: The innervation of the pelvic and adjoining viscera. V. Position of the nerve cells on the course of the efferent nerve fibers. *J. Physiol. (London)* **19**:131–139 (1896).

Libet, B.: Generation of slow inhibitory and excitatory postsynaptic potentials. *Fed. Proc. Fed. Am. Soc. Exp. Biol.* **29**:1945 (1970).

Lichtman, J. W.: The reorganization of synaptic connexions in the rat submandibular ganglion during post-natal development. *J. Physiol.* **273**:155–177 (1977).

Mayer, M. L., Higashi, H., Shinnick-Gallagher, P, and Gallagher, J. P.: A hyperpolarizing GABA response associated with a conductance decrease. *Brain Res.* **222**:204–208 (1981).

Mayer, M. L., Higashi, H., Gallagher, J. P., and Shinnick-Gallagher, P.: On the mechanism of action of GABA in pelvic vesical ganglia: Biphasic response voked by two opposing actions on membrane conductance. *Brain Res.* **260**:233–148 (1983).

Nakamura, T., Yoshimura, M., Shinnick-Gallagher, P., Gallagher, J. P., and Akasu, T.: Alpha$_2$ and alpha$_1$ adrenoreceptors mediate opposing actions on parasympathetic neurons. *Brain Res.* **323**:349–353 (1984).

Shinnick-Gallagher, P., Nakamura, T., Yoshimura, M., and Gallagher, J.: Norepinephrine-induced hyperpolarization and depolarizations in parasympathetic neurons are mediated through alpha$_2$ and alpha$_1$ adrenoceptors, respectively. *Soc. Neurosci. Abstr.* **9**:1000 (1983).

Shinnick-Gallagher, P., Akasu, T., and Gallagher, J. P.: P_2 and P_1 purinoceptors mediate neuronal depolarization and hyperpolarization in parasympathetic neurons. 9th International Congress of Pharmacology, IUPHAR Meeting, London **9**:1835P (1984).

Simonds, W. F., Booth, A. M., Thor, K. B., Ostrowski, N. C., Nagel, J. R., and deGroat, W. C.: Parasympathetic ganglia: naloxone antagonizes inhibition by leucine-enkephalin and GABA, *Brain Research* **271**:365–370 (1983).

Skok, V. I.: *Physiology of Autonomic Ganglia*. Igaku Shoin, Tokyo (1973).

Suzuki, T., and Kusano, K.: Hyperpolarizing potentials induced by Ca-mediated K-conductance increase in hamster submandibular ganglion cells. *J. Neurobiol.* **9**:367–392 (1978).

Suzuki, T, and Volle, R. L.: Nicotinic, muscarinic and adrenergic receptors in a parasympathetic ganglion. *J. Pharmacol. Exp. Ther.* **211**:252–256 (1979).

Tum Suden, C., and Marrazzi, A. S.: Synaptic inhibitory action of adrenaline at parasympathetic synapses. *Fed. Proc. Fed. Am. Soc. Exp. Biol.* **10**:138 (1951).

16

Transmission in Enteric Ganglia

J. P. HODGKISS and G. M. LEES

I. INTRODUCTION

There can be few topics in physiology that have attracted such fluctuating interest as the neural control of intestinal motility. Recently, there has been much renewed interest in this deceptively simple nervous system, mainly as a result of the successful application of electrophysiological techniques (see below) and histochemical and immunohistochemical methods (see Chapter I). Moreover, different aspects of the subject have formed the basis of several thoughtful reviews, to which the reader is referred (Hirst, 1979; Gershon, 1981; Holman, 1981; Wood, 1981; North, 1982a,b; Szerb, 1982; Llewellyn-Smith *et al.*, 1983).

A detailed study of the different types of neurons of the enteric nervous system is the focus of this chapter, since a knowledge of their properties will help to explain intrinsic reflexes and drug action in the intestine. Almost all the information currently available relates to the characteristics of enteric neurons of the small intestine (duodenum, jejunum, and ileum). Caution should therefore be exercised in extrapolating the information to the neurons of the large intestine, since it is known, for example, that the organization of inhibition in the small intestine differs from that in the large bowel (Costa and Furness, 1976, 1982). Furthermore, recent preliminary electrophysiological studies of the enteric

---

J. P. HODGKISS • Department of Ethology, Agricultural and Food Research Council-Poultry Research Centre, Roslin, Midlothian, Scotland. G. M. LEES • Department of Pharmacology, Marischal College, University of Aberdeen, Aberdeen, Scotland.

neurons in parafascicular ganglia of the stomach have failed to reveal the presence of an AH-type (see below) neuron (King and Szurszewski, 1984).

Since the enteric neurons play a key role in the control of peristalsis (Kosterlitz, 1968), which is itself essential for health, it is pertinent to collate information concerning the activity of individual neurons. This is necessary in order to identify pathways within the myenteric (Auerbach's) and submucous (Meissener's) plexuses involved in integrated neuronal activity associated with particular physiological functions of the gut. As in the central nervous system, an essential preliminary is the establishment of the functions and synaptic connections of individual neurons or groups of neurons. The problem of identifying the functions of particular enteric neurons is formidable, and the architecture of the plexuses gives no clues to their likely connections. However, together with information obtained from clsssic experimental procedures (Kosterlitz and Lees, 1964; Kosterlitz, 1968; Kottegoda, 1969; Frigo *et al.*, 1972; Costa and Furness, 1976, 1979, 1982, Szerb, 1982), the recognition of different classes of neurons and of their specific electrophysiological, neurochemical, and pharmacological properties will provide a sound basis for future experiments that are needed to define functional pathways involved in reflex responses.

II. ELECTROPHYSIOLOGICAL CLASSIFICATIONS OF NEURONS

A. Extracellular Recording

Extracellular studies of myenteric and submucous plexus neurons have shown that many enteric neurons are spontaneously active (Wood, 1970; Ohkawa and Prosser, 1972a,b; Nozdrachev and Vataev, 1981). In these experiments, individual ganglia were visualized by staining the preparation with methylene blue; however, this vital stain is said not to affect the electrical properties (Wood, 1970) or to exert an inhibitory effect on these neurons (Nozdrachev and Vataev, 1981). Furthermore, spontaneous activity is also seen in the absence of methylene blue (North and Williams, 1977), but North and Williams (1977) provided evidence that the recorded "spontaneous" electrical activity had been induced by suction pressure. For a summary of excitatory and inhibitory effects of gut peptides on the firing of myenteric neurons studied with this method, the reader should consult North *et al.* (1980b).

Wood (1975) devised a classification of myenteric plexus neurons based on their pattern of discharge. Three types of units can be recognized: (1) units firing in distinct bursts either regularly or irregularly, sometimes with more than one unit contributing to a burst (Wood *et al.*, 1979); (2) units firing single spikes irregularly at a low frequency; and (3) mechanosensitive units that may be either rapidly or slowly adapting and that respond to mechanical displacement of the ganglion within which their receptive field lies. As a further subdivision of category (3), there are tonic-type neurons that exhibit a fixed-pattern, prolonged discharge of action potentials in response to a transient distortion of the ganglion.

Burst and single-spike unit activity can also be recorded from the submucous plexus, where about 50% of neurons are spontaneously active *in vitro* (Nozdrachev and Vataev, 1981).

Comparisons have been made between the discharge patterns recorded extracellularly and those recorded with intracellular microelectrodes (Wood and Mayer, 1978a). It is thought that AH-neurons exhibiting slow excitatory postsynaptic potentials (see Section II. B for nomenclature) are equivalent to the tonic-type neurons seen with extracellular recording methods and that, as are the latter, they are activated by inputs from mechanoreceptors. Such neurons may be important in gating excitation in the myenteric plexus (Wood and Mayer, 1979a).

B. Intracellular Recording

With the increasing application of the skills associated with intracellular recording to the autonomic sympathetic ganglia, it was not unexpected that there should be simultaneous, independent investigations of myenteric plexus neurons with intracellular microeelectrode techniques. Two main categories of neurons were identified by means of microelectrode methods in the myenteric plexus of guinea pig duodenum (Holman *et al.*, 1972). In about two thirds of excitable cells, an excitatory postsynaptic potential (fast EPSP) could be evoked by transmural electrical stimulation of the isolated preparation; increasing stimulus intensity evoked a series of fast EPSPs that tended to summate and to initiate an action potential (Figure 1). In this type of neuron, the action potential was followed only by a brief (40-msec) positive afterpotential. The remainder of the excitable cells had no demonstrable fast EPSPs, but exhibited a property unique among autonomic ganglion neurons, namely, a secondary hyperpolarization of up to 20 sec in duration consequent and dependent on the firing of a soma action potential; this long hyperpolarization followed the initial, brief positive afterpotential. After further investigations confirmed these preliminary observations, these two types of neurons were termed appropriately S-cells or S-neurons, to account for their synaptic (S) input, i.e., the presence of fast EPSPs, and AH-cells or AH-neurons, thus called by virtue of the long-lasting after hyperpolarization (Hirst *et al.*, 1974).

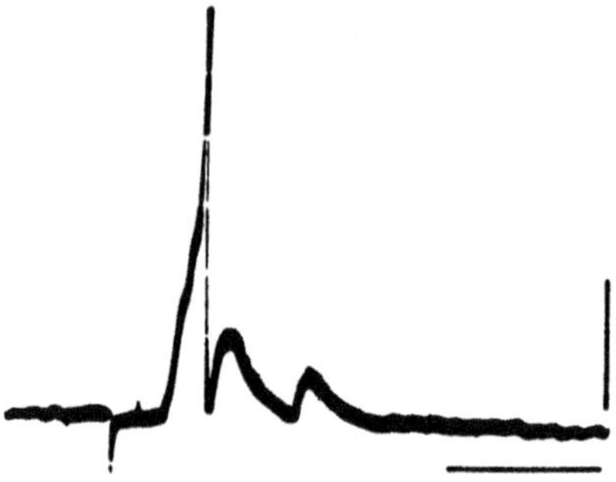

Figure 1. Responses of an S-neuron to a single indirect (transmural) stimulus applied at a point orad to the ganglion in which the soma was located. This myenteric plexus neuron from guinea pig ileum had at least three excitatory synaptic inputs. Calibration bars: 20 mV and 50 msec.

At the same time, Nishi and North (1973) also described two types of myenteric plexus neurons in the guinea pig ileum. The first type exhibited a relatively high membrane excitability that allowed it to generate many action potentials at a high frequency during a maintained suprathreshold depolarization (Nishi and North, 1973; North and Nishi, 1974). Accommodation was not a pronounced feature of this class of neuron; the threshold current ($1–3 \times 10^{-10}$ A) (Hirst *et al.*, 1974) was increased for only 100 msec following a single direct action potential. Remarkably, recovery from repetitive firing of 150–200 action potentials at up to 150 Hz (induced by strong depolarizing current pulses) took only 2–3 min (Nishi and North, 1973). Associated with the firing of each soma action potential was a brief (100-msec) positive afterpotential, which was also seen following antidromic spikes. In the latter case, the afterpotentials were up to 15 mV in amplitude and were due to an increased potassium conductance (g) (Nishi and North, 1973); posttetanic hyperpolarizations of up to 15 mV and 100 msec in duration were also recorded following antidromic invasion of the soma by focal stimulation applied at a distance of up to 150 μm from the soma of the cells in question. Nishi and North (Nishi and North, 1973; North and Nishi, 1974) referred to these cells as type 1 cells.

The second major class of neurons described by Nishi and North (Nishi and North, 1973; North and Nishi, 1974) included neurons that were much less excitable, and exhibited somewhat different passive membrane properties as compared to type 1 neurons (see below); further, unlike most type 1 cells, many (about 65%) of these type 2 cells exhibited a long-lasting afterhyperpolarization (slow AH) following a direct or antidromically evoked soma spike and its positive afterpotential (Figure 2). As concluded by North and Nishi (1976), this kind of neuron appears to be identical with the AH-neurons.

The designations "S" and "AH" are preferable to the designations type "type 1" and "type 2" for classifying enteric neurons because among the criteria (Nishi and North, 1973; North and Nishi, 1974, 1976) used to

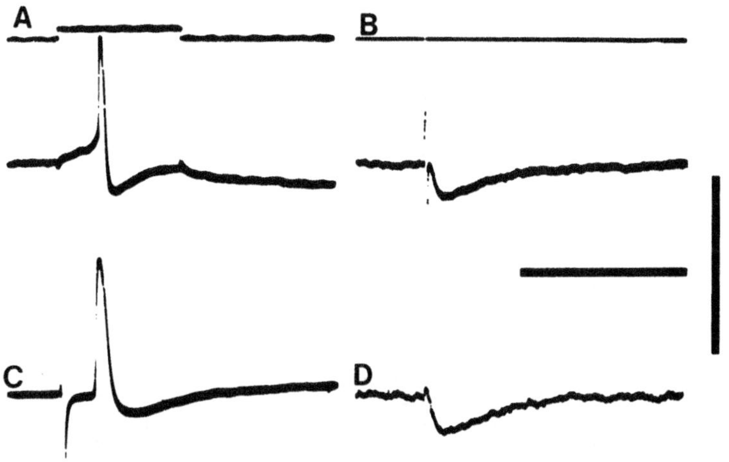

Figure 2. Intracellular records from an AH-neuron of the myenteric plexus of guinea pig ileum (E_m −60 mV) showing soma action potentials elicited by either direct depolarization of the soma (A) or by antidromic invasion (B). In this latter case, the transmural stimulating electrodes were 7 mm orad to the impaled neuron. In both cases, the action potential was followed by a slow AH. Calibration: vertical bar: 100 mV (A, C); 50 mV (B, D); 1.25 nA (A, B); horizontal bar: 25 msec (C); 50 msec (A), 2.5 sec (B, D).

distinguish type 1 from type 2 cells, the occurrence of fast EPSPs and the absence of the slow AH were not explicitly stated or obligatory.

It should be added that some investigators feel that the various types of cells actually represent various states or phases of excitability rather than different neurons [see Section III.G and Grafe *et al.* (1978)]. It appears that there is no intramural AH-neuron in the stomach (King and Szurszewski, 1984), although it is not altogether clear that the slow inhibitory potential described by these authors is not the slow AH.

III. MYENTERIC PLEXUS NEURONS

A. Electrical Properties

The apparent resting membrane potential (E_m) of AH-neurons was either the same as that of S-neurons (Hirst *et al.*, 1974) or, more likely, higher (Nishi and North, 1973; Wood and Mayer, 1978a; Grafe *et al.*, 1980; Hodgkiss and Lees, 1983 (see Table 1). The high membrane potential of AH-neurons may be due to a calcium-associated g_K (Grafe *et al.*, 1980). In practical terms, however, any one neuron cannot be typed on the basis of either its apparent resting potential or its membrane time constant (τ_m)

Table 1. Comparison of Electrophysiological Characteristics of Myenteric Plexus Neurons

Investigators	Cell type	E_m (mV)	τ_M (msec)	R_{in} (MΩ)	C_m Mean total (pF)	C_m Specific membrane (F/cm^2)
Nishi and North (1973)	1	−45 to −60	5.2	58 (12–140)	107	—
	2	−55 to −75	2.0	21	—	—
Hirst *et al*. (1974), Hirst and McKirdy (1974a)	S AH	−40 to −60	20	125–250	—	—
Wood and Mayer (1978a), Hodgkiss (1981)	All	−35 to −84	5.0	10–185	—	—
	Cateogry 3 (equivalent to type 1)	(−35 to −45)	—	(60–180)	—	—
	Category 1 (equivalent to type 2)	(−60 to −65)	—	Not lower	—	—
	AH	−59	—	Lower	—	—
	AH	E_m −51.3 (mean)				
Hodgkiss and Lees (1983 and unpublished)	S	−34 to −80 (mean −51)	7.9	153 (43–368)	73	18
	AH	−36 to −84 (mean −55)[a]	12.6[a]	150+ (55− 368)	91[b]	28[b]

[a] Significantly different, $p < 0.05$. [b]Not significantly different.

(this constant represents the time required by small hyperpolarizing current pulses to reach 1/e of the value of the displacement of the resting potential). The time constant of AH-neurons has been variously reported as being the same as (Hirst *et al.*, 1974), smaller than (Nishi and North, 1973; Wood and Mayer, 1978a), or significantly larger than (Hodgkiss and Lees, 1983) that of S-neurons (see Table 1). The large disparity between the reported values of τ_m is not easily explained by differences in equipment used, composition of the saline solution employed, or temperature of the bathing medium. Since isoprenaline does not affect the passive membrane properties of myenteric neurons (Hirst and Spence, 1973; Hirst *et al.*, 1974), the occasional use by Hirst *et al.* (1974) and by Hodgkiss and Lees (1983) of isoprenaline (employed to prevent the underlying longitudinal muscle from contracting excessively either spontaneously or on transmural electrical stimulation) cannot explain the high values of τ_m reported by these investigators. A consistent finding in our laboratory has been that the time constant of an AH-neuron increases in the course of the experiment. Thus, the earlier reports of an exceedingly brief time constant may refer to τ_m measurements carried out too soon after impalement.

Likewise, there have been discrepancies with respect to the estimates of the input resistance (R_{in}) of AH- and S-neurons, much lower values being initially reported for AH-neurons (see Table 1). The discrepancies in values of τ_m and R_{in} are probably closely related to the problem of recognition of AH-neurons, that is, of distinguishing them from type 3 cells, which may not be neurons at all, but rather glial cells (Nishi and North, 1973) (see below).

An important practical consideration is the nonlinearity of the current–voltage relationship. The neurons in question show anomalous rectification (Holman *et al.*, 1972; Hirst *et al.*, 1974; Nishi and North, 1973); they may also show delayed rectification (Nishi and North, 1973). Since the current–voltage relationship is linear only for small currents (less than 1 nA) producing a ± 10 mV change in E_m, changes in R_{in} should be assessed by means of hyperpolarizing electrotonic potentials of not more than 10 mV in amplitude.

While all authors are agreed that the excitability of S-neurons is relatively high, there is an apparent lack of unanimity concerning the excitability of other cells, and this has led to a certain measure of difficulty in the identification of the type of cell impaled. Although an S-neuron can fire many action potentials in reponse to continued, direct depolarization of its membrane, this must not be used as the sole criterion for its classification. Initially, it was reported that certain cells are impaled two to three times more frequently than excitable cells; the cells in question were characterized by a very high resting potential (-70 to -90 mV), a very low R_{in} (about 9 MΩ), and a very short τ_m (0.5 msec); also, these cells did not show any excitation in response to anodal break and did not generate action potentials in response to direct or indirect stimulation with single or repetitive stimuli (Nishi and North, 1973). The cells in question were probably glial cells, although there is a faint possibility of their being nonspiking interneurons (Shepherd, 1981).

In the myenteric plexus, estimates by various investigators of the fraction of impaled cells exhibiting a slow AH (and therefore being by definition AH-neurons) vary from 6 to 62% (see Table 2). One of the reasons for this variability may be that some AH-neurons do not show action potentials and slow afterhyperpolarizations for many minutes after impalement (Hodgkiss and Lees, 1978, 1980, 1983). Another reason may be that for some neurons, E_k is very near or equal to E_m; a slow AH in such cells would not be readily recognized, although it could be seen if E_m were shifted away from E_K or if E_K were increased by lowering the extracellular potassium ion concentration (Nishi and North, 1973).

A number of interesting variations in the amplitude and time–course of the slow AH have been reported. Although the amplitude may be constant over about 90 min (Hirst *et al.*, 1974), marked changes are sometimes

Table 2. Frequency Distribution of Different Electrophysiological Cell Types in Myenteric Plexus

Investigators	Cell type[a]	Proportion of neurons (%)	Proportion of all cells (%)
Holman *et al.* (1972)	S	66	—
Hirst *et al.* (1974)	AH	33	—
Nishi and North (1973), North (1973),	1	60	15–20
North and Nishi (1974)	2	40	10–13
	AH	25	6–8
	3 (no spikes)	—	66–75
Hirst and McKirdy (1974a)	S	88.6	—
	AH	11.4	—
Hodgkiss and Lees (1978, 1983a)	S	34.1	—
	AH	61.7	—
	No S, no AH[b]	4.2	—
	3 (no spikes)	—	11
Wood and Mayer (1978a)	AH	60.3	31.5
	No spikes	—	24.3

[a] Neurons were typed S not on the basis of showing fast EPSPs, but on the ability to generate action potentials in response to direct stimulation with 20 to 60-msec pulses at 1 Hz for 20–30 sec.
[b] Cells fired spikes, but these were never followed by a slow AH and there were no demonstrable fast EPSPs.

observed following stable, good impalements, in the absence of a detectable alteration in resting potential and in the absence of drugs (Hodgkiss and Lees, 1980). Furthermore, the initiation of the slow AH may occur before (Nishi and North, 1973) (Figure 3A 1–4) or 45–50 msec following (Wood and Mayer, 1978a (Figure 3C 1–4) recovery of E_m from the level of the positive afterpotential. The slow AH reaches the maximum amplitude of 15 mV (Nishi and North, 1973) or 20 mV (Holman *et al.*, 1972; Hirst *et al.*, 1974; Hodgkiss and Lees, unpublished observations) in 100–200 msec (Wood and Mayer, 1978a; Hirst *et al.*, 1974) to 1 sec (Nishi and North, 1973) or even 20 sec (Holman *et al.*, 1972; Hirst *et al.*, 1974) after the action potential. The input resistance of the soma membrane falls

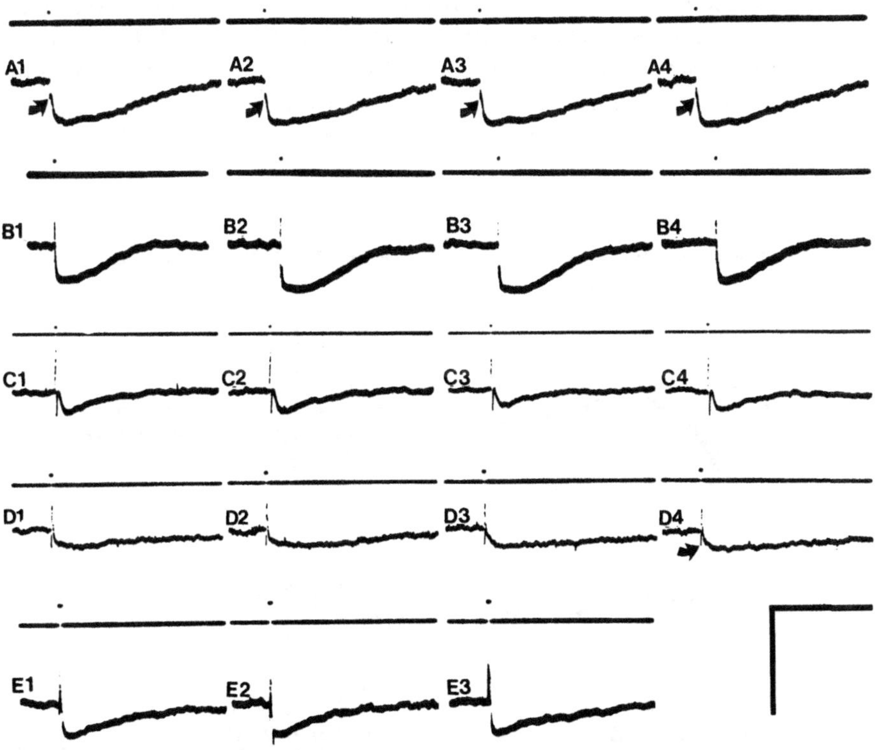

Figure 3. Slow afterhyperpolarizations following a single (A–D) or several action potentials (E) in an AH-neuron of myenteric plexus of guinea pig ileum. The upper trace in each pair of records represents the suprathreshold depolarizing constant current pulse passed across the soma membrane. The responses from each neuron, which are not successive, were obtained at a stimulation rate of 0.09 Hz over a period of 1–2 min. There was no change in the apparent membrane potential during this time. In some neurons, there was little variation in the amplitude or the course of the slow AH (A). In others, the variation was obvious (C, D) or striking (B). The resting potentials of the neurons were −60 (A), −70 (B), −80 (C), and −40 mV (E). Calibration: vertical bar: 50 mV (A–E); 1.25 nA (B–D); 2.5 nA (A, E); horizontal bar: 2.5 sec (C–E); 5 sec (A, B).

during the slow AH and gradually increases again as the membrane potential returns to E_m. The ionic basis for the slow AH is a selective increase in a calcium-dependent potassium conductance (Hirst and Spence, 1973; North, 1973; North and Nishi, 1976; Grafe *et al.*, 1980). The essential calcium influx occurs during the action potential, but is not inhibited by the so-called calcium antagonist D-600 (Grafe *et al.*, 1980).

Although the amplitude of the slow AH has been reported as never exceeding that of the positive afterpotential (Nishi and North, 1973), this is not always the case (see Figure 4C in Hirst *et al.*, 1974); for example, it was found in this laboratory that in some AH neurons, the amplitude of the slow AH may exceed that of the positive afterpotential by up to 14 mV (see Figure 3A4, D4).

It has also been our consistent finding that an individual AH-neuron may display marked variations in the duration of the slow AH, irrespective of the frequency at which it is made to fire action potentials, although with faster driving rates, slow AH tends to shorten. The range of responses is illustrated in Figure 3, in which the responses of five AH-neurons to

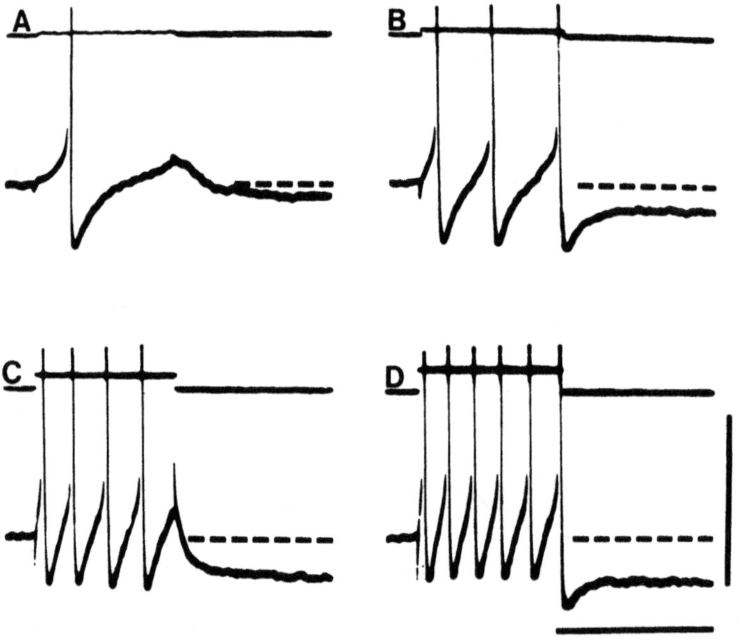

Figure 4. Effect of increasing the intensity of the depolarizing current pulse passed across the soma membrane of AH-neuron (E_m −40 mV) of myenteric plexus of guinea pig ileum (same neuron as illustrated in Figure 3E). As the intensity was increased (A–D), additional action potentials were generated, which were followed by corresponding increases in the amplitude of the accompanying slow AH (not seen in full at sweep speed employed). Calibration bar: vertical bar: 50 mV, 2.5 nA; horizontal bar: 100 msec.

suprathreshold depolarizations of the soma membrane are shown; the slow AH that followed a single soma action potential is shown in four neurons, while the slow AH that followed seven action potentials is shown in the other neuron. In some neurons, there is indeed very little if any variation in the amplitude or time–course of the slow AH following one (Figure 3A1–4) or many (Figure 3C1–3) action potentials; however, in others, despite appearing relatively constant, the slow AH exhibits marked variation of both amplitude and duration, sometimes by as much as approximately 30 and 65%, respectively (Figure 3B) (see also Hodgkiss and Lees, 1980). The amplitude of the slow AH grows larger with increases in the number of preceding action potentials (Figure 4) (Hirst *et al.*, 1974), although the slow AH of large amplitude that follows many action potentials in one neuron may not necessarily have a longer duration than a smaller-amplitude response that follows a single action potential in another neuron (compare Figure 3D1–4 with C1–3).

The slow AH may also summate following repetitive antidromic excitation, to reach values of 25–30 mV (Nishi and North, 1973; Morita *et al.*, 1982a). In most cells, however, the antidromic spike is likely to be blocked before such a high value is attained; instead, a "proximal process potential," i.e., a fast-rising, all-or-nothing potential change (5–25 mV) is likely to be recorded because the fractionation of antidromic spikes into one or more, all-or-nothing fast responses is common in myenteric neurons (Nishi and North, 1973; North and Nishi, 1974).

Since a calcium-dependent increase in g_K underlies the slow AH, it is possible that changes in the amplitude and duration of the slow AH reflect variations in either the intracellular sequestration of Ca^{2+} or the synaptic modulation of a Ca^{2+}-dependent g_K (Grafe *et al.*, 1980). Certainly, the presence of the slow AH greatly influences the excitability of AH neurons; in some, only a single action potential can be evoked by direct stimulation, whereas in others two to four spikes can be evoked before firing is inhibited. Generally, however, the excitability is low. There is much electrophysiological evidence (Nishi and North, 1973; North and Nishi, 1974, 1976) in favor of a predominantly bidirectional distribution of processes of AH-neurons. Transmission from one process to another on the opposite side of the soma is likely to be restricted by the occurrence of the slow AH and greatly facilitated when the slow AH is suppressed during activation of slow excitatory synaptic inputs to these neurons (see below). It should be noted that the action potential of the soma but not of the processes of AH-neurons is resistant to the blocking effect of tetrodotoxin (Nishi and North, 1973; Hirst and Spence, 1973; North, 1973; North and Nishi, 1976), the charge carrier being calcium, as well as sodium.

Some AH-neurons also have a fast cholinergic synaptic input (Nishi and North, 1973; Mayer *et al.*, 1978; Grafe *et al.*, 1979b; Johnson *et al.*,

1980a; Hodgkiss and Lees, unpublished observations). To claim that such fast synaptic potentials are unlikely to be of physiological significance (Johnson *et al.*, 1980a; Szerb, 1982) seems unwarranted at present, particularly since many AH-neurons can undergo dramatic changes in input resistance and excitability as a result of the activation of their slow excitatory synaptic input (see Section III.C).

B. Fast Excitatory Postsynaptic Potentials

The outstanding feature of S-neurons is the occurrence of fast EPSPs in response to focal stimulation of the surface of the ganglion and interganglionic fiber tracts, or transmural electrical stimulation (Holman *et al.*, 1972; Nishi and North, 1973; Hirst *et al.*, 1974). Fast EPSPs take about 3–5 msec to peak and decay with a time constant considerably longer than the τ_m of the same cell (Nishi and North, 1973). The reversal potential lies between 0 and -25 mV, probably at about -15 mV (Nishi and North, 1973). Repetitive activation of the input leads to a marked tetanic "rundown," which is not due to densensitization of the postsynaptic membrane (Nishi and North, 1973). With focal stimulation, summation of EPSPs is seen only occasionally, whereas it is a common finding in preparations in which the ganglion is not isolated from its neighbors and transmural stimulation is applied. Even cells located up to 1.2 cm away from the stimulating (transmural) electrodes could be synaptically activated; many neurons could also be excited antidromically (Hirst *et al.*, 1974); orthodromic activation may be polysynaptic. These observations are in sharp contrast to the findings of Hirst *et al.* (1974) with regard to AH-neurons, which never responded to transmural excitation except antidromically and then only when they were located within 1 mm of the electrodes. These electrophysiological observations are supported by the results obtained with intracellular application of dyes employed to reveal the distribution of processes (see Chapter 2).

As expected, the fast EPSPs, which can be mimicked by iontophoretic application of acetylcholine (ACh) to the cell soma, are abolished by hexamethonium (10–400 μM) or tubocurarine (0.14–1.4 μM) and unaffected by atropine (14 μM) and phenoxybenzamine (3 μM) (Hirst *et al.*, 1974; Nishi and North, 1973); they are depressed, especially the first in a train, by norepinephrine (1–10 μM) and epinephrine (0.1–1 μM) (Nishi and North, 1973; North and Nishi, 1974) and by periarterial nerve stimulation (Hirst and McKirdy, 1974b). After treatment with physostigmine (10 μM), the fast EPSP is augmented in amplitude and its duration is prolonged more than 10-fold, resulting in repetitive firing (North and Nishi, 1974).

The EPSPs are reduced in amplitude by 5-hydroxytryptamine (5-HT), which is itself without effect on the resting potential, the R_{in}, or the τ_m of these neurons (North and Henderson, 1975; North *et al.*, 1980a; Hodgkiss and Lees, unpublished observations); thus, the action of 5-HT on S-neurons must be presynaptic. The inhibitory effect of morphine on this type of neuron is brought about by a postsynaptic hyperpolarization associated with an increased conductance of the membrane (for reviews, see North, 1979, 1982a,b; North and Egan, 1983). Substance P (10–300 nM) applied in the bathing medium does not affect the fast EPSPs (Katayama *et al.*, 1979).

Recently, it has been established that about one third or more of S-neurons show enkephalin like immunoreactivity (Bornstein *et al.*, 1983, 1984a).

C. Slow Excitatory Postsynaptic Potentials

Although undetected in early intracellular investigations of myenteric plexus neurons (Holman *et al.*, 1972; Nishi and North, 1973; Hirst *et al.*, 1974, 1975; Hirst and McKirdy, 1974a), a variety of slow synaptic potentials were recorded from myenteric neurons in the course of more recent studies; these potentials may occur both spontaneously and in response to electrical stimulation (Katayama and North, 1978; Wood and Mayer, 1978b; Grafe *et al.*, 1979a, 1980; Johnson *et al.*, 1980a,b; Hodgkiss, 1981; Hodgkiss and Lees, 1984). As yet, slow synaptic excitation of myenteric neurons of the extrinsically denervated gut has not been examined to rule out excitation by extrinsic nerves.

Excitation of the surface of a myenteric ganglion or fiber tract by means of focal microelectrode or transmural stimulation of the myenteric plexus–longitudinal muscle preparation with repetitive pulses results in a slow depolarization (slow EPSP) in about 40% of S-neurons and in 19% (Johnson *et al.*, 1980a), 38% (Wood, 1981), or 58% (Hodgkiss and Lees, 1984) of AH-neurons (see Figures 5, 6, 10, and 11). When only a single pulse is used as the stimulus, the proportions of S-neurons responding is 20%, while AH neurons fail to respond [although in the Summary of their study, Johnson *et al.* (1980a), reported that 20% of cells are responding].

The slow EPSP evoked by repetitive stimulation at 0.5–20 Hz for 2–10 sec was found to have a mean amplitude of 9.7 mV (Hodgkiss, 1981) or 12 mV (Wood and Mayer, 1979a), with a range of 4–22 mV (Wood and Mayer, 1979a) or 8–20 mV (Wood and Mayer, 1978b). Johnson *et al.* (1980a) reported maximum amplitudes as 14 mV (E_m, −61 mV) and 20 mV (E_m, −52 mV) for the slow EPSPs arising from single and repetitive pulses, respectively. The duration of the slow EPSP ranged from 10 sec (Katayama and North, 1978) to 455 sec (Wood and Mayer, 1979a), with a mean value of 88 sec (Wood and Mayer, 1979a).

The slow EPSP is associated with a fall in membrane conductance (G_m) and an increased excitability of the neuron (Figure 5). The reversal potential for the slow EPSP was the same as that for the slow AH in those neurons that exhibited both characteristics and was reduced to less negative values by raising the extracellular K^+ concentration (Johnson *et al.*, 1980a). These results, together with the finding that the slow EPSP can be mimicked by superfusion of neurons with solutions containing reduced Ca^{2+} (Wood *et al* 1979) or elevated Mg^{2+} and reduced Ca^{2+} concentrations (Grafe *et al.*, 1980), strongly suggest that the ionic basis of the slow EPSP is an inactivation of a resting Ca^{2+}-dependent G_K of myenteric neurons.

The slow EPSP may sometimes reach the threshold for the initiation of action potentials (see Figures 5 and 9) (Katayama and North, 1978; Wood and Mayer, 1978b, 1979a) that can discharge at frequencies of up to 100 Hz for from 5 to 42 sec (Grafe *et al.*, 1978; Wood and Mayer, 1979a) or even 120 sec (Hodgkiss and Lees, 1984) after cessation of stimulations.

In AH-neurons, the characteristic slow AH is much reduced or abolished during the slow EPSP, and such cells, which normally can fire only one to four action potentials when the soma is depolarized with long-duration constant-current pulses (see below), acquire the ability to discharge continuously at high rates throughout such pulses delivered within the duration of the slow EPSP (Wood and Mayer, 1978b, 1979a). It is interesting to note that while the slow AH is often abolished throughout the duration of the slow EPSP (see Figures 6 and 10), this is not always the case (see Figure 5) (see also Figure 1a in Grafe *et al.*, 1979a).

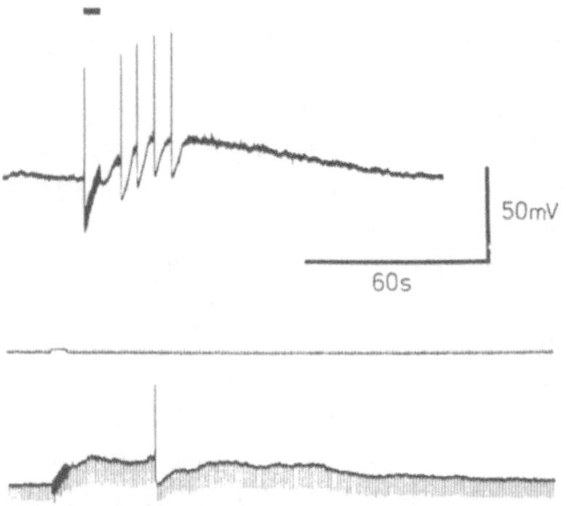

Figure 5. Slow EPSP of each of two AH-neurons of myenteric plexus of guinea pig ileum. Both slow EPSPs were evoked by transmural stimulation at 10 Hz for 5 sec, 3 mm aborad (upper record; E_m −65 mV) and 7 mm orad (lower record; E_m −42 mV) from the point of recording; the period of stimulation is indicated by the bar (upper record) and the signal in the timer trace (lower record). The slow EPSPs were accompanied by the firing of action potentials and their subsequent slow AHs, the latter being only very slightly reduced during the slow EPSP. In the lower record, changes in membrane input resistance were monitored by passing hyperpolarizing constant-current pulses across the soma membrane. The G_m that is characteristically increased during the course of the slow AH is shown to be preceded and succeeded by a period of reduced conductance before it returned to control values. Due to limitations in the frequency response of the pen, the full amplitude of the action potentials was not recorded.

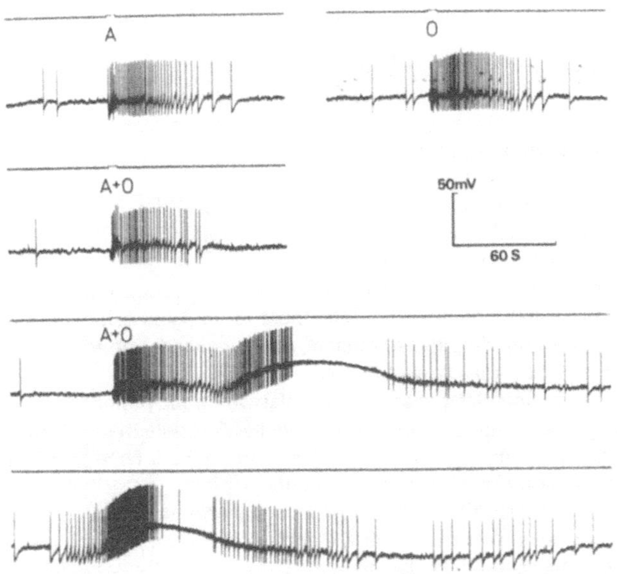

Figure 6. Evoked and spontaneously occurring slow EPSPs in an AH-neuron (E_m −48 mV) of myenteric plexus of guinea pig ileum. Slow EPSPs were evoked by transmural stimulation at 10 Hz for 5 sec, 3 mm orad (O) and/or 7.5 mm aborad (A) from the point of recording. Note the greatly increased excitability of this spontaneously firing AH-neuron and the variability of the slow AH, which is completely suppressed during the initial stages of the slow EPSPs and during the recovery phase of the two spontaneous slow EPSPs despite the low frequency of firing. Due to limitations in the frequency response of the pen, the full amplitude of the action potentials was not recorded.

In the cells illustrated in Figure 5 as in other AH neurons, the slow AH that follows the action potentials fired during the slow EPSP may course at or below the level of the resting potential.

Quite clearly, electrophysiologically as well as morphologically (Hodgkiss and Lees, 1980, 1983) (see Chapter 2), the AH-neuron does not represent a single uniform class of cells. Although its functions are unknown, it is not likely that the majority of AH-neurons are afferent neurons, as has been suggested (Holman *et al.*, 1972; Nishi and North, 1973; Hirst *et al.*, 1974; Hirst, 1979; Holman, 1981). It has also been suggested (Nishi and North, 1973) that neurons showing the slow AH may be the nonadrenergic inhibitory neurons that are present in this preparation (Kosterlitz and Lydon, 1969). For further discussion of this point, see Section V.

In view of the variety of slow synaptic events tht can be recorded in myenteric neurons, a simple checklist of the properties and responses that an enteric neuron may show (Table 3) may be useful. It should be appreciated that it is essential that the event and the properties of the neuron

Table 3. Checklist of Information Required for Characterization of Enteric Cells

1. Location of ganglion in preparation.
2. Position of cell in ganglion.
3. Value of apparent resting potential; constant or variable?
4. Value of input resistance of cell; constant or variable?
5. Value of membrane time constant.
6. Occurrence of an action potential at any time during intracellular recording; if not, the cell is probably not a neuron. Likelihood of impaled cell being a neuron greater than 90% in myenteric plexus; frequency of impalement of nonneuronal cells in the submucous plexus not known but rare.
7. Occurrence of a long-lasting afterhyperpolarization (or prolonged period of increased membrane conductance) following the soma action potential (are amplitude and time–course constant?). If so, it is an *AH-neuron* (about 60% probability in the myenteric plexus, about 2% in the submucous plexus).
8. Occurrence of a fast EPSP in response to transmural or to indirect, focal stimulation of interconnecting fiber strands or the surface of the ganglion (is it sub- or suprathreshold, mono- or polysynaptic?) If so, it is an *S-neuron* (about 30–35% probability in the myenteric plexus, occasionally up to 66% in some ganglia; probability more than 80% in the submucous plexus). If no fast EPSP demonstrable and no slow AH detectable yet action potentials can be generated, cell is one of the rare *no S-, no AH-neurons* (about 4–5% in both plexuses). Fast EPSPs in AH-neurons are rare.
9. Development of slow EPSPs in response to transmural or indirect, focal stimulation. The slow EPSP may be cholinergic (muscarinic-receptor-mediated) or noncholinergic in origin. In myenteric plexus, about 60% of AH-neurons show noncholinergic slow EPSPs, whereas such slow EPSPs may be seen in up to 40% of S-neurons. Muscarinic synaptic potentials are found in about 25% of myenteric plexus S-neurons, but are not commonly seen in AH-neurons, in which they are small (see North and Tokimasa, 1982). About 30% of submucous plexus neurons show a slow EPSP in response to a single indirect stimulus.
10. Development of IPSPs in response to transmural or indirect, focal stimulation; probability less than 12% in myenteric plexus neurons, 40–45% in submucous plexus neurons.
11. Development of biphasic slow potentials in response to transmural or indirect, focal stimulation; probability 2–8% in myenteric plexus neurons.
12. Location of transmural and focal stimulating electrodes should be noted in relation to the cell.
13. Occurrence of "spontaneous" changes in resting potential (rare).
14. Presence of any drugs and of their effects on any of these responses: similarly, the responsiveness of the cell to putative neurotransmitter substances should be noted.
15. Morphology of the cell in question as revealed by an intracellular marker, e.g., lucifer yellow, noting the soma shape, location with respect to other distinctive landmarks, orientation, and number of short and long soma (use a modified Dogiel's Classification I, II, or III) (see Bornstein *et al.*, 1984a).
16. Determination of neurochemical characteristics of the cell and its surrounding neural elements, e.g., by immunohistochemistry.

studied be adequately defined; otherwise, confusion may arise. For example, in some published records (e.g., Figure 4A in Johnson *et al.*, 1980a), the slow "depolarizations" appear to be biphasic potential changes, a distinct hyperpolarization following the initial depolarizing response.

1. Transmitter Candidates for the Slow EPSP

North and his colleagues (Morita *et al.*, 1982b,c; North and Tokimasa, 1982) have recently found that activation of muscarinic receptors leads to slow depolarizations with characteristics very similar to those of the slow EPSP. A controversy exists at present, however, concerning the identity of the transmitter mediating the slow EPSP. The two most likely candidates seem to be the undecapeptide substance P and the indolalkylamine 5-HT. The evidence for and against the role of these substances as transmitters is presented in detail elsewhere (Katayama and North, 1978; Wood and Mayer, 1978b, 1979a–c; Grafe *et al.*, 1979a; Katayama *et al.*, 1979; Johnson *et al.*, 1980b, 1981; Morita *et al.*, 1980; Wood *et al.*, 1980; Gershon, 1982).

Briefly, the evidence for 5-HT being the transmitter generating the slow potential is as follows: (1) 5-HT mimics the changes in membrane potential and conductance that occur during the slow EPSP (however, see below); (2) the reversal potential for the depolarizing response to 5-HT is the same as that of the slow EPSP; (3) methysergide blocks both the slow EPSP and the action of 5-HT (but not that of substance P); and (4) desensitization to 5-HT reversibly blocks the slow EPSP. It should be restated, however, that methysergide is not a selective antagonist of the action of 5-HT on enteric neurons (Costa and Furness, 1979; Wallis, 1981) and has depressant effects of its own.

The arguments favoring the transmitter role of substance P are as follows: (1) Substance P mimics the slow EPSP in its action on membrane potential and conductance (Figure 7); (2) the reversal potentials for the slow EPSP and for substance P are identical (Y. Katayama, S. Mihara, and S. Nishi, personal communication); (3) the proteolytic enzyme chymotrypsin reversibly abolishes the slow EPSP and reduces the substance P potential, but leaves unaltered the action of 5-HT on the same neuron; and (4) desensitization to substance P reversibly depresses the response to substance P and the presynaptically evoked slow EPSP. Curiously, although, in the experiments performed in connection with point (3), the control response to nerve stimulation was much more intense than the response to iontophoretically applied substance P, chymotrypsin completely abolished the response to nerve stimulation but only partially reduced the response to the application of substance P (Morita *et al.*, 1980 see their Figure 2). Also, it must be noted that 5-HT can depolarize or hyperpolarize myenteric neurons that exhibit a slow EPSP; this leads to

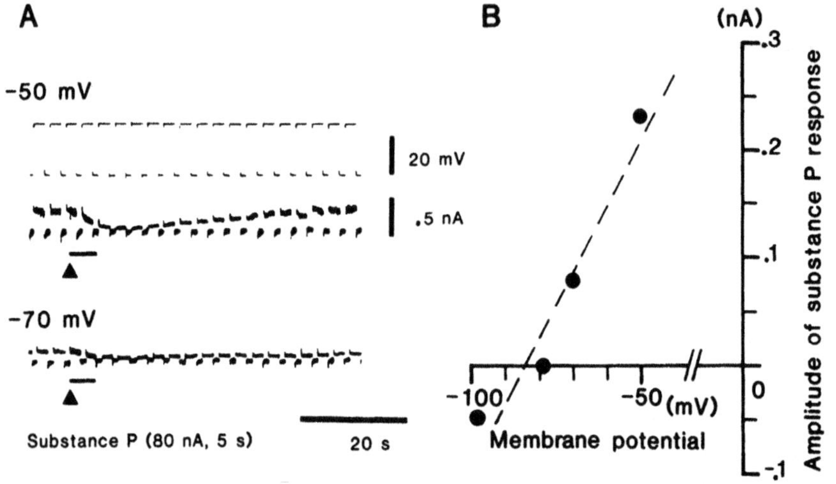

Figure 7. Records from a myenteric neuron that was voltage-clamped at two levels alternately (top trace) by applying commanding pulses (duration 1 sec) of 28 mV, repeated at 0.3 Hz. Middle trace: Substance P [applied by iontophoresis, (▲)] caused an inward current of about 0.25 nA at −50 mV (upper envelope), almost no current at −78 mV (upper envelope), and an outward current of 0.05 nA at −98 mV (lower envelope). Middle and bottom records demonstrate that substance-P-induced currents were associated with a reduction of the membrane conductance, which was determined from the decrease in current pulses for jumping the membrane between the two potential levels. (B) Relationship between the membrane potential (abscissa) and the amplitude of the substance-P-induced response (upper direction of ordinate: inward current). The substance-P-induced response was reversed in polarity between −80 and −90 mV. Figure kindly supplied by Professor Y. Katayama.

the conclusion that 5-HT is unlikely to be the transmitter mediating the slow EPSP in neurons capable of generating the latter. On the other hand, the slow AH is not always abolished during the slow EPSP (see Figure 5), whereas it is greatly reduced or abolished by 5-HT (Wood and Mayer, 1979b) and substance P (Katayama *et al.*, 1979; North *et al.*, 1982).

Furthermore, since some extrinsic, possibly sensory, nerves to the gut contain substance P (Hökfelt *et al.*, 1980; Costa *et al.*, 1980; Dalsgaard, *et al.*, 1982), it is possible that antidromic impulses induced by transmural or focal stimulation in sensory terminals may cause a nonsynaptic release of this peptide.

A more powerful argument against a role for 5-HT in the generation of the slow EPSP comes from recent experiments in which circumferential cuts have been made in the wall of the intestine (myotomy), an operation that completely interrupts the myenteric plexus. In these experiments, slow EPSPs were recorded from neurons located in ganglia in which 5-HT immunoreactivity was absent but substance P immunoreactivity was present (Bornstein *et al.*, 1984b).

On the basis of available evidence, a more convincing case exists for substance P as the mediator of the slow EPSP in the great majority of myenteric neurons, but a less extensive role for 5-HT in slow excitation cannot be ruled out at present.

2. Effect of γ-Aminobutyric Acid and Muscarinic Agonists on Myenteric Neurons

Recently, attention has been focused again on γ-aminobutyric acid (GABA) as a transmitter in the enteric nervous system. The evidence, though incomplete in fulfilling the criteria for a transmitter role (Furness and Costa, 1982b), is sufficiently compelling to warrant further study. Thus, it is known that a population of enteric neurons possess high-affinity uptake sites for [^{3}H]-GABA (Krantis and Kerr, 1981) and that intact ganglia can synthesize [^{3}H]-GABA from its precursor (Jessen *et al.*, 1979). GABA stimulates intrinsic inhibitory and excitatory nerves in the guinea pig intestine, but interestingly does not interfere with either ascending excitation or descending inhibition of the peristaltic reflex (Krantis *et al.*, 1980). Furthermore, it was demonstrated that depolarization induced release of [^{3}H]-GABA from segments of large intestine containing the myenteric plexus and from isolated, cultured ganglia (Jessen *et al.*, 1983).

Intracellular studies of the action of GABA provide at this time no direct evidence for a transmitter role for GABA. GABA causes a depolarization of AH- (but not S-) neurons that is the result of a conductance increase. At first, GABA causes chloride activation of AH-neurons, this effect being subject to rapid desensitization; subsequently, a nondesensitizing depolarization ensues (Cherubini and North, 1984a). A reduction in the amplitude of the fast EPSP and of the cholinergic and noncholinergic slow EPSPs is also seen in the presence of GABA (Cherubini and North, 1984b). This accords with the suggestion made by Jessen *et al.* (1983) that GABAergic neurons function to modulate presynaptically the release of transmitter from neighboring axons.

Much study has also been directed at the muscarinic action of ACh within the myenteric plexus. It appears that muscarinic agonists can inhibit presynaptically the release of transmitters responsible for the fast EPSP and the slow EPSP (Morita *et al.*, 1982c). Muscarinic receptors are also present postsynaptically (Morita *et al.*, 1982b) and are activated by single-pulse stimulation of presynaptic elements within the plexus. The event recorded postsynaptically is a slow EPSP resulting from inactivation of a potassium conductance (North and Tokimasa, 1982).

Muscarinic agonists also cause excitation by depressing the slow AH of myenteric neurons, even when applied after the calcium entry (during the spike) has occurred. Thus, ACh does not interfere with calcium entry, but only with events consequent on its entry (North and Tokimasa, 1983), perhaps by reducing the association of calcium with an intracellular membrane site (North and Tokimasa, 1984a,b).

There is evidence that the long latency and slow recovery after exposure of myenteric neurons to muscarinic agonists is the result of events set in motion by the initial agonist–receptor interaction. These events may include the movement of calcium ions from one intracellular site to another (North and Tokimasa, 1984b) or activation of adenylate cyclase (Nemeth *et al.*, 1984).

It is becoming more and more apparent that not only ACh but also a number of peptides and other compounds exert their inhibitory or excitatory influence by altering the potassium conductance(s) of the membrane (Tokimasa, *et al.*, 1981; Morita and North, 1981; North and Egan, 1982; Cherubini *et al.*, 1984; North and Tokimasa, 1984a).

D. Slow Hyperpolarizing Inhibitory Postsynaptic Potentials

Slow hyperpolarizations following single (Wood and Mayer, 1978a) and repetitive stimulus pulses (Johnson *et al.*, 1980a; Hodgkiss and Lees, 1984) (Figure 8) are recorded in about 8% of neurons (Johnson *et al.*, 1980a) and range in amplitude from 3 to 7 mV (Wood and Mayer, 1978a). The amplitude of the slow inhibitory postsynaptic potential (IPSP) varied with the number as well as the frequency of preceding stimuli up to a maximum of 10 mV; the duration of the hyperpolarizations ranged from 2 to 40 sec (Johnson *et al.*, 1980a). The slow IPSP that was recorded in S- and AH-neurons appeared to be very often associated with an increase in G_m as determined by passing constant-current pulses across the soma membrane (but see the current trace of Figure 5 in Johnson *et al.*, 1980a). In AH-neurons, the slow IPSP and the slow AH were found to reverse at the same level of membrane potential. Thus, it seems likely that the slow IPSP is the result of the activation of G_K (Johnson *et al.*, 1980a; Y. Katayama, S. Miharo, and S. Nishi, personal communication).

A portion of S- and AH-neurons are hyperpolarized, with a concomitant rise in G_m, by iontophoretic application of substance P and 5-HT. Similar changes in membrane potential and conductance are also seen in some myenteric neurons in response to the application of norepinephrine, somatostatin, and enkephalin (North and Henderson, 1975; North *et al.*, 1980b; North, 1979). Thus, these substances are obvious transmitter candidates for the slow IPSP. The case for 5-HT is strong, since desensitization with 5-HT (300 nM) abolishes the slow IPSP, although this result is not dependent on the effect of 5-HT on membrane potential; this difficulty can be overcome by postulating the existence of extrasynaptic 5-HT receptors that gate ionophores different from those involved in the

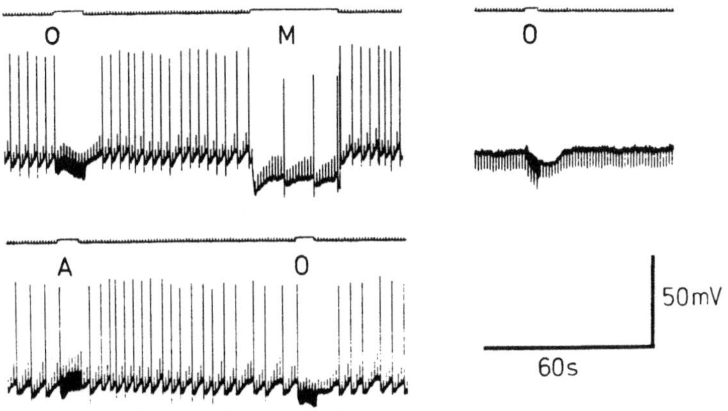

Figure 8. Slow hyperpolarizing IPSPs in an AH-neuron (E_m −55 mV) of myenteric plexus of guinea pig ileum. The neuron was driven by passing suprathreshold depolarizing constant-current pulses across the soma membrane at 1 Hz. When the preparation was stimulated transmurally at 10 Hz for 11 sec, orad (O; upper trace) to the point of recording, there was an increase in membrane potential associated with a fall (20%) in R_{in} (see also right-hand trace) and a cessation of spiking. This inhibition was not mimicked even by hyperpolarization of the cell (M). An inhibitory response in the same AH sec, neuron was evoked by aborad stimulation (A; lower trace) at 10 HZ for 7 sec, but it was less effective and shorter-lasting than when evoked by orad stimulation (O; lower trace) under the same conditions. Due to limitations in the frequency response of the pen, the full amplitude of the action potentials was not recorded.

generation of the slow IPSP. The slow IPSP is also blocked by methysergide (10–30 μM) (Johnson *et al.*, 1981).

E. Slow Depolarizing Inhibitory Postsynaptic Potentials

Not all inhibitory potentials recorded in myenteric neurons involve hyperpolarization associated with a rise in membrane conductance. In about 5–10% of AH-neurons (Hodgkiss, 1981; Hodgkiss and Lees, 1984), transmural stimulation is followed by a depolarization of 2.0–12.7 mV associated with a rise in G_m as detected by fall in input resistance (Figure 9); in fact, it is this rise in G_m that renders this potential functionally inhibitory, since the increased G_m serves to short-circuit the membrane of the cells in question. This depolarization is clearly not the same event as the slow IPSP in AH-neurons because this potential is still depolarizing in neurons with a detectable slow AH (Hodgkiss, 1981), whereas the slow IPSP has its reversal potential at E_K. Of interest is the interaction between this depolarizing inhibitory potential and the slow EPSP, since it is possible to modulate the intensity and duration of the slow EPSP by timing appropriately the occurrence of the slow depolarizing inhibitory potential

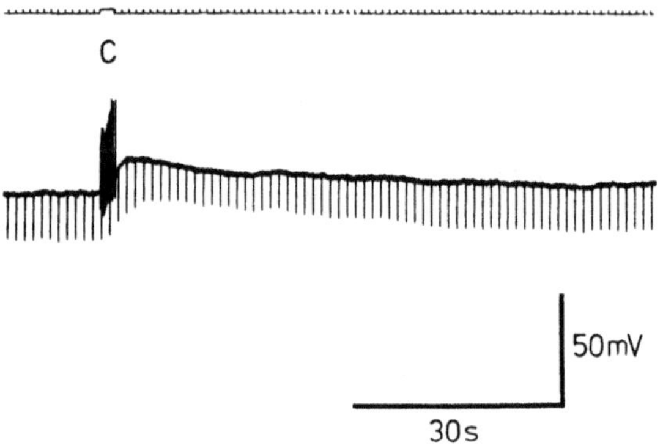

Figure 9. Slow depolarizing IPSP in an AH-neuron (E_m −53 mV) of myenteric plexus of guinea pig ileum. Membrane conductance was monitored by passing hyperpolarizing constant-current pulses (60 msec) across the soma membrane at 0.9 Hz. Focal stimulation at 10 Hz for 2 sec with an extracellular electrode placed on an interconnecting strand circumferential (C) to the point of recording resulted in antidromic excitation during the stimulation period, followed by a slow depolarization associated with a transient small rise followed by a fall (20–25%) in R_{in}, which declined gradually over a period of about 40 sec. In this neuron, transmural stimulation, with electrodes located 8 mm orad from the point of recording, had no effect on membrane potential and little if any effect on R_{in}.

(Hodgkiss, 1981) (Figure 10); indeed, this interplay between these two depolarizing potentials emphasizes the inhibitory nature of the slow depolarizing inhibitory potential.

The presence of a slow depolarizing inhibitory potential means that not all depolarizations in myenteric neurons should be assumed to have the nature of slow EPSPs; thus, it is important that the conductance change accompanying all slow depolarizing events be determined before the latter are classified as slow EPSPs or slow IPSPs.

A further point of interest is the possible occurrence of a presynaptic inhibitory mechanism that subserves the modulation of the slow EPSP (Figure 11).

F. Biphasic Synaptic Potentials

Biphasic potentials in the course of which a hyperpolarization precedes or follows a depolarization have been recorded in a small proportion [2–3% (Johnson *et al.*, 1980a); 8% (Hodgkiss and Lees, unpublished observations)] of myenteric neurons. These potential changes are thought to represent the release of at least two transmitter substances (Johnson *et al.*, 1980a), although an activation of a different population of receptors by diffusion of a single released transmitter cannot be ruled out.

G. Spontaneous Activity

Intracellular studies have revealed the presence of spontaneous activity in the form of fast EPSPs, which sometimes reach threshold; in

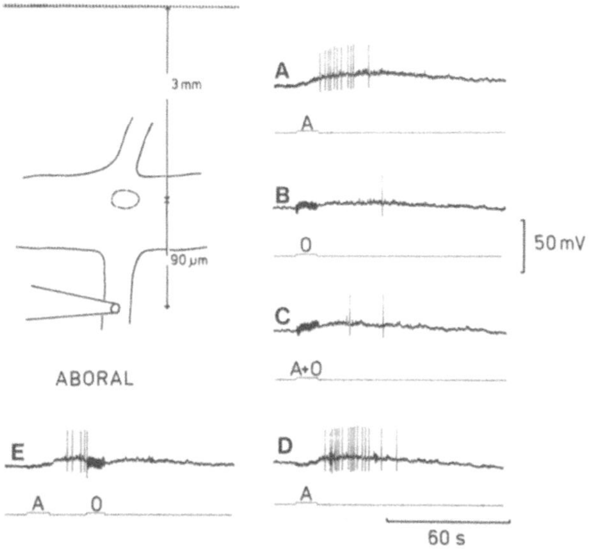

Figure 10. Effect on the membrane potential of an AH-neuron (E_m −48 mV) of the stimulation of the preparation with an aborad (A)-placed microelectrode (A, D) and an orad (O)-placed pair of transmural electrodes (B). The slow EPSP evoked by fiber tract stimulation at 10 Hz for 10 sec was greatly reduced (C) or abbreviated [(E) descending inhibition] by oral transmural stimulation, which alone produced a small, slow depolarization (B). Due to limitations in the frequency response of the pen, the full amplitude of the action potentials was not recorded. Reproduced from Hodgkiss (1981) with permission.

addition, unitary spikes and bursts of action potentials, not accompanied by fast EPSPs but followed by a slow AH, have been recorded in a small number of myenteric neurons (Nishi and North, 1973; Hirst *et al.*, 1974). These observations were confirmed by Wood and Mayer (1978a), who also described spontaneous potentials that appeared to be slow IPSPs. Spontaneous changes in membrane potential and spike discharge similar to those that can be seen during the course of the slow EPSP were also recorded (Wood and Mayer, 1978a; Johnson *et al.*, 1981; Hodgkiss and Lees, unpublished observations) (See Figure 6).

When maintained in tissue culture, myenteric neurons that are effectively isolated under tissue-culture conditions also exhibit spontaneous electrical activity consisting of fast-EPSP-like potentials and action potentials (Jessen *et al.*, 1978; Hanani *et al.*, 1982). It should be noted that cultured AH-neurons (about 4% of the population) can fire repetitively at high frequency in response to a depolarizing stimulus.

These observations suggest that spontaneous activity occurs in S- and AH-neurons, although it has been suggested that these neurons constitute but two excitability states of the same neuronal type (see Section II. B and Grafe *et al.*, 1978).

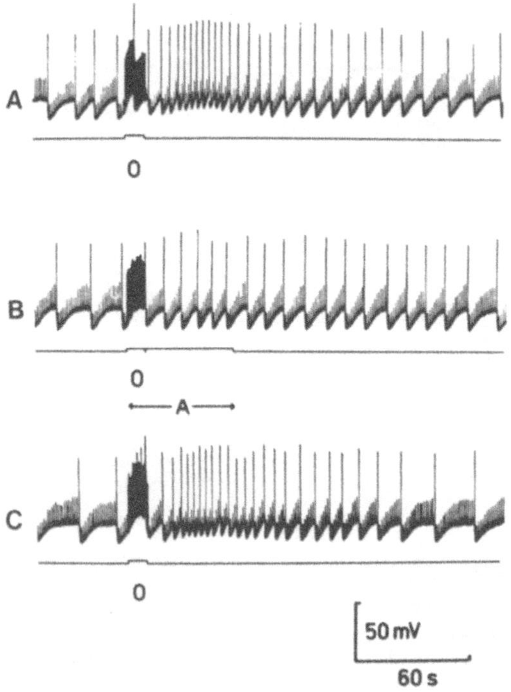

Figure 11. Ascending presynaptic inhibition of the slow EPSP in an AH-neuron (E_m −50 mV) of myenteric plexus of guinea pig ileum. The cell was driven with suprathreshold depolarizing constant-current pulses delivered at 0.9 Hz. Transmural orad stimulation (O) evoked a slow EPSP except when aborad stimulation was applied. Note the absence of a change in resting potential during aborad stimulation.

In view of the rather variable effect, from cell to cell, of the slow EPSP on the slow AH, it was of interest to evaluate variations in the amplitude of the slow AH of spontaneously active neurons. In some, action potentials were followed by a very small AH (Figure 12) that was on occasions undetectable, while in others the slow AH had a large amplitude; indeed, sometimes when two or three action potentials discharged in close sequence, they were followed by a summated slow AH (Figure 12) (see also Hirst *et al.*, 1974, p. 306).

IV. SUBMUCOUS PLEXUS NEURONS

Submucous plexus neurons of guinea pig small intestine have been studied extensively by Hirst and his colleagues, who have identified several important qualitative and quantitive differences between these and myenteric plexus neurons with respect to their responses to orthodromic excitation and the frequency distribution of these responses. The vast majority (77–90%) of neurons exhibited a fast excitatory postsynaptic input (Hirst and McKirdy, 1975; Hirst and Silinsky, 1975; Surprenant,

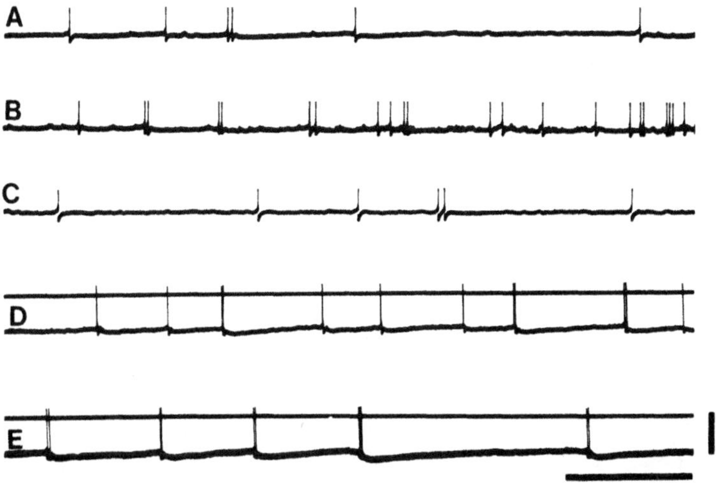

Figure 12. Spontaneous activity recorded in two AH-neurons. In one cell (E_m −50 mV), the spontaneous activity occurred as either single spikes (A, C) or bursts of spikes (B). The bursts were recorded at a time of increased membrane "noise." (A,B) Recorded 15 min, (C) 60 min after impalement. Note that the slow AH was very small. In the other neuron (E_m −45 mV), the action potentials occurred as singlets, doublets (D), or triplets (E) and were followed by a pronounced slow AH, which increased in amplitude with increases in the number of preceding action potentials. Calibration: vertical bar: 120 mV (A–C), 60 mV (D, E); horizontal bar: 2 sec (A–C), 4 sec (D,E).

1984a) that could reach threshold for the generation of an action potential. Increasing the intensity of a single transmural stimulus increased the number of excitatory inputs and shortened the latency of the appearance of the response, provided the cell was located within 6 mm of the transmural electrodes. The firing of an action potential due to synaptic activation by a single transmural stimulus initiated, in about 40% of neurons, a long-lasting (1–5 sec) hyperpolarizing response that followed the excitatory response; this hyperpolarization was not evoked by direct excitation of the soma membrane (Figure 13) and therefore was not comparable to the slow response of AH-neurons. Occasionally, this hyperpolarization was evoked at stimulus intensities too low to activate sufficient excitatory inputs to generate an action potential. Furthermore, when tubocurarine was used to abolish fast EPSPs, this slow hyperpolarization remained present and constituted an all-or-none phenomenon (Figure 14B). A fall in input resistance occurred during the hyperpolarization; this fall was associated solely with an increased potassium conductance, which was not secondary to an increased calcium ion influx; a change in chloride conductance was not involved (Hirst and Lang, 1976). It was concluded, therefore, that most submucous neurons receive several excitatory synaptic inputs, which are probably cholinergic, and that a significant pro-

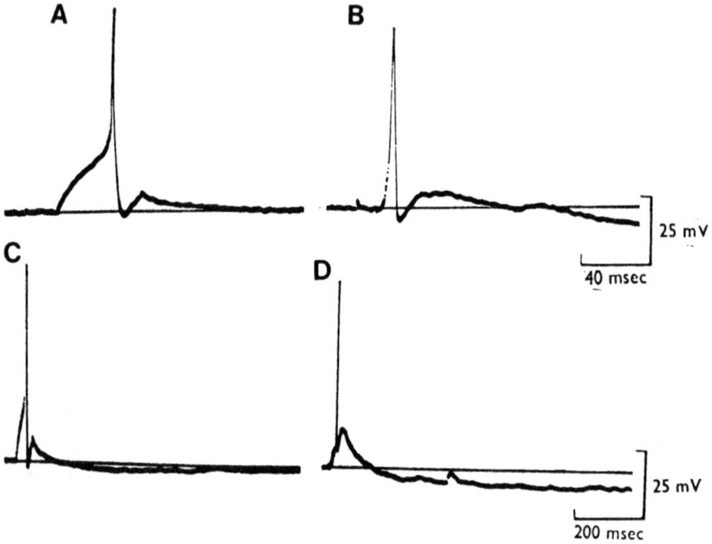

Figure 13. Comparison of responses evoked in a submucous plexus S-neuron by direct and indirect (transmural) stimulation. (A,C) Depolarizing constant-current pulses (duration 50 msec, 8×10^{-10} A) were injected through the recording electrode. (B,D) The transmural stimulus strength was 11 V (pulse width 0.5 msec). Note that the early responses evoked only by transmural stimuli were followed by long-lasting increases in membrane potential. The level of resting potential has been drawn in, and action potentials were retouched. The upper calibration bars apply to (A) and (B), the lower calibrations bars to (C) and (D). Reproduced from Hirst and McKirdy (1975) with permission.

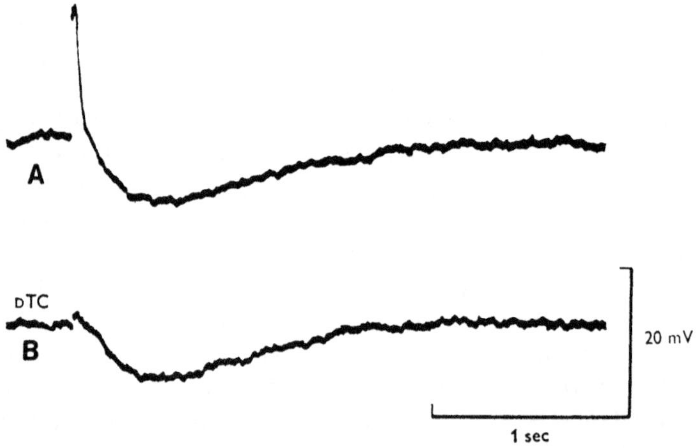

Figure 14. Synaptic responses recorded from a neuron lying in a flap of submucous plexus taken from a segment of small intestine that had its extrinsic supply cut 11 days previously. Traces (A) and (B) represent intracellular records made from the same S-neuron before (A) and 8 min after (B) addition of tubocurarine (1×10^{-4} g/ml) to the perfusion fluid. Reproduced from Hirst and McKirdy (1975) with permission.

portion of them receive both a multiple, excitatory and a single, inhibitory input (which evoked the slow IPSP). The remainder of the neurons are likely to show only a slow AH [up to 13% (Surprenant, 1984b)] or else neither fast EPSPs nor a slow AH [5% (Hirst and Silinsky, 1975)].

The IPSP had a latency of 30–80 msec, rose to a peak in about 100–200 msec, and declined slowly [half-life of 700–1600 msec at 35°C, 450–700 msec at 37°C (Surprenant, 1984a)]. Summation of IPSPs occurred when pairs of stimuli were used, 100 msec apart, and the membrane potential increased then to a value that was 16–43 mV more negative than the apparent resting potential (at E_m levels ranging from −40 to −70 mV). The polarity of the IPSP was readily reversed by strong hyperpolarization of the cell, the reversal potential being 20–40 mV negative with respect to E_m. An additional interesting finding was that the decrease in amplitude of the electronic potential used to test input impedance was always greater than the reduction in magnitude of the fast EPSP; furthermore, the depression of the EPSP outlined the decrease in input resistance of the neuron (Edwards *et al.*, 1976).

Extrinsic denervation of a midileal segment was found not to change the incidence of IPSPs. From these and other observations, Hirst and McKirdy (1975) concluded that the inhibition is not likely to result from the release of norepinephrine, which is present in extrinsic sympathetic nerves. Furthermore, Hirst and Silinsky (1975) provided evidence that the IPSP is most unlikely to be due to 5-HT. Further work is required to establish unequivocally the identity of the neurotransmitter and the type of receptor involved.

About 30% of submucous neurons exhibited a slow EPSP (duration 15–20 sec) that could be elicited by a single stimulus. The slow EPSP, like the IPSP, was an all-or-none phenomenon, and it was suggested that the former is the result of inactivation of a g_K, while the latter is the result of its activation (Surprenant, 1984a). In addition, about 20% of submucous neurons did not exhibit any demonstrable synaptic input. Of these, one group were similar to AH-neurons of the myenteric plexus; these neurons may be sensory in nature. The other group exhibited sponteneous action potentials endowed with a characteristic ionic basis (Surprenant, 1984b).

V. INTERCONNECTIONS AMONG ENTERIC NEURONS IN RELATION TO MOVEMENTS OF THE GUT

There have been many attempts to model the interconnections among neurons within the enteric plexuses to account for the motility patterns seen in, or recorded from, the gut (Kosterlitz, 1967; Crema, 1970; Furness and Costa, 1973; Hirst and McKirdy, 1974a; Hirst *et al.*, 1975; Franco *et al.*, 1979; Hirst, 1979; Wood and Mayer, 1979a; Yokoyama and Ozaki, 1978, 1980; North, 1982a; Wood, 1984). These studies have so far been confined to the small and large intestine of the guinea pig, rabbit, and cat.

Three factors or parameters seem to be important for the integrity of the peristaltic reflex. The first is a relaxation of smooth muscle ahead of the bolus (descending inhibition), the second is a contraction orad to the bolus (ascending excitation), and the third is the maintenance of an orad-to-aborad wave of contraction (descending excitation). In the models of the functioning of the gut that relate these factors to possible neuronal pathways, the classic elements of reflex circuitry are combined in such a way that output from a sensory neuron impinges on one or more interneurons that project for a variable length along the gut before synapsing either with another interneuron or with a final motor neuron to the smooth muscle, which is usually the circular muscle layer.

A. Ascending Excitation

This component of the peristaltic reflex was demonstrated simply and elegantly by Costa and Furness (1976) for the isolated guinea pig large intestine. The technique used allowed small lengths of intestine to be stretched, and the mechanical response of the circular muscle to such stimuli was recorded at points orad and/or aborad with respect to the stimulated portion. Such experiments demonstrated that the integrity of the myenteric plexus but not that of the submucous plexus was essential for ascending excitation. It seems, therefore, that the sensory neurons involved in this reflex arc are not located in the submucous plexus, as had been proposed earlier (Bülbring *et al.*, 1958; Crema, 1970; Kottegoda, 1970). This finding and the similar results obtained by Hirst *et al.* (1975) make it unlikely that the facilitatory effect of 5-HT on peristalsis, when the drug is applied to the mucosa, is due to stimulation of distension-sensitive receptors of the mucosa (Bülbring and Lin, 1958). This should not be interpreted as evidence against the presence of sensory neurons in the submucous plexus, since recent studies suggest a sensory role for submucous AH-neurons that receive no synaptic input (Surprenant, 1984b). Such neurons may play a role in desceding excitation, for which the integrity of the submucous plexus is obligatory (Hirst *et al.*, 1975).

On the basis of a pharmacological analysis of ascending excitation, Costa and Furness (1976) concluded that the ascending excitatory pathway consists of a cholinergic and a noncholinergic neuron running either in series or in parallel in the myenteric plexus. Since in the presence of low concentrations of atropine, contractions of the circular muscle take place above and below the bolus, excitatory pathways may also project aborally as well as orally in this plexus (Crema, 1970; Frigo *et al.*, 1972). Alternatively, excitatory and inhibitory influences on the smooth muscle may be expected to propagate in both directions along the gut, the predominating direction being determined by the orientation of the projection of neural processes in the myenteric plexus. Intracellular recordings made from S-neurons of the myenteric plexus showed that evoked synaptic activity is seldom seen when the point of stimulation is located aborally; thus, long orally directed pathways are thought to be uncommon (Hirst and McKirdy, 1974a). We have confirmed electrophysiologically as well as morphologically (by means of intracellular staining of neurons) that orally directed processes are very uncommon and short. The greatest distances over which S- and AH-neurons projected orad were 260 and 520 μm, respectively (Hodgkiss and Lees, 1978, 1980, 1983a) (see Chapter 2); since the ascending excitatory reflex contraction was strongest 5 mm orad, the pathway in question may be polysynaptic (Costa and Furness, 1976).

B. Descending Inhibition

As in the case of the ascending excitatory reflex, all the neural elements for the descending inhibitory reflex are located outside the submucous plexus (Hirst *et al.*, 1975; Costa and Furness, 1976). In the colon, distension of the intestinal wall resulted in a relaxation, aborad from the point of stimulation, that could be recorded up to 90 mm from the locus of distension; the relaxation was, however, maximal 10 mm aborad from the stimulus (Costa and Furness, 1976). This suggests that the descending inhibitory pathway is also polysynaptic, since the processes of AH-and S-neurons in the ileum appear to extend aborally for up to only 240 and 630 μm, respectively, as revealed by intracellular dye injection (Hodgkiss and Lees, 1983); however, when immunohistochemical techniques are used, soma processes of intrinsic neurons appear to be very much longer (Furness and Costa, 1980). In contrast to the findings of Bülbring and Tomita (1967) and of Bornstein *et al.* (1982) suggesting that the processes of intrinsic inhibitory neurons are only a few millimeters long, preliminary observations by Hirst and McKirdy (1974a) indicated that the processes of inhibitory neurons extend longitudinally for about 1.5 cm, the neurotransmitter being released only from the final 500 μm of the processes. Furthermore, with distances greater than 5 mm, a cholinergic (nicotinic) synapse is possibly present in the reflex pathway (Costa and Furness, 1976).

Intracellular recordings made from S-neurons of small intestine provide evidence that the descending pathways are polysynaptic (Hirst and McKirdy, 1974a). One of the descending inputs to S-neurons situated aborally from the locus of distension is activated by distension after a short latency (about 1 sec) and is always of short duration. It occurs at about the same time as the inhibitory junction potential that can be detected in the circular but not the longitudinal muscle. This short latency response is thought to be part of the descending inhibitory mechanism (Hirst *et al.*, 1975). Descending inhibition in the small intestine is, however, quite different from that in the large intestine, since sustained distension of the colon results in sustained relaxation aborad from the point of stimulation (Furness and Costa, 1976, 1977). Therefore it could be predicted that a population of S-neurons in the myenteric plexus of the colon would respond to distension, after a short delay, with a maintained generation of fast EPSPs.

Studies of the interaction between ascending and descending influences in the intestine show that descending inhibition is capable of overcoming ascending excitation when both act on the same region of the intestine (Costa and Furness, 1976). Electrophysiological studies of slow synaptic potentials in AH-neurons revealed that in a small proportion of these myenteric neurons, there is a descending inhibitory input that is capable of attenuating the slow EPSP (see Figure 10) (Hodgkiss, 1981). Unfortunately, it is not possible at present to correlate these observations.

The available evidence suggests that the slow EPSP is not associated with the enteric inhibitory neuron (Vermillion *et al.*, 1979). An input to the enteric inhibitory neuron involving 5-HT is, however, thought to be present (Gershon and Dreyfus, 1977; Gershon, 1981). The recent interesting findings of extensive aborad projections of intrinsic neurons showing 5-HT-like immunoreactivity may be of importance in this context (Furness and Costa, 1982a; Costa *et al.*, 1982).

Taken together, recent studies suggest that descending inhibition itself can be divided into two descending components, one acting primarily on the nonadrenergic, noncholinergic inhibitory neuron to produce inhibition of smooth muscle, the other acting on interneurons that block excitation of cholinergic excitatory motor neurons to smooth muscle (Julé, 1980).

C. Descending Excitation

Descending inhibition is essential for peristalsis to occur, but following the relaxation, a contraction is necessary to move the intestinal contents to the adjacent aborad region of relaxed intestine. It is thought that rebound excitation may play a role in this sequence of events (see Furness and Costa, 1973).

Electrophysiological studies of S-neurons of the myenteric plexus showed that in addition to those cells that receive a short-latency synaptic input in response to distension, there is another population of neurons that exhibit a much longer latency. In this latter group, the latency varied from 2 to 11 sec and was apparently synchronized with excitatory junction potentials of the smooth-muscle layers. These long latency responses occurred in the circular muscle after a distinct delay, following the cessation of the inhibitory junction potential; they were atropine-sensitive. This smooth-muscle excitation is therefore not of the secondary or rebound type. Hirst *et al.* (1975) proposed that this delayed excitatory response is part of the descending inhibitory mechanism, because it is obviously linked to descending inhibition and thus provides a means of ensuring the onward movement of intestinal contents by complementing the secondary excitation.

D. Circumferential Excitation and Inhibition

In review of the predominant circumferential orientation of processes of myenteric neurons (Hodgkiss and Lees, 1978, 1980, 1983), it was of interest to find that in AH-neurons, slow excitatory or inhibitory responses did not show any symmetry, nor did they show any obvious correlation with the type of slow synaptic input activated (Hodgkiss and Lees, 1984).

VI. CONCLUSIONS

Having reviewed the early literature on intrinsic intestinal reflexes, Kosterlitz and Lees (1964) concluded:

> Although the intestine is perhaps the organ in which pharmacological methods have been most frequently used to elucidate function, the organization of its tissue is so complex that the success does not seem commensurate with the effort expended. While the research of recent years has answered some question, it has posed new, and more difficult, problems.

What frequently happens to investigators active in this field is that after spending many hours in the laboratory, they have obtained no useful information; they may then be tempted to agree with this quotation. Yet, the concentrated efforts of a number of groups, distributed worldwide, have in recent years led to a burgeoning of essential information.

Furthermore, the results obtained with immunohistochemical techniques are providing at this time exciting new insights into the function, organization, and pharmacology of the enteric nervous system at a rate comparable to that of the advances in the physiology and pharmacology of the central nervous system that stemmed from the application of fluoresence histochemistry to the elucidation of the central catecholaminergic and serotonergic pathways. These developments encourage us to hope that the greatly improved understanding of the neural control of gastrointestinal motility that will surely follow will soon be somewhat more commensurate with the effort expanded than it was in earlier decades.

ACKNOWLEDGMENTS. Our published and unpublished work was generously supported by a Medical Research Council Project Grant. The assistance of Mrs. Margaret Gray is gratefully acknowledged.

REFERENCES

Bornstein, J. C., Land, R. J., Costa, M., and Furness, J. B.: An electrophysiological investigation of the projections of the enteric inhibitory nerves in the guinea-pig small intestine. *Proc. Aust. Physiol. Pharmacol. Soc.* **13**:202P (1982).

Borntein, J. C.; Costa, M., Furness, J. B., and Lees, G. M.: Correlation of electrophysiology, morphology and enkephalin-immunoreactivity of neurones in guinea-pig myenteric plexus. *J. Physiol. (London)* **341**:59P (1983).

Bornstein, J. C., Costa, M., Furness, J. B., and Lees, G. M.: Electrophysiology and enkephalin immunoreactivity of identified myenteric plexus neurones of guinea-pig small intestine. *J. Physiol. (London)* **351**:313–325 (1984a).

Bornstein, J. C., North, R. A., Costa, M., and Furness, J. B.: Excitatory synaptic potentials due to activation of neurones with short projections in the myenteric plexus. *Neuroscience* **11**:723–731 (1984b).

Bülbring, E., and Lin, R. C. Y.: The effect of intraluminal application of 5-hydroxytryptamine and 5-hydroxytryptophan on peristalsis; the local production of 5-HT and its release in relation to intraluminal pressure and propulsive activity. *J. Physiol (London)* **140**:381–407 (1958).

Bülbring, E., and Tomita, T.: Properties of the inhibitory potential of smooth muscle as observed in the response to field stimulation of the guinea-pig coli taenia, *J. Physiol (London)* **189**:299–315 (1967).

Bülbring, E., Lin, R. C. Y., and Schofield, G.: An investigation of the peristaltic reflex in relation to anatomical observations. *Q. J. Exp. Physiol.* **140**:381–407 (1958).

Cherubini, E., and North, R. A.: Actions of γ-aminobutyric acid on neurones of guinea pig myenteric plexus. *Br. J. Pharmacy* **82**:93–100 (1984a).

Cherubini, E., and North, R. A.: Inhibition of calcium spikes and transmitter release by γ-aminobutyric acid in guinea-pig myenteric plexus. *Br. J. Pharmacol.* **82**:101–105 (1984b).

Cherubini, E., Morta, K., and North, R. A.: Morphine augments a calcium-dependent potassium conductance in guinea pig myenteric neurones. *Br. J. Pharmacol.* **81**:617–622 (1984).

Costa, M., and Furness, J. B. The peristaltic reflex: An analysis of nerve pathways and their pharmacology. *Naunyn-Schmiedeberg's Arch. Pharmacol.* **294**:47–60 (1976).

Costa, M., and Furness, J. B. The sites of action of 5-hydroxytryptamine in nerve–muscle preparations from guinea-pig small intestine and colon. *Br. J. Pharmacol.* **65**:237–248 (1979).

Costa, M., and Furness, J. B.: Nervous control of intestinal motility, in: *Handbook of Experimental Pharmacology* Vol. 59, (G. Bertaccini, ed.), *Mediators and Drugs in Gastrointestinal Motility I. Morphological Basis and Neurophysiological Control*, pp. 279–382. Springer-Verlag, Berlin, Heidelberg, and New York (1982).

Costa, M., Cuello, A. C., Furness, J. B., and Franco, R.; Distribution of enteric neurons showing immunoreactivity for substance P in the guinea-pig ileum. *Neuroscience* **5**:323–331 (1980).

Costa, M., Furness, J. B., Cuello, A. C., Verhofstad, A. A. J., Steinbusch, H. W. J., and Elde, R. P.: Neurons with 5-hydroxytryptamine-like immunoreactivity in the enteric nervous system: Their visualization and reactions to drug treatment. *Neuroscience* **7**:351–363 (1982).

Crema, A.: On the polarity of the peristaltic reflex in the colon, in: *Smooth Muscle: An Assessment of Current Knowledge* (E. Bulbring, A. F. Brading, A. W. Jones, and T. Tomita, eds.), pp. 542–548, Edward Arnold, London (1970).

Dalsgaard, C. J., Hökfelt, T., Elfvin, L.-G., Skirboll, L., and Emson, P.: Substance P containing primary sensory neurons projecting to the inferior mesenteric ganglion: Evidence from combined retrograde tracing and immunohistochemistry. *Neuroscience* **7**:647–654 (1982).

Edwards, F. R., Hirst, A. D. S., and Silinsky, E. M.: Interaction between inhibitory and excitatory synaptic potentials at a peripheral neurone. *J. Physiol. (London)* **259**:647–663 (1976).

Franco, R., Costa, M., and Furness, J. B.: The presence of a cholinergic excitatory input to substance P neurons in the intestine. *Proc. Aust. Physiol. Pharmacol. Soc.* **10**:255P (1979).

Frigo, G. M., Torsoli, A., Lecchini, S., Falaschi, C. F., and Crema, A.: Recent advances in the pharmacology of peristalsis. *Arch. Pharmacodyn. Ther.* **196**(Suppl):9–24 (1972).

Furness, J. B., and Costa, M.: The nervous release and the action of substances which affect intestinal muscle through neither adrenoreceptors nor cholinoreceptors. *Philos. Trans. R. Soc. London Ser. B* **265**:123–133 (1973).

Furness, J. B., and Costa, M.: Ascending and descending enteric reflexes in the isolated small intestine of the guinea pig. *Proc. Aust. Physiol. Pharmac. Soc.* **7**:172P (1976).

Furness, J. B., and Costa, M.: The participation of enteric inhibitory nerves in accommodation of the intestine to distension. *Clin Exp. Pharmacol. Physiol.* **4**:37–41 (1977).

Furness, J. B., and Costa, M. Types of nerves in the enteric nervous system. *Neuroscience* **5**:1–20 (1980).

Furness, J. B., and Costs, M. Neurons with 5-hydroxytryptamine-like immunoreactivity in the enteric nervous system: Their projections in the guinea-pig small intestine. *Neuroscience* **7**:341–349 (1982a).

Furness, J. B., and Costa, M.: Identification of gastrointestinal neurotransmitters, in: *Handbook of Experimental Pharmacology* Vol. 59, (G. Bertaccini, ed.), *Mediators and Drugs in Gastrointestinal Motility. I. Morphological Basis and Neurophysiological Control*, pp. 383–462, Springer-Verlag, Berlin, Heidelberg, and New York (1982b).

Gershon, M. D.: The enteric nervous system. *Annu. Rev. Neurosci.* **4**:227–272 (1981).

Gershon, M. D.: Serotonergic neurotransmission in the gut, in: *Structure of the Gut* (J. M. Polak, S. R. Bloom, N. A. Wright, and M. J. Daly, eds.), pp. 205–219 Glaxo, Ware, United Kingdom (1982).

Gershon, M. D., and Dreyfus, C. F.: Serotonergic neurons in the mammalian gut, in: *Nerves and the Gut* (F. P. Brooks, and P. W. Evers, eds.), pp. 197–206, Charles Slack, Thorofare, New Jersey (1977).

Grafe, P., Wood, J. D., and Mayer, C. J.: Long duration excitability change in guinea-pig myenteric plexus. *Pfluegers Arch. Suppl.* **377**:R30, (1978).

Grafe, P., Mayer, C. J., and Wood, J. D.: Evidence that substance P does not mediate slow synaptic excitation within the myenteric plexus. *Nature (London)* **279**:720–721 (1979a).

Grafe, P., Wood, J. D., and Mayer, C. J.: Fast excitatory postsynaptic potentials in AH (Type 2) neurons of the guinea pig myenteric plexus. *Brain Res.* **163**:349–352 (1979b).

Grafe, P., Mayer, C. J., and Wood, J. D.: Synaptic modulation of calcium-dependent potassium conductance in myenteric neurons in the guinea-pig. *J. Physiol. (London)* **305**:235–248 (1980).

Hanani, M., Baluk, P., and Burnstock, G.: Myenteric neurons express electrophysiological and morphological diversity in tissue culture. *J. Auton. Nerv. Syst.* **5**:155–164 (1982).
Hirst, G. D. S.: Mechanisms of peristalsis. *Br. Med. Bull.* **35**:263–268 (1979).
Hirst, G. D. S., and Lang, R. J.: Permeability change during synaptic inhibition at an enteric neuron. *Proc. Aust. Physiol. Pharmacol. Soc.* **7**:165P (1976).
Hirst, G. D. S., and McKirdy, H. C.: A nervous mechanism for descending inhibition in guinea-pig small intestine. *J. Physiol (London)* **238**:129–143 (1974a).
Hirst, G. D. S., and McKirdy, H. C.: Presynaptic inhibition at mammalian peripheral synapse? *Nature (London)* **250**:430–431 (1974b).
Hirst, G. D. S., and McKirdy, H. C.: Synaptic potentials recorded from neurons of the submucous plexus of guinea-pig small intestine. *J. Physiol. (London)* **249**:369–385 (1975).
Hirst, G. D. S., and Silinsky, E. M.: Some effects of 5-hydroxytryptamine, dopamine and noradrenaline on neurons in the submucous plexus of guinea-pig small intestine. *J. Physiol. (London)* **251**:817–832 (1975).
Hirst, G. D. S., and Spence, I.: Calcium action potentials in mammalian peripheral neurons. *Nature (London) New Biol.* **243**:54–56 (1973).
Hirst, G. D. S., Holman, M. E., and Spence, I.: Two types of neurons in the myenteric plexus of duodenum in the guinea-pig. *J. Physiol. (London)* **236**:303–326 (1974).
Hirst, G. D. S., Holman, M. E., and McKirdy, H. C.: Two descending nerve pathways activated by distension of guinea-pig small intestine. *J. Physiol. (London)* **244**:113–127 (1975).
Hodgkiss, J. P.: Slow synaptic potentials in AH-type myenteric plexus neurons. *Pfluegers Arch.* **391**:331–333 (1981).
Hodgkiss, J. P., and Lees, G. M.: Correlated electrophysiological characteristics of myenteric plexus neurons. *J. Physiol. (London)* **285**:19–20P (1978).
Hodgkiss, J. P., and Lees, G. M.: Morphological features of guinea-pig myenteric plexus neurons, in: *Gastrointestinal Motility* (J. Christensen, ed.), pp. 111–117, Raven Press, New York (1980).
Hodgkiss, J. P., and Lees, G. M.: Morphological studies of electrophysiologically identified myenteric plexus neurons of the guinea pig ileum. *Neuroscience* **8**:593–608 (1983).
Hodgkiss, J. P., and Lees, G. M.: Slow intracellular potentials in AH-neurons of the myenteric plexus evoked by repetitive activation of synaptic inputs. *Neuroscience* **11**:255–261 (1984).
Hökfelt, T., Johansson, O., Ljungdahl, A., Lundberg, J. M., and Schultzberg, M.: Peptidergic neurons. *Nature (London)* **284**: 515–521 (1980).
Holman, M. E.: The intrinsic innervation and peristaltic reflex of the small intestine, in: *Smooth Muscle: An Assessment of Current Knowledge.* (E. Bulbring, A. F. Brading, A. W. Jones, and T. Tomita, eds.), pp. 311–338, Edward Arnold, London (1981).
Holman, M. E., Hirst, G. D. S., and Spence, I.: Preliminary studies of the neurons of Auerbach's plexus using intracellular microelectrodes. *Aust. J. Exp. Biol.* **50**:795–801 (1972).
Jessen, K. R., McConnell, J. D., Purves, R. D., Burnstock, G., and Chamley-Campbell, J.: Tissue culture of mammalian enteric neurons. *Brain Res.* **152**:573–579 (1978).
Jessen, K. R., Mirsky, R., Dennison, M. E., and Burnstock, G.: GABA may be a transmitter in the vertebrate peripheral nervous system. *Nature (London)* **281**:71–74 (1979).
Jessen, K. R., Hills, J. M., Dennison, M. E., and Mirsky, R.: γ-Aminobutyrate as an autonomic neurotransmitter: Release and uptake of [^{3}H]-γ-aminobutyrate in guinea pig large intestine and cultured enteric neurones using physiological methods and electron microscope autoradiography. *Neuroscience* **10**:1427–1442 (1983).
Johnson, S. M., Katayama, Y., and North, R. A.: Slow synaptic potentials in neurons of the myenteric plexus. *J. Physiol (London)* **301**:505–516 (1980a).
Johnson, S. M., Katayama, Y., and North, R. A.: Multiple actions of 5-hydroxytryptamine on myenteric neurons of the guinea-pig ileum. *J. Physiol. (London)* **304**:459–470 (1980b).
Johnson, S. M., Katayama, Y., Morita, K., and North, R. A.: Mediators of slow synaptic potentials in the myenteric plexus of the guinea-pig ileum. *J. Physiol. (London)* **320**:175–186 (1981).

Julé, Y.: Nerve-mediated descending inhibition in the proximal colon of the rabbit. *J. Physiol. (London)* **309**:487–498 (1980).

Katayama, Y., and North, R. A.: Does substance P mediate slow synaptic excitation within the myenteric plexus? *Nature (London)* **274**:387–388 (1978).

Katayama, Y., North, R. A., and Williams, J. T.: The action of substance P on neurons of the myenteric plexus of the guinea-pig small intestine. *Proc. R. Soc. London Ser. B* **206**:191–208 (1979).

King, B. F., and Szurszewski, J. H.: Intracellular recordings from vagally innervated intramural neurons in opossum stomach. *Am. J. Physiol.* **246**:G209–212 (1984).

Kosterlitz, H. W.: Intrinsic intestinal reflexes. *Am. J. Dig. Dis.* **12**:245–254 (1967).

Kosterlitz, H. W.: Intrinsic and extrinsic nervous control of motility of the stomach and the intestines, in: *Handbook of Physiology*, Section 6, *Alimentary Canal*, Vol. IV, *Motility* (C. F. Code, eds.), pp. 2147–2171, The American Physiological Society, Washington D. C. (1968).

Kosterlitz, H. W., and Lees, G. M.: Pharmacological analysis of intrinsic intestinal reflexes. *Pharmacol. Rev.* **16**:301–339 (1964).

Kosterlitz, H. W., and Lydon, R. J.: Spontaneous electrical activity and nerve-mediated inhibition in the innervated longitudinal muscle strip of the guinea-pig ileum. *J. Physiol. (London)* **200**:126–128P (1969).

Kottegoda, S. R.: An analysis of possible nervous mechanisms involved in the peristaltic reflex. *J. Physiol (London)* **200**:687–712 (1969).

Kottegoda, S. R.: Peristalsis of the small intestine, in: *Smooth Muscle* (E. Bülbring, A. F. Brading, A. Jones, and T. Tomita, eds.), pp. 525–541, Edward Arnold, London (1970).

Krantis, A., and Kerr, D. I. B.: Autoradiographic localization of ^{3}H-gamma aminobutyric acid in the myenteric plexus of the guinea-pig small intestine. *Neurosci. Lett.* **23**:263–268 (1981).

Krantis, A., Costa, M., Furness, J. B., and Orbach, J.: Gamma-aminobutyric acid stimulates intrinsic inhibitory and excitatory nerves in the guinea-pig intestine. *Eur. J. Pharmacol.* **67**:461–468 (1980).

Llewellyn-Smith, I. J., Furness, J. B., Wilson, A. J., and Costa, M.: Organization and fine structure of enteric ganglia, in: *Autonomic Ganglia* (L.-G. Elfvin, ed.), pp. 145–182, John Wiley, Chichester and New York (1983).

Mayer, C. J., Wood, J. D., and Grafe, P.: Synaptic activation of so-called AH neurons in guinea-pig myenteric plexus. *Pfluegers Arch. Suppl.* **377**:R30 (1978).

Morita, K., and North, R. A.: Enkephalin reduces excitability of cellular processes of myenteric neurones. *Neuroscience* **6**:1943–1951 (1981).

Morita, K., North, R. A., and Katayama, Y.: Evidence that substance P is a neurotransmitter in the myenteric plexus. *Nature (London)* **287**:151–152 (1980).

Morita, K., North, R. A., and Tokimasa, T.: The calcium-activated potassium conductance in guinea-pig myenteric neurons. *J. Physiol. (London)* **329**:341–354 (1982a).

Morita, K., North, R. A., and Tokimasa, T.: Muscarinic agonists inactivate potassium conductance in guinea-pig myenteric neurons. *J. Physiol. (London)* **333**:125–139 (1982b).

Morita, K., North, R. A., and Tokimasa, T.: Muscarinic presynaptic inhibition of synaptic transmission in myenteric plexus of guinea-pig ileum. *J. Physiol. (London)* **333**:141–149 (1982c).

Nemeth, P. R., Zafirov, D., and Wood, J. D.: Forskolin mimics slow synaptic excitation in myenteric neurons. *Eur. J. Pharmacol.* **101**:303–304 (1984).

Nishi, S., and North, R. A.: Intracellular recording from the myenteric plexus of the guinea-pig ileum. *J. Physiol. (London)* **231**:471–491 (1973).

North, R. A.: The calcium-dependent slow after-hyperpolarization in myenteric plexus neurons with tetrodotoxin-resistant action potentials. *Br. J. Pharmacol.* **49**:709–711 (1973).

North, R. A.: Opiates, opioid peptides and single neurons (minireview). *Life Sci.* **24**:1527–1546 (1979).

North, R. A.: Electrophysiology of the enteric nervous system (commentary). *Neuroscience* **7**:315–325 (1982a).

North, R. A.: Electrophysiology of the enteric neurons, in: *Handbook of Experimental Pharmacology* Vol. 59, (G. Bertaccini, ed.), *Mediators and Drugs in Gastrointestinal Motility. I. Morphological Basis and Neurophysiological Control*, pp. 145–179, Springer-Verlag, Berlin, Heidelberg, and New York (1982b).

North, R. A., and Egan, T. M.: Electrophysiology of peptides in the peripheral nervous system. *Br. Med. Bull.* **38**:291–296 (1982).

North, R. A., and Egan, T. M.: Action and distributions of opioid peptides in peripheral tissues. *Br. Med. Bull.* **39**:71–75 (1983).

North, R. A., and Henderson, G.: Action of morphine on guinea-pig myenteric plexus and mouse vas deferens studied by intracellular recording. *Life Sci.* **17**;63–66 (1975).

North, R. A., and Nishi, S.: Properties of the ganglion cells of the myenteric plexus of the guinea-pig ileum determined by intracellular recording, in: Proc. Fourth Int. Symp. Gastrointestinal Motility (E.E. Daniel, ed.), pp. 667–676, Mitchell Press, Vancover (1974).

North, R. A., and Nishi, S.: The soma spike in myenteric plexus neurons with a calcium-dependent after-hyperpolarization, in: *Physiology of Smooth Muscle* (E. Bulbring and M. F. Shuba, eds.), pp. 303–307, Raven Press, New York (1976).

North, R. A., and Tokimasa, T.: Muscarinic synaptic potentials in guinea-pig myenteric plexus neurons. *J. Physiol. (London)* **333**:151–156 (1982).

North, R. A., and Tokimasa, T.: Depression of calcium dependent potassium conductance of guinea pig myenteric neurones by muscarinic agonists. *J. Physiol. (London)* **342**:253–266 (1983).

North, R. A., and Tokimasa, T.: Muscarinic suppression of a calcium activated potassium conductance. *Trends Pharmacol. Sci. Suppl.* Jan., pp. 35–38 (1984a).

North, R. A., and Tokimasa, T.: The time course of muscarinic depolarization of guinea-pig myenteric neurones. *Br. J. Pharmacol.* **82**:85–91 (1984b).

North, R. A., and Williams, J. T.: Extracellular recording from the myenteric plexus of the guinea-pig ileum and the action of morphine. *Eur. J. Pharmacol.* **45**:23–33 (1977).

North, R. A., Henderson, G., Katayama, Y., and Johnson, S. M.: Electrophysiological evidence for presynaptic inhibition of acetylcholine release by 5-hydroxytryptamine in the enteric nervous system. *Neuroscience* **5**:581–586 (1980a).

North, R. A., Katayama, Y., and Williams, J. T.: Action of peptides on enteric neurons, in: *Neural Peptides and Neuronal Communication* (E. Costa and M. Trabucchi, eds.), pp. 83–91, Raven Press, New York (1980b).

North, R. A., Morita, K., and Tokimasa, T.: Peptide actions on autonomic nerves, in: *Systemic Role of Regulatory Peptides* (S. R. Bloom, J. M. Polak, and E. Lindenlaub, eds.), Symposia Medica Hoechst, 18, pp. 78–87, F. K. Scattauer, Stuttgart (1982).

Nozdrachev, A. D., and Vataev, I.: Neuronal electrical activity in the submucosal plexus of the cat small intestine. *J. Auton. Nerv. Syst.* **3**:45–53 (1981).

Ohkawa, H., and Prosser, C. L.: Electrical activity in myenteric and submucous plexuses of cat intestine. *Am. J. Physiol.* **222**:1412–1419 (1972a).

Ohkawa, H., and Prosser, C. L.: Functions of neurons in enteric plexuses of cat intestine. *Am. J. Physiol.* **222**:1420–1426 (1972b).

Shepherd, G. M.: Introduction: The nerve impulse and the nature of nervous function, in: *Neurons without Impulses* (A. Roberts and B. M. H. Bush, eds.), Society for Experimental Biology Seminar Series, No. 6, pp. 1–27, Cambridge University Press, Cambridge (1981).

Surprenant, A.: Slow excitatory synaptic potential recorded from neurones of guinea pig submucous plexus. *J. Physiol. (London)* **351**:343–361 (1984a).

Surprenant, A.: Two types of neurones lacking synaptic input in the submucous plexus of guinea-pig small intestine. *J. Physiol. (London)* **351**:378 (1984b).

Szerb, J. C.: Correlation between acetylcholine release and neuronal activity in the guinea-pig ileum myenteric plexus; Effects of morphine. *Neuroscience* **7**:327–340 (1982).

Tokimasa, T., Morita, K., and North, R. A.: Opiates and clonidine prolong calcium dependent afterhyperpolarizations. *Nature (London)* **249**:162–163 (1981).

Vermillion, D. L., Gillespie, J. P., Cooke, A. R., and Wood, J. D.: Does 5-hydroxytryptamine influence "purinergic" inhibitory neurons in the intestine? *Am. J. Physiol.* **237**:E198–E202 (1979).

Wallis, D.: Neuronal 5-hydroxytryptamine receptors outside the central nervous system (minireview). *Life Sci.* **29**:2345–2355 (1981).

Wood, J. D.: Electrical activity from single neurons in Auerbach's plexus. *Am. J. Physiol.* *219*:159–169 (1970).

Wood, J. D.: Neurophysiology of Auerbach's plexus and control of intestinal motility. *Physiol. Rev.* **55**:307–324 (1975).

Wood, J. D.: Intrinsic neural control of intestinal motility. *Annu. Rev. Physiol.* **43**:33–51 (1981).

Wood, J. D.: Enteric neurophysiology. (Editorial review), *Am. J. Physiol.* **247**:G585–G598 (1984).

Wood, J. D., and Mayer, C. J.: Intracellular study of electrical activity of Auerbach's plexus in guinea-pig small intestine. *Pfluegers Arch.* **374**:265–275 (1978a).

Wood, J. D., and Mayer, C. J.: Slow synaptic excitation mediated by serotonin in Auerbach's plexus. *Nature (London)* **276**:836–837 (1978b).

Wood, J. D., and Mayer, C. J.: Intracellular study of tonic-type enteric neurons in guinea pig small intestine. *J. Neurophysiol.* **42**:569–581 (1979a).

Wood, J. D., and Mayer, C. J.: Serotonergic activation of tonic-type enteric neurons in guinea pig small bowel. *J. Neurophysiol.* **42**:582–592 (1979b).

Wood, J. D., and Mayer, C. J.: Adrenergic inhibition of serotonin release from neurons in the guinea pig Auerbach's plexus. *J. Neurophysiol.* **42**:594–603 (1979c).

Wood, J. D., Mayer, C. J., Ninchoji, T., and Erwin, D. N.: Effects of depleted calcium on electrical activity of neurons in Auerbach's plexus. *Am. J. Physiol.* **236**:C78–C86 (1979).

Wood, J. D., Grafe, P., and Mayer, C. J.: Comparison of the action of 5-hydroxytryptamine and substance P on intracellularly recorded electrical activity of myenteric neurons, in: *Gastrointestinal Motility* (J. Christensen, ed.), pp. 131–138, Raven Press, New York (1980).

Yokoyama, S., and Ozaki, T.: Polarity of effects of stimulation of Auerbach's plexus on longitudinal muscle. *Am. J. Physiol.* **235**:E345–E353 (1978).

Yokoyama, S., and Ozaki, T.: Effects of gut distension on Auerbach's plexus and intestinal muscle. *Jpn. J. Physiol.***30**:143–160 (1980).

IV

Spinal and Reflex Activities of the Ganglia

17

The Pharmacology of Sympathetic Preganglionic Neurons

PATRICIA SHINNICK-GALLAGHER, JOEL P. GALLAGHER, and M. YOSHIMURA

I. INTRODUCTION

The sympathetic preganglionic neurons (SPNs) are found in the intermediolateral (IML) cell column of the thoraco lumbar spinal cord. In cats, presynaptic fibers that originate in the medulla and higher centers approach the IML cell column primarily from the lateral funiculus (Rethelyi, 1972). Three types of presynaptic vesicles suggesting the presence of several different transmitters have been identified in the terminals (Rethelyi, 1972). Dahlstrom and Fuxe (1965) found that in rats, a high density of monoamine-containing bulbospinal fibers terminated in close proximity to SPNs. A similar distribution of monoamine neurons was reported for cats, and fibers associated with these terminals were located in the lateral funiculus (Coote and Macleod, 1974b; Coote *et al.*, 1981). Descending adrenergic (Soller, 1977) and serotonergic (Anden *et al.*, 1964; Mensah *et al.*, 1974) pathways have also been described for amphibia.

Recent investigations have identified the presence of leu- and met-enkephaline-, substance-P-, and somatostatin-immunoreactive fibers that

PATRICIA SHINNICK-GALLAGHER and JOEL P. GALLAGHER • Department of Pharmacology, University of Texas Medical Branch, Galveston, Texas. M. YOSHIMURA • Department of Physiology, Kurume University School of Medicine, Kurume, Japan.

surround and appear to contact SPNs of cats (Holets and Elde, 1983), rats (Ljungdahl *et al.*, 1978; Holets and Elde, 1982), and hamsters (Hancock, 1982). Somatostatin as well as oxytocin and neurophysin immunoreactivity was comparatively sparse (Holets and Elde, 1982, 1983). These peptidergic fibers appeared to arise primarily from descending tracts, since met-enkephalin, neurophysin, oxytocin, somatostatin, and substance P reactivity in the IML was totally depleted after high (C_6) spinal transection (Holets and Elde, 1982, 1983). It is also interesting to note that the enkephalin immunoreactivity coincided with regions containing SPNs and resembled the ladderlike distribution of SPNs in the spinal transection (Romagnano and Hamill, 1984). In addition, thyrotropin-releasing hormone (TRH)-immunoreactive fibers and terminals probably originating from descending inputs have also been localized in the lateral horn (Gilbert *et al.*, 1982).

This chapter reviews investigations in the course of which SPN activity was measured with either intra- or extracellular microelectrodes or was monitored by recording spinal sympathetic reflexes (SSRs). The use of SSRs as an index of sympathetic preganglionic activity seems justified, since Sato and Schmidt (1971) concluded that any afferent input has a twofold action on the sympathetic nervous system, a local action involving sympathetic reflexes at the segmental level and a more generalized action via supraspinal sympathetic reflex centers. It is important to note that pools of SPNs may be activated discretely by various afferents and may respond differentially to neurotransmitters.

II. APPROACHES

Since in general neurotransmitter candidates penetrate the blood–brain barrier poorly and since they induce widespread systemic effects, systemic administration of these substances does not offer the best avenue for their study. Two other approaches to the study of neurotransmitter effects have been used *in vivo*, namely, iontophoresis and precursor loading. Both approaches have inherent disadvantages.

There are several problems in evaluating data obtained with iontophoretic methods. It has been shown that neuronal responses to iontophoretically applied norepinephrine (NE) and serotonin [5-hydroxytryptamine (5-HT)] may vary depending on pH (Frederickson *et al.*, 1971; Jordan *et al.*, 1972). In addition, since only a small fraction of the total number of SPNs can be sampled, entire pools of SPNs may be overlooked. Other limitations of or problems with iontophoretic techniques that are generally characteristic for this methodology wherever it is applied also pertain when it is used for the study of SPNs, to wit: (1) Does the drug applied iontophoretically interact with specific receptors on the cell? (2) If it does, are these receptors functionally important? And (3) what is the concentration of the applied drug?

The second approach used to overcome the problem posed by the blood–brain barrier is precursor loading. This technique relies on several assumptions: (1) The precursors, L-3,4-dihydroxyphenylalanine (L-dopa), L-5-hydroxytryptophan (5-HTP), and L-tryptophan, do not exhibit direct effects on SPNs; (2) these substances will be selectively taken up into the appropriate terminals and subsequently metabolized to NE or 5-HT; and (3) as a result of increased synthesis, the functional activity of the loaded terminal is increased, resulting in an overflow of monoamines from the terminal (cf. Anderson and Shibuya, 1966; Baker and Anderson, 1970). Furthermore, several investigators have questioned the reliability of this technique because some precursors have been shown to affect other transmitter systems in addition to those present in their target neurons (Moir and Eccleston, 1968; Neumayr *et al.*, 1974).

Two *in vitro* preparations have been developed that eliminate some of the problems encountered *in vivo*. With the cat cord slice preparation (Yoshimura *et al.*, 1981) and the frog (Shinnick-Gallagher, 1979) and neonatal rat (McKenna and Schramm, 1983) hemisected spinal cord, known concentrations of drugs can be applied to SPNs. Furthermore, common problems, such as those due to anesthesia, movement artifact, unwanted sensory input, and changing cardiovascular status, that plague *in vivo* preparations are eliminated when the hemisected spinal cord or slice preparations are used. However, there are also some disadvantages associated with the *in vitro* preparations. Discrete stimulation of specific afferents is not possible in the case of the slice preparation and is limited to whole root afferents in that of the hemisected cord, so that an observed effect cannot be correlated with a particular physiological function. Another problem with the slice preparation is identification of the neuron, since both SPNs and interneurons may be found in the IML cell column. Furthermore, when a population response is being recorded, as in the case of the frog hemisected cord preparation, it is difficult to determine the particular site of drug action. It is necessary, therefore, to utilize several types of approaches in studying the pharmacology of SPNs.

III. PHARMACOLOGY

A. Serotonin

The cell bodies for the descending serotonergic neurons are located in caudal raphe nuclei (Oliveras *et al.*, 1977; Basbaum *et al.*, 1978; Tohyama *et al.*, 1979; Bowker *et al.*, 1981a) and in the central gray and midbrain reticular formation (Bowker *et al.*, 1981b). Loewy and colleagues (Loewy, 1981; Loewy and McKellar, 1981) showed that serotonin (5-HT) neurons from the raphe project to the IML cell column, and biochemical analysis indicated that the highest serotonin concentration in the spinal gray matter is encountered in the IML cell column (Zivin *et al.*, 1975). Finally, serotonin was localized in the IML using fluorescent histochemical techniques (Dahlstrom and Fuxe, 1965) and immunofluorescent techniques (Steinbusch, 1981; Holets and Elde, 1982; 1983).

In 1967, DeGroat and Ryall (1967) studied the effect of 5-HT applied iontophoretically to the SPNs. Application of 5-HT excited about 76% and had no effect on 24% of spontaneously active neurons (for the matter of spontaneous vs. silent neurons, see Chapter 18). In a later study, Coote *et al.* (1981) found that 5-HT excited 52%, inhibited 14%, and had no effect on 34% of SPNs. All types of SPNs, cardiac or noncardiac, spontaneously active or silent neurons (which could be activated by DL-homocysteic acid or glutamate), were affected by 5-HT. Coote *et al.* (1981) suggested that the inhibition was due to an action of the creatinine salt of 5-HT, since it was not observed with the bimaleate salt. These investigators also suggested that 5-HT causes considerable depolarization of the SPNs, since 5-HT decreased the amplitude of extracellularly recorded action potentials.

Recently, Kadzielawa (1983b) observed effects of iontophoretically applied 5-HT hydrochloride that were similar to those reported by Coote *et al.* (1981) and found that the excitatory effect of 5-HT exhibited tachyphylaxis and was antagonized by the 5-HT-receptor antagonists methysergide and cinanserin. Kadzielawa (1983b) also reported that 5-HT exhibited a biphasic effect, an inhibition being followed by an excitation. In his hands, there was no difference in excitation induced by either hydrochloride or creatinine salts of 5-HT; however, 5-HT-induced inhibition was recorded in 12.5% of SPNs. It is possible, on the basis of the available data, that a small percentage of SPNs may indeed be inhibited by 5-HT (cf. Higashi, 1977).

On the other hand, McCall (1983) recently found only excitatory effects of 5-HT applied as the creatinine salt on SPNs; these effects could be observed at low iontophoretic ejection currents. These data obtained *in vivo* are supported by data obtained from *in vitro* recordings. In the case of intracellular recordings obtained from IML horn cells in a cat cord slice preparation, Yoshimura and Nishi (1982) found that 5-HT (10–30 μM) depolarized the membrane and increased the membrane resistance in 68% of the neurons; 32% of the neurons did not respond to 5-HT. Thus, the results of the studies in which 5-HT was applied directly to SPNs indicate that 5-HT may act as a depolarizing, excitatory compound.

Results obtained with 5-HT-receptor antagonists support the concept of an excitatory action for 5-HT. It was found that methysergide and cinanserin decrease spontaneous (McCall and Humphrey, 1981; McCall, 1983) and reflex (Gilbey *et al.*, 1981) sympathetic discharges. The effect of these antagonists was abolished following depletion of 5-HT by *p*-chlorophenylalanine (PCPA) in intact (McCall and Humphrey, 1981) as well as spinalized animals (McCall, 1983). These data suggest that endogeneous 5-HT excites the SPN directly or indirectly via an interneuron.

Conflicting results have also been reported with the precursor 5-HTP. Hare *et al.* (1972) and Neumayr *et al.* (1974) observed that 5-HTP as well as L-tryptophan depressed SSRs. In addition, Coote and Macleod (1974b) found that 5-HTP inhibited spinal spontaneous and supraspinal sympathetic reflex activity. On the basis of those findings and of studies in which localized stimulation of the medulla was combined with fluorescence localization of serotonergic cell bodies, Coote and Macleod (1974b) suggested that serotonergic fibers descend into the spinal sympathetic neuron pool from the nucleus raphe pallidus. Subsequently, Gilbey *et al.* (1981) found that stimulation of the raphe inhibited sympathetic reflex activity and that this effect was antagonized by lysergic acid diethylamide. In contrast, the results of a study of Haeusler (1977), who used high intravenous doses of 5-HTP, suggested an excitatory role for this innervation. DeGroat *et al.* (1975) also reported that spinal vesicosympathetic reflexes were enhanced or unaffected by 5-HTP administered in doses similar to those employed in the study of Coote and Macleod (1974b); however, DeGroat *et al.* (1975) also found that 5-HTP used at lower doses depressed somatosympathetic reflexes. Coote *et al.* (1981) and Gilbey *et al.* (1981) suggested that 5-HT may normally be released, not onto the SPNs, but onto inhibitory interneurons close to the SPNs, since it seemed that 5-HT exerted primarily an excitatory action on SPNs, while 5-HTP exhibited mainly an inhibitory action on spinal reflexes, perhaps through a bulbospinal pathway.

In summarizing the information concerning the effects of 5-HT on the SPNs, it is important to note that dose–response analyses were not performed in the studies discussed above. Despite the lack of that information, two consistent findings suggest a physiologically relevant excitatory role for 5-HT: (1) Direct application of 5-HT primarily excites SPNs and (2) 5-HT antagonists decrease spontaneous firing of SPNs and reflex sympathetic discharge.

B. Norepinephrine

The noradrenergic input to SPNs is thought to originate in the A5 area of the brainstem in rats (Loewy *et al.*, 1979) and the A11 cell group in cats (Fleetwood-Walker and Coote, 1981). There is also biochemical and histochemical evidence to suggest the presence of separate dopaminergic (Commissiong *et al.*, 1978; Blessing and Chalmers, 1979; Bjorklund and Skagerberg, 1979) and adrenergic inputs to the sympathetic preganglionic cell column (Hökfelt *et al.*, 1974; Zivin *et al.*, 1975).

In an early study carried out with spinal cats, Hongo and Ryall (1966) found that iontophoretic application of NE had no effect on antidromically activated ($N = 12$), spontaneously firing ($N = 3$), or synaptically activated ($N = 2$) SPNs of the seventh thoracic segment. Even cells excited by DL-homocysteic acid ($N = 3$) were not inhibited by NE. In another study, NE depressed 4 of 10 spontaneously firing SPNs (DeGroat and Ryall, 1967). Coote *et al.* (1981) found that NE inhibited 30 of 46 neurons tested, 16 neurons were unaffected, and none was excited; all types of SPNs were depressed, including cardiac and noncardiac neurons, synaptically active cells, and silent neurons activated by DL-homocysteic acid or glutamate. Coote *et al.* (1981) also reported that epinephrine depressed 2 of 16 SPNs, while dopamine depressed 3 of 11 neurons. Recently, Kadzielawa (1983a) observed that NE applied iontophoretically inhibited the majority of SPNs, although it exerted no effect on some SPNs; however, in contrast to the results of Coote *et al.* (1981), Kadzielawa (1983a) found that a small percentage of SPNs (14%) were excited by NE and that epinephrine induced effects that were similar to those observed with NE. The receptor involved in the inhibitory effects of iontophoretically applied catecholamines has been characterized in an *in vivo* study by Guyenet and Cabot (1981). These authors quantitated the inhibitory effect of drugs by calculating the amount of charge necessary to decrease firing to 50% of the control level. Using that value as an index of drug potency, Guyenet and Cabot (1981) determined the following rank order of inhibition: clonidine > α-methyl-norepinephrine > epinephrine > NE > phenylephrine. Since α_2-antagonists, yohimbine and piperoxan, but not the α_1-antagonist prazosin, antagonized the effects of clonidine and NE, Guyenet and Cabot (1981) concluded that the inhibitory effects of catecholamines and clonidine on SPNs are mediated through α_2-adrenoceptors. Recording intracellularly *in vitro* from IML horn cells, Yoshimura *et al.*, (1981) found that superfusion of NE (10–30 μM) depolarized the membrane and increased membrane resistance in 57% of the neurons; a hyperpolarization accompanied be decreased membrane resistance as well as a biphasic effect (i.e., hyperpolarization followed by depolarization) were also observed in 10 and 14% of the neurons, respectively.

There may be numerous reasons for the lack of consistency of the results obtained by several investigators with direct application of NE to SPNs. The pH-dependent variation in NE effects (Jordan *et al.*, 1972) and small sampling of SPNs may account for some inconsistencies. It is possible that different pools of SPNs or neurons other than SPNs were sampled in the respective studies, since an NE-induced depolarization could be recorded in the majority of IML neurons *in vitro*, while relatively few neurons were excited by NE applied iontophoretically *in vivo*.

The results obtained in studies utilizing systemic rather than localized application of NE are also controversial. Hare *et al.* (1972) reported that L-dopa (15–50 mg/kg) produced a biphasic effect on SSRs in spinal cats pretreated with phenoxybenzamine. After an initial depression (30 min), the reflexes returned to control level in 1 hr and increased in size after $1\frac{1}{2}$–2 hr. Hare and colleagues suggested that the depression of reflexes was due to release of 5-HT (Ng *et al.*, 1970) and the enhancement of reflexes to released NE, since the depression was attenuated by pretreatment with PCPA, while the enhancement was reversed by chlorpromazine. Furthermore, L-dopa (15–50 mg/kg) depressed discharges of SPNs evoked by stimulation of a descending excitatory pathway and facilitated the discharges in animals pretreated with PCPA (Neumayr *et al.*, 1974).

In contrast, Coote and Macleod (1974b), using a similar experimental protocol, found that L-dopa (40–100 mg/kg) plus a peripheral dopa decarboxylase inhibitor, MK 486, only depressed SSRs. These authors suggested that a descending noradrenergic system originating in the ventrolateral medulla inhibits SPNs. They argued that there was no evidence to suggest that maximal somatosympathetic reflexes could be facilited by activation of supraspinal pathways and that no correlation could be made between position of descending sympathoexcitatory tracts in the spinal cord and descending NE-containing axons. Coote and Macleod (1975) have also suggested that bulbospinal noradrenergic pathways may be involved in baroreceptor inhibition at the spinal level.

It is important to note that the concentrations of L-dopa that reached SPNs in the study of Coote and Macleod (1974b) were probably much higher than those that reached SPNs in the studies of Hare *et al.* (1972) and Neumayr *et al.* (1974). It is possible that higher concentrations of L-dopa depress and much lower concentrations facilitate SSRs. Finally, the facilitation of SSRs may be due to indirect effects of the peripheral metabolites of L-dopa, NE, dopamine, or epinephrine. The effects of increased synthesis and release of these metabolites in the periphery may account for the biphasic effect observed by Hare *et al.* (1972), since a dopa decarboxylase inhibitor was not used to prevent the breakdown of peripheral L-dopa in those studies.

Other investigators employed the centrally active α-adrenergic agonist clonidine to mimic the effect of NE on sympathetic reflexes. Baum and Shropshire (1977) found that clonidine (1–32 μg/kg) inhibited spontaneous sympathetic outflow activity as well as supraspinal and spinal reflexes. The spontaneous activity was most sensitive to inhibition by clonidine, whereas supraspinal and spinal reflexes were less sensitive. In contrast, clonidine did not attenuate baroreceptor-mediated inhibition of sympathetic outflow (cf. Coote and Macleod, 1975). Similarly, Dembowsky *et al.* (1981) found that clonidine (20 μg/kg) depressed the amplitude of a somatosympathetic spinal reflex during cold blockade; this effect was reversed by the α-adrenoceptor antagonist yohimbine. These authors found that there was a tonic sympathetic inhibition of SSRs that was independent of baroreceptor input. A variation in sensitivity of reflex pathways to clonidine was reported by other investigators. Vesicosympathetic spinal reflexes were depressed by L-dopa (15–30 mg/kg), by p-methoxyphenylethylamine, an agent thought to release NE from nerve terminals (DeGroat and Lalley, 1973), and by clonidine (DeGroat and Douglas, 1975; DeGroat *et al.*, 1975); furthermore, L-dopa depressed this reflex in animals with an intact spinal cord, but not in chronic spinal animals (DeGroat *et al.*, 1975). However, clonidine had no effect on lumbar sympathetic inhibitory outflow to the large intestine (Krier *et al.*, 1979), which is thought to be generated by local circuits in the spinal cord and which is somewhat influenced by baroreceptor inhibition generated by stimulation of certain afferents (DeGroat and Krier, 1979). Franz *et al.* (1978) also observed that clonidine depressed SSRs as well as descending reflex pathways. In keeping with their hypothesis that 5-HT inhibits and NE excites SPNs, these authors suggested that the clonidine-induced depression of spinal reflexes was due to stimulation of inhibitory 5-HT receptors on SPNs, since effects of clonidine and 5-HTP were blocked by tolazaline, an α-adrenoceptor antagonist, but not by chlorpromazine. It is difficult to evaluate results obtained using tolazaline as an experimental tool to discriminate between NE and 5-HT receptors, since tolazaline is known to block 5-HT as well as NE effects.

Franz and his colleagues (C. F. Sangdee and Franz, 1983) have recently modified their original hypothesis to account for the direct depressant effect of NE and epinephrine on SPNs. Their data showed that inhibitors of phenylethanolamine-*n*-methyltransferase, the enzyme that converts NE to epinephrine in neurons, enhanced an intraspinal reflex involving descending epinephrine and NE inputs to SPNs by selectively depleting epinephrine from the epinephrine pathways; they inferred that epinephrine pathways are inhibitory to SPNs and suggested that the inhibitory somal α_2-adrenoceptors (which are activated preferentially during iontophoretic application) are normal substrates for bulbospinal epinephrine pathways and that those epinephrine inputs along with excitatory NE pathways terminating on the dendrites (Glazen and Ross, 1980) regulate the excitability of SPNs.

Results obtained in studies employing neurochemical measurements and intracisternally administered neurotoxin, 6-hydroxydopamine (6-OHDA), (Chalmers, 1975; Chalmers and Reid, 1972; Chalmers and Wurtman, 1971) suggested that bulbospinal catecholaminergic nerves facilitate sympathetic outflow. Anatomical evidence correlated with physiological data led Hökfelt and Fuxe (Fuxe *et al.*, 1975) to suggest that central noradrenergic pathways may exert an excitatory and adrenergic pathways an inhibitory effect on SPNs. Coote and Macleod (1977), who administered 6-OHDA intraspinally, obtained data indicating that bulbospinal neurons convey inhibitory effects of baroreceptors on spinal sympathetic activity. Results obtained with 6-OHDA are difficult to interpret because intraspinal or intracerebral administration of 6-OHDA leads to its diffusion over many nerves of a particular transmitter type, rather than to a differential effect on the specific tracts being studied; also, 6-OHDA may have unwanted nonspecific effects on other neurons (Butcher *et al.*, 1974; Poirier *et al.*, 1973). Altogether, biochemical as well as anatomical data obtained with 6-OHDA must be extrapolated, and it must be remembered that the region affected may contain many subpopulations of monaminergic neurons having different or opposite functions.

We have studied the effects of catecholamines in the isolated hemisected spinal cord of the frog (Undesser *et al.*, 1981). NE (0.56 nM to 56 μM) inhibited spinal sympathetic reflexes elicited by supramaximal but not by submaximal stimulation in a concentration-dependent manner. This inhibition was mimicked by other α-adrenoceptor agonists (clonidine) and antagonized by phentolamine. Epinephrine and also isoproterenol facilitated reflex activity; however, in the presence of propranolol, an inhibitory effect of epinephrine was observed. The high concentrations of dopamine employed (100 μM) inhibited SSRs; this effect could be antagonized by phentolamine. In addition, L-dopa (20 μM) superfusion produced a delayed-onset, long-lasting inhibition of SSRs that was prevented by pretreatment with R0-4-4602, a decarboxylase inhibitor. These data suggest that the effect of L-dopa was due to its metabolites such as dopamine, NE, or epinephrine. It was concluded that in the frog, SSRs could be inhibited by activation of α_2-adrenoceptors and facilitated by activation of β_2-adrenoceptors (Undesser *et al.*, 1981). On the basis of these data as well as the results obtained in previous studies, it seems likely that NE has a direct inhibitory effect on SPNs and that facilitatory adrenergic receptors may be present as well; these receptors may not ordinarily be functionally important, but may prove useful therapeutically.

Since there are excitatory descending pathways from several regions of the brain, some of which pass through the medulla and influence spinal sympathetic neurons (Coote and Macleod, 1974a), it is possible that neurotransmitters other than catecholamines or 5-HT may also influence sympathetic outflow.

C. Acetylcholine

Acetylcholine (ACh) was applied iontophoretically to a total of 25 SPNs *in vivo* and was found to be inactive (Coote *et al.*, 1981; DeGroat and Ryall, 1967; Hongo and Ryall, 1966). More than 50% of the IML horn cells tested by means of intracellular recording methods were depolarized by superfusion of ACh (1 mM); this ACh-induced depolarization was prevented by D-tubocurarine (50 μM) (Yoshimura and Nishi, 1982). Similar effects of ACh were observed in recordings obtained from IML horn neurons that were clearly identified as SPNs. The depolarization was often followed by a long-lasting burst of excitatory postsynaptic potentials that was blocked by atropine (50 μM) (Yoshimura and Nishi, 1982). Altogether, it appears that some SPNs lack cholinoceptors, although IML neurons and other spinal cells (such as dorsal horn and ventral horn neurons including the Renshaw cells) can respond muscarinically or nicotinically (Ryall, 1983; Nowak and MacDonald, 1983).

D. Amino Acids

DL-Homocysteic acid applied *in vivo* by iontophoresis usually excited SPNs (Backman and Henry, 1983a; Coote *et al.*, 1981; DeGroat and Ryall, 1967; Hongo and Ryall, 1966). Similarly, glutamate (Coote *et al.*, 1981) as well as aspartate (Backman and Henry, 1983a) applied iontophoretically to SPNs excited the latter *in vivo*. The majority of SPNs from which *in vivo* records were obtained appeared to show equipotent sensitivities to glutamate and aspartate applied iontophoretically; however, about 30% of SPNs could not be induced to fire during application of the excitatory amino acids (Backman and Henry, 1983a). The inhibitory amino acids γ-aminobutyric acid (GABA) and glycine inhibited spontaneous activity and blocked antidromic invasion of the soma–dendritic region of SPNs; these effects were antagonized by bicuculline and strychnine, respectively (Backman and Henry, 1983b). Glycine has been implicated in possible recurrent inhibition of these neurons, since strychnine abolished the inhibitory effect of antidromic activation of one pool of SPNs on the orthodromic activation of another pool (Lebedev *et al.*, 1980). Furthermore, baclofen (*p*-chlorophenyl GABA), a drug that is used clinically as an antispastic agent, inhibited at a concentration of 10 μM both spontaneously firing and reflexly activated SPNs in the isolated neonatal rat spinal cord (McKenna and Schramm, 1984).

In vitro intracellular recordings from IML horn cells showed that glutamate (0.5 mM) markedly depolarized 80% of these neurons, while only about 50% of the cells were depolarized by aspartate (0.5 mM). Glutamate-induced depolarization was usually accompanied by decreased membrane resistance, but occasionally an increased membrane resistance was recorded. On the other hand, aspartate depolarizations were always associated with an increased resistance (Yoshimura and Nishi, 1982). Glycine (0.5 mM) and GABA (0.5 mM) usually hyperpolarized IML horn cells and decreased membrane resistance (Yoshimura and Nishi, 1982).

Thus, it seems that amino acids may be candidates for the status of excitatory or inhibitory neurotransmitters, or both, to SPNs.

Some immunoreactive fibers and terminals in the lateral horn are thought to arise from neurons in the caudal raphe nuclei known to contain, in addition to serotonin, other substances including substance P (Chan-Palay *et al.*, 1979; Hökfelt *et al.*, 1978; Johansson *et al.*, 1981) and TRH (Johansson *et al.*, 1981). This possibility is supported by observations that 5,6- or 5,7-dihydroxytryptamine (neurotoxins that destroy 5-HT-containing neurons) causes depletion of serotonin-, substance-P-, and TRH-immunoreactive fibers and terminals in the spinal cord (Gilbert *et al.*, 1982; Hökfelt *et al.*, 1978; Johansson *et al.*, 1981). When applied *in vivo* by iontophoresis, substance P and TRH exerted weak excitatory effects on 67 and 20% of the SPNs tested, respectively (Backman and Henry, 1984). These results suggested that these peptides may be chemical mediators of synaptic transmission in the IML.

Enkephalin-immunoreactive fibers and terminals are present in the IML column in ladderlike rungs in the spinal cord and disappear on transection of the spinal cord at the level of the sixth cervical vertebra (Romagnano and Hamill, 1984). Franz *et al.* (1982) showed that opiates as well as clonidine depressed intraspinal transmission through SPNs in cats. They suggested that the ability of clonidine to depress the symptoms of opiate withdrawal that are characterized by sympathetic hyperactivity may occur at the spinal level. Thus, it seems that although the physiological relevance of the enkephalinergic fibers and terminals in the IML is not known, their presence does have therapeutic significance.

IV. CONCLUSIONS

It is obvious that the pharmacology of spinal sympathetic neurons constitutes an incomplete story. In general, it appears that NE inhibits and 5-HT excites the SPNs, but the pertinent data are frequently inconsistent. It seems also that some SPNs, i.e., neurons innervating the cholinergic ganglia, may not be cholinoceptive, although the data are not sufficiently extensive to permit definitive statements on this matter.

It is of increasing interest that SPNs appear to respond to a number of peptides and amino acids and that some of the latter may subserve SPN inhibitions.

However, in many of these areas, inconsistent and controversial results have been obtained. To clarify these matters and to define the pharmacology and physiology of these neurons *in vivo* and *in vitro*, intracellular recordings of SPNs will have to be combined with appropriate labeling techniques.

ACKNOWLEDGMENTS. We thank Dr. Betty Williams and Dr. Alexander G. Karczmar for their careful review of this manuscript.

REFERENCES

Anden, N., Carlsson, A., Hillarp, N., and Magnussen, T.: Hydroxytryptamine release by nerve stimulation of the spinal cord. *Life Sci.* **3**:473–478 (1964).

Anderson, E. G., and Shibuya, T.: The effects of 5-hydroxytryptophan and 1-tryptophan on spinal synaptic activity. *J. Pharmacol. Exp. Ther.* **153**:352–360 (1966).

Backman, S. B., and Henry, J. L.: Effects of glutamate and aspartate on sympathetic preganglionic neurons in the upper thoracic intermediolateral nucleus of the cat. *Brain Res.* **277**:370–374 (1983a).

Backman, S. B., and Henry, J. L.: Effects of GABA and glycine on sympathetic preganglionic neurons in the upper thoracic intermediolateral nucleus of the cat. *Brain Res.* **277**:365–369 (1983b).

Backman, S. B., and Henry, J. L.: Effects of substance P and thyrotropin-releasing hormone on sympathetic preganglionic neurones in the upper thoracic intermediolateral nucleus of the cat. *Can. J. Physiol. Pharmacol* **62**:248–251 (1984).

Baker, R. G., and Anderson, E. G.: The effects of L-3,4-dihydroxyphenylalanine on spinal reflex activity. *J. Pharmacol. Exp. Ther.* **173**:212–223 (1970).

Basbaum, A. I., Clanton, C. H., and Fields, H. L.: Three bulbospinal pathways from the rostral medulla of the cat: An autoradiographic study of pain modulating systems. *J. Comp. Neurol.* **178**:209–224 (1978).

Baum, T., and Shropshire, A. T.: Susceptibility of spontaneous sympathetic outflow and sympathetic reflexes to depression by clonidine. *Eur. J. Pharmacol.* **44**:121–129 (1977).

Bjorklund, A., and Skagerberg, G.: Evidence for a major spinal cord projection from the diencephalic A11 dopamine cell groups in the rat using transmitter-specific fluorescent retrograde tracing. *Brain Res.* **177**:170–175 (1979).

Blessing, W. W., and Chalmers, J. P.: Direct projection of catecholamine (presumably dopamine)-containing neurons from hypothalamus to spinal cord. *Neurosci. Lett.* **11**:35–40 (1979).

Bowker, R. M., Steinbusch, H. W. M., and Coulter, J. D.: Serotoninergic and peptidergic projections to the spinal cord demonstrated by a combined retrograde HRP histochemical and immunocytochemical staining method. *Brain Res.* **211**:412–417 (1981a).

Bowker, R. M., Westlund, K. N., and Coulter, J. D.: Origins of serotonergic projections to the spinal cord in the rat: An immunocytochemical–retrograde transport study. *Brain Res.* **226**:187–199 (1981b).

Butcher, L. L., Eastgate, S. M., and Hodge, G. K.: Evidence that punctate intracerebral administration of 6-hydroxydopamine fails to produce selective neuronal degeneration. *Naunyn-Schmiedeberg's Arch. Pharmacol.* **285**:31–70 (1974).

Chalmers, J. P.: Brain amines and models of experimental hypertension. *Circ. Res.* **36**:469–480 (1975).

Chalmers, J. P., and Reid, J. L.: Participation of central noradrenergic neurones in arterial baroreceptor reflexes in the rabbit: A study with intracisternally administered-6-OHDA. *Circ. Res.* **31**:789–804 (1972).

Chalmers, J. P., and Wurtman, R. J.: Participation of central noradrenergic neurones in arterial baroreceptor reflexes in the rabbit. *Circ. Res.* **28**:480–491 (1971).

Chan-Palay, V., Johnsson, G., and Palay, S.: Combined imunocytochemistry and autoradiography after *in vivo* injections of monoclonal antibody to substance P and ^{3}H-serotonin: Co-existence of two putative neurotransmitters in single raphe cells and fiber plexuses. *Anat. Embryol.* **156**:241–254 (1979).

Commissiong, J. W., Galli, C. L., and Neff, N. H.: Differentiation of dopaminergic and noradrenergic neurons in rat spinal cord. *J. Neurochem.* **30**:1095–1099 (1978).

Coote, J. H., and Macleod, V. H.: Evidence for the involvement in the baroreceptor reflex of a descending inhibitory pathway. *J. Physiol.* (*London*) **241**:477–496 (1974a).

Coote, J. H., and Macleod, V. H.: The influence of bulbospinal monaminergic pathways on sympathetic nerve activity. *J. Physiol.* (*London*) **241**:453–475 (1974b).

Coote, J. H., and Macleod, V. H.: The spinal route of sympathoinhibitory pathways descending from the medulla oblongata. *Pfluegers Arch.* **359**:335–347 (1975).

Coote, J. H., and Macleod, V. H.: The effect of intraspinal microinjections of 6-hydroxydopamine on the inhibitory influence exerted on spinal sympathetic activity by the baroceptors. *Pfluegers Arch.* **371**:271–277 (1977).

Coote, J. H., Macleod, V. H., Fleetwood-Walker, S., and Gilbey, M. P.: The response of individual sympathetic preganglionic neurones to microelectrophoretically applied endogenous monoamines. *Brain Res.* **215**:135–145 (1981).

Dahlstrom, A., and Fuxe, K.: Evidence for the existence of monoamine neurons in the central nervous system. II. Experimentally induced changes in the intraneuronal amine levels in bulbospinal neuron systems. *Acta Physiol. Scand.* **64** (Suppl. 245):7–85 (1965).

DeGroat, W. C., and Douglas, J. W.: The effect of L-5-hydroxytryptophan (5-HTP) and clonidine on central sympathetic and parasympathetic reflex pathways to urinary bladder of the cat. *Fed. Proc. Fed. Am. Soc. Exp. Biol.* **34**:795 (1975).

DeGroat, W. C., and Krier, J.: The central control of the lumbar sympathetic pathway to the large intestine of the cat. *J. Physiol.* (*London*) **289**:449–468 (1979).

DeGroat, W. C., and Lalley, P. M.: Depression by p-methoxyphenylethylamine of sympathetic reflex firing elicited by electrical stimulation of the carotid sinus nerve or pelvic nerve. *Brain Res.* **64**:460–465 (1973).

DeGroat, W. C., and Ryall, R. W.: An excitatory action of 5–hydroxytryptamine on sympathetic preganglionic neurons. *Exp. Brain Res.* **3**:229–305 (1967).

DeGroat, W. C., Douglas, J., and Lalley, P.: Effects of clonidine, L-DOPA and L-5-HTP on spinal and supraspinal sympathetic reflexes elicited by viseral and somatic afferent stimulation. Abstract presented at the Sixth International Congress of Pharmacology (IUPHAR), Helsinki, Finland (1975).

Dembowsky, K., Lackner, K., Czachurski, J., and Seller, H.: Tonic catecholaminergic inhibition of the spinal somato-sympathetic reflexes originating in the ventrolateral medulla oblongata. *J. Auton. Nerv. Syst.* **3**:277–290 (1981).

Fleetwood-Walker, S. M., and Coote, J. H.: The contribution of brain stem catecholamines cell groups to the innervation of the sympathetic lateral cell column. *Brain Res.* **205**:141–155 (1981).

Franz, D. N., Hare, B. D., and Neumayr, R. J.: Depression of sympathetic preganglionic neurons by clonidine: Evidence for stimulation of 5-HT receptors. *Clin. Exp. Hyperten.* **1**:115–140 (1978).

Franz, D. N. Hare, B. D., and McCloskey, K. L.: Spinal sympathetic neurons: Possible sites of opiate-withdrawal suppression by clonidine. *Science* **215**:1643–1645 (1982).

Frederickson, R. C. A., Jordan, L. M., and Phillis, J. W.: The action of noradrenaline on cortical neurons: Effects of pH. Brain Res. **35**:556–560 (1971).

Fuxe, K., Hökfelt, T., Blome, P., Goldstein, M., Johansson, O., Jonsson, G., Lidbrink, P., Ljungdahl, A., and Sachs, C. H.: The topography of central catecholamine pathways in relation to their positive role in blood pressure control, in: *Central Action of Drugs in Blood Pressure Regulation* (D. S. Davies and J. L. Reid, eds.), pp. 8–23, University Park Press, Baltimore (1975).

Gilbert, R. F. T., Emson, P. C., Hunt, S. P., Bennett, G. W., Marsen, C. A., Sandberg, B. E. B.,

Steinbusch, H. W. M., and Verhofstad, A. A. J.: The effects of monoamine neurotoxins on peptides in the rat spinal cord. *Neuroscience* **7**:69–87 (1982).

Gilbey, M. P., Coote, J. H., Macleod, V. H., and Peterson, D. F.: Inhibition of sympathetic activity by stimulating in the raphe nuclei and the role of 5-hydroxytryptamine in this effect. *Brain Res.* **226**:131–142 (1981).

Glazer, E. J., and Ross, L. L.: Localizations of noradrenergic terminals in sympathetic preganglionic nuclei of the rat: Demonstration by immunocytochemical localization of dopamine-β-hydroxylases. *Brain Res.* **185**:39–49 (1980).

Guyenet, P. G., and Cabot, J. B.: Inhibition of sympathetic preganglionic neurons by catecholamine and clonidine: Mediation by an α-adrenergic receptor. *J. Neurosci.* **1**:908–917 (1981).

Haeusler, G.: Neuronal mechanism influencing transmission in the baroreceptor reflex, in: *DeJong's Hypertension and Brain Mechanism*, (W. deJong, A. P. Provost, and A. P. Shapiro, eds.), pp. 95–109, Elsevier, Amsterdam (1977).

Hancock, M. B.: Leu-enkephalin, substance P, and somatostatin immunohistochemistry combined with the retrograde transport of horseradish peroxidase in sympathetic preganglionic neurons. *J. Auton. Nerv. Syst.* **6**:263–272 (1982).

Hare, D. H., Neumayr, R. J., and Franz, F. N.: Opposite effects of L-DOPA and 5-HTP on spinal sympathetic reflexes. *Nature (London)* **239**:336–337 (1972).

Higashi, H.: 5-Hydroxytryptamine receptors on visceral primary afferent neurons in the nodose ganglion of the rabbit. *Nature (London)* **267**:448–449 (1977).

Hökfelt, T., Fuxe, K., Goldstein, M., and Johansson, O.: Immunohistochemical evidence for the existence of adrenaline neurons in the rat brain. *Brain Res.* **66**:235–251 (1974).

Hökfelt, T., Ljungdahl, A., Steinbusch, H., Verhofstad, A., Nilsson, G., Bradin, E., Pernau, B., and Goldstein, M.: Immunohistochemical evidence of substance P-like immunoreactivity in some 5-hydroxytryptamine-containing neurons in the rat central nervous system. *Neuroscience* **3**:517–538 (1978).

Holets, V., and Elde, R.: The differential distribution and relationship of serotoninergic and peptidergic fibers to sympathoadrenal neurons in the intermediolateral cell column of the rat: A combined retrograde axonal transport and immunofluorescence study. *Neuroscience* **7**:1155–174 (1982).

Holets, V., and Elde, R.: Sympathoadrenal preganglionic neurons: Their distribution and relationship to chemically-coded fibers in the kitten intermediolateral cell column. *J. Auton. Nerv. Syst.* **7**:149–163 (1983).

Hongo, T., and Ryall, R. W.: Electrophysiological and microelectrophoretic studies on sympathetic preganglionic neurones in the spinal cord. *Acta Physiol. Scand.* **68**:96–104 (1966).

Johansson, O., Hökfelt, T., Pernou, B., Jeffcoate, S. L., White, W., Steinbusch, W. M., Verhofstad, A. J., Emson, P. C., and Spindel, E.: Immunohistochemical support for three putative transmitters in one neurone: Coexistence of 5-hydroxytryptamine, substance P, and thyrotropin releasing hormone-like immunoreactivity in medullary neurons projecting to the spinal cord. *Neuroscience* **6**:1857–1881 (1981).

Jordan, L. M., Frederickson, R. C. A., Phillis, J. W., and Lake, N.: Microelectrophoresis of 5-hydroxytryptamine: A clarification of its action on cerebral cortical neurones. *Brain Res.* **40**:552–558 (1972).

Kadzielawa, K.: Inhibition of the activity of sympathetic preganglionic neurones and interneurones activated by visceral afferents by alpha-methylnoradrenaline and endogenous catecholamines. *Neuropharmacology* **22**:3–17 (1983a).

Kadzielawa, K.: Antagonism of the excitatory effects of 5-hydroxytryptamine on sympathetic preganglionic neurones and neurones activated by visceral afferents. *Neuropharmacology* **22**:19–27 (1983b).

Krier, J., Thor, K. B., and DeGroat, W. C.: Effects of clonidine on the lumbar sympathetic pathways to the large intestine and urinary bladder of the cat. Eur. *J. Pharmacol.* **59**:47–53 (1979).

Lebedev, V. P., Petrov, V. I., and Skobelev, V. A.: Do sympathetic preganglionic neurones have a recurrent inhibitory mechanism? *Pfluegers Arch.* **383**:91–92 (1980).

Ljungdahl, A., Hokfelt, T., and Nilsson, G.: Distribution of Substance P-like immunoreactivity in the central nervous system of the rat–I. Cell bodies and nerve terminals. *Neuroscience* **3**:861–943 (1978).

Loewy, A. D.: Raphe pallidus and raphe obscurus projections to the intermediolateral cell column in the rat. *Brain Res.* **222**:129–133 (1981).

Loewy, A. D., and McKellar, S.: Serotoninergic projections from the ventral medulla to the intermediolateral cell column in the rat. *Brain Res.* **211**:146–152 (1981).

Loewy, A. D., McKellar, S., and Saper, C. B.: Direct projection from the A5 catecholamine cell group to the intermediolateral cell column. *Brain Res.* **174**:309–314 (1979).

McCall, R. B., and Humphrey, S. J.: Central serotoninergic neurons facilitate sympathetic nervous discharge (SND). *Soc. Neurosci. Abstr.* **7**:365 (1981).

McCall, R. B.: Serotonergic excitation of sympathetic preganglionic neurons: A microiontophoretic study. *Brain Res.* **289**:121–127 (1983).

McKenna, K. E., and Schramm, L. P.: Sympathetic preganglionic neurons in the isolated spinal cord of the neonatal rat. *Brain Res.* **269**:201–210 (1983).

McKenna, K. E., and Schramm, L. P.: Baclofen inhibits sympathetic preganglionic neurons in an isolated spinal cord preparation. *Neurosci. Lett.* **47**:85–88 (1984).

Mensah, P. L., Glangman, D. L., Levy, W. B., and Thompson, R. F.: The effects of 5,6-dihydroxytryptamine in the amphibian spinal cord using silver techniques. *Brain Res.* **78**:255–261 (1974).

Moir, A. T. B., and Eccleston, D. E.: THe effects of precursor loading on the cerebral metabolism of 5-hydroxyindoles. *J. Neurochem.* **15**:1093–1108 (1968).

Neumayr, R. J., Hare, B. S., and Franz, D. N.: Evidence for bulbospinal control of sympathetic preganglionic neurones by monoaminergic pathways. *Life Sci.* **14**:793–806 (1974).

Ng, K. Y., Chase, T. N., Colburn, R. W., and Kopin, I. J.: L-DOPA induced release of cerebral monoamines. *Science* **170**:76–77 (1970).

Nowak, L. M., and MacDonald, R.: Ionic mechanism of muscarinic cholinergic depolarization of mouse spinal cord neurons in cell culture. *J. Neurophysiol.* **49**:792–803 (1983).

Oliveras, J. L., Bourgoin, S., Hery, F., Besson, J. M., and Hamon, M.: The topographical distribution of serotoninergic terminals in the spinal cord of the cat: Biochemical mapping by the combined use of microdissection and microassay procedures. *Brain Res.* **178**:393–406 (1977).

Poirier, L. J., Langelier, P., Roberge, A., Boucher, R., and Kitsikis, A.: Nonspecific histopathological changes induced by the intracerebral injection of 6-hydroxydopamine. *J. Neurol. Sci.* **16**:401–416 (1973).

Rethelyi, M.: Cell and neuropil architecture of the intermediolateral (sympathetic) nucleus of cat spinal cord. *Brain Res.* **46**:203–213 (1972).

Romagnano, M. A., and Hamill, R. W.: Spinal sympathetic pathway: An enkephalin ladder. *Science* **225**:737–739 (1984).

Ryall, R. W.: Cholinergic transmission in the spinal cord, in: *Handbook of the Spinal Cord* (R. A. Davidoff, ed.), pp. 203–240, Marcel Dekker, New York (1983).

Sangdee, C., and Franz, D. N.: Evidence for inhibition of sympathetic preganglionic neurons by bulbospinal epinephrine pathways. *Neurosci. Lett.* **37**:167–173 (1983).

Sato, A., and Schmidt, R. F.: Spinal and supraspinal components of the reflex discharges into lumbar and thoracic white rami. *J. Physiol.* (London) **212**:839–850 (1971).

Shinnick-Gallagher, P.: An *in vitro* preparation for the recording of spinal sympathetic reflexes. *J. Pharmol. Methods* **2**:187–191 (1979).

Soller, R. W.: Monoaminergic inputs to frog motoneurons: An anatomical study using fluorescence histochemical and silver degeneration techniques. *Brain Res* **122**:445–448 (1977).

Steinbusch, H. W. N.: Distribution of serotonin immunoreactivity in the central nervous system of the rat—cell bodies and terminals. *Neuroscience* **6**:557–618 (1981).

Tohyama, M., Sakai, K., Touret, M., Salvert, D., and Jouvet, M.: Spinal projections from the lower brain stem in the cat as demonstrated by the horseradish peroxidase technique. II. Projections from the dorsolateral pontine tegmentum and raphe nuclei. *Brain Res.* **176**:215–231 (1979).

Undesser, E. K., Shinnick-Gallagher, P., and Gallagher, J. P.: Catecholamine modulation of spinal sympathetic reflexes. *J. Pharmacol. Exp. Ther.* **217**:170–176 (1981).

Yoshimura, M., Katayama, Y., and Nishi, S.: Chemosensitivities of lateral horn cells in cat spinal cord. Eighth Int. Cong. Pharmacol., Tokyo, **8**:604 (1981).

Yoshimura, M., and Nishi, S.: Intracellular recordings from alteral horn cells of the spinal cord *in vitro*. *J. Auton. Ner. Syst.* **6**:5–11 (1982).

Zivin, J. A., Reid, J. L., Saavedra, J. M., and Kopin, I. J.: Qualitative localization of biogenic amines in the spinal cord. *Brain Res.* **99**:293–301 (1975).

18

Spontaneous and Reflex Activities: General Characteristics

V. I. SKOK

I. GENERAL CHARACTERISTICS

A. Origins

As used in this chapter, the term "spontaneous activity" refers to electrical potentials that are not induced directly by the investigator; "reflex activity," to potentials induced by special afferent stimulation. It should be noted, however, that spontaneous activity is also usually controlled by afferent input, even if it is not specially induced by the latter. For example, the cardiac rhythm generated by spontaneous sympathetic activity is due to the baroreceptor control of sympathetic centers. Furthermore, one can observe the spontaneous activity in sympathetic ganglion neurons only if their preganglionic fibers remain connected to their centers or if their afferent fibers of peripheral origin remain connected to appropriate peripheral organs. The neurons with the autogenic type of intracellular spontaneous activity resulting from the properties of their soma, similar to the spontaneous activity observed in invertebrate pacemaker neurons (cf. Tauc, 1955; Hagiwara and Bullock, 1957), have not been found so far in the vertebral autonomic ganglia, except the enteric system (see Chapter 6; Wood, 1975; Wood and Mayer, 1978) and the inferior mesenteric ganglion

---

V. I. SKOK • Bogomoletz Institute of Physiology, Kiev, U.S.S.R.

(Jule and Szurszewski, 1983; Jule *et al.*, 1983). In all cases, the spontaneous intracellular spikes recorded from the neuronal soma appear to be triggered by excitatory postsynaptic potentials (EPSPs), by electrotonic invasions of the soma by spikes originating in the processes of the cell, or by an endogenous depolarizing mechanism (see below).

B. Preganglionic Neurons

Records obtained from single sympathetic preganglionic neurons show that only a fraction of these neurons (21–63%, depending on the author) are spontaneously active. The other neurons are "silent"; many of these cells do not respond to afferent stimuli, and their physiological significance is unclear (Polosa, 1968; Lebedev, 1972; Coote and Westbury, 1979; Janig and Szulczyk, 1981). A small proportion of spontaneously active neurons discharge at a rhythm corresponding to the cardiac or respiratory rhythm (Iggo and Vogt, 1960) or other rhythms of intrinsic central origin (Gebber and Barman, 1980). The remainder of these cells display an irregular firing pattern (Polosa, 1968; Coote and Westbury, 1979). Each sympathetic system (vasomotor, pilomotor, or sudomotor) has its own typical pattern of spontaneous or evoked neuronal activity (Janig and Szulczyk, 1981), while the parasympathetic preganglionic neurons show an irregular spontaneous firing pattern (DeGroat and Krier, 1979).

The mean frequency of spontaneous firing ranges from 0.2 to 9/sec, the maximal frequency of the burst being 30/sec (Iggo and Vogt, 1960; Polosa, 1968; Coote and Westbury, 1979). The range of conduction velocities that were recorded in spontaneously active preganglionic fibers innervating the superior cervical ganglion (5–18 m/sec) (Skok and Remizov, 1977) is narrower than that recorded when these fibers are stimulated electrically (0.8–31 m/sec) (Skok, 1973; Skok and Heeroog, 1975). This is consistent with the observation that preganglionic fibers which exhibit the highest excitation threshold are not spontaneously active (Coote and Westbury, 1979).

Vasomotor preganglionic fibers are smaller than sudomotor or pilomotor fibers (Janig and Szulczyk, 1981; see Skok, 1973). However, exact correlation between the size of preganglionic fibers and their function is limited or not possible due to the convergence of fibers of different sizes on the same neuron of the ganglion, as for instance in the case of the vasomotor neurons (Lebedev *et al.*, 1978).

C. Efferent Neurons of Autonomic Ganglia

The spontaneous firing of sympathetic ganglionic neurons is driven by their preganglionic fibers. This is shown by the finding that there is a close correlation between the spontaneous discharges of preganglionic and postganglionic fibers of superior cervical ganglion, the discharges in question exhibiting the cardiac rhythm (Skok and Mirgorodski, 1971; Skok, 1973).

Only vasoconstrictor and sudomotor neurons exhibit resting activity, as shown by studies carried out in the cat and man. Vasoconstrictors for the muscle and cutaneous vasoconstrictor neurons are under strong and weak control, respectively, of arterial baroreceptors (Janig *et al.*, 1983b). Thus, cardiac rhythm is usually observed in sympathetic vasoconstrictor neurons subserving skeletal muscles in humans (Hagbarth and Vallbo, 1968; Delius *et al.*, 1972; Hagbarth *et al.*, 1972; Vallbo *et al.*, 1979; Skok *et al.*, 1984) and in animals (Gregor *et al.*, 1977), while only irregular firing that is strongly influenced by afferent stimuli is observed in the case of the sympathetic cutaneous vasoconstrictor neurons (Hagbarth *et al.*, 1972). Again, the enteric system differs from the autonomic system, marked variability having been observed with regard to the patterns of spontaneous discharges of the single neurons of the enteric plexus (Wood, 1975; Nozdrachev and Vataev, 1981; Nozdrachev *et al.*, 1981). The mean frequency of spontaneous firing observed in sympathetic ganglion neurons is 3.4/sec, and the maximal frequency observed during the burst is 15/sec (Mirgorodski and Skok, 1969; Skok, 1973), which is probably much lower than the maximal activity obtainable on stimulation (Andersson, 1983).

Figure 1 illustrates examples of the spontaneous activity recorded with intracellular electrodes from the neurons of several autonomic ganglia. The activity was evoked by spontaneous excitatory synaptic input from either preganglionic or peripheral afferent neurons (Figure 1A and B, respectively), by excitatory or inhibitory (Figure 1C and E, respectively) synaptic inputs from the unidentified (peripheral afferent or internuncial) neurons, by electrotonic invasion of the soma by spikes generated in the cell's processes (Figure 1D), and probably by an endogenous depolarizing mechanism (Figure 1F). It can be seen that in all synaptically driven cells, each spontaneous spike is triggered by separate high-amplitude EPSP.

It is of particular interest that in the cat inferior mesenteric ganglion, two types of spontaneously active neurons have been observed, in addition to commonly known efferent ganglionic neurons: the regularly discharging neurons (Figure 1F), which are probably intraganglionic, and efferent irregularly discharging neurons, which are provided with the synaptic inputs by the regularly discharging ones (Jule and Szurszewski, 1983; Jule *et al.*, 1983). An important finding was that the peripheral afferent input can be blocked by sympathetic stimulation; this finding provides evidence for the existence of a negative feedback (Crowcroft *et al.*, 1971).

The responses evoked in the neurons of autonomic ganglia by reflex stimulation are illustrated in Figure 2. The excitatory stimulation results in the appearance of high-amplitude EPSPs evoked by preganglionic fi-

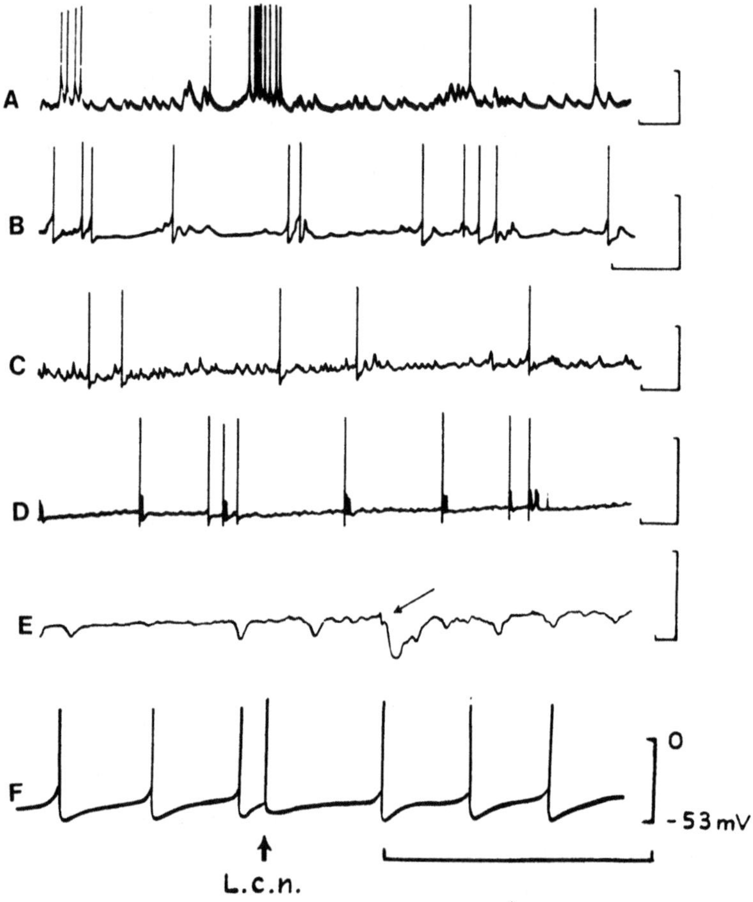

Figure 1. Spontaneous activity in the neurons of autonomic ganglia. (A) Superior cervical ganglion of the rabbit provided with intact preganglionic input. From Mirgorodski and Skok (1969). (B) Inferior mesenteric ganglion of the guinea pig provided with intact afferent input from the colon. From Crowcroft *et al.* (1971). (C–E) Myenteric plexus of the guinea pig small intestine; the activity was evoked by excitatory synaptic input (C), by electrotonic invasion of the soma by spikes generated in the cell's processes (D), and by inhibitory synaptic input (E). The arrow in (E) indicates the application of the single stimulus to the fiber tract that evoked hyperpolarization of the cell. Modified from Wood and Mayer (1978). (F) Regular discharges in an endogenously active neuron of the inferior mesenteric ganglion of the cat; at the arrow, suprathreshold stimulation was applied to the lumbar colonic nerve (L.c.n.) Modified from Jule and Szurszewski (1983). Intracellular recordings. Calibration: 50 mV, 0.5 sec.

bers. Many of the EPSPs trigger spikes (Figure 2A). Similar excitatory reactions are evoked in autonomic neurons by afferent fibers of peripheral origin (Crowcroft *et al.*, 1971). The inhibitory stimulation is followed by a disappearance of spontaneous EPSPs and spikes, although inhibitory postsynaptic potentials (IPSPs) do not appear (Figure 2B).

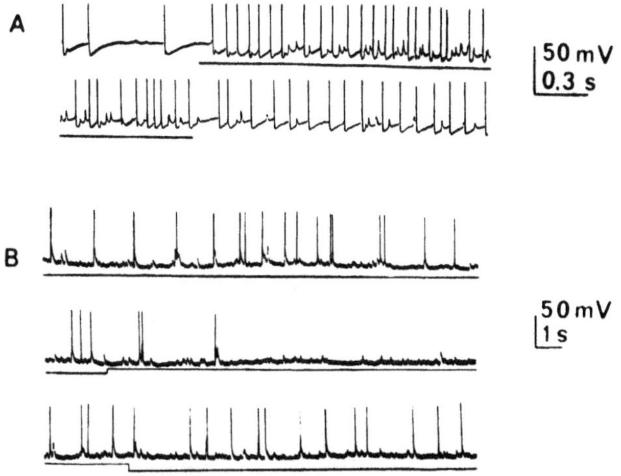

Figure 2. Reflex responses in the neurons of autonomic ganglia provided with intact preganglionic input for afferent stimulation. (A) Excitatory response of the cat ciliary ganglion neuron to illumination of the eye. (B) Inhibitory response of the neuron of the rabbit superior cervical ganglion to stimulation of the depressor nerve. The time of stimulation is indicated by a horizontal line in (A) and by a rise in the horizontal line in (B). Portions of the recordings originally intervening between the top and bottom tracings shown in (A) (2.2 sec) and between the middle and bottom tracings shown in (B) (9.0 sec) are omitted. Intracellular recordings. From Melnichenko and Skok (1970) and Skok (1980).

D. Afferent Neurons of Autonomic Ganglia

There are afferent neurons in the enteric plexus that establish synaptic contacts with the adjacent neurons or with the neurons located in the remote ganglia (in the inferior mesenteric ganglion and in the solar plexus), providing a pathway for a peripheral reflex (see Skok, 1973). It can be suggested that the spontaneous activity recorded from the soma of the afferent neuron of the enteric plexus should be similar to that shown in Figure 1D, since it resembles the activity recorded from the soma of the identified afferent neurons of the dorsal root ganglia (cf. Svaetichin, 1958); however, the evidence for this hypothesis is lacking at present.

Some autonomic ganglia possess afferent nerve terminals of cerebrospinal origin (Kolosov and Milokhin, 1963; Milokhin, 1967; Koval *et al.*, 1981). The spontaneous firing recorded from these afferent fibers is strongly affected by a number of drugs (Bulygin, 1983). The physiological role of these afferent terminals remains unclear.

II. NEURONAL AND CHEMICAL MECHANISMS OF NATURAL ACTIVITY

As illustrated in Figure 1, each spike of the soma membrane is triggered by a separate fast EPSP that is commonly accepted to be mediated by nicotinic transmission. Besides such effective EPSPs, there are many ineffective EPSPs of lower amplitudes that are not followed by cell discharge. It is well known that as many as 30 preganglionic fibers (Blackman, 1974) organized in two to seven groups differing in their thresholds and conduction velocities may converge on the same neuron of the sympathetic ganglion (Skok and Heeroog, 1975) and that a spike produced in the neuron by a single preganglionic stimulus results from synchronous firing of at least several presynaptic fibers (see Skok, 1973). This led to the suggestion that spontaneous effective EPSPs are due to synchronous firing of several converging presynaptic fibers, while the ineffective EPSPs are due to firing of only a few such fibers.

That synaptic inputs from at least several presynaptic fibers may be involved in producing spontaneous firing in each neuron of a sympathetic ganglion is indicated by the finding that the spontaneous activity recorded from the neurons of the inferior mesenteric ganglion gradually diminishes following sequential acute sections of colonic nerves that provide the afferent input to this ganglion (Crowcroft *et al.*, 1971).

Furthermore, using partial anodal block of conduction in the cervical sympathetic nerve, it was possible to compare the excitatory effect on the neurons of the superior cervical ganglion of spontaneous and evoked discharges evoked in the same group of preganglionic fibers (Mirgorodski and Skok, 1970). The current that selectively blocked the initial low-threshold component in the compound orthodromic response depressed, but did not completely eliminate, the spontaneous activity. In other neurons of this ganglion, the spontaneous activity could be completely eliminated by the current that blocked selectively either the initial low-threshold component or the late high-threshold component of the evoked response. Thus, in some neurons, the spontaneous activity is due to the activity of preganglionic fibers forming a single group type, whereas in other neurons, the preganglionic fibers belonging to several groups may interact. These results confirm the suggestion (see above) that many converging presynaptic fibers may participate in producing the postsynaptic spontaneous activity of the ganglion neuron. However, these results do not answer the question whether postsynaptic spontaneous spikes can be triggered only by summated EPSPs or may also be evoked by a single preganglionic fiber similarly to spikes evoked in singly innervated neurons, as for example in the amphibian sympathetic ganglion (Nishi, 1974).

To study this point, the spontaneous activity of a sympathetic neuron was compared recently with the responses evoked in the same cell at various membrane potential levels by orthodromic stimuli of graded intensities. This approach revealed that there are two types of spontaneous postsynaptic spikes. As shown in Figure 3A, the spikes of these two types are markedly different as to their rising phase, which is steep and does not show any inflections in the case of the first type of spike, in contrast to the time–course of the second type of spike.

Strong hyperpolarization is necessary to obtain the EPSP normally generating the first type of spike (Figure 3C), since a gradual decrease of stimulus intensity is not effective in producing this EPSP (Figure 3B). The EPSP obtained in this way has an all-or-none character and is similar in amplitude and shape to that appearing spontaneously (Figure 3C). In contrast, the EPSP normally leading to the second type of spike can be easily obtained at resting membrane potential level using a gradual decrease of stimulus intensity and consists of at least four components differing as to their threshold (Figure 3D). These results suggest that there are two distinct synaptic inputs to a mammalian sympathetic ganglion neuron: the dominant input, which is provided by a single preganglionic fiber that triggers the postsynaptic spike with a high safety factor, and the accessory input provided by several preganglionic fibers that can trigger the postsynaptic spike only through summation of their EPSPs (Skok and Ivanov, 1983).

It is essential that the great majority of the neurons of the sympathetic ganglia possess both synaptic inputs working together to produce the

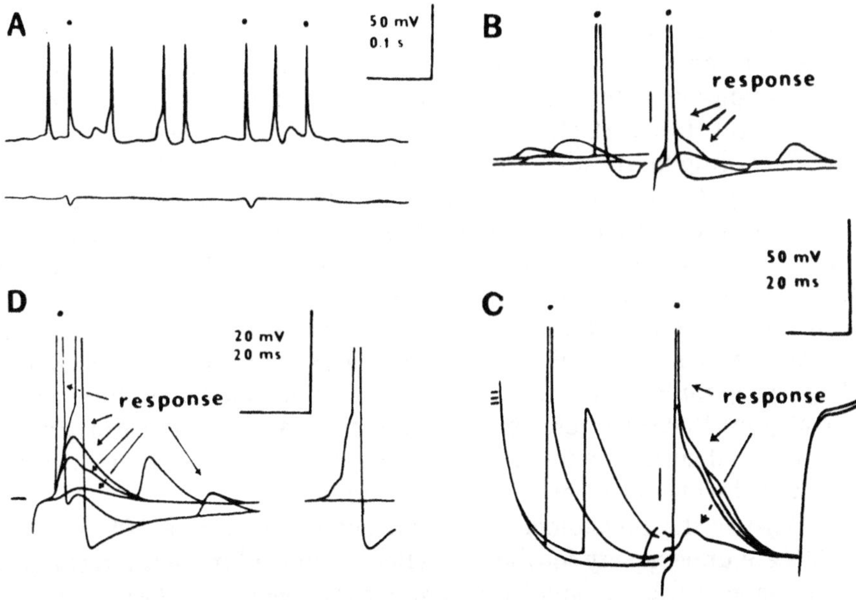

Figure 3. Two types of spontaneous and evoked spikes in the neurons of the superior cervical ganglion of the rabbit. (A) Spontaneous activity. Dominant spikes are labeled by dots. (B–D) Dominant spikes (B,C) and accessory spikes (D) are evoked by a single orthodromic stimulus at the resting membrane potential level (B,D) and at the hyperpolarized membrane potential level (C). Multiple sweeps with gradual increase of stimulus intensity. Arrows show the components of the response (spikes and EPSPs). Spontaneous activity (spikes and EPSPs) is also shown in the same sweeps (B–D) and in a separate sweep (D). Records A, B and C, and D were obtained from three different neurons, respectively. From Skok and Ivanov (1983).

postsynaptic spontaneous activity of the neuron. The pertinent inputs do not differ as to their preganglionic conduction velocities and their modalities, but do differ markedly with respect to their postsynaptic interspike histograms. The analysis of the histograms suggests that dominant spikes, accessory spikes, and ineffective EPSPs that participate in the spontaneous activity of the ganglion neuron are each produced by separate central generators and that the accessory generator may be influenced by both dominant and ineffective EPSP generators (Skok and Ivanov, 1983). The sympathetic preganglionic neurons probably do not derive dominant excitatory inputs from the driving cells, but instead discharge whenever the activity of their several synaptic inputs is synchronized sufficiently to summate and depolarize the membrane to the threshold level (Coote and Westburg, 1979).

Dominant and accessory synaptic inputs probably differ with respect to the localization of their preganglionic terminals on the neuronal surface. The obvious similarity between the effectiveness of the dominant input and of the single preganglionic fiber innervating amphibian sympathetic ganglionic neurons suggests that the dominant fiber terminates on the soma close to the site of the axonal initiation of the response, whereas accessory fibers terminate on the dendrites. This conclusion is consistent with the recent finding that the rate of spontaneous synaptic activity tends to be progressively higher in neurons endowed with complex geometries (Purves and Hume, 1981; D. A. Johnson and Purves, 1983).

According to the stochastic theory (see Blackman, 1974), the preganglionic impulses arrive at a neuron of the sympathetic ganglion at irregular intervals, and the summation necessary for spike initiation occurs by chance. However, in view of the perfect synchrony that is achieved during the stimulation of a neuron by several converging, tonically active preganglionic fibers, synchronous firing of these fibers seems necessary, and the occurrence of such firing by chance appears to be unlikely. Rather, the preganglionic fibers contributing to each summated EPSP form a functional group that is triggered by a common presynaptic impulse originating in the spinal cord. The number of firing preganglionic fibers in the group may vary by chance from one burst to another.

What is the physiological significance of ineffective EPSPs? These EPSPs are unnecessary for the final effect produced in their target organs by autonomic ganglia, since they are not followed by spikes. A possible physiological role of the ineffective EPSPs may be that they represent a side effect of a highly specialized neuronal organization that ensures selective firing of certain ganglion cells within a large neuronal population which is induced by a small number of preganglionic fibers. According to this hypothesis (Skok, 1980), to fire a particular, small group of neurons, a preganglionic volley must be conveyed to the ganglion via a specific combination of preganglionic fibers, all these fibers converging on each neuron of the neuronal group in question and all these fibers being necessary to evoke an effective EPSP. Other neurons of the ganglion that receive some but not all fibers that constitute an effective combination would respond with an ineffective EPSP. This hypothesis is consistent with the observation that the increase in the number of converging preganglionic fibers parallels the increase in the ratio of the ganglionic neurons to preganglionic fibers; it is also consistent with the finding that the preganglionic fiber evoking a noneffective EPSP in one neuron of the rabbit superior cervical ganglion may trigger discharge in another neuron (V. I. Skok and A. Y. Ivanov, unpublished observation). An alternative explanation is that ineffective EPSPs may produce a trophic effect in the ganglionic neurons (Black and Green, 1973; Chalazonitis and Zigmond, 1980).

Another interesting aspect of the spontaneous activity recorded intracellularly from the neurons of the autonomic ganglia concerns IPSPs. A fast IPSP, i.e., an IPSP with a time–course similar to that of the fast EPSP, has never been observed in the course of natural activity of autonomic ganglia, whether spontaneous or evoked (see Chapter 16), in contrast to the central nervous system. On the other hand, spontaneous slow IPSPs were recorded from myenteric neurons (See Figure 1E). Their putative synaptic transmitter is a catecholamine or an enkephalin (Hirst and Silinsky, 1975; S. M. Johnson *et al.*, 1980), and these IPSPs are apparently due to an increase in membrane potassium conductance (Hirst and McKirdy, 1975; Hirst and Silinsky, 1975; Wood, 1981). Many investigators have observed slow IPSPs (see Chapter 6) in orthodromically stimulated neurons of mammalian sympathetic ganglia. Thus, it can be expected that a slow IPSP should follow strong inhibition of spontaneous sympathetic activity, eg., inhibition caused by stimulation of depressor nerves. However, no signs of a slow IPSP could be observed in this case (see Figure 2B). The possibility remained, however, that a slow IPSP might appear in the neurons that were not available for intracellular electrodes because of their small size. To test this possibility, the spontaneous activity of a sympathetic ganglion was recorded by means of the sucrose-gap method. The results of such an experiment are shown in Figure 4. Although stimulation of the depressor nerve produced electropositivity in the ganglion, this electropositivity was not affected by atropine employed in the dose that blocked the slow IPSP evoked by orthodromic stimulation (Figure 4A). The electropositivity caused by stimulation of the depressor nerve appeared to be due to a decrease in electronegativity of the ganglion maintained by spontaneous preganglionic firing (see Figure 4B), rather than to a slow IPSP (Skok *et al.*, 1974; Skok, 1976).

Thus, it has not been possible as yet to observe slow IPSPs in the course of the spontaneous activity or reflex responses of the sympathetic and some parasympathetic (Melnichenko and Skok, 1970) neurons, while such potentials could be recorded from enteric neurons. Similarly, there

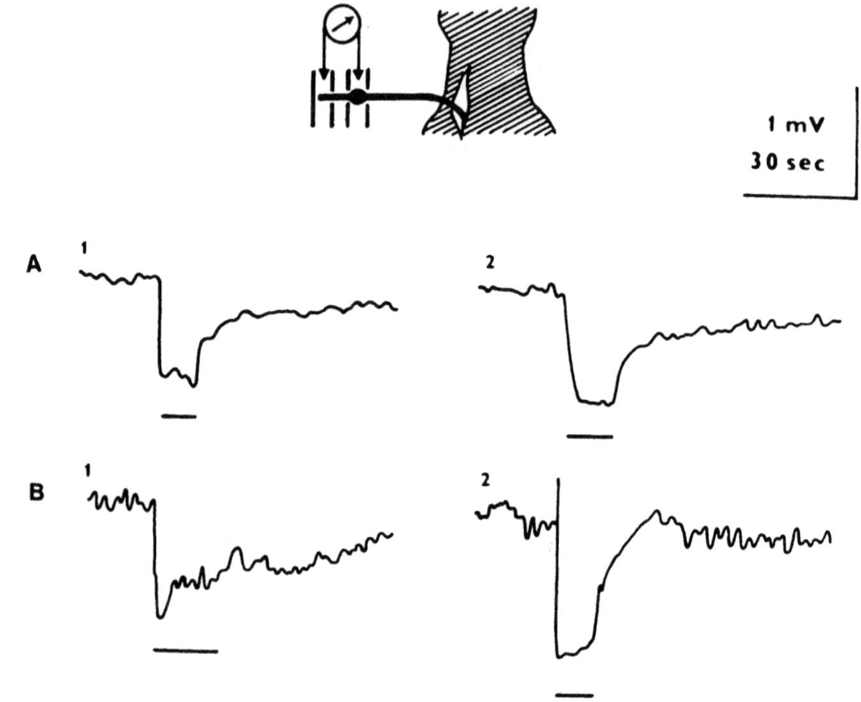

Figure 4. Effects of stimulation applied to the rabbit depressor nerve and of transient block of preganglionic synaptic input on the spontaneous activity recorded with the sucrose-gap technique from rabbit superior cervical ganglion. (A) Responses to stimulation of the depressor nerve before (1) and after (2) perfusion of the ganglion with atropine (1×10^{-6} M). (B) Response to stimulation of the depressor nerve (1) and the effect of blockade of conduction produced by the application of a constant current to the cervical sympathetic nerve (2). The records were obtained from two preparations (A,B). Horizontal bars indicate the time period of repetitive stimulation of the depressor nerve (A1 and 2; B1) and of the application of a constant current that blocked conduction in the cervical sympathetic nerve (B2). The scheme at top shows the experimental arrangement. Modified from Skok *et al.* (1974).

have been no reports as yet as to the presence of spontaneous slow EPSPs, although the slow EPSP has been commonly evoked in mammalian ganglia by orthodromic stimulation that activates muscarinic or 5-hydroxytryptamine receptors of the neurons (see Skok, 1973, 1981; Nishi, 1974; Kuba and Koketsu, 1978; Wood, 1981), as described in detail in Chapter 7. Moreover, it has been shown recently that the frequency of spontaneous spikes recorded with intracellular electrodes from the rabbit superior cervical ganglion provided with intact preganglionic input is not affected by intravenous or intraarterial injections of atropine in doses sufficient to block slow synaptic processes of this ganglion (Ivanov and Savchenko, 1982). Thus, the role of the slow EPSPs with respect to the resting spontaneous activity of autonomic ganglionic neurons remains unclear. Some data, however, suggest that noncholinergic mechanisms may be involved in the transmission through the sympathetic ganglion of reflex activity evoked by stimulation of arterial chemoreceptors (Janig *et al.*, 1983a) (see also Chapters 7, 8, 13 and 19).

III. CONCLUSIONS

Together, the findings described above indicate that the spontaneous activity recorded in the neurons of the autonomic ganglia is due to synaptic inputs from preganglionic neurons or from peripheral afferent neurons, rather than representing autogenic spontaneous activity, except in the case of the inferior mesenteric ganglion and the enteric plexus, which probably posses autogenically active neurons. It also appears that each spontaneous spike is triggered by a separate fast EPSP. The spontaneous activity of each neuron in the mammalian sympathetic ganglion is triggered by double preganglionic input: a single preganglionic fiber of high potency that evokes an EPSP high enough for spike initiation and a group of converging preganglionic fibers of low potency that trigger the postsynaptic spike only if they fire simultaneously. Spontaneously active neurons of autonomic ganglia did not exhibit fast IPSPs. In fact, slow IPSPs, and slow EPSPs, although they appear characteristically in many neurons of sympathetic ganglia following their orthodromic stimulation (see Chapters 7, 8, 9, and 13), have not been observed as yet in the course of their spontaneous or reflex activities; again, spontaneous slow IPSPs were observed in the case of myenteric neurons.

Finally, the results of studies of the natural physiological activity of autonomic ganglia suggest that in the case of most ganglia, there is an enlargement of the centrifugal impulse outflow, this system including a feedback mechanism generated in their target organs via the central nervous system; in addition, there is a feedback impinging directly on the ganglia in the case of some abdominal sympathetic ganglia, as well as in that of the enteric plexus, this feedback constituting peripheral reflexes.

REFERENCES

Andersson, P. O.: Comparative vascular effects of stimulation continuously and in bursts of the sympathetic nerves to cat skeletal muscle. *Acta Physiol. Scand.* **118**:343–348 (1983).

Black, I. B., and Green, S. C.: Trans-synaptic regulation of adrenergic neuron development: Inhibition by ganglionic blockade. *Brain Res.* **63**:291–302 (1973).

Blackman, F. G.: Function of autonomic ganglia, in: *The Peripheral Nervous System* (J. I. Hubbard, ed.), pp. 257–276, Plenum Press, New York (1974).

Bulygin, I. A.: A consideration of the general principles of organization of sympathetic ganglia. *J. Auton. Nerv. Syst.* **8**:303–330 (1983).

Chalazonitis, A., and Zigmond, R. E.: Effects of synaptic and antidromic stimulation on tyrosine hydroxylase activity in the rat superior cervical ganglion. *J. Physiol.* (*London*) **300**:525–538 (1980).

Coote, J. H., and Westbury, D. R.: Functional grouping of sympathetic preganglionic neurones in the third thoracic segment of the spinal cord. *Brain Res.* **179**:367–372 (1979).

Crowcroft, P. J., Holman, M. E., and Szurszewski, J. H.: Excitatory input from the distal colon to the inferior mesenteric ganglion in the guinea-pig. *J. Physiol.* (*London*) **219**:443–461 (1971).

DeGroat, W. C., and Krier, J.: The central control of the lumbar sympathetic pathway to the large intestine of the cat. *J. Physiol.* (*London*) **289**:449–468 (1979).

Delius, W., Hagbarth, K. E., Hongell, A., and Wallin, B. G.: General characteristics of sympathetic activity in human muscle nerves. *Acta Physiol. Scand.* **84**:65–81 (1972).

Gebber, G. L., and Barman, S. M.: Basis for 2–6 cycle/s rhythm in sympathetic nerve discharge. *Am. J. Physiol.* **239**:R48–R56 (1980).

Gregor, M., Jänig, W., and Wiprich, L.: Cardiac and respiratory rhythmicities in cutaneous and muscle vasoconstrictor neurones to the cat's hindlimb. *Pflügers Arch.* **370**:299–302 (1977).

Hagbarth, K. E., and Vallbo, A. B.: Pulse and respiratory grouping of sympathetic impulses in human muscle nerves. *Acta Physiol. Scand.* **74**:96–108 (1968).

Hagbarth, K. E., Hallin, R. G., Hongell, A., Torebjork, H. E., and Wallin, B. G.: General characteristics of sympathetic activity in human skin nerves. *Acta Physiol. Scand.* **84**:164–176 (1972).

Hagiwara, S., and Bullock, T. H.: Intracellular potentials in pacemaker and integrative neurons of the lobster cardiac ganglion. *J. Cell Comp. Physiol.* **50**:25–47 (1957).

Hirst, G. D. S., and McKirdy, H. C.: Synaptic potentials recorded from neurones of the submucous plexus of guinea-pig small intestine. *J. Physiol.* (*London*) **249**:369–385 (1975).

Hirst, G. D. S., and Silinsky, E. M.: Some effects of 5-hydroxytryptamine, dopamine and noradrenaline on neurones in the submucous plexus of guinea-pig small intestine. *J. Physiol.* (*London*) **251**:817–832 (1975).

Iggo, A., and Vogt, M.: Preganglionic sympathetic activity in normal and in reserpine-treated cats. *J. Physiol.* (*London*) **150**:114–133 (1960).

Ivanov, A. Ya., and Savchenko, V. I.: Effects of nicotinic and muscarinic blocking agents on tonic activity of mammalian sympathetic ganglion neurones (in Ukrainian). Abstracts of the I Ukrainian Physiological Society Meeting, Naukova Dumka, Kiev, p. 172 (1982).

Jänig, W., and Szulczyk, P.: The organization of lumbar preganglionic neurons. *J. Auton. Nerv. Syst.* **3**:177–191 (1981).

Jänig, W., Krauspe, R., and Wiedersatz, G.: Reflex activation of postganglionic vasoconstrictor neurones supplying skeletal muscle by stimulation of arterial chemoreceptors via nonnicotinic synaptic mechanisms in sympathetic ganglia. *Pflügers Arch.* **396**:95–100 (1983a).

Jänig, W., Sundlöf, G., and Wallin, B. G.: Discharge patterns of sympathetic neurons supplying skeletal muscle and skin in human and cat. *J. Auton. Nerv. Syst.* **7**:239–256 (1983b).

Johnson, D. A., and Purves, D.: Tonic and reflex synaptic activity recorded in ciliary ganglion cells of anaesthetized rabbits. *J. Physiol.* (*London*) **339**:599–613 (1983).

Johnson, S. M., Katayama, Y., and North, R. A.: Slow synaptic potentials in neurones of the myenteric plexus. *J. Physiol.* (*London*) **301**:505–516 (1980).

Jule, Y., and Szurszewski, J. H.: Electrophysiology of neurones of the inferior mesenteric ganglion of the cat. *J. Physiol.* (*London*) **344**:277–292 (1983).

Jule, Y., Krier, J., and Szurszewski, J. H.: Patterns of innervation of neurones in the inferior mesenteric ganglion of the cat. *J. Physiol.* (*London*) **344**:293–304 (1983).

Kolosov, N. G., and Milokhin, A. A.: Afferent innervation of the ganglia in the autonomic nervous system (in Russian) *Arkh. Anat. Gistol. Embriol.* **44**:13–23 (1963).

Koval, L. M., Skibo, G. G., and Skok, V. I.: Ultrastructural study of interneuronal connections in the superior cervical sympathetic ganglion of the cat (in Russian). *Neirofiziologiya* **13**:299–306 (1981); see also *Neurophysiology* (*USSR*) **13**:220–229 (1981).

Kuba, K., and Koketsu, K.: Synaptic events in sympathetic ganglia. Prog. *Neurobiol.* **II**:77–169 (1978).

Lebedev, V. P.: Some patterns of lateral horn sympathetic neurons responses to anti- and orthodromic stimulation (in Russian), in: *Interneuronal Transmission in the Autonomic Nervous System* (P. G. Kostyuk, ed.), pp. 76–90, Naukova Dumka, Kiev (1972).

Lebedev, V. P., Syromyatnikov, A. V., and Skok, V. I.: Convergence of preganglionic fibres on vasomotor neurons of the sympathetic ganglia in cats (in Russian). *Neirofiziologiya* **9**:592–597 (1978); see also *Neurophysiology* (*USSR*) **9**:448–451 (1979).

Melnichenko, L. V., and Skok, V. I.: Natural electrical activity in mammalian parasympathetic ganglion neurones. *Brain Res.* **23**:277–279 (1970).

Milokhin, A. A.: *Afferent Innervation of Autonomic Neurons* (in Russian). Nauka, Leningrad (1967).

Mirgorodski, V. M., and Skok, V. I.: Intracellular potentials recorded from a tonically active mammalian sympathetic ganglion. *Brain Res.* **15**:570–572 (1969).

Mirgorodski, V. M., and Skok, V. I.: The role of different preganglionic fibres in tonic activity of mammalian sympathetic ganglion. *Brain Res.* **22**:262–263 (1970).

Nishi, S.: Ganglionic transmission, in: *The Peripheral Nervous System* (J. I. Hubbard, ed.), pp. 225–255, Plenum Press, New York (1974).

Nozdrachev, A. D., and Vataev, S. I.: Neuronal electrical activity in the submucosal plexus of the cat small intestine. *J. Auton. Nerv. Syst.* **3**:45–53 (1981).

Nozdrachev, A. D., Kachalov, Y. P., and Sanin, G. Yu.: Neuronal activity of submucosal plexus of pyloric and ileocecal spincteric regions of the cat gastrointestinal tract. *J. Auton. Nerv. Syst.* **4**:33–42 (1981).

Polosa, C.: Spontaneous activity of sympathetic preganglionic neurons. *Can. J. Physiol. Pharmacol.* **46**:887–896 (1968).

Purves, D., and Hume, R.: The relation of postsynaptic geometry to the number of presynaptic axons that innervate autonomic ganglion cells. *J. Neurosci.* **1**:441–452 (1981).

Skok, V. I.: *Physiology of Autonomic Ganglia*. Igaku Shoin, Tokyo (1973).

Skok, V. I.: Convergence of preganglionic fibres in autonomic ganglia (in Russian), in: *Mechanisms of Neuronal Integration of Nervous Centers* (P. G. Kostyuk, ed.), pp. 27–34, Nauka, Leningrad (1974).

Skok, V. I.: On the physiological role of slow inhibitory postsynaptic potential in the neurons of sympathetic ganglia, in: *Electrobiology of Nerve, Synapse and Muscle* (I. P. Reuben, D. P. Purpura, M. V. L. Bennett, and E. R. Kandel, eds.), pp. 123–128, Raven Press, New York (1976).

Skok, V. I.: Ganglionic transmission: Morphology and physiology, in: *Pharmacology of Ganglionic Transmission* (D. A. Kharkevich, ed.), pp. 9–39, *Handb. Exp. Pharmacol.*, Vol. 53, Springer-Verlag, Berlin, Heidelberg, and New York (1980).

Skok, V. I.: Introductory remarks: New approaches in the study of transmission in autonomic ganglia, in: *Physiology of Excitable Membranes* (J. Salanki, ed.), pp. 295–304, *Advances Physiological Sciences* Vol. 4, Pergamon Press, New York; Akademiai Kiado, Budapest (1981).

Skok, V. I., and Heeroog, S. S.: Synaptic delay in superior cervical ganglion of the cat. *Brain Res.* **87**:343–353 (1975).

Skok, V. I., and Ivanov, A. Y.: What is the ongoing activity of sympathetic neurone? *J. Auton. Nerv. Syst.* **7**:263–270 (1983).

Skok, V. I., and Mirgorodski, V. N.: Supraspinal influence upon the activity of the preganglionic sympathetic neurons sending their axons in cervical sympathetic nerve (in Russian), in: *Mechanisms of Descending Control of Spinal Cord Activities* (P. G. Kostyuk, ed.), pp. 181–185, Nauka, Leningrad (1971).

Skok, V. I., and Remizov, I. N.: Automatic multi-channel analyser for the electrical activity of mixed visceral nerve. Proceedings of the XVIII International Congress of Neurovegetative Research, pp. 100–102, Tokyo (1977).

Skok, V. I., Bogomoletz, V. I., Ivanov, A. Ya., and Mirgorodski, V. N.: Electrical activity of sympathetic ganglia during depressor reflex (in Russian). *Neirofiziologiya* **6**:519–524 (1974); see also *Neurophysiology (USSR)* **6**:410–414.

Skok, V. I., Bogomoletz, V. I., Melnichenko, L. V., Purnyn, S. L., and Gerzanich, V. V.: Human sympathetic activity recorded with skin surface electrodes. *J. Auton. Nerv. Syst.* **12**:261–266 (1984).

Svaetichin, G.: Component analysis of action potentials from single neurons. Exp. *Cell Res. Suppl.* **5**:234–261 (1958).

Tauc, L.: Les divers modes d'activité du soma neuronique ganglionnaire de l'Aplysie et de l'Escargot, in: *Microphysiologie Comparée des Elements Excitables*, p. 91, Centre National de la Recherche Scientifique, Paris (1955).

Vallbo, A. B., Hagbarth, K. E., Torebjork, H. E., and Wallin, B. G.: Somatosensory, proprioceptive, and sympathetic activity in human peripheral nerves. *Physiol. Rev.* **59**:919–957 (1979).

Wood, J. D.: Neurophysiology of Auerbach's plexus and control of intestinal motility. *Physiol. Rev.* **55**:307–324 (1975).

Wood, J. D.: Synaptic interactions in the enteric plexuses. *J. Auton. Nerv. Syst.* **4**:121–133 (1981).

Wood, J. D., and Mayer, C. J.: Intracellular study of electrical activity of Auerbach's plexus in guinea-pig small intestine. *Pfluegers Arch.* **374**:265–275 (1978).

19

Chemosensitivity of Visceral Primary Afferent Neurons: Nodose Ganglia

HIDEHO HIGASHI

I. INTRODUCTION

The sensory nerve terminal consists of the generator region and the regenerative region, which are involved in, respectively, the production of receptor potential and the generation of afferent nerve impulses. The latter region is believed to be endowed with receptors susceptible to various bioactive substances, such as acetylcholine (ACh), 5-hydroxytryptamine (5-HT), and histamine (HST) (see Paintal, 1964). While it is important to investigate the nature of these receptors at the membrane level, such a study is almost impossible to conduct because of the technical difficulties due to the small size of the axon terminal.

Using respiratory and cardiovascular reflexes as the criteria, Jacobs and Comroe (1971) demonstrated that injection of 5-HT into the carotid artery activates cat nodose ganglion cells. The results of the experiments of Sampson and Jaffe (1974), in which extracellular recordings were carried out along the infra- and supranodose vagus nerve, confirmed the findings of Jacobs and Comroe (1971). Using intracellular recording methods, Higashi (1977) showed that low concentrations of 5-HT (≥ 1 μM) depolarize the soma membrane of a majority of rabbit nodose ganglion cells. Other substances, such as ACh, veratrum alkaloids, and phenyldi-

---

HIDEHO HIGASHI • Department of Physiology, Kurume University School of Medicine, Kurume, Japan.

guanide, stimulate the sensory nerve terminals and depolarize the visceral sensory neurons, i.e., the nodose ganglion cells (Sampson and Jaffe, 1974; Blackman *et al.*, 1975; Higashi *et al.*, 1982b; Wallis *et al.*, 1982). This evidence suggests that the somata of primary afferent neurons are endowed with receptors similar to those located on the regenerative region of the primary afferent fibers. If this is the case, the nature and functional characteristics of the peripheral receptors can be elucidated by characterizing the receptors of the visceral primary afferent neurons.

A. Electrical Properties of Nodose Ganglion Cells

The nodose ganglion of the vagus nerve contains the cell bodies of afferent fibers innervating the respiratory, cardiovascular, and gastrointestinal systems. Approximately 80% of the 30,000 fibers in the cervical vagus are sensory, and of these, 70% are unmyelinated afferent fibers originating in abdominal viscera (Evans and Murray, 1954; Agostoni *et al.*, 1957; see also Leek, 1972). The cell bodies in the nodose ganglion are devoid of synapses (Cajal, 1909).

In 1976, Jaffe and Sampson (1976) described the electrical properties of nodose ganglion cells of the cat and rabbit. They did not, however, identify the impaled cells in terms of the nature of their peripheral processes. Thus, their measurements probably refer to a mixed population of type A myelinated and type C unmyelinated neurons. Gallego and Eyzaguirre (1978) measured the passive and active electrical properties of neurons (Table 1) identified as type A or C on the basis of conduction velocities of their axons. Type A and C neurons exhibited similar resting membrane potentials (MP) and input resistance (R_0). They differed, however, in several respects; as compared to type C neurons, type A neurons were characterized by small time constants (τ) and capacitances (C_0) and shorter and smaller action potentials with less overshoot. The ratio of type A to type C neuronal populations was in good agreement with the ratio of myelinated to nonmyelinated sensory fibers in the vagus. Moreover, it has been also suggested that the membrane characteristics of the sensory ganglion cells are related to the properties of the peripheral nerves to which they give rise (see Gallego and Eyzaguirre, 1978).

Our results obtained in studies of the membrane properties of rabbit nodose ganglion cells were generally in agreement with those of Gallego and Eyzaguirre (1978); an additional, major difference between type A and C neurons was that type C neurons exhibited a regenerative Ca^{2+} system (see also Chapter 4) in addition to the Na^+ spike-producing system, while type A neurons lacked the former, as indicated by the finding that about 80% of type C neurons generated an action potential characterized by having a shoulder on its repolarizing phase that disappeared in a low-Ca^{2+} (0.5 mM)/high-Mg^{2+} (5.5 mM) medium and in solutions containing

Table 1. Electrical Properties of A and C Neurons in Nodose Ganglia[a]

	Membrane Constants			
Cells	MP (mV)	R_0 (MΩ)	τ (msec)	C_0 (pF)
A	53 ± 2.1 (9)	20 ± 6.2 (5)	2.0 ± 0.2[b] (5)	122 ± 19.8[b] (5)
C	54 ± 1.1 (53)	15 ± 1.2 (39)	3.4 ± 0.2[b] (28)	249 ± 5.2[b] (28)
	Action potentials			
Cells	Amplitude (mV)	Overshoot (mV)	AHP (mV)	Duration (msec)
A (9)	67 ± 3.4[b]	14 ± 3.0[b]	10 ± 2.1	1.0 ± 0.2[b]
C (45)	85 ± 1.2[b]	31 ± 1.1[b]	12 ± 0.7	2.8 ± 0.1[b]

[a] From Gallego and Eyzaguirre (1978). Values are means ± S.E. Figures in parentheses are numbers of cells.
[b] Significant difference ($P < 0.001$) between A and C cells.

cadmium (0.2 mM) or cobalt (0.5–3.0 mM). Furthermore, the action potential of type C neurons was lowered but not abolished in a Krebs solution containing tetrodotoxin (TTX) (0.3 μM), and the addition of tetraethylammonium (10 mM) enhanced the TTX-inhibited action potential, so that it was heightened to an overshooting level of 20 mV or more as well as prolonged, exhibiting a plateau of some tens or hundreds of milliseconds. Type A neurons, on the other hand, completely lost their excitability in a TTX-containing solution (Higashi *et al.*, 1982a; Ito, 1982).

In approximately 40% of type C neurons, the action potential was followed by a long-lasting (90 msec to 4 sec) afterhyperpolarization. Tetanic stimulation of such neurons produced an extremely prolonged afterpotential [posttetanic afterhyperpolarization (PTH)]. The PTH was reversed at a membrane potential of about −90 mV. This reversal level was shifted by approximately 54 mV when the external K^+ concentration was increased 10-fold. The PTH was abolished in a Ca^{2+}-free solution or a solution containing Co^{2+} (1 mM); this suggested that the PTH was due to a Ca^{2+}-activated K^+ conductance (Morita *et al.*, 1982; Higashi *et al.*, 1984) (see also Chapter 4).

II. CHEMOSENSITIVITY OF NODOSE GANGLION CELLS

Visceral afferent A and C nerve terminals differ in their chemosensitivity to various endogenous substances (see Paintal, 1963, 1964). Furthermore, Neto (1978) demonstrated that 5-HT depolarizes nonmyelinated C fibers but not myelinated A fibers of the rabbit cervical vagus and sciatic nerves. Nerve terminals of vagal nonmyelinated fibers can be activated by 5-HT or ACh (Douglas and Ritchie, 1957; Armett and Ritchie, 1961). As shown in Table 2, the somata of type C neurons are depolarized by substances that evoke pain on skin injection or application to appropriate nerve terminals [such as bradykinin (BK) or histamine (HST)] and by other bioactive autacoids, while type A neurons are less sensitive to these compounds (Higashi, 1980; Higashi *et al.*, 1982b). Together, these results indicate that type C somata and C axons of visceral primary afferent neurons are excited by algesics, while type A neurons exhibit little or no sensitivity to algesics.

Recent studies disclosed that approximately 10% of the vagal afferent fibers exhibit physiological and morphological properties of C fibers in their distal part and properties of B (myelinated) fibers in their proximal part (Duclaux *et al.*, 1976) and that some neurons of type A are sensitive to the depolarizing action of ACh and 5-HT (Higashi, 1980; Higashi *et al.*, 1982b). This seems to indicate that the differences in the sensitivities of type A and type C neurons to endogenous bioactive substances are due to the relative differences in the densities of pertinent receptors, rather than to the absolute existence or absence of certain receptors.

The following sections describe the responses of nodose ganglion cells to several endogenous substances.

A. 5-Hydroxytryptamine

It is well known that the cell bodies of vagal primary afferents that are located in the nodose ganglion are excited and discharge action potentials in response to application of 5-HT (Sampson and Jaffe, 1974; DeGroat and Simonds, 1976), and a direct depolarizing action of 5-HT has been reported (Higashi, 1977; Higashi, 1980; Simonds and DeGroat, 1980; Wallis *et al.*, 1982). In a study by Higashi and Nishi (1982), the susceptibilities of type A and type C neurons to 5-HT appeared to be as follows: In response to 5-HT applied by superfusion ($\geq 1\ \mu M$) or by iontophoresis (≥ 5 nA, 50 msec), the majority of type C neurons showed a rapid depolarization of 20–30 mV in amplitude that was followed by a hyperpolarization of a few millivolts. Both the initial depolarization and the afterhyperpolarization were associated with a reduction in membrane resistance. The initial depolarization induced by 5-HT was abolished by Na^+-free Krebs solution and was reduced by a few millivolts in K^+-free Krebs solution. The response in normal Krebs solution was reversed at a membrane potential level of $+7.3$ mV. The afterhyperpolarization dis-

Table 2. Sensitivity of A and C Neurons in Nodose Ganglia to Endogenous Substances (10 μM)[a]

Neurons	5-HT-sensitive	BK-sensitive	ACh-sensitive	HST-sensitive	GABA-sensitive	
A neurons (14)	2 (14%)	0	1 (7%)	0	14 (100%)	
C neurons (42)	39[b] (93%)	15 (36%)	14 (33%)	10 (24%)	33 (79%)	
	Depolarization (mV)					RP (mV)
Neurons	5-HT	BK	ACh	HST	GABA	
A neurons	4.5 ± 2.1	0	7.0	0	14.4 ± 8.7	−58.0 ± 5.5
C neurons	23.8 ± 10.9	8.4 ± 5.7	12.5 ± 6.9	6.7 ± 7.0	6.6 ± 7.0	−57.7 ± 5.5

[a] From Higashi *et al.* (1982b).
[b] Of these 39 neurons, 34 were depolarized and 5 hyperpolarized. Each response is the mean ± S.D. in mV of the responses tabulated above.

appeared in Na^+-free or Ca^{2+}-free Krebs solution, while it was markedly enhanced the K^+-free Krebs solution. This afterhyperpolarization was reversed in normal Krebs solution at a membrane potential of -88.7 mV; it was abolished at membrane potentials more positive than -20 mV. These results suggest that the initial depolarization induced by 5-HT was due mainly to simultaneous increases in Na^+ and K^+ conductances, while the afterhyperpolarization was brought about by an increase of K^+ conductance that is triggered by a voltage-dependent influx of Ca^{2+}. Finally, a majority of type A neurons did not respond to 5-HT, whether 5-HT was applied by superfusion or iontophoretically; the few type A neurons that did respond to 5-HT showed a depolarization of only a few millivolts (Higashi and Nishi, 1982; Higashi *et al.*, 1982b; Stansfeld and Wallis, 1984).

Wallis *et al.* (1982) investigated the depolarizing action of 5-HT analogues on the rabbit nodose ganglion and disclosed that for activation of 5-HT receptors of nodose ganglion cells, the ligand must possess a hydroxyl moiety at position 5 of the indole nucleus. Absence from the ligand of a hydroxyl group, as in tryptamine, or the presence of a bulky substituent group, as in 5-methoxytryptamine, leads to a marked reduction in activity. Among the compounds tested in which the side chain was *N*-methylated, only bufotenine showed substantial depolarizing ability. Psilocin, which is *N*-methylated but exhibits the hydroxyl group in position 4 of the nucleus, was inactive. Methysergide and LSD_{25}, both recognized as specific 5-HT antagonists, did not produce any significant antagonism of 5-HT depolarization (Higashi and Nishi, 1982; Wallis *et al.*, 1982). Neto (1978) also reported that methysergide does not affect 5-HT depolarization of rabbit cervical vagal fibers, and Nishi (1975) noted that LSD_{25} or methysergide does not alter the excitatory effect of 5-HT on the cat carotid body that causes firing recorded from the carotid nerve. On the other hand, quipazine, which is reputed to act centrally by mimicking the action of 5-HT, reduced 5-HT depolarization without exhibiting any agonist activity on its own (Wallis *et al.*, 1982). D-Tubocurarine (D-TC) depressed, in a reversible manner, 5-HT depolarization of nodose neurons without altering their resulting membrane properties and, in a concentration range of 0.2–5 μM, caused a parallel shift to the right of the concentration response curves for 5-HT. Although the peak amplitude of the 5-HT-induced inward currents was depressed by D-TC, there was no alteration in the time–course of the falling phase of these currents. At concentrations greater than 10 μM, D-TC caused a nonparallel shift to the right and depression of the maximum peak response to 5-HT. In addition, at these higher concentrations, the time–course of the 5-HT currents was altered significantly; i.e., the falling phase was frequently shortened and became biphasic (Higashi and Nishi, 1982). D-TC also blocked the depolarizing action of 5-HT on myenteric plexus neurons of the guinea pig (Hirst and Silinsky, 1975). Besides antagonizing the 5-HT effect on nodose ganglion cells, D-TC (0.2–5 μM) blocked both ACh- and γ-aminobutyric acid (GABA)-induced depolarization. Picrotoxin (1–10 μM) also blocked 5-HT- as well as GABA- and ACh-induced depolarizations (see below). Picrotoxin was, however, 6–10 times less potent than D-TC in blocking ACh- and 5-HT-induced depolarizations (Higashi *et al.*, 1982b). Also, picrotoxin blocked *in vivo* 5-HT- as well as GABA-induced depolarizations of the cat nodose ganglion (DeGroat and Lalley, 1973; Saum and DeGroat, 1973; Simonds and DeGroat, 1980). Higashi *et al.* (1982b) suggested that picrotoxin may act on the receptor complex rather than the neuronal excitable membrane, since resting membrane properties were not affected by picrotoxin. However, the precise mechanism of picrotoxin blocking action remains unknown.

Recently, Stansfeld and Wallis (1984), using a sucrose-gap method, demonstrated that 5-HT-induced depolarization in rabbit nodose ganglia was not only markedly potentiated but also changed in character in a Ca^{2+}/Mg^{2+}-free medium; the dose dependence between quantity of 5-HT and response magnitude was lost under these circumstances, and 5-HT appeared to trigger the response in an all-or-none fashion. They suggested that this unusual response is due to a depolarization that is initiated in a few cells and spreads through the ganglion; they further indicated that K^+ efflux probably plays a role in this process, since the increase in external K^+ concentration during superfusion with Ca^{2+}/Mg^{2+}-free medium initiated the responses. The responses were dependent on external Na^+ concentration, but were not blocked by TTX. These observations suggest that voltage-sensitive Na^+ channels that are TTX-insensitive are responsible for this unusual response.

In a small number of type C neurons, 5-HT applied by superfusion at concentrations of 0.1 μM or with iontophoretic currents of 0.5–20 nA (50-msec duration) evoked a simple hyperpolarizing response of 3–8 mV without any associated pre- and postdepolarization (Higashi, 1977; Higashi and Nishi, 1982; Stansfeld and Wallis, 1984). Since removal of K^+ enhanced this hyperpolarization and since displacement of the membrane potential to a level more negative than -90 mV reversed its polarity, voltage-independent K^+ conductance may constitute the ionic mechanism responsible for this hyperpolarization. When the same neurons were superfused with a higher concentration (10 μM) of 5-HT or when 5-HT was applied iontophoretically with larger currents (50–100 nA, 50 msec), the hyperpolarizing response was preceded by a rapid depolarization of 5–15 mV; both the hyperpolarizing and the depolarizing component were associated with a decrease in membrane resistance. Finally, a small number of type C neurons did not show any response to 5-HT applied either by iontophoresis or by superfusion.

B. Acetylcholine

In response to ACh applied by superfusion (≥10 μM) or by iontophoresis (≥5 nA, 50 msec), approximately 30% of type C neurons showed a rapid depolarization amounting to about 10 mV, which was accompanied by a decreased input resistance (Higashi, 1980). ACh depolarization disappeared in low-Na^+ (11.7 mM) Krebs solution and reversed its polarity at 0 mV (Higashi et al., 1980b). These data indicate that the ACh-induced depolarization may be due to a simultaneous increase in Na^+

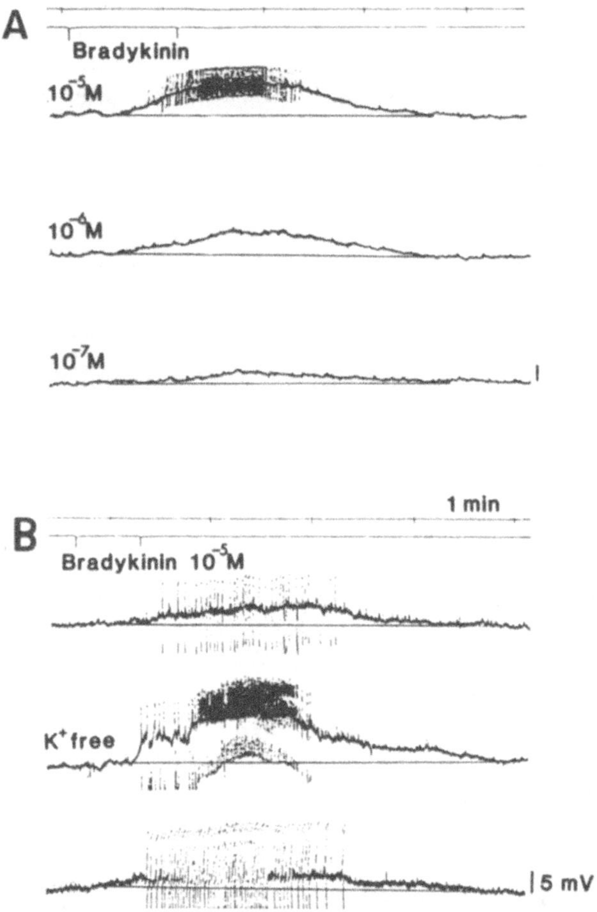

Figure 1. BK-induced depolarization at various concentrations of BK (A) and in K^+-free Krebs solution (B). The top pair of tracings in (A) and (B) depicts time marks (1 min) (upper) and on–off marks of BK application (lower). (A) BK (10^{-7}–10^{-5} M) produces dose-related depolarizations in normal Krebs solution. (B) BK (10^{-5} M) induces depolarization in normal Krebs solution, K^+-free solution, and normal Krebs solution (top to bottom).

and K^+ conductance. ACh depolarization was blocked by D-TC (10–50 μM), but unaffected by atropine employed at the same concentrations (Higashi, 1980; Higashi *et al.*, 1982b); thus, this ACh depolarization must have been generated by nicotinic receptors. In agreement with these findings, Wallis *et al.* (1982) reported that a nicotinic receptor ligand, 1,1-dimethyl-4-phenylpiperazinium iodide, depolarizes nodose ganglion cells of the rabbit.

C. Bradykinin

BK (0.1–10 μM) depolarized the membrane and increased input resistance in about one third of type C neurons. The BK depolarization was long-lasting and dose-dependent (Figure 1A). BK depolarization was reduced in amplitude as the membrane potential was increased by anodal currents applied to the cell membrane. The BK depolarization was enhanced in a K^+-free Krebs solution (Figure 1B). These findings indicate that the BK depolarization is brought about mainly by a decrease in K^+ conductance (Higashi, 1980). In the case of a few type C neurons sensitive to BK, the slow depolarization was preceded by a rapid hyperpolarization of 4–10 mV that was associated with increased conductance (Higashi *et al.*, 1982b).

Some of the effects of BK on the sympathetic autonomic ganglia are of interest in this context. Although in many instances BK had no significant effect on the membrane of sympathetic ganglion cells (Haefely, 1970; Wallis and Woodward, 1974) (see also Cahpter 13), BK-induced depolarization was also noticed, as in the case of the effects of concentrations of BK ranging from 10 to 100 μM (Yonemura *et al.*, 1982); these concentrations produced dose-dependent depolarization of guinea pig sympathetic ganglia (see also Figure 9 concerning the rabbit superior cervical ganglion in Wallis and Woodward, 1974) (see also Chapter 13).

D. Histamine

About 25% of C neurons respond to the depolarizing action of HST (Table 2). The behavior of HST-induced depolarization at different levels of membrane potential and in a K^+-free medium is similar to that of BK-induced depolarization, suggesting that both share the same ionic mechanism; i.e., both are due to reduction in K^+ conductance (Higashi *et al.*, 1982b). It is still unknown whether HST_1 or HST_2 receptors are involved in the HST-induced decrease of K^+ conductance.

It is of interest that HST receptors are also present in mammalian sympathetic ganglia (Brimble and Wallis, 1973); both pre- and postsynaptic effects may be involved (Snow *et al.*, 1980) (see also Chapters 11 and 13). It should be added that two types of receptors may be present in these ganglia, H_1 and H_2 receptors being involved in excitatory and blocking effects of histamine, respectively (Lindl, 1983).

E. γ-Aminobutyric Acid

A direct depolarizing action of GABA on nodose ganglion cells has been described by DeGroat (1970) and Simonds and DeGroat (1980). In response to GABA applied by superfusion ($\geq 5\ \mu M$) or by iontophoresis (≥ 5 nA, 50 msec), all type A neurons and the majority of type C neurons of rabbit nodose ganglia show a rapid depolarization associated with decreased input resistance. The observation that the reversal potential for GABA responses (-15 to -30 mV) obtained with a K_2SO_4 electrode is similar to that of cat dorsal root ganglion cells suggests that increased Cl^- conductance is involved in the GABA depolarization. D-TC as well as picrotoxin at relatively low concentrations depress GABA responses (Higashi *et al.*, 1982b). These effects of GABA on rabbit nodose ganglion cells resemble those on the primary afferent neurons of the lumbar spinal ganglia of the bullfrog (Nishi *et al.*, 1974) and the cat (Gallagher *et al.*, 1978). In the studies of Nishi *et al.* (1974) and Gallagher *et al.* (1978), the effects of GABA and their mechanism, i.e., an increased chloride conductance, were similar in the case of both terminals and primary afferent neurons.

These effects of GABA on the cells of nodose and of spinal ganglia are comparable to the action of GABA on the neurons of autonomic ganglia. GABA depolarized the cell membranes of cat and rat superior cervical sympathetic ganglion (DeGroat, 1970; Bowery and Brown, 1974), and this depolarization was attributed to an increased Cl^- conductance (Adams and Brown, 1975) (see also Chapters 12 and 13); it is of interest that picrotoxin blocked GABA- as well as ACh- and 5-HT-induced depolarization and that D-TC exhibited similar blocking actions on ACh-, GABA-, and 5-HT-induced depolarizing responses. Picrotoxin was 6–10 times less potent than D-TC in blocking ACh and 5-HT responses (Higashi *et al.*, 1982b).

F. Endogenous Opiatelike Substances

Autoradiographic techniques indicate that tritiated etorphine and diprenorphine become localized along the course of the intramedullary vagus nerve following systemic administration (Atweh and Kuhar, 1977); these binding sites are reduced following vagotomy (Atweh *et al.*, 1978). This suggested that the vagal cell bodies in the nodose ganglion and perhaps the infranodose vagal fibers may also have opiate-binding sites.

Indeed, Shefner *et al.* (1981) found stereospecific binding sites for tritiated dihydromorphine and Leu^5-enkephalin in both the nodose ganglion and the vagus nerve of the rabbit, but they failed to observe any electrophysiological effect of these ligands on the neuron. However, this negative finding may have resulted from a form of desensitization, since much lower concentrations (1–100 nmol/liter) both hyperpolarize the resting membrane of C-type neurons and also prolong the calcium-dependent after-hyperpolarization that follows a burst of action potentials or application of 5-HT (North and Egan, 1982; Higashi *et al.*, 1982; Higashi, unpublished observations). Higashi *et al.* (1982) demonstrated that in approximately 50% of type C neurons, bath application of morphine (1–100 nM) causes a hyperpolarization of a few millivolts in amplitude. The hyperpolarization was associated with a decreased input resistance and antagonized by a prior exposure of the preparation to naloxone (10 nM). The apparent reversal potential for this morphine-induced hyperpolarization was estimated by employing constant-current anelectrotonic pulses; its average value was −86.3 mV, which is similar to the reversal level of the afterhyperpolarization of soma spikes; this suggests that the hyperpolarizing effect of morphine may be due to selective increase of K^+ conductance. Furthermore, Higashi *et al.* (1982a) observed that morphine (1–100 nM) initially augments and subsequently depresses Ca^{2+}-dependent spikes and that this effect of morphine can be prevented by an equal concentration of naloxone. It is of interest to compare these effects of opioids on nodose ganglia with their effect on pre- and postsynaptic sites of the sympathetic ganglion (see Chapters 11 and 13).

G. Other Substances

Many of the primary afferents of the vagus have been shown to contain substance P (Lundberg *et al.*, 1978). Since it has been shown that pericellular plexuses of substance-P-immunoreactive fibers are present in rabbit nodose ganglia, substance P may have a direct effect on neuronal cell bodies of the nodose ganglia (Katz and Karten, 1980). However, Wallis *et al.* (1982), using a sucrose-gap method, demonstrated that bath application of substance P to the rabbit nodose ganglion produced no detectable change in resting membrane potential.

Chemoreceptor nerve endings in rabbit or cat carotid bodies appear to be hyperpolarized by dopamine (Sampson and Vidruk, 1977). Since the nodose ganglion contains some afferents that originate in aortic bodies, the effect of dopamine on the ganglion was tested (Wallis *et al.*, 1982); dopamine produced no change in resting membrane potential in the case of many preparations and a slight depolarization in the case of a minority of ganglia. The effect of noradrenaline was also tested, and it was found that noradrenaline exhibits a medium to low depolarizing ability (Wallis *et al.*, 1982). Our intracellular recordings showed that noradrenaline and dopamine (0.1–50 μM) do not significantly depolarize the majority of type C neurons. Employed at low concentrations, these catecholamines augment the long-lasting slow afterpolarization and PTH, which are both due to an increase in a Ca^{2+}-activated K^+ conductance; at high concentrations, they depress this Ca^{2+}-dependent hyperpolarization, block the Ca^{2+}-dependent shoulder of the action potential, and eliminate or shorten the Ca^{2+}-dependent spikes without affecting the resting membrane properties (Higashi, unpublished observations). Similar observations have been made with respect to sympathetic ganglion cells (Minota and Koketsu, 1977; Horn and McAfee, 1980), and it was proposed that activation by catecholamines of an α-adrenoceptor inhibits a voltage-dependent Ca^{2+} conductance (Horn and McAfee, 1980) (see also Chapters 4 and 13).

III. INTERACTIONS BETWEEN ALGESIC AND ANALGESIC SUBSTANCES AND THEIR MECHANISMS

The observations described above are compatible with the hypothesis that the cell bodies of vagal afferent neurons are endowed with algesic receptors (such as those responding to 5-HT, BK, ACh, and HST) that may be similar to those located along their axons and on their axon terminals. A similar receptor distribution has been described for other neuronal elements, such as GABA receptors on the afferent terminals and somata of primary afferent neurons (see Nishi *et al.*, 1974; Gallagher *et al.*, 1978; Curtis, 1979) and the cholinoceptors and adrenoceptors on the somata and axon terminals of the sympathetic nerves to the heart (see Westfall, 1977). Stimulation of low-threshold vagal afferents produces primary afferent depolarization (PAD), resulting in presynaptic inhibition (Eccles *et al.*, 1963) within the solitary tract nucleus receiving sensory projections from nodose ganglion cells of the vagus (Rudomin, 1967; Dubbledam *et al.*, 1979). The time–course and physiological properties of these PADs are similar to those evoked in the spinal cord. They may result from the action of GABA released by interneurons forming axonal synapses with primary afferents (see Curtis, 1979). Thus, it seems likely that the nerve terminals of the visceral afferent fibers within the solitary tract nucleus may be endowed with GABA receptors similar to those located on the nodose ganglion cells.

The algesic-induced depolarizations lend themselves to the following classification: The first type concerns 5-HT- and ACh-induced response concomitant with simultaneously increased Na^+ and K^+ conductance; this depolarization is characterized by rapid onset and short duration.

This concept is consistent with results obtained in studies of the rabbit superior cervical ganglion (Skok and Selyanko, 1979) (see also Chapters 6, 12, and 13). The second type of response is that due to BK and HST, in which case K^+ conductance is decreased and the depolarization has a slow onset and a long-lasting course. In several studies (Mense and Schmidt, 1974; Nishi, 1975; Fock and Mense, 1976: Foreman *et al.*, 1979) in which extracellular recordings were obtained with respect to peripheral mammalian afferent nerves, intraarterial injection of 5-HT or ACh evoked a short burst of repetitive firing followed by a supression of spontaneous firing, while the application of BK and HST produced long-lasting repetitive firing. These different patterns of firing evoked by various algesics may be due to different ionic mechanisms underlying their depolarizations. The analgesics such as morphine produced a slight hyperpolarization of the afferent nerves arising from an increased K^+ conductance. In this preparation, however, the conductance increase induced by morphine was small, and occasionally morphine produced no noticeable hyperpolarization, while it markedly inhibited Ca^{2+}-dependent spikes. It is therefore possible that at high concentrations (10–100 nM), morphine depresses Ca^{2+} entry into the nerve terminal as well as the somata of the visceral primary afferent neurons. Again, these actions of opioids on afferent neurons should be compared to their actions on pre- and postsynaptic sites in the autonomic ganglia (see Chapters 11 and 13).

Since ACh and 5-HT depolarizations are generated by increases in Na^+ and K^+ conductances and are blocked by D-TC or picrotoxin, there is a possibility that ACh and 5-HT may share the same receptor–ionophore complex. In the case of nodose ganglia, however, this possibility is unlikely, since in this preparation (1) hexamethonium blocks ACh but not 5-HT depolarization and (2) ACh does not desensitize 5-HT depolarization (Higashi and Nishi, 1982). In addition, there appeared to be no interaction between 5-HT and BK receptors, although BK depressed the afterhyperpolarizing phase of 5-HT responses (Higashi *et al.*, 1982b). A similar lack of cross-desensitization among 5-HT, BK, and HST was reported for the endings of muscular group IV afferent units (Hiss and Mense, 1976) and unmyelinated afferents from the skin (Beck and Handwerker, 1974). On the other hand, 5-HT exerted a long-lasting amplifying effect on BK responses (Sicuteri, 1968). The depression by BK of the 5-HT-induced afterhyperpolarization may explain the results of Sicuteri (1968).

IV. SUMMARY

It appears that particularly the type C cells of mammalian nodose ganglia exhibit multiple chemosensitivity to several transmitters and autacoids. Altogether, the somata of type C neurons in the rabbit nodose ganglion are endowed with receptor sites for 5-HT, BK, ACh, HST, catecholamines, opiates, and GABA. The effects of these substances vary from compound to compound, as do their mechanisms. 5-HT and ACh produce rapid depolarization associated with an increased membrane conductance, most likely to Na^+ and K^+, while BK and HST elicit slow depolarization accompanied by a decreased membrane conductance, probably to K^+. Since these substances are endowed with algesic properties, their effects on afferent neurons can be considered as teleological, yet it is of interest that they exert actions on autonomic ganglia as well (see Chapters 8, 11, and 13); this may illustrate the parsimony principle as exerted by nature. At high concentrations, catecholamines and morphine depress a voltage-dependent Ca^{2+} conductance. GABA induces a rapid depolarization associated with an increased membrane conductance to, presumably, Cl^-, as does afferent-terminal depolarization involved in presynaptic inhibition. In contrast, type A neurons are relatively insensitive to the four algesic agents, but respond to GABA. The nature of the receptors involved remains unclear, since the results of pertinent pharmacological analysis are inconclusive. D-TC or picrotoxin at relatively low concentrations blocks ACh-, 5-HT-, and GABA-induced depolarizations without affecting the membrane properties, while hexamethonium blocks ACh but not 5-HT responses. In addition, 5-HT and ACh or BK responses were not subject to cross-desensitization, which suggests that the depolarizing effect of these substances on the visceral neurons may be exerted via different receptor sites.

REFERENCES

Adams, P. R., and Brown, D. A.: Action of γ-aminobutyric acid on sympathetic ganglion of cells. *J. Physiol. (London)* **250**:85–120 (1975).

Agostoni, E., Chinnock, J. E., Deburgh Daly, M., and Murray, J. G.: Functional and histological studies of the vagus nerve and its branches to the heart, lungs and abdominal viscera in the cat. *J. Physiol. (London)* **135**:182–205 (1957).

Armett, C. J., and Ritchie, J. M.: The action of acetylcholine and some related substances on conduction in non-myelinated fibers. *J. Physiol. (London)* **155**:372–384 (1961).

Atweh, S. F., and Kuhar, M. J.: Autoradiographic localization of opiate receptors in rat brain. I. Spinal cord and lower medulla. *Brain Res.* **124**:53–67 (1977).

Atweh, S. F., Murrin, L. C., and Kuhar, M. J.: Presynaptic localization of opiate receptors in the vagal and accessory optic systems: An autoradiographic study. *Neuropharmacology* **17**:65–71 (1978).

Beck, P. W., and Handwerker, H. O.: Bradykinin and serotonin effects on various types of cutaneous nerve fibres. *Pfluegers Arch.* **347**:207–222 (1974).

Blackman, J. G., Borison, H. L., and Milne, R. J.: Intracellular recording of after-discharge induced by veratrum alkaloids in the guinea-pig nodose ganglion. *Brain Res.* **98**:369–372 (1975).

Bowery, N. G., and Brown, D. A.: Depolarizing actions of γ-aminobutyric acid and related compounds on rat superior cervical ganglion *in vitro*. *Br. J. Pharmacol.* **50**:205–218 (1974).

Brimble, M. J., and Wallis, D. I.: Histamine H_1 and H_2 and receptors at a ganglionic synapse. *Nature* (*London*) **246**:156–158 (1973).

Cajal, R. Y.: *Histologie du Systeme Nerveux de l'Homme et des Vertébrés, p 732 pp. A. Malonine, Paris (1909).*

Curtis, D. R.: Pre- and postsynaptic action of GABA in the mammalian spinal cord, in: *Advances in Pharmacology and Therapeutics*, Vol. 2, *Neuro-Transmitters* (P. Simon, ed.), pp. 281–298, Pergamon Press, Oxford (1979).

DeGroat, W. C.: The actions of γ-aminobutyric acid and related amino acids on mammalian autonomic ganglia. *J. Pharmacol. Exp. Ther.* **172**:384–396 (1970).

DeGroat, W. C., and Lalley, P. M.: Interaction between picrotoxin and 5-hydroxytryptamine in the superior cervical ganglion of the cat. *Br. J. Pharmacol.* **48**:233–244 (1973).

DeGroat, W. C., and Simonds, W.: Antagonism by picrotoxin of 5-hydroxytryptamine induced depolarization and excitation of the nodose ganglion of the cat. *Neurosci. Abstr.* *2*:780 (1976).

Douglas, W. W., and Ritchie, J. M.: On excitation of nonmedullated afferent fibres in the vagus and aortic nerves by pharmacological agents. *J. Physiol* (*London*) **138**:31–43 (1957).

Dubbledam, J. D., Burs, E. R., Menken, S. B. J., and Zeilstra, S.: The central projections of the glossopharyngeal and vagus ganglia in the mallard, *Anas platyrhynchos* L. *J. Comp. Neurol.* **183**:149–168 (1979).

Duclaux, R., Mei, N., and Ranieri, F.: Conduction velocity along the afferent vagal dendrites: A new type of fibre. *J. Physiol* (*London*) **260**:487–495 (1976).

Eccles, J. C.: Ionic mechanism of postsynaptic inhibiton, in: *Les Prix Nobel, 1963*, pp. 261–283, Nobel Foundation, Stockholm (1964).

Eccles, J. C., Schmidt, R. F., and Willis, W. D.: Pharmacological studies on presynaptic inhibition. *J. Physiol* (*London*) **168**:500–530 (1963).

Evans, D. H. L., and Murray, J. G.: Histological and functional studies on the fibre composition of the vagus nerve of the rabbit. *J. Anat.* (*London*) **88**:320–337 (1954).

Fock, S., and Mense, S.: Excitatory effect of 5-hydroxytryptamine, histamine and potassium ions on muscular group IV afferent unit: A comparison with bradykinin. *Brain Res.* **105**:459–469 (1976).

Foreman, R. D., Schmidt, R. F., and Willis, W. D.: Effects of mechanical and chemical stimulation of fine muscle afferents upon primate spinothalmic tract cells. *J. Physiol* (*London*) **286**:215–231 (1979).

Gallagher, J. P., Higashi, H., and Nishi, S.: Characterization and ionic basis of GABA-induced depolarizations recorded *in vitro* from cat primary neurons. *J. Physiol* (*London*) **275**:263–282 (1978).

Gallego, R., and Eyzaguirre, C.: Membrane and action potential characteristics of A and C nodose ganglion cells studies in whole ganglia and in tissue slices. *J. Neurophysiol.* **41**:1217–1232 (1978).

Haefely, W.: Some actions of bradykinin and related peptides on autonomic ganglion cells, in: *Advances in Experimental Medicine and Biology*, Vol. 8, *Bradykinin and Related Kinins* (F. Sicuteri, M. Rocha e Silva, and N. Back, eds.), pp. 591–599, Plenum Press, New York (1970).

Higashi, H.: 5-Hydroxytryptamine receptors on visceral primary afferent neurones in the rabbit. *Nature* (*London*) **267**:448–450 (1977).

Higashi, H. Chemoreceptors of visceral primary afferent neurons in the rabbit. *Biomed. Res.* (*Suppl.*) **1**:98–101 (1980).

Higashi, H., and Nishi, S.: 5-Hydroxytryptamine receptors of visceral primary afferent neurones on rabbit nodose ganglion. *J. Physiol (London)* **323**:543–567 (1982).

Higashi, H., Shinnick-Gallagher, P., and Gallagher, J. P.: Facilitatory and inhibitory effects of morphine on Ca^{++} dependent responses in visceral afferent neurons. *Brain Res.* **251**:186–191 (1982).

Higashi, H., Ueda, N., Nishi, S., Gallagher, J. P., and Shinnick-Gallagher, P.: Chemoreceptors for serotonin (5-HT), bradykinin (BK), histamine (H) and γ-aminobutyric acid (GABA) on rabbit visceral afferent neurons. *Brain Res. Bull.* **8**:23–32 (1982b).

Higashi, H., Morita, K., and North, R. A.: Calcium dependent afterpotentials in visceral afferent neurones of the rabbit. *J. Physiol. (London)* **355**:479–492 (1984).

Hirst, G. D. S., and Silinsky, E. M.: Some effects of 5-hydroxytryptamine, dopamine and noradrenaline on neurons in the submucous plexus of guinea-pig small intestine. *J. Physiol. (London) 251*:817–832 (1975).

Hiss, E., and Mense, S.: Evidence for existence of different receptor sites for algesic agents at the endings of muscular group IV afferent units. *Pfluegers Arch.* **362**:141–146 (1976).

Horn, J. P., and McAfee, D. A.: Alpha-Adrenergic inhibition of calcium-dependent potentials in rat sympathetic neurons. *J. Physiol. (London)* **301**:191–204 (1980).

Ito, H.: Evidence for initiation of calcium spikes in C-cells of the rabbit nodose ganglion. *Pfluegers Arch.* **394**:106–112 (1982).

Jacobs, L., and Comroe, J. H., Jr.: Reflex apnea, bradycardia, and hypotension produced by serotonin and phenyldiguanide acting on the nodose ganglia of the cat. *Circ. Res. 29*:145–155 (1971).

Jaffe, R. A., and Sampson, S. R.: Analysis of passive and active electrophysiologic properties of neurons in mammalian nodose ganglia maintained *in vitro*. *J. Neurophysiol.* **39**:802–815 (1976).

Katz, D. M., and Karten, H. J.: Substance P in the vagal sensory ganglia: Localization in cell bodies and pericellular arborizations. *J. Comp. Neurol.* **193**:549–564 (1980).

Leek, B. F.: Abdominal visceral receptors, in: *Handbook of Sensory Physiology*, Vol. IV/1, *Enteroceptors* (E. Neil, ed.), pp. 114–160, Springer-Verlag, New York, Berlin, and Heidelberg (1972).

Lindl, T.: Effects of histamine agonists and antagonists (H_1 and H_2) on ganglionic transmission and on accumulation of cyclic neurotides (CAMP and CGMP) in rat superior cervical ganglion *in vitro*. *Neuropharmacology* **22**:203–211 (1983).

Lundberg, J., Hökfelt, T., Nilsson, G., Terenius, L., Rehfeld, J., Elde, R., and Said, S.: Peptide neurons in the vagus, splanchnic and sciatic nerves. *Acta Physiol. Scand.* **104**:499–501 (1978).

Mense, S., and Schmidt, R. F.: Activation of group IV afferent units from muscle by algesic agents. *Brain Res.* **72**:305–310 (1974).

Minota, S., and Koketsu, K.: Effects of adrenaline on the action potential of sympathetic ganglion cells in bullfrogs. *Jpn. J. Physiol.* **27**:353–366 (1977).

Morita, K., Tokimasa, T., Higashi, H., North, R. A., and Nishi, S.: The calcium activated potassium conductance in rabbit nodose ganglion cells. *J. Physiol. Soc. Jpn.* **44**:342 (1982).

Neto, F. R.: The depolarizing action for 5-HT on mammalian non-myelinated nerve fibres. *Eur. J. Pharmacol.* **49**:351–356 (1978).

Nishi, K.: The action of 5-hydroxytryptamine on chemoreceptor discharges on the cat's carotid body. *Br. J. Pharmacol.* **55**:27–40 (1975).

Nishi, S., Minota, S., and Karczmar, A. G.: Primary afferent neurons: The ionic mechanism of GABA-mediated depolarization. *Neuropharmacology* **13**:215–220 (1974).

North, R. A., and Egan, T. M.: Electrophysiology of peptides in the peripheral nervous system. *Br. Med. Bull.* **38**:291–296 (1982).

Paintal, A. S.: Vagal afferent fibres. *Ergeb. Physiol.* **52**:74–156 (1963).

Paintal, A. S.: Effects of drugs on vertebrate mechanoreceptors. *Pharmacol. Rev.* **16**:341–380 (1964).

Rudomin, P.: Presynaptic inhibition induced by vagal afferent volleys. *J. Neurophysiol.* **30**:964–981 (1967).

Sampson, S. R., and Jaffe, R. A.: Excitatory effect of 5-hydroxytryptamine, veratridine and phenyldiguanide on sensory ganglion cells of the nodose ganglion of the cat. *Life Sci.* **15**:2157–2165 (1974).

Sampson, S. R., and Vidruk, E. H.: Hyperpolarizing effects of dopamine on chemoreceptor nerve endings from cat and rabbit carotid bodies *in vitro*. *J. Physiol.* (London) **268**:211–221 (1977).

Saum, W. R., and DeGroat, W. C.: The actions of 5-hydroxytryptamine on the urinary bladder and on vesical autonomic ganglia in the cat. *J. Pharmacol.* Exp. Ther. **185**:70–83 (1973).

Shefner, S. A., North, R. A., and Zukin, R. S.: Opiate effects on rabbit vagus nerve: Electrophysiology and radioligand binding. *Brain Res.* **221**:109–116 (1981).

Sicuteri, F.: Sensitization of nociceptors by 5-hydroxytryptamine in man, in: *Pharmacology of Pain* (R. K. S. Lim, D. Armstrong, and E. Pardo, eds.), pp. 57–86, Pergamon Press, Oxford (1968).

Simonds, W. F., and DeGroat, W. C.: Antagonism by picrotoxin of 5-hydroxytryptamine induced excitation of primary afferent neurons. *Brain Res.* **192**:592–597 (1980).

Skok, V. I., and Selyanko, A. A.: Acetylcholine and serotonin receptors in mammalian sympathetic ganglion neurons, in: *Integrative Functions of the Autonomic Nervous System* (C. McC. Brooks, K. Koizumi, and A. Sato, eds.), pp. 248–258, University of Tokyo Press, Tokyo; Elsevier/North-Holland, Amsterdam and New York (1979).

Snow, W., Stein, C., and Weinreich, D.: Histamine modulates synaptic transmission in rat superior cervical ganglion and at frog neuromuscular junction via two pharmacologically distinct receptors (H_1 & H_2). *Soc. Neurosci. Abstr.* **6**:184 (1980).

Stansfeld, C. E., and Wallis, D. I.: Generation of an usual depolarizing response in rabbit primary afferent neurones in the absence of divalent cations. *J. Physiol.* (London) **352**:49–72 (1984).

Wallis, D. I., and Woodward, B.: The facilitatory actions of 5-hydroxytryptamine and bradykinin in the superior cervical ganglion of the rabbit. *Br. J. Pharmacol.* **51**:521–531 (1974).

Wallis, D. I., Stansfeld, C. E., and Nash, H. L.: Depolarizing responses recorded from nodose ganglion cells of the rabbit evoked by 5-hydroxytryptamine and other substances. *Neuropharmacology* **21**:31–40 (1982).

Westfall, T. C.: Local regulation of adrenergic neurotransmission. *Physiol. Rev.* **57**:660–728 (1977).

Yonemura, K., Morimoto, K., Mtsuoka, Y., and Tanaka, I.: Effects of bradykinin on the guinea-pig sympathetic ganglion cells *in vitro*. *J. Physiol. Soc. Jpn.* **44**:463 (1982).

V

Clinical and CNS-Related Aspects of Ganglionic Transmission

20

Autonomic Disease and Clinical Applications of Ganglionic Agents

ALEXANDER G. KARCZMAR

I. INTRODUCTION

This chapter will summarize pertinent clinical facts concerning the autonomic ganglia, including ganglionic malfunction and disease, as well as clinical uses of ganglionic agents. The purpose is to complete the subject of ganglionic autonomic transmission with respect to its clinical relevance and to point out certain general aspects of the subject, rather than to present a full clinical description of certain disease states—such as hypertension—and their treatment.

Since this book concerns autonomic and enteric ganglia rather than postganglionic nerve endings and effector sites, this chapter will focus on the ganglia. The extensive use of drugs in the treatment of peripheral autonomic disease and the pharmacology of particularly muscarinic effector sites is beyond the scope of this book; thus, the extensive and popular use of anticholinesterases and cholinomimetics in the treatment of bladder atomy and paralytic ileus and of glaucoma will not be discussed.

With respect to the ganglia themselves, it should be stressed that soon after Acheson's (Acheson and Moe, 1946) recommendation for the use of ganglionic blockers in the treatment of hypertension, ganglionic blockers

---

ALEXANDER G. KARCZMAR • Department of Pharmacology and Experimental Therapeutics, Stritch School of Medicine, Loyola University Medical Center, Maywood, Illinois 60153.

became very popular in the therapy of this condition (see this chapter, Section IV). However, this use peaked in the 1960s and 1970s, and the employment of these agents today, while important, is limited, at least in the United States (Karczmar, 1979), although it seems to be much more extensive elsewhere (see Section IV). This current limitation of the use of ganglionic drugs must be juxtaposed with their past, current, and future importance as tools employed in the analysis of basic phenomena of autonomic transmission. On the other hand, the autonomic ganglia are of primary importance in a number of clinical conditions, such as familial dysautonomia, idiopathic orthostatic hypotension (IOH), and, indeed, hypertension. In many cases, the role of the ganglia in these conditions is still poorly understood, and much interesting work lies ahead. Furthermore, several new concepts and speculations have emerged recently as to the role and/or diagnostic significance of autonomic ganglia in conditions in which such a role was heretofore not expected, e.g., mental disease. This implication of autonomic ganglia in a number of disease states and the need for the development of useful ganglionic drugs constitute the current significance of this topic.

II. AUTONOMIC DISEASE

A. Introduction

The failure of autonomic ganglia has been suspected or suggested for many years now as underlying a number of diseases; this is true, for example, for neurogenic or idiopathic orthostatic hypotension (Bradbury and Eggleston, 1925), several ocular disorders such as Horner's syndrome and Argyll–Robertson pupil, and neuropathies resulting from diabetes and syphilis or drug use and abuse (as in alcoholism). On the other hand, this concept has been expanded markedly in the last 15 years to include, for example, autonomic failure of the Shy–Drager type; in a related context, autonomic function may be abnormal in mental disease (Zahn, 1977; Karczmar and Richardson, 1985).

B. Familial Dysautonomia

This is an autosomal recessive disease (Riley–Day syndrome) (see Riley *et al.*, 1949), also referred to as a type of hereditary motor–sensory neuropathy (see Schaumburg *et al.*, 1983) that occurs predominantly in Ashkenazic (Jewish) infants or children. While its manifestations include sensory, reflex, and motor disturbances, the majority of symptoms focus on the autonomic system and on autonomic ganglia since the symptoms include abnormal responses to norepinephrine (Smith and Dancis, 1964) and to cholinomimetics (Smith *et al.*, 1965), labile blood pressure, drooling, postural hypotension, change in skin color, faulty bowel activity, diminished lacrimation, poor temperature control, and sweating (Van Woert, 1976; Johnson and Spaulding, 1974; McLeod, 1982). Biochemical studies are consistent with the concept of autonomic including adrenal medulla malfunction, since metabolism of catecholamines is impaired, perhaps due to deficiency of cholinergic innervation of the adrenal medulla or sympathetic malfunction or both (Van Woert, 1976) and since acetylcholine (ACh) synthesis may be faulty (Brunt and McKusik, 1970). There is a reduction of the volume of several ganglia and the number of preganglionic neurons (Dyck *et al.*, 1978), and the response to methacholine is exaggerated. The biochemical analysis in question has been directed at secretions and certain superficial tissues, since the ganglia could not be readily analyzed (see also Van Woert, 1976). However, certain pathological findings are consistent with autonomic malfunction, since they indicate loss of autonomic neurons (Pearson *et al.*, 1971).

It is of interest that this serious or fatal disease—many patients die in the first year of life (usually as a result of repeated aspiration pneumonia) (McLeod, 1982; Johnson and Spaulding, 1974)—may be treated, with some degree of success, with cholinomimetic drugs, such as urecholine, that tend to normalize the patient's parasympathetic functions (cf. Van Woert, 1976); in this context, it should be stressed that ganglionic stimulants and drugs or compounds that might be expected to improve release or synthesis of ACh have not so far been employed in this condition.

C. Congenital Megacolon (Hirschsprung's Disease)

Congenital megacolon involves the enteric nervous system. This disease constitutes a birth defect, presumably hereditary and ontogenic in nature, since it results from a defect in neural-crest development (Bodian, 1952; Bolande, 1975). Infants and children suffering from this condition show extreme constipation, bleeding, vomiting and malnutrition, and, secondarily, megacolon (Gershon *et al.*, 1979). The disease is due to the absence of both myenteric and submucosal ganglion cells (Gannon *et al.*, 1969) at sites of initial colonic constriction, although intrinsic ganglion cells are present in the dilated portion of the colon, proximal to the constriction, and although adrenergic and cholinergic (preganglionic?) fibers are present (Kamijo *et al.*, 1953; see Gershon *et al.*, 1979). It is of interest that a hereditary congenital megacolon condition exists in several strains of laboratory mice (Derrick and St. George-Grambauer, 1957; see Gershon *et al.*, 1979). There is currently no pharmacological therapy for the disease, which is treated surgically.

D. Ocular Disorders

Several ocular conditions are listed frequently as dysautonomic in nature (see Johnson and Spaulding, 1974; McLeod, 1982; Schaumburg *et al.*, 1983), although direct evidence that autonomic ganglionic lesion is involved is not always available. Among such conditions may be included Horner's syndrome, with resulting miosis and ptosis; Argyll–Robertson pupil, leading to miosis, ptosis, and poor light reflexes; and Holmes–Adie (tonic pupil) syndrome, which is associated with loss of tendon reflexes, dilated pupil and poor light reflexes, and generally, but not always, orthostatic hypotension.

In some of these conditions, there is documentation for the actual presence of autonomic lesion, as in the case of the marked abnormality of the ciliary ganglion in a subject suffering from Holmes–Adie syndrome (Harriman and Garland, 1968) and since the pupil of the patients in question is hypersensitive to instillation of cocaine or methacholine or both. The tonic pupil may result from undue response of the denervated iris sphincter muscle to ACh released during the process of accommodation (Pallis, 1978). In the case of Argyll–Robertson pupil, there appears to be an interruption of sympathetic innervation; the lesion may be presynaptic or supraspinal or occur at any site between the effector, the postganglionic or preganglionic nerves, and the spinal cord (Johnson and Spaulding, 1974). However, the ganglion may be the structure that is least likely to be involved, (Miller, 1982).*

E. Metabolic, Idiopathic, and Postinfectious, Long-Course Neuropathies

Several conditions induced by metabolic disease, viral conditions or inflammation, and neuronal atrophy of idiopathic nature, involve autonomic ganglia. Thus, in amyloidosis, there may be amyloid infiltration of the sympathetic nerves or autonomic ganglia or both; orthostatic hypotension and additional symptoms of autonomic disturbance may be not present initially, but result eventually (Johnson and Spaulding, 1974;

*Argyll–Robertson syndrome was pathognomonic in the past for tertiary syphilis (tabes dorsalis). For instance, Guy de Maupassant who died in 1893 from syphilitic general paresis suffered for many years from Argyll–Robertson syndrome. He also complained of ocular pain ("yeux grinçants") (Lanoux, 1967) which does not characterize Argyll–Robertson syndrome.

Schaumburg *et al.*, 1983; McLeod, 1982). Similarly, as is well known, chronic diabetic and porphyric polyneuropathy may result in hypotension and other autonomic disturbances, including Argyll–Robertson syndrome, and pathologies of autonomic ganglia in these conditions have been reported (Johnson and Spaulding, 1974; Schaumberg *et al.*, 1983; McLeod, 1982).

Inflammatory neuropathies are classically represented by Landry–Guillain–Barré syndrome (Landry's paralysis or acute inflammatory polyradiculoneuropathy), which may be postinfectious in nature. This rare disease resembles porphyric neuropathy, since hypotension and sensory and reflex involvements are present in either case. The differentiation may depend on the CSF analysis, since the protein level is increased in Guillain–Barré syndrome, but not in porphyric neuropathy, and since slowing of nerve conduction is again predominant in Guillain–Barré syndrome. In addition, porphyric neuropathy presents elevated urinary levels of delta amino levulinic acid and porphobilinogen (Low and Stevens, 1984). Landry–Guillain–Barré syndrome may occasionally assume the form of acute pandysautonomia (Low and Stevens, 1984). In this case, autonomic imbalance is particularly prominent, the patients exhibiting severe hypotension, total anhidrosis, extreme dryness of the eyes and of nasal and oral mucousal membranes, nonreactive pupils, reduced bowel sounds, and atonic bladder (Low and Stevens, 1984).

Guillain–Barré syndrome—particularly its variant, acute pandysautonomia of the Low–Stevens type—resembles in many of its symptoms idiopathic orthostatic hypotension or Shy–Drager syndrome, which is a chronic failure of the autonomic nervous system (see Johnson and Spaulding, 1974; Bannister, 1978). Shy–Drager syndrome differs from the Guillain–Barré condition because of the chronic and progessive character of the former; Shy–Drager syndrome is a disease of middle age, and it leads to death within 7–8 years after the symptomatic onset of the disease (McLeod, 1982). As already stated, this condition resembles the types of dysautonomia that were already mentioned; the prominent symptoms include loss of sweating and sphincter function, impotence, postural hypotension with dizziness, and weakness on standing or walking. In a large proportion of cases, the nonautonomic neurological manifestations are either present or develop subsequently. These conditions include parkinsonism, characterized, however, by absence of ataxia, and mutiple system atrophy with clinical features of extrapyramidal, pyramidal, and cerebellar degeneration (see Bannister, 1978). The pathology of this condition includes lesions of the intermedial—lateral columns with a loss of cells (Johnson and Spaulding, 1974); the abnormalities of the autonomic ganglia are relatively slight and not always present, although the presence of Lewy bodies has been described (Rajput and Rozdilsky, 1976). There is also neurotransmitter malfunction, since plasma noradrenaline and, perhaps, dopamine-β-hydroxylase levels are decreased (Bannister *et al.*, 1977;

McLeod, 1982). Furthermore, certain maneuvers—such as tilt—that evoke significant rise of norepinephrine plasma levels in normals fail to do so in patients suffering from Shy–Drager dysautonomia (as well as from other types of dysautonomia) (Sever, 1978; Bannister, 1978). There may be also change in plasma renin and aldosterone levels. The relationship of this condition to that presumably caused by postganglionic degeneration (Bonnin *et al.*, 1961) is not clear.

F. Acute Autonomic Neuropathy

Contrary to the Shy–Drager condition, this condition resembles the inflammatory autonomic neuropathies, since the onset of the disease is rapid and recovery usually occurs in weeks or months. The symptoms are relatively unexceptional, resembling those already described for the various dysautonomias. The disease presents impotence, hypotension, blurred vision, decreased tear formation, loss of sweating, constipation, and urinary retention or incontinence; the pupils are fixed, and light reflex is poor. Additional symptoms may occur, such as lethargy and tiredness. Sometimes the condition becomes cholinergic in nature rather than presenting a combined failure of sympathetic and parasympathetic function. In this case, the symptomatology is similar to that of acute pandysautonomia, but there is absence of postural hypotension. It is of interest that in either case, the condition is clearly autonomic in nature, since there are no abnormalities of mental function, motor performance, sensation, reflexes, or ocular movements.

The etiology of the disease is relatively unclear, since the few biopsies that have been performed do not appear to be eventful (McLeod, 1982). However, the tests reveal neurotransmitter dysfunction, since there was generally denervation supersensitivity to norepinephrine as well as abnormalities of pupilary responses including those to methacholine, findings that are consistent with the concept that preganglionic or ganglionic damage has occurred. The lack of pathology may be consistent with the reversible nature of the disease.

G. Old-Age Dysautonomia

It is of interest to refer in this context to the circulatory disorders, impotence, orthostatic hypotension and poor response to Valsalva's maneuver, and bladder atony that are prone to occur in the elderly; this condition is frequently accompanied in the aged population by hypothermia, which may also occur on its own (see Johnson and Spaulding, 1974; Low and Stevens, 1984). Blood pressure disorders (cf. Roth and Morrissey, 1952) including sympathetic insufficiency have frequently been described. Anecdotally, these problems in the regulation of blood pressure and temperature are related to abnormal sweating and salivation (or their absence) and to episodes of falling, and the condition characterized by such multiple presentations could be referred to as old-age dysautonomia (W. Wood, personal communication). Actually, the condition in question seems to be related to cerebrovascular aging, lesions whether spinal or supraspinal, and degenerative disease such as a parkinsonian condition (see also above); degenerative conditions or lesions of the autonomic ganglia have occasionally been inculpated in these syndromes. Indeed, senility-related degenerative changes in ganglion cells have been described (Botar, 1966); these include accumulation of several types of granules, including sudanophilic granules, and decreased neuronal counts. Furthermore, Botar (1966) described decreased innervations of the adrenal medulla in aged individuals. Botar (1966) stated that a number of pathogenic conditions, including nutritional conditions, seemed to contribute to the changes in question, but he did not relate these changes to senile hypotension (or hypertension) and considered them to be nonspecific and age-related. It may be added that neuronal loss and transmitter deficits occur in the ciliary ganglion of the aged chick (Marchi *et al.*, 1980); of course, this information should not be unduly extrapolated to man.

It must be added that elderly hypertensives, as is well known, are unduly responsive to hypotensive drugs, and ganglionic blockers are counterindicated in this population (see below); this may indicate ganglionic abnormality in aged hypertensives, but may have no bearing on elderly hypotensives.

H. Drug-Induced Autonomic Neuropathies

As already indicated, many drugs and their use and abuse may lead to chronic or acute autonomic malfunction and neuropathies; the signs and symptoms include hypotension or hypertension, ocular and gastrointestinal disturbances, and impotence. This is true generally for autonomic or centrally acting drugs used in hypertension such as monoamine oxidase inhibitors, methyldopa, ganglionic blockers, and others (see also below), as well as such agents as phenothiazines and certain antibiotics such as chloramphenicol. Furthermore, certain alkaloids used in the treatment of leukemias and lymphomas may produce dysautonomia manifested by hypotension and constipation; this is true, for instance, for vincristine and vinblastine. It should be emphasized that while neurological disorders that occur with many drugs and toxins include primarily motor dysfunction, paresthesias, and sensory disturbances, some degree of dysautonomia occurs with these drugs as well. In this context, it is of particular interest that there is a full documentation of autonomic damage, including ganglionic lesions, in alcoholism (chronic alcoholic neuropathy) (Appenzeller and Richardson, 1966). In this case, avitaminosis and poor nutrition may also constitute etiological factors, and it is frequently found that in alcoholism, besides ganglionic cell damage, axonal and myelin changes appear to be involved (Schaumburg *et al.*, 1983) A somewhat similar picture obtains with regard to chronic smoking and resulting nicotinic neuropathy.

I. Hypertension

In hypertensive disease, vasoconstriction is present, ganglionic blockade produces—at least temporarily—vasodilation and fall of blood pressure, and ganglionic therapy is utilized; yet today it is not considered generally that autonomic sympathetic ganglia are primarily involved, even though some information indicates that norepinephrine plasma levels are increased (vide below).

It must be stressed, of course, that hypertension, as is well known, is a heterogeneous condition; in fact, the remarks in this section are addressed mainly to essential hypertension, and it further complicates the matter that this condition is, by definition, of unknown or uncertain etiology. In the case of essential hypertension, as in that of several diseases or conditions outlined above, there is no indication of a change in the ganglionic neurons, although some ganglionic inclusions or neuronal loss or both have been noticed in conditions unrelated to hypertension (see above) (see also Schaumburg *et al.*, 1983; Johnson and Spaulding, 1974). Indeed, while since the early days of the study of hypertension (see Barker and Cole, 1924; Page, 1951) many mechanisms have been postulated or even demonstrated to play a part in hypertension, and although the neurotransmitters or modulators that have been discussed in this book as being involved in ganglionic transmission, such as prostaglandins, angiotensins and other peptides, kinins, and Na^+ (see Chapters 8 and 13), as well as, in fact, norepinephrine, have been and are considered to be factors that participate in the complex etiology of hypertension, still the ganglia or ganglionic imbalance (Gellhorn, 1959) probably do not constitute the primary mechanism underlying the etiology of hypertension. For example, much evidence has been brought forward indicating that plasma norepinephrine levels are elevated in subjects suffering from essential hypertension (see Kopin *et al.*, 1981), and, as is well known, β-adrenergic blockers are effective in essential and malignant hypertension (see, Menard *et al.*, 1981). However, the data are generally inconsistent, and the elevation of plasma norepinephrine may be more characteristic for younger, particularly black, hypertensives (Sever, 1978) than for other patients. At any rate, the elevation of blood catecholamines may not depend primarily on sympathetic ganglia or even on the adrenal medulla, since it may be generated centrally. This does not signify, of course, that ganglia do not serve as relay stations in hypertension, that they may not generate vasoconstriction in response to centrally originating stress, and that they may not be involved in renin-mediated hypertension; in fact, autonomic sympathetic ganglia are generally included in diagrams illustrating the genesis of hypertension (Kaplan, 1979), and indeed they are components of what can be rather loosely referred to as "increased sympathetic system activity" (Kaplan and Lieberman, 1982). Also, the foregoing does not signify that speculations that include ganglionic genesis of essential hypertension are not being raised; for example, it has been speculated in these laboratories that hypofunction or abnormal absence of the slow IPSP (and/or SIF cells or other sources of interganglionic generation of catecholamines) (see Chapter 13) may lead to hypertension; preliminary studies do not seem to support this speculation (N. J. Dun, unpublished observations).

III. MENTAL DISEASE AND THE AUTONOMIC NERVOUS SYSTEM

The involvement of autonomic ganglia in mental disease generally and in schizophrenia particularly is an interesting concept that is only infrequently appreciated by students of ganglionic transmission. Speculatively, several investigators have proposed that schizophrenics exhibit autonomic hyperresponsiveness as measured in terms of, for example, galvanic skin response (GSR) and cardiac rate change (for references, see Karczmar and Richardson, 1985). The data are not always consistent, and there is also some evidence of the occurrence in schizophrenics of autonomic hypo- rather than hyperarousal (Kornetsky and Hecht-Orzack, 1978; see also Karczmar and Richardson, 1985). Additional evidence as to the inconsistency of data concerning the GSR and related responses as well as the cardiovascular response of schizophrenics is adduced by Zahn (1977). We may actually be dealing with an improper autonomic response of schizophrenics to stress and environmental stimuli, rather than increased or decreased response.

One of the problems with this concept is that what is generally referred to as autonomic imbalance or autonomic hyperarousal in schizophrenics cannot be readily related to the autonomic ganglia. In fact, the GSRs cannot be clearly related to the autonomic function, although they are referred to in the psychiatric literature as reflecting autonomic phenomena (see Zahn, 1977). Furthermore, the autonomic hyperarousal that allegedly may exist in schizophrenics could depend on improper autonomic responses due primarily not to the autonomic ganglia but to the cortico- and-subcorticospinal connections of the ganglia that may involve the limbic system (see Chapters 17 and 18) or the hypothalamico–pituitary axis. This may particularly be so as there are a number of concepts and speculations that a general hyperarousal or central imbalance may be present in the schizophrenic conditions (see Karczmar and Richardson, 1985). In this context, it is important to emphasize that this general hyperarousal in affective and schizophrenic disease and the possible improper response to stress of schizophrenic patients may be mediated by the cholinergic system and by undue release or synthesis of ACh (Karczmar, 1976; Karczmar and Richardson, 1985; Janowsky and Risch, 1984). What must be particularly emphasized, however, is that the speculations as to the autonomic imbalance or hyperarousal in schizophrenia are not accompanied by any evidence showing directly that autonomic ganglia are malfunctioning or exhibit abnormalities or lesions.

IV. CLINICAL USE OF GANGLIONIC DRUGS

A. General and Past Uses of Ganglionic Drugs

As already indicated in the Introduction, cholinergic agonists and antagonists are used today in a number of conditions that involve, particularly, parasympathetic muscarinic and neuromyal nicotinic sites (see Gilman *et al.*, 1980; Bowman and Rand, 1980). On the other hand, ganglionic drugs, particularly nicotinic ganglionic blocking agents, are somewhat passé today. Yet they were employed in a variety of conditions even a few years ago, and their use in essential hypertension was much more frequent and the number of drugs available much more extensive than it is today (see below). It must be added that even today, ganglionic drugs are used in some countries (Bunatian and Meshcherjakov, 1980; Erina, 1980) much more extensively than in the United States.

With regard to the use of ganglionic blocking drugs in conditions other than chronic or essential hypertension, U.S.S.R. writers list a number of conditions that are generally not considered in the United States as primary indicators for the use of ganglionic blockers. According to Erina (1980) and Bunatian and Meshcherjakov (1980), these conditions include peptic ulcer and chronic gastritis, angioneurotic "eczema," and vascular conditions such as pulmonary edema and endarteritis obliterans. Furthermore, ganglionic blockers seem to be used in the U.S.S.R. relatively frequently in conjunction with anesthesia and surgery. Thus, they seem to be used in the course of surgery to induce "artificial hypotension" ("controlled hypotension") for prevention of bleeding and decreasing the load on pulmonary circulation and cardiac return; they are also employed to produce vasodilation and to facilitate surgical intervention and mechanical manipulation of blood vessels in surgical treatment of coarctation of the aorta, patent arterial duct, and neurosurgery of the brain. In addition, they seem to be employed relatively regularly during bypass procedures, including cardiopulmonary bypass (Bunatian and Meshcherjakov, 1980). Finally, postoperative and intraoperative hypertensive crises or malignant hypertension or both seem to be treated in the U.S.S.R. with ganglionic blockers, rather than with directly acting smooth muscle depressants.

In other countries, these uses of ganglionic blockers are generally considered ineffective as compared to the use of other drugs (see below) and inconvenient or even dangerous in view of the side effects and toxicities readily produced by ganglionic blockers (Cottrell, 1983). In fact, this viewpoint was expressed as early as in the 1970s (Goldberg, 1975). It is of interest that even in the case of surgery of patients suffering from hypertension, ganglionic drugs are not added to their regimen, although the patients are generally maintained during anesthesia and surgery on their antihypertensive therapy, which, similarly, does not include ganglionic blockers (Kaplan and Lieberman, 1982). However, there is still in the United States sporadic use of such drugs as trimethaphan in emergency treatment of pulmonary edema in patients suffering from pulmonary hypertension associated with systemic hypertension, for controlled hypotension, and particularly for acute aortic dissection (McFarland *et al.*, 1972); in the latter case, they are employed to prevent reflex rise in blood pressure that may occur during aortic manipulation, although even then a nitroprusside–β-blocker combination may be preferred (see below for their use in severe or malignant hypertension).

B. Use of Ganglionic Blockers in Essential or Chronic Hypertension

1. General Statement

Historically, and still today, the major therapeutic use of ganglionic blocking agents has been in the management of essential or chronic hypertension and malignant hypertension. Some 25 years ago, five or six ganglionic agents were habitually listed in the Western pharmacopeias or texts (see, for example, Beckman, 1958; Sollmann, 1957). Today, only one or two such drugs are listed in United States pharmacopeias and related sources (see *USP Drug Information*, 1983; *AMA Drug Evaluations*, 1983; Michaels and Brown, 1985).

The original clinical use of ganglionic blockers was initiated by Acheson's (Acheson and Moe, 1946) experimental studies of tetraethylammonium (TEA) (Etamon), which led Acheson to propose its use in hypertension. The studies of Acheson and his associates actually expanded on the earlier and by then forgotton work of Marshall (1913) and Burn and Dale (1915) indicating potent "paralysis" by TEA of ganglionic stimulation by nicotine.

The original findings of clinical effectiveness of Etamon led to development of additional, clinically more potent or, it was hoped, more selective (see below) ganglionic blockers. Drugs of two types can induce block, namely, depolarizers and nondepolarizers (Paton and Perry, 1953; Paton and Zaimis, 1949; Zaimis, 1963) (see also Chapter 13; for additional subdivision of ganglionic blockers, see Chapter 6). Indeed, depolarizing ganglioactive drugs such as nicotine, ACh (particularly in the presence of an anticholinesterase), tetramethylammonium, dimethyl-phenyl-piperazinium, and anticholinesterases (Koppanyi and Karczmar, 1951) produce ganglionic block following initial depolarizing excitatory effect; sometimes, their block may occur following, or independently of, their depolarizing actions (Brown, 1980) (see also Chapter 13). For obvious reasons, these drugs could not be exploited clinically, only nondepolarizers drugs being useful in this regard.

As discussed in detail by Volle (1966) and Brown (1980), the nondepolarizing drugs such as mecamylamine, pentolinium, TEA, C_6, chlorisondamine, and the newer drugs such as trimethaphan exert their main therapeutic effect by their postsynaptic action; the presynaptic effects are minimal (as with mecamylamine) or absent (Brown, 1980). Generally, they do not affect neuronal excitability, their effect being competitive in nature with respect to depolarizing drugs or ACh; however, they may also bind nicotinic receptors noncompetitively (Blackman, 1970) and/or act on Ca^{2+} or Na^+ current. Some, if not all, of these actions resemble those of curarimimetic neuromyal blockers, and indeed the drugs in question exert some neuromyal blocking actions, although they show differential affinity for the ganglionic receptor (Brown, 1980; Gyermek, 1980).

It is of particular practical importance that, generally, these drugs affect both parasympathetic and sympathetic ganglia. While structure–activity studies (Gyermek, 1980; Trcka, 1980) indicate that differentiation between the two sites is possible, this differentiation has not led, so far, to clinical exploitation, and the drugs currently used in the United States (mecamylamine and trimethaphan), as well as pentolinium and chlorisondamine, still employed in Europe, block all autonomic ganglia as therapeutic doses.

The two blockers currently used in the United States differ chemically. Trimethaphan, contrary to some of the older and classic drugs, is not a quaternary onium compound but a congener of triethylsulfonium salts; however, similarly to the onium compounds, it essentially does not pass the blood–brain barrier. On the other hand, mecamylamine is a secondary amine.

One of the problems with the lack of specificity for sympathetic ganglia of the agents in question is that of their side effects, which will be reviewed below. Let it be stressed at this time that even in therapeutic doses, they impede or depress the function of all visceral organs, of the visual system, and of the endocrine system; thus, the patient treated with these drugs may not be able to see, coit, defecate, or stand up without fainting. In addition, the secondary nitrogenous compound mecamylamine may produce prominent central effects. An additional problem with these drugs is that they produce tolerance to their therapeutic effects, although not necessarily to their side effects.

2. Use of Mecamylamine and Trimethaphan

Both these drugs may be used in the treatment of moderately severe to severe, essentially uncomplicated hypertension and uncomplicated malignant hypertension; they are never considered as primary agents, but are "step 4" drugs (see *USP Drug Information*, 1983; *AMA Drug Evaluations*, 1983). Their excretion is renal and their metabolism prior to excretion is minimal [trimethaphan may be partially metabolized by plasma (butyryl) cholinesterase]. Mecamylamine is given orally, since it is well absorbed; trimethaphan is given intravenously. Even mecamylamine should not be given on an outpatient basis.

If and when these drugs are given in the treatment of malignant hypertension and other emergency conditions, their side effects are not of primary importance. When they are used (particularly mecamylamine) in severe and/or intractable essential hypertension, these side effects must be taken into consideration. That the side effects are frequent and prominent has already been indicated. The list of actual effects of mecamylamine (*as compiled from USP Drug Information*, 1983) follows: dizziness or lightheadedness, especially when getting up from a lying or sitting position (postural hypotension); confusion or unusual excitement or mental depression or trembling; drowsiness and tiredness or weakness; shortness of breath; difficult urination and constipation, which may be preceded by small, frequent liquid stools and rarely lead to paralytic ileus; blurred vision; decreased sexual ability or impotence; dry mouth; enlarged pupils; loss of appetite; nausea and vomiting.

Trimethaphan exerts similar side effects except for dizziness and lightheadedness. Furthermore, as already mentioned, these drugs may be contraindicated (or should be used carefully) in the elderly and should be used with caution in children; the drugs are also capable of adverse interaction with several other drugs and conditions.

V. CONCLUSIONS

Ganglionic agents, although important, are of limited use clinically, at least in the United States. Accordingly, this chapter stresses certain unfinished business and presents several items for potential exploitation. Thus, the limited current use of ganglionic blockers in hypertension may constitute only a nadir of the spectacular cycle of early interest in, and present disappointment with, these drugs. The cycle may continue; further structure-activity relationship studies of pertinent compounds may lead to drugs with differential effects on sympathetic vs. parasympathetic ganglia and/or ganglionic actions that may interfere with the renin–angiotensin system, thus bringing about a dramatic revival in the clinical use of these drugs.

Another aspect of the clinical relevance of the autonomic ganglia is that several disease conditions involve dysautonomias. Since final proof of ganglionic etiology of some of these conditions is missing or incomplete, further work is indicated, and confirmed etiology may then lead to novel therapy of these conditions as well as to a better understanding of the role of the ganglia in health and disease. This is particularly important, since many of these diseases—e.g., familial dysautonomia and Shy–Drager syndrome—are progressive or fatal in nature.

REFERENCES

Acheson, G. H., and Moe, G. K.: The action of tetraethylammonium ion on the mammalian circulation. *J. Pharmacol. Exp. Ther.* **87**:220–236 (1946).

AMA Drug Evaluations, 5th ed. AMA, Chicago (1983).

Appenzeller, O., and Richardson, E. P.: The sympathetic chain in patients with diabetic and alcoholic polyneuropathy. *Neurology* **16**:1205–1209 (1966).

Bannister, R.: The diagnosis and management of chronic autonomic failure, in: *Neurotransmitter Systems and Their Clinical Disorders (N. J. Legg, ed.),pp. 65–83, Academic Press, London (1978).*

Bannister, R., Sever, P., and Grots, M.: Cardiovascular reflexes and biochemical responses in progressive autonomic failure. *Brain* **100**:327–344 (1977).

Barker, L. F., and Cole, N. B.: *Blood Pressure: Cause, Effect and Remedy*. Appleton and Co., New York (1924).

Beckman, H.: *Drugs—Their Nature, Action and Use*. W. B. Saunders, Philadelphia (1958).

Blackman, J. G.: Dependence on membrane potential of the blocking action of hexamethonium at a sympathetic ganglionic synapse. *Proc. Univ. Otago Med. Sch.* **48**:4–5 (1970).

Bodian, M.: Chronic constipation in children, with particular reference to Hirschsprung's disease. *Practitioner* **169**:517 (1952).

Bolande, R. P.: Animal model of human disease: Hirschsprung's disease, aganglionic or hypoganglionic megacolon. *Am. J. Pathol.* **79**:189–172 (1975).

Bonnin, M., Skinner, S. L., and Whelan, R. F.: Holmes–Adie syndrome with progressive autonomic degeneration. *Aust. Ann. Med.* **10**:303–307 (1961).

Botar, J.: *The Autonomic Nervous System.* Akademiai Kiado, Budapest (1966).

Bowman, W. C., and Rand, M. J.: *Textbook of Pharmacology, 2nd ed. Blackwell, Oxford (1980).*

Bradbury, S., and Eggleston, C.: Postural hypotension: A report of three cases. *Am. Heart J.* **1**:73–86 (1925).

Brown, D. A.: Focus and mechanism of action of ganglion-blocking agents, in: *Pharmacology of Ganglionic Transmission* (D. A. Kharkevich, ed.), pp. 185–235, Springer-Verlag, New York (1980).

Brunt, P. W., and McKusik, V. A.: Familial dysautonomia: A report of genetic and clinical studies with review of the literature. *Medicine (Baltimore)* **49**:343–374 (1970).

Bunatian, A. A., and Meshcherjakov. Ganglion-blocking agents in anesthesiology, in: *Pharmacology of Ganglionic Transmission* (D. A. Kharkevich, ed.), pp. 439–463, Springer-Verlag, New York (1980).

Burn, J. H., and Dale, H. H.: The action of certain quarternary ammonium bases. *J. Pharmacol. Exp. Ther.* **6**:417–438 (1915).

Cottrell, J. E.: Deliberate hypotension, in: "Thirty-fourth Annual Refresher Course Lectures." *Am. Soc. Anesthesiol. Lect.* **112**:1–7 (1983).

Derrick, E. H., and St. George-Grambauer, B. M.: Megacolon in mice. *J. Pathol.* **73**:569–571 (1957).

Dyck, P. J., Kawamura, Y., Low, P. A. and Shimoro, M.: The number and sizes of reconstructed peripheral autonomic, sensory and motor neurons in a case of dysautonomia. *J. Neuropath. Exp. Neuron.* **37**:741–755 (1978).

Erina, E. V.: Ganglion-blocking agents in internal medicine, in: *Pharmacology of Ganglionic Transmission* (D. A. Kharkevich, ed.), pp. 417–438, Springer-Verlag, New York (1980).

Gannon, B. J., Noblett, H. R., and Burnstock, G.: Adrenergic innervation of bowel in Hirschsprung's disease. *Br. Med. J.* **3**:338–340 (1969).

Gellhorn, E.: Hypothalamus, bladder and gut: A contribution to the understanding of autonomic integration. *Acta Neuroveg.* **19**:235–249 (1959).

Gershon, M. D., Dreyfus, D. F., and Rothman, T. P.: The mammalian enteric nervous system: A third autonomic division, in: *Trends in Autonomic Pharmacology (S. Kalsner, ed.), pp. 59–101, Urban and Schwarzenberg, Munich (1979).*

Gilman, A. G., Goodman, L. S., and Gilman, S. (eds.): *Goodman & Gilman's The Pharmacological Basis of Therapeutics*, 6th ed. Macmillan, New York (1980).

Goldberg, L. I.: Current therapy of hypertension: A pharmacologic approach. *Am. J. Med.* **58**:489–494 (1975).

Gyermek, L.: Methods for the examination of ganglion-blocking activity, in: *Pharmacology of Ganglionic Transmission* (D. A. Kharkevich, ed.), pp. 63–121, Springer-Verlag, New York (1980).

Harriman, D. G., and Garland, H.: The pathology of Adie's syndrome. *Brain* **91**:401–418 (1968).

Janowsky, D. S., and Risch, S. C.: Cholinomimetic and anticholinergic drugs used to investigate an acetylcholine hypothesis of affective disorders and stress. *Drug Dev. Res.* **4**:125–142 (1984).

Johnson, R. H., and Spaulding, J. N. K.: *Disorders of the Autonomic Neurons System.* Contemporary Neurology Series, F. H. Davis, Philadelphia, (1974).

Kamijo, K. I., Hiatt, R. B., and Koelle, G. B.: Congenital megacolon: A comparison of the spastic and hypertrophied segments with respect to cholinesterase activities and sensitivities to acetylcholine, DFP, and barium ion. *Gastroenterology* **24**:173–185 (1953).

Kaplan, N. M.: The Goldblatt Memorial Lecture. Part II. The role of the kidney in hypertension. *Hypertension* **1**:456–461 (1979).

Kaplan, N. M. and Lieberman, E.: *Clinical Hypertension*. Williams and Wilkins, Baltimore (1982).

Karczmar, A. G.: Central actions of acetylcholine, cholinomimetics and related drugs, in: *Biology of Cholinergic Function* (A. M. Goldberg and I. Hanin, eds.), pp. 395–449, Raven Press, New York (1976).

Karczmar, A. G.: New roles for cholinergics in CNS disease. *Drug Ther.* **4**:31–42 (1979).

Karczmar, A. G., and Richardson, D. L.: Cholinergic mechanisms, schizophrenia, and neuropsychiatric adaptive dysfunctions, in: *Central Cholinergic Mechanisms and Adaptive Dysfunctions* (M. M. Singh, D. M. Warburton, and H. Lal, eds.), pp. 193–221, Plenum Press, New York (1985).

Kopin, I. J., Goldstein, D. S., and Feuerstein, G. Z.: Position paper: the sympathetic nervous system and hypertension. in: *Frontiers of Hypertension Research* (J. H. Laragh, F. R. Buhler, and D. W. Seldin, eds.), pp. 283–289, Springer-Verlag, New York (1981).

Koppanyi, T., and Karczmar, A. G.: Contribution to the study of the mechanism of action of cholinesterase inhibitors. *J. Pharmacol. Exp. Ther.* **101**:327–343 (1951).

Kornetsky, C., and Hecht-Orzack, M.: Physiological and behavioral correlates of attention dysfunction in schizophrenic patients. *J. Psychiatr. Res.* **4**:69–79 (1978).

Lanoux, A.: *Maupassant le Bel Ami*. Librairie Artheme Fayard, Paris (1967).

Low, P. A., and Stevens, J. C.: Peripheral and autonomic neuropathy, in: *Neurology* (J.N. Whishant, ed.), *Clinical Medicine*, Vol. 11, pp. 1-31, Harper and Row, Philadelphia (1984).

Marchi, M., Hoffman, D. W., and Giacobini, E.: Aging of the peripheral nervous system: Age dependent changes in cholinergic synapses, in: *Neural Regulatory Mechanisms during Aging*, (R. C. Adelman, J. Roberts, C. T. Baker III, S. I. Baskin, and V. J. Cristofalo, eds.), pp. 215–220 Alan Liss, New York (1980).

Marshall, C. R.: Studies on the pharmaceutical action of tetra-alkyl-ammonium compounds. *Trans. R. Soc. Edinburgh*, **1**:17–40 (1913).

McFarland, J., Willerson, J. T., Dinsmore, R. E., Austen, G. W., Buckley, M. J., Sanders, C. A., and DeSanctis, R. W.: The medical treatment of dissecting aortic aneurysms. N. Engl. *J. Med.* **26**:115–119 (1972).

McLeod, J. G.: Peripheral autonomic failure, in: *Disorders of Neurohumoral Transmission* (T. J. Crow, ed.), pp. 45–82, Academic Press, London (1982).

Menard, J., Plouin, P. F., Thibonnier, M., Alhenc-Gels, F., Deqoulet, P., and Corvol, P.: Renin activity and the response to beta-blockade, in: *Frontiers in Hypertension Research* (J. H. Laragh, F. R. Buhler, and D. W. Seldin, eds.), pp. 450–453, Springer-Verlag, New York (1981).

Michaels, R. M., and Brown, G. R.: *Drug Consultant, 1985–1986*. John Wiley, Chichester and New York (1985).

Miller, R. N.: *Walsh and Hoyt's Clinical Neuro-Ophtholmology*. 4th ed. William and Wilkins, Baltimore (1982).

Page, I. H.: Treatment of essential and malignant hypertension. *J. Am. Med. Assoc.* **147**:1311–1318 (1951).

Pallis, C.: The pupil as an indicator of denervation hypersensitivity in the autonomic nervous system, in: *Neurotransmitter Systems and Their Clinical Disorders (N. J. Legg, ed.), pp. 85–97, Academic Press, London (1978).*

Paton, W. D. M., and Perry, W. L. M.: The relationship between depolarisation and block in the cat's superior cervical ganglion. *J. Physiol. (London)* **119**:43–57 (1953).

Paton, W. D. M., and Zaimis, E. J.: The pharmacological actions of polymethylene bistrimethyl-ammonium salts. *Br. J. Pharmacol* **4**:381–400 (1949).

Pearson, J., Budzilovich, G., and Finegold, M. J.: Sensory, motor and autonomic dysfunction: The nervous system in familial dysautonomia. *Neurology (Minneapolis)* **21**:486–493 (1971).

Rajput, A. H., and Rozdilsky, B.: Dysautonomia in parkinsonism: A clinicopathological study. *J. Neurol. Neurosci. Psychiatry* **39**:1092–1100 (1976).

Riley, C. M., Day, R. I., Greeley, D. M., and Langford, W. S.: Central autonomic dysfunction with defective lacrimination: Report of 5 cases. *Pediatrics* **3**:468–478 (1949).

Roth, M., and Morrissey, J. D.: Problems in diagnosis and classification of mental disorders in old age. *J. Ment. Sci.* **98**:66–80 (1952).

Schaumburg, H. H., Spencer, P. S., and Thomas, P. K.: *Disorders of Peripheral Nerves*. Contemporary Neurology Series, F. A. Davis, Philadelphia (1983).

Sever, P. S.: Hypertension, hypotension and the autonomic nervous system, in: *Neurotransmitter Systems and Their Clinical Disorders* (N. J. Legg, ed.), pp. 53–66, Academic Press, London (1978).

Smith, A. A., and Dancis, J.: Exaggerated response to infused norepinephrine in familial dysautonomia. *N. Engl. J. Med.* **270**:704–707 (1964).

Smith, A. A., Dancis, J., and Breinin, G.: Ocular responses to autonomic drugs in familiar dysautonomia. *Invest. Ophthalmol.* **4**:358–361 (1965).

Sollmann, T.: *The Manual of Pharmacology*, 8th ed. W. B. Saunders, Philadelphia (1957).

Trcka, V.: Relationship between chemical structure and ganglion-blocking activity, in: *Pharmacology of Ganglionic Transmission* (D. A. Kharkevich, ed.), pp. 155–184, Springer-Verlag, New York (1980).

USP Drug Information The United States Pharmacopeial Convention, Rockville, Maryland (1983). Distributed by the C. V. Mosby Co., St. Louis.

Van Woert, M. H.: Myasthenia gravis, Eaton Lambert syndrome, and familial dysautonomia, in: *Biology of Cholinergic Function* (A. M. Goldberg and I. Hanin, eds.), pp. 567–581, Raven Press, New York (1976).

Volle, R. L.: Muscarinic and nicotinic stimulant actions at autonomic ganglia, in: *Ganglionic Blocking and Stimulating Agents* (A. G. Karczmar, ed.), International Encyclopedia of Pharmacology and Therapeutics, Pergamon Press, Oxford, Section 12, **1**:1–110 (1966).

Zahn, T. P.: Autonomic nervous system characteristics possibly related to a genetic predisposition to schizophrenia. *Schizophr. Bull.* **3**:49–60 (1977).

Zaimis, E. J.: Actions at autonomic ganglia, in: *Cholinesterase and Anticholinesterase Agents* (G. B. Koelle, ed.), pp.530–569, *Hndbuch. der experimentullen Pharmakologie, Ergänzungswerk.*, Vol. 15, Springer-Verlag, Berlin (1963).

21

Ganglionic Transmission as a Model for CNS Function

ALEXANDER G. KARCZMAR

I. INTRODUCTION

It should be obvious from perusal of the preceding chapters that a number of ganglionic and central nervous system (CNS) phenomena are similar and even identical. It is well known that shortly after chemical, cholinergic transmission was demonstrated at the periphery, i.e., at the parasympathetic, motor, and ganglionic synapses (cf. Chapter 1), Dale (cf. Dale, 1938) and others proposed that cholinergic transmission obtains in the CNS, a proposal that was supported by much indirect evidence at the time and that was finally demonstrated in 1954 for the synapse between the motor nerve collateral and the Renshaw cell by Eccles *et al.* (1954). Since 1954, the presence of cholinergic synapses at other central sites has been amply demonstrated (see Krnjevic, 1974; Karczmar, 1976; McGeer *et al.*, 1978, 1984). While the central cholinergic system is ubiquitous, however, anatomical identification and localization of some of the pathways is not yet certain (McGeer *et al.*, 1978, 1984; Kimura *et al.*, 1981; Karczmar, 1986). Thus, the cholinergic ganglion with its proven cholinergic axosomatic synapses (see Chapters 1, 2, and 3) is both homologous and analogous with respect to identified central cholinergic synapses and to those that will be identified in the future.

Furthermore, certain types of studies of cholinergic synapses that cannot be readily carried out in the CNS can be carried out on the ganglion,

ALEXANDER G. KARCZMAR • Department of Pharmacology and Experimental Therapeutics, Stritch School of Medicine, Loyola University Medical Center, Maywood, Illinois 60153.

which can then serve as a model for the central cholinergic synapses. There are several reasons for the feasibility of such studies in the ganglion rather than in the CNS; these reasons were stated as follows (Nishi *et al.*, 1978):

> Autonomic cells exhibit long lasting, steady and consistent responses to presynaptic input and to iontophoretic application of drugs and of transmitters. Thus, they lend themselves to the studies of the mechanisms of transmitter action and their ionic basis; in fact, the voltage clamp method, which is a potent tool in such investigations, can be used only at the periphery. Similarly, enduring changes involved in long term modulations can be studied advantageously only in the ganglia. Finally, several ganglia—sympathetic, myenteric and pelvic—may be considered as polysynaptic systems, and they offer obvious advantages as compared to the central circuits for the study of synaptic and neurotransmitter interactions.

What are the specific questions that are important for the central cholinergic system and that can be answered or at least investigated using the ganglion as a model?

First, it is remarkable that central cholinoceptive neurons, similarly to ganglionic neurons, respond either nicotinically or muscarinically, or show both responses (Krnjevic, 1974; McGeer *et al.*, 1978); however, for the reasons already enumerated, the characteristics of these responses cannot be readily studied in the CNS, and the autonomic ganglion can serve for elucidation of the characteristics of these central cholinoceptive responses. This elucidation includes the investigation of the involvement of second messengers in synaptic responses and conductance changes; in fact, the ganglion was used in the studies of Greengard and his associates (see Greengard, 1978) (see also Chapters 12 and 13) that led to his proposal concerning the general role of cyclic nucleotides in synaptic responses.

Second, there is the question of the responses of central neurons to such transmitters as dopamine, amino acids, and peptides. Again, the characteristics of these responses, particularly their ionic mechanisms, cannot be readily studied in the CNS, but can be heuristically investigated in the ganglia.

Third, there is the question of modulation. It is assumed and, to a certain extent, demonstrated that responses of central neurons—whether to acetylcholine (ACh) or to other than ACh transmitters—may be modulated postsynaptically by a number of mechanisms, including the regulating action of other neurotransmitters (cf., e.g., Puil and Krnjevic, 1978). Again, postsynaptic regulation by transmitters and modulators (see Chapter 3 for definitions) occurs and can be readily studied in the ganglia. The pertinent studies concerning the interplay between ACh and other neurotransmitters or the interplay between ACh and modulators have been presented in Chapters 12 and 13; an example of these studies is the study of the interplay between cholinergic and peptidergic transmissions (see

Chapter 8). Still in the context of modulation, modulation may be exerted presynaptically, via control of the release of the transmitter, in this case ACh, and such modulation, which has barely been studied in the CNS (see Szerb, 1977; Ben-Ari *et al.*, 1981), can be readily studied, in all its complexity, in the ganglion (see Chapters 2, 10, and 11). It may be added that transsynaptic regulation of transmitter synthesis, a phenomenon described by Costa and his associates (Wood *et al.*, 1981) that can be readily demonstrated with respect to ganglia, constitutes a special example of synaptic modulation.

Still another aspect of the ganglion as a model for the CNS is constituted by its function as a reflex system. This system may involve the classic reflex arc that includes the spinal cord (see Chapters 17, 18 and 19) or may present certain novel, sometimes extraspinal aspects (see Chapters 8 and 13).

In sum, the ganglion is most appropriate for in-depth investigation of transmitter mechanisms and of interplay between transmitters and modulators; this may then throw light on the mechanisms underlying moment-to-moment regulation and fine tuning that the neuron is capable of. This knowledge that can be derived from ganglion studies may serve for even further goals; thus, the mechanisms of ganglionic response regulation are considered by some investigators (Libet *et al.*, 1975) as appropriate models for the study of such CNS-related phenomena as memory, learning, and cognitive functions.

This chapter will thus consider briefly the contribution of ganglionic research to the elucidation of certain aspects of central, particularly cholinergic, transmission.

II. MUSCARINIC (SLOW) AND NICOTINIC (FAST) EXCITATORY POTENTIALS

While the excitatory response of cholinoceptive CNS neurons, as already indicated, either is muscarinic in nature or exhibits both nicotinic and muscarinic components (Krnjevic, 1974, 1976) (for inhibitory cholinoceptive response, see Section III), it is of interest that there are only a few central synapses that respond similarly to the ganglionic synapses, i.e., primarily nicotinically. The classic example is the Renshaw cell (Eccles *et al.*, 1954; Eccles, 1969); its postsynaptic membrane is generally considered as nicotinic and its muscarinic response may not be within the general ken, yet the Renshaw interneuron shows a delayed, long muscarinic response (Eccles, 1969; Ryall and Haas, 1975). At any rate, most central cholinoceptive neurons respond muscarinically or show a mixed response, while only a few show a typical fast and sharp nicotinic response that is blocked by D-tubocurarine, dihydro-β-erythroidine, or large doses of nicotine (cf. Karczmar, 1976, 1979; McGeer *et al.*, 1978; Karczmar and Dun, 1978; Krnjevic, 1974, 1976).

Are the central and ganglionic muscarinic and nicotinic actions equivalent? Do the ionic mechanisms exhibited by the ganglionic responses constitute a good model for their central counterparts? The generation of the fast EPSP depends on the inward current directed across the subsynaptic site and increased Na^+ and K^+ conductance (Koketsu, 1969) (cf. Chapter 5); this is true for the neuromyal endplate potential (EPP) as well. It is of interest that while pharmacological analysis clearly indicates the nicotinic nature of the fast response of central cholinoceptive neurons, it is not clearly established whether its ionic mechanism is similar to that of the ganglionic response (cf., e.g., Krnjevic, 1976; McGeer *et al.*, 1978); it can only be surmised that this is so. Thus, since the neuromyal and ganglionic generation of the nicotinic response has been studied most dependably by means of several procedures, including the voltage-clamp method (see Chapters 5 and 6), it must be assumed that the increase of the conductance of Na^+ and K^+ channels underlies both peripheral and central nicotinic responses.

Strangely enough, the ionic mechanisms of the muscarinic central response have been studied in more depth than the ionic mechanisms of the central nicotinic response. In particular, this is the case with the septohippocampal cholinergic pathway and the slow cholinoceptive response of the hippocampal pyramidal neurons. It was initially proposed that this effect, a slow depolarization, is mediated by a reduction in K^+ conductance and a rise in membrane resistance (Adams *et al.*, 1981; cf. Krnjevic and Ropert, 1982), similar to the slow depolarization induced by ACh in the neocortical cholinoceptive neurons (Krnjevic *et al.*, 1971) (it should be added that in this latter case, contrary to the hippocampal neurons, we are not dealing with a definitely identified cholinergic synapse). This decrease in conductance is an unusual finding for the CNS (see McGeer *et al.*, 1978). More recently, Cole and Nicoll (1983) proposed that two types of K^+ conductance are blocked in the course of the hippocampal response, Ca^{2+}-activated and the M current (see Chapter 7), and these investigators consider this type of response as nonclassic in nature. Actually, this type of response is characteristic for some ganglionic neurons (see Chapter 7), and in this case it has been studied in depth; thus, this mechanism for the generation of muscarinic potentials is well documented. It must be stressed that ganglion cell populations exhibit several types and several ionic mechanisms of muscarinic response (cf. Chapter 7), and the response in question is only one of them. However, most recent studies of Adams et al. (1981; 1986) indicate that there is a number of similarities between ionic mechanisms that underlie ganglionic and hippocampal muscarinic responses; interestingly enough, several types of M current that characterize ganglionic muscarinic response (see Chapter 7) are pertinent as well for the response of pyramidal hippocampal cells, but their relative contribution to the response differs in the two cases. It must be added that Cole and Nicoll (1983) stressed that in their hippocampal slice preparation, the ACh-generated slow response reduced the afterhyperpolarization induced by a train of action potentials; this was also true in the course of the generation of the slow ganglion potential (Chapter 7).

It is also of interest that studies carried out with respect to ganglionic inhibitory responses [the slow IPSP (see Chapters 9 and 13)] shed light on the inhibitory central responses to catecholamines; this subject is discussed below.

III. INHIBITORY RESPONSE

The central inhibitory cholinergic phenomenon may be generally little appreciated and is in fact relatively rare in appearance (Krnjevic, 1974, 1976; McLennan and York, 1967). The mechanism and indeed the authenticity of this cholinergic IPSP response are not quite clear. Of pertinence in this context are the results of studies of inhibitory ganglionic transmission. Indeed, these studies suggest that cholinergic transmission may result in inhibition via one of the two mechanisms: disynaptically or as a consequence of direct action of ACh (see Chapters 3, 9, 12, and 13). It is possible that there is present in the CNS an interneuron that can be exicted cholinergically, may release an inhibitory, hyperpolarizing transmitter, and is located at strategically appropriate sites. Catecholamine involvement in this hypothesized disynaptic mechanism is not likely, even though hyperpolarizing action is exhibited by catecholamines both in the CNS and in the ganglia (see below). While central catecholamines may be released cholinergically (cf. Glisson *et al.*, 1974), pontine and midbrain localization of the somata of the catecholaminergic pathways and their long axonal nature precludes the possibility that catecholaminergic release may be activated by iontophoretic application of ACh at the forebrain and other brain sites where IPSPs were recorded (Krnjevic, 1976). Local, short-axon circuitry involving other transmitters is a possible explanation of the effects in question; thus, the SIF-cell system of the ganglion may constitute an appropriate model and basis for further studies (cf. Chapters 9 and 12). On the other hand, it is tempting to ascribe central cholinergic inhibition to a direct inhibitory action of ACh (such as described in Chapters 9 and 13) or an action similar to that exerted by ACh at cardiac vagal endings; again, autonomic sites may serve as a model for the study of the central inhibitory phenomena, particularly to ascertain whether the ionic mechanisms of the pertinent central and ganglionic phenomena are identical.

IV. GANGLIONIC RESPONSES TO TRANSMITTERS OTHER THAN ACETYLCHOLINE AND THEIR INTERACTIONS

A. Responses to Noncholinergic Neurotransmitters

1. Peptides

While it has been well known since the investigations of Hökfelt and his associates (Hökfelt *et al.*, 1975; see also Pernow, 1963) that many peptides are present in the CNS, it is less appreciated that the same investigators almost simultaneously established the presence of these peptides in the autonomic ganglia (Hökfelt *et al.*, 1977; cf. Chapters 2, 3, 8, and 13). Soon thereafter, it was proposed that substance P and luteinizing-hormone-releasing hormone (LH-RH) act as neurotransmitters responsible for the noncholinergic late slow EPSP in the inferior mesenteric ganglion of the guinea pig and in bullfrog sympathetic ganglia, respectively (see Chapters 8 and 13) (North and Egan, 1982; Sejnowski, 1982). What must be also stressed is that while in the case of the central synapses not much more can be said as to the effects of these peptides than that they are excitatory and depolarizing in nature and involve an increase in the membrane conductance (cf., e.g., Nicoll *et al.*, 1980; Kelly, 1982; Hökfelt *et al.*, 1984), in the case of the ganglia, the mechanisms in question could be described rather readily. It appears (cf. Chapter 8) that substance P exhibits two types of responses, depending on the cell population of the inferior mesenteric ganglion of the guinea pig, the type of the response resulting from the interaction between increase in G_{Na} and decrease in G_K. Similarly, several types of responses characterize the excitatory action of LH-RH on bullfrog sympathetic neurons, G_K, G_{Na}, and in fact also the M current being involved (see Chapter 7). The pertinent data were obtained by means of dependable methods, such as the voltage-clamp method; thus, the current conclusion—that, in the case of central neurons, these two peptides increase membrane conductance (Kelly, 1982)—appears to be simplistic and will have to be further evaluated in light of the pertinent studies of the autonomic ganglia.

One additional point merits attention. In the case of both the autonomic ganglia and the CNS, the pathways for substance P and LH-RH have not been firmly established (see Chapter 8) (see also Kelly, 1982).

However, it appears that substance P may act, in both cases, as the transmitter for sensory, afferent cells; peripheral reflex involving gastrointestinal activity may activate afferent fibers to the inferior mesenteric ganglion and induce the release of substance P (see Chapter 7), while substance P may act as the transmitter agent for the primary afferent sensory cells in vertebrate dorsal ganglia (cf. McGeer *et al.*, 1978). Thus, the results obtained both at the periphery and in the spinal cord reinforce each other as to the sensory role of substance P.

Again, in the case of enkephalins, dependable data obtained with the autonomic ganglia are at this time the only definitive results that are pertinent as to the possible action of enkephalins on the CNS and as to its mechanism. In the CNS, particularly the spinal cord, enkephalins were originally considered as depressant in action, since they inhibit neuronal firing at several sites (cf. Nicoll *et al.*, 1980; Kelly, 1982). However, excitatory effects of opioids including enkephalins were observed as well (cf., e.g., Baker *et al.*, 1978; Phillis and Kirkpatrick, 1980; see also Kelly, 1982). Actually, there may be no inconsistency between these two sets of findings, since Nicoll (cf. Nicoll *et al.*, 1980) considered the excitatory actions of opioids to be dependent on enkephalin-induced disinhibitions resulting from either postsynaptic hyperpolarization of inhibitory interneurons or block of presynaptic inhibition, i.e., antagonism by enkephalins of inhibition of transmitter release. It should be noted in keeping with enkephalin-induced inhibition of transmitter release in the CNS that the results obtained in the ganglia indicate that while enkephalins do not affect the potential or the sensitivity of the postsynaptic membrane, they prevent transmitter release (see Chapter 7).

The mechanisms of the central actions of angiotensin II and bradykinin are also not clear. These actions appear to be excitatory or depolarizing or both, although disinhibition rather than postsynaptic effect may be involved (Haas *et al.*, 1980); the excitatory actions of bradykinin have been more consistently reported. The results obtained with the autonomic ganglia may be helpful with respect to angiotensin, which exerted consistently potent stimulant, depolarizing actions in the cat and rat superior cervical ganglion, changes in G_{Na} and G_K underlying these effects, respectively; however, this peptide exerted no effects on other ganglia (see Chapter 8). The results obtained with bradykinin were less clear-cut (see Chapter 8).

2. Amino Acids, Purines, Monoamines, and Cyclic Nucleotides

It is interesting that the action of γ-aminobutyric acid (GABA) on autonomic neurons resembles the effects exerted by GABA on central afferent nerve terminals, rather than its effects on central postsynaptic membranes. As is well known, the depolarizing effect of GABA on the afferent terminals is the origin of Ecclesian presynaptic depolarizing inhibition (see below). It has been demonstrated that the same depolarizing effect of GABA with the same underlying ionic mechanism can occur at the postsynaptic site in the case of autonomic ganglia (see Chapter 12); what is even more noteworthy is that in the ganglia, as in the CNS, this depolarizing action also occurs presynaptically, leading to a decrease in release of the transmitter (see Chapter 11).

It is well known how variable the affects of *serotonin* (5-HT) on central neurons are, although the more general effect seems to be reduction in firing and excitability (see McGeer *et al.*, 1978), particularly in the case of cells with identified serotonergic input (Aghajanian *et al.*, 1975). It may be reinforcing for this concept to note that applied to the sympathetic ganglia of the bullfrog, 5-HT exerted a similar depressant action (see Chapter 12). This "pharmacological" depression was ascribed to activation of an electrogenic sodium pump. On the other hand, 5-HT was described as the transmitter responsible for the noncholinergic slow EPSP of the coeliac mammalian ganglion, the depolarization in this case presumably being due to inactivation of K^+ conductance (see Chapter 13).

Finally, it should be pointed out that sertonin seems to depress the sensitivity of the cholinergic receptor to ACh (see Chapter 12). There is no evidence as yet concerning the central counterpart of this effect.

It is generally accepted that such *catecholamines* as norepinephrine and dopamine essentially depress and hyperpolarize the postsynaptic membranes of central neurons, although these findings are not consistent, since it is sometimes suggested that α-adrenergic —particularly α_1—agonists exert excitatory effects; biphasic actions have also been observed (McGeer *et al.*, 1978; Bloom, 1975; York, 1975; Minneman *et al.*, 1981). The ionic mechanisms are not clear. On the other hand, it is frequently hypothesized that at least the hyperpolarizing effects of these compounds relate to membrane protein phosphorylation generated by catecholamine-induced formation of cyclic nucleotides via activation of adenylate cyclase (cf., e.g., Greengard, 1978; McGeer *et al.*, 1978; Bloom, 1975; Nestler *et al.*, 1984). This seems to be particularly well documented for basal ganglia (cf. Bloom, 1975; York, 1975; Walaas and Greengard, 1984) and the cerebellum (Hoffer *et al.*, 1973; cf. McGeer *et al.*, 1978).

Some of this evidence is corroborated by results obtained in studies of the autonomic ganglia. Thus, dependable methods showed that several catecholamines produce a hyperpolarization of ganglion cells, which is possibly coupled with a decrease in membrane resistance; accordingly, an increase in G_K may be involved (see Chapters 9, 12, and 13). Also, catecholamines may exert a potent ganglionic block via their presynaptic action and inhibition of transmitter release (see Chapter 8). It should be stressed that there is still another mechanism by means of which catecholamines may depress ganglionic synaptic transmission, and that is by reducing receptor sensitivity (see Chapter 12). At this time, it is not known whether or not a similar mechanism obtains in the CNS, and it seems very worthwhile to investigate this possibility.

It must be added in this context that the concept of the role of *cyclic AMP* in the generation of the dopamine-induced hyperpolarization of the central neurons is not substantiated at all in the case of the ganglia (see Chapters 9 and 13); it must further be stressed in this context that this concept could be investigated in depth in the case of the ganglia. It should also be added that the hyperpolarizing action of dopamine used as a drug may not necessarily relate to the concept of the disynaptic nature of the slow IPSP and of the role of the SIF cell and its dopamine in the generation of this inhibitory potential (see Chapters 9, 12, and 13). Nevertheless, autonomic ganglia constitute an ideal system to test the concept that differential phosphorylations of specific phosphoproteins and related cyclic nucleotide phenomena underlie ionic, synaptic movements and neuronal function, whether at the postsynaptic (Nestler *et al.*, 1984; Snyder, 1984) or presynaptic (Nestler and Greengard, 1982) sites.

ATP and related compounds, *purines*, and *pyrimidines* may act as transmitters of "purinergic" neurons (Burnstock, 1972). The evidence, and particularly their effects on central neurons, are diversified and may include depression (e.g., Perkins and Stone, 1980, 1983) as well as excitatory pre- and postsynaptic actions (McGeer *et al.*, 1978). Again, ganglionic studies are important in this context, since pertinent voltage-clamp experiments could be carried out on the ganglia (e.g., Akasu *et al.*, 1983) (see Chapter 12). It appears that in the ganglia, ATP may depolarize the membrane via action on K^+ and Ca^{2+} currents (see Chapter 12). Furthermore, as in the case of catecholamines, ATP, adenosine, and its congeners may act on the nicotinic receptor independently of their effect on membrane potential, and augment its ACh sensitivity; this effect may or may not be mediated by Ca^{2+} currents (Henon and McAfee, 1983) (Chapter 12).

3. Phosphatidylinositol Turnover

The role of cyclic nucleotides and related activators of protein phosphorylation (Nestler *et al.*, 1984) as second messengers involved "in communicating the influence of hormones, neurotransmitters, and other mediators to the interiors of" . . . neurons (Snyder, 1984) was stressed above and in Chapter 13. It is in fact of interest that results obtained with the ganglia constitute a portion of the evidence obtained by Greengard and his associates (Greengard, 1978; Nestler and Greengard, 1982; Nestler *et al.*, 1984) in support of this concept.

Phosphatidylinositol turnover and resulting phosphorylated metabolites (polyphosphoinositides) constitute a system analogous in its function to the second-messenger system of the nucleotides and related substances, and in fact there is a link between the two systems (Nishizuka, 1984). While the phosphatidylinositol phenomena are readily activated by muscarinic rather than nicotinic signals (Fisher *et al.*, 1984; Brown and Masters, 1984) (cf. Chapter 13), the effects of pertinent metabolites on the nicotinic synapses have been demonstrated (see Chapter 13), and again, ganglia constitute an ideal system for further study of this particular second-messenger system.

B. Neurotransmitter Interactions

1. Postsynaptic Site

To state the problem in a simple way, synaptic potentials elicited by various transmitters may interact postsynaptically, since hyperpolarization will decrease depolarization and may render excitatory, depolarizing postsynaptic responses ineffective, while depolarizing responses may interact additively or potentiatively, leading to initiation of transmissive spikes. In several instances, such interactions were demonstrated for the CNS (see McGeer *et al.*, 1978; Puil and Krnjevic, 1978); they frequently result from feed-forward or feedback circuitries that involve interneurons and result in inhibitions; however, this interaction may be generalized and include interactions that operate solely via neurons (Puil and Krnjevic, 1978) and lead to both facilitations and inhibitions. Thus, Puil and Krnjevic (1978) adduced as examples the depression, by dopamine, of the firing of cortical neurons induced by ACh or glutamate and the facilitation, by glutamate, of firing induced by substance P. Similar interactions between a number of transmitters occur, classically, in the striate (e.g., Javoy *et al.*, 1978).

It is of interest that ACh itself may be involved in these interactions. Indeed, Krnjevic (1969, 1976) proposed early that slow muscarinic responses that characterize the central cholinoceptive neurons are modulatory and facilitatory in nature, rather than transmissive (see also above). Supportive evidence was recently added to this concept by Krnjevic and Ropert (1982) (see also Glavinovic *et al.*, 1983), who found that ACh as well as stimulation of the septum facilitate the spike response of the CA1 hippocampal pyramidal neurons evoked by commissural stimulation; these investigators commented that the slow cholinergic response or the ACh potential "potentiates other excitatory synaptic inputs" via inactivation of K^+ channels and/or "reducing the potency of the GABA-ergic system" (Krnjevic and Ropert, 1982).

Such interactions abound in the case of the ganglia. In many cases, they are analogous to the central events, though sometimes the interactions in question lead to effects opposite to those that occur centrally. With regard to the former case, it was shown early by Koketsu and Nishi (1967) that in the course of the IPSP, the excitability of the sympathetic ganglia is decreased, as evidenced by the block by the IPSP of the generation of ganglionic afterdischarges by repetitive stimulation and by the block of the slow (muscarinic) EPSP; these effects could be mimicked by dopamine (Koketsu and Nishi, unpublished data). These results were confirmed recently by Dodd and Horn (1983); related results are described in Chapter 8. It is noteworthy that the diminution of excitability by the IPSP or dopamine or both does not suffice to block the fast EPSP, although it can block repetitive responses; the latter are related to the increase of ganglionic excitability that is induced by slow excitatory potentials (Nishi, 1974; Nishi *et al.*, 1978). Indeed, the noncholinergic late show EPSP induces repetitive firing, as does the application of neurotransmitters responsible for the response, i.e., substance P in the case of the inferior mesenteric ganglion of the guinea pig (Dun and Jiang, 1982) and LH-RH in that of bullfrog sympathetic ganglia. Furthermore, there is possibly an interaction between the slow IPSP and the fast (nicotinic) as well as the slow (muscarinic) potential, and this interaction makes difficult the evaluation of the time of actual initiation of the slow potentials.

These findings are analogous to the antagonistic interaction between dopamine and ACh described for cortical neurons by Puil and Krnjevic (1978) (see above). On the other hand, it appears that in the ganglia, exogenous dopamine, and perhaps dopamine generated by SIF cells (Libet, 1970) (see, however, Chapter 13), may enhance certain excitatory responses, such as the muscarinic slow EPSP (Libet and Tosaka, 1970; Libet, 1981; Ashe and Libet, 1981; Kobayashi, 1982). It appears, moreover, that this modulation by dopamine is not mimicked by other catecholamines and that it may occur with concentrations of dopamine below those that cause hyperpolarization (cf. also Chapter 13). Furthermore, it was proposed that the dopamine-induced potentiation of the slow EPSP is mediated by cyclic AMP (cAMP) in its role as second messenger (Kobayashi, 1982). It should be pointed out that the facilitation of the slow EPSP by dopamine or by the slow IPSP cannot be readily explained in terms of their ionic responses, since IPSP or dopamine responses are due to an increase in G_K, while the slow EPSP or slow ACh potential results mainly from G_K inactivation (see Chapters 7, 9, and 13); on the other hand, a AMP-dependent mechanism appears consistent with the slow EPSP–slow IPSP facilitatory interaction. Kobayashi (1982) stressed, parenthetically, that AMP mediation of the facilitation of the muscarinic EPSP is not inconsistent with the evidence that cAMP may not be involved in the generation of the IPSP (see also Chapters 9 and 13). It must be added that cAMP and dopamine as well as other catecholamines may also facilitate the noncholinergic, late slow EPSP (Katayama and Nishi, 1980).

It should be stressed that a number of substances or neurotransmitters that are effective both centrally and at the ganglion have not been studied in depth as to their central postsynaptic interaction with each other and with ACh. Enkephalins and endorphins are in this category. Their effects, *per se*, at the ganglion and in the CNS were already discussed and compared above, as well as in Chapters 8, 9, and 13 (see also North and Egan, 1982). On the postsynaptic level, the hyperpolarizing effects of enkephalins deserve some attention, since these effects may constitute their role as putative transmitters involved in the generation of the IPSP in the myenteric plexus (see Chapter 16) and since these postsynaptic inhibitory effects have been observed in the case of the other ganglia as well (see Chapter 13). As putative generators of the IPSP or of related actions, enkephalins should be important, similarly to dopamine (see below), in postsynaptic modulation, whether in the CNS or at the ganglia, although there is little direct evidence on this matter. Also, such regulations may obtain presynaptically, and this will be referred to subsequently.

Serotonin (5-HT) also seems to belong to the category of neuronally active substances (for its ganglionic effects, see above and Chapters 12 and 13) that have not been studied as to their central or ganglionic interactions. It is therefore important to point out that since 5-HT may decrease sensitivity or affinity of the cholinergic receptor for ACh (see Chapter 12), it appears worthwhile to ascertain whether 5-HT interacts with cholinergic transmission in the CNS. An important aspect of this matter is that content (or synthesis) of serotonin in the sympathetic ganglia may be regulated via muscarinic signals; this example of "trans-synaptic modulation" (Hadjiconstantinou *et al.*, 1982) may be important in the context of ganglionic serotonin–ACh interaction (see also Chapters 12 and 13).

In sum, while it must be expected on theoretical grounds that both centrally and peripherally, neurotransmitters interact postsynaptically to provide subtle and precise modulatory tuning, there is a paucity of pertinent evidence with respect to the CNS; the evidence derived from ganglionic studies is more extensive, and it may serve as a good basis for the analogous studies of the CNS.

2. Presynaptic Site

The presynaptic site is most strategic with regard to the modulation of autonomic transmission, since this site regulates the release of the transmitter (see Chapters 3, 10, and 11). It has been shown definitively that many endogenous substances, including transmitters or putative transmitters, as well as drugs act presynaptically at the ganglia and affect ACh release (see Chapters 11 and 13), yet relatively little work of an analogous nature has been carried out with the CNS. There is apparently clear-cut evidence concerning enkephalins. The inhibitory nerve-terminal action of these compounds has already been stressed in this chapter and in Chapters 10 and 11, and in fact some evidence has been adduced that indicates that endogenous enkephalins may be involved in physiological regulation of ACh release in the myenteric plexus (see also North and Egan, 1982). It appears at this time that a similar effect obtains centrally as well. In fact, there is good evidence indicating that enkephalins and opioids block the presynaptic inhibition in the frog spinal cord. The principle of presynaptic inhibition was established for the CNS by Eccles (1964). This inhibition, which is due to depolarization of the afferent fibers and diminution of the release of excitatory transmitter (Eccles, 1964), is evoked by liberation of GABA from appropriate afferents abutting on the primary afferents in question (cf. Nicoll *et al.*, 1980); block of this inhibition—which results in disinhibition—may be induced by enkephalins via hyperpolarization of the terminals of spinal afferents (Nicoll *et al.*, 1980). Related effects of enkephalins have been described at other sites such as the hippocampus (Nicoll *et al.*, 1980; cf. Kelly, 1982). While these effects of enkephalins involve release of GABA, they are of course related in principle to their effects on release of ACh from the ganglia.

In this context, it should be stressed that the central presynaptic, depolarizing action of GABA is analogous to its presynaptic ganglionic action as discussed in Chapters 10 and 11 and in this chapter. In either case, the effect of GABA on G_{Cl^-} is involved; this has been definitively demonstrated on ganglionic preparations, although in this case, spinal rather than autonomic ganglia were used (Nishi *et al.*, 1974). It should be added that since the soma of the afferent neurons of the spinal ganglia reacts similarly to the spinal afferent nerve terminals (Nishi *et al.*, 1974), spinal ganglia constitute a good model for the study of the central ionic and neurotransmitter mechanisms underlying afferent functions; Chapter 19 should be viewed in this light.

An interesting aspect of presynaptic action involves ACh itself. In the 1960s, Koelle (1961, 1963) proposed that ACh released from cholinergic synapses acts retrogradely on the terminal, facilitating its own liberation via a "percussive" action; he assigned this action to certain central sites as well as to peripheral synapses such as autonomic ganglia. Actually, as discussed in Chapters 10, 11, and 13, both a muscarinic and a nicotinic cholinoceptive site seem to be present at autonomic presynaptic terminals; the activation of either site causes nerve-terminal depolarization. It is a moot point whether or not the activation of the nicotinic site facilitates ACh release (cf. Chapters 10 and 13) (see also Nishi, 1974). On the other hand, it is quite clear that the depolarizing muscarinic effect leads to depression of ACh release (see Chapter 10). Again, there is a central counterpart of this effect, since Szerb (1977) has demonstrated that muscarinic activation and, in fact, retrograde action of endogenous ACh block ACh release at several central sites; as in the case of ganglionic terminals, atropine disinhibits the terminal from block due to exogenous or endogenous muscarinic action (Szerb, 1977). In the case of either the CNS or ganglia, the ionic mechanism of this effect is not clear, although Ca^{2+} may be involved (Nordstrom *et al.*, 1981). On the other hand, there is little if any evidence that central nicotinic receptors are present on the central cholinergic terminals. In view of the ubiquity of the central cholinergic terminals (Karczmar, 1979), it is important to continue investigating the feedback, nerve-terminal action of ACh. Furthermore, since the central cholinergic pathways abut on a number of central noncholinergic pathways, including GABAergic, catecholaminergic, and serotonergic pathways, and since the cholinergic pathways or drugs or both affect the levels and/or turnover of GABA, 5-HT, and catecholamines (Karczmar, 1981a,b), it is worthwhile to investigate whether or not these interactions are due to presynaptic actions of ACh.

3. Coexistence of Transmitters and Peptides

That such a possibility exists was mentioned to this author by B. B. ("Steve") Brodie in the 1950s, long before any evidence on this matter was available. Today, this possibility is at least implied in the case of peripheral tissues including para- and prevertebral ganglia (see Chapters 3, 8, and 11) (Hökfelt *et al.*, 1977; Simmons, 1985) and suggested for the CNS (Hökfelt *et al.*, 1984). Spontaneous release of neurotransmitters that may induce either similar (excitatory, as in the case of substance P and ACh) or dissimilar (excitatory and inhibitory, as in the case of dopamine and ACh) postsynaptic actions leads to particularly telling interactions, and it must be emphasized that these interactions may also occur presynaptically (see above). Again, autonomic ganglia constitute a most convenient system for both establishing and evaluating the significance of simultaneous release of more than one neurotransmitter.

It should be added in this context that ganglionic activity may result in presynaptic modulatory effects that depend, not on neurotransmitter interaction, but on ionic phenomena. A case in point is ganglionic long-term potentiation (Briggs *et al.*, 1985). This potentiation does not appear to be mediated by postsynaptic interaction between either various types of cholinergic responses or catecholamines and ACh; rather, it is presynaptic in nature and may be due to increased flux of Ca^{2+} into the terminals. This presynaptic mechanism of potentiation that leads to increased release of ACh is known to operate at the skeletal neuromyal junctions. This potentiation may be pertinent for the relevance of the ganglion as a model for higher CNS function (see below).

C. A Special Instance of Interactions: Rat Pelvic Ganglia

This section serves to alert the reader to the research potential of a unique autonomic system constituted by the male rat pelvic ganglion, which is ideal as a model of functional regulation resulting from circuitry and neurotransmitter interplay; while only preliminary pharmacological data are available concerning this material, these data as well as pertinent morphological and neurochemical information clearly indicate such a potential for the system in question. Uniquely among the mammals, in the male rat, the pelvic innervation is organized as the major pelvic ganglion (Purington *et al.*, 1973); it is noteworthy that this structure does not appear within the pelvic innervation and its plexuses in the female rat or other mammalian species which have been investigated so far (Simmons, 1985). The rat system includes cholinergic and catecholaminergic neurons, as well as noradrenergic SIF cells (Dail *et al.*, 1975; Dail, 1976); more recently, the presence of peptidergic neurons, including neurons that probably release vasoactive intestinal polypeptide, and enkephalins was demonstrated (Dail *et al.*, 1983; Owman *et al.*, 1983; DeGroat *et al.*, 1986) (see also Chapter 13). Of particular interest is that first of all, cholinergic cells of the major pelvic ganglion are enclosed by a plexus of adrenergic nerve endings that originate in the adrenergic ganglion cells (Dail, 1976). Second, the SIF cells of this ganglion appear to make several synaptic connections with the ganglion cells. The SIF cells seem to be cholinergically innervated, since incoming nerve terminals, containing numerous small, clear vesicles of the cholinergic type, make synaptic contacts with the perikarya of the SIF cells. Moreover, the SIF cells make efferent, presumably adrenergic connections with ganglion cells, since nerve terminals containing dense-cored vesicles that originate from the SIF cell appear to make synaptic contacts with cell processes of principal ganglion cells (Dail *et al.*, 1975). Finally, it appears that the major pelvic ganglion of the male rat exhibits slow and fast excitatory potentials (DeGroat and Saum, 1976), as well as inhibitory potentials evoked by sympathetic preganglionic stimulation and applied catecholamines or enkephalins (DeGroat and Saum, 1971, 1972; Akasu *et al.*, 1984; DeGroat *et al.*, 1986).

This preparation appears, then, to be a convenient model for the study of complex circuitry and neurotransmitter interaction, at both the pre- and postsynaptic level.

V. SPECULATIONS ON SPECIAL RELEVANCE OF AUTONOMIC GANGLIONIC PHENOMENA TO HIGHER CNS FUNCTION

As stated in Chapter 1, already in the mid-1850s, investigators of the autonomic system visualized the complexities of the ganglionic system clearly enough to refer to the ganglia as "little brains." The preceding chapters of this book serve to illustrate the truth of this statement. Furthermore, ganglionic events seem to offer special mechanisms that may be pertinent to higher central function.

First, the negative feedback mechanisms studied in depth with respect to the ganglion and demonstrably present in the CNS obviously serve to provide a safety margin that ensures maintenance of function even in unfavorable conditions such as those resulting from hyperactivity or chemical damage (Karczmar *et al.*, 1972); this is particularly true with respect to presynaptic modulations (see Chapters 3, 11, 12, 13, 17, and 18). It is just possible that the alleged participation of autonomic ganglia in schizophrenia (cf. Chapter 20) relates to the homeostatic role of the ganglia.

Furthermore, postsynaptic processes interact vectorally (see above) to provide moment-to-moment tuning and regulation; this is particularly true with regard to slow potentials, and it was suggested (Karczmar *et al.*, 1972) that, again, these processes may be important with regard to maintenance of function, whether in the ganglia or centrally, under precarious conditions.

Second, specific proposals as to the relationship of these and other phenomena to high (mental) CNS function have been made by a number of investigators.

One such proposal concerns long-term ganglionic potentiation, which has been discussed above and elsewhere in this book; this potentiation may constitute a phenomenon that is either analogous or homologous with phenomena that underlie such higher functions as memory. Indeed, Briggs *et al.* (1985) emphasized the long-lasting nature of this potentiation and its similarity to the posttetanic potentiation that occurs in several sites, including invertebrate synapses (see, for example, Baxter *et al.*, 1985) and has already been implicated in learning.

Another proposal was presented by Libet and his associates. Libet (1981) stressed the long-lasting character of the slow potentials, particularly those due to dopamine; indeed, in dramatic contrast to the fast EPSP lasting milliseconds and even to the muscarinic slow EPSP lasting seconds, dopamine evokes what Libet refers to as "contingent synaptic function" (Libet, 1981, 1984) (this term corresponds to the concept of modulatory action as espoused in Chapter 3). As described by Libet, in the course of this phenomenon, "a brief exposure to one transmitter induces a persisting change (lasting for hours) in response to another transmitter." It must be pointed out that a similarly long duration of action characterizes the peptidergic, noncholinergic slow responses (see Chapter 8 and above). It should be emphasized that besides dopamine, other substances that are demonstrably released at the ganglia and generate ganglionic potentials [such as the slow noncholinergic EPSP (see Chapter 8)] provoke long-lasting effects on ganglionic responses and ganglionic excitability (see Chapters 4, 12, and 13). Cyclic nucleotides and phosphatidylinositol may exert similar actions (see Chapter 13).

It is interesting in the context of Libet's proposal to refer to results of Koshtojantz *et al.* (1979, 1980). These investigators showed that stimulation of the catecholaminergic locus coeruleus of the rat induced changes in the responses of hippocampal neurons (presumably pyramidal cells?) to ACh; it must be noted, however, that the effect was variable (the response could increase or decrease), that it was measured only shortly following the stimulation of the locus coeruleus (1 min after the stimulus), and that responses to dopamine were affected as well.

An important aspect of the role of the slow potentials is that their function may constitute a model for centrally occurring memory processes, and Libet (Libet *et al.*, 1975; Libet, 1981) presented important heuristic speculation on this subject. He proposed that the "contingent" action (see above) of dopamine results in still another phenomenon, i.e., the "consolidation" of dopamine action, mediated by cAMP, and the concomitant phosphorylation of the membrane (see above and Chapter 13) (see also Greengard, 1978). The earlier version of this hypothesis read as follows (Libet *et al.*, 1975):

> A heterosynaptic interaction takes place between two types of synaptic inputs to the same neuron; the memory trace is initiated by a brief (dopaminergic) input in one synaptic line, while "read-out" of the memory consists simply in the enhanced ability of the postsynaptic unit to produce its specific response to another (cholinergic) synaptic input.

This arrangement provides for a "learned" change in the response to one input as a result of an "experience" previously carried by way of the other input. It should be added that Libet *et al.* (1975) also showed that cGMP abolishes dopamine-generated enhancement of muscarinic sensitivity, the two nucleotides acting on the memory processes in terms of a yin–yang control.

A comment should be added. In Chapters 9 and 13, the role of nucleotides in the generation of the slow excitatory (muscarinic EPSP) and the slow inhibitory potential (IPSP) was presented as essentially controversial, particularly with regard to the generation of the IPSP. However, the evidence for the generation by dopamine of cAMP is well documented (see Chapters 9 and 13) (see also Greengard, 1978; Nestler *et al.*, 1984; Karczmar and Dun, 1978), and it is entirely conceivable that the "consolidatory" or "storage" effect of dopamine via the cyclic nucleotide mechanism is not related to and is independent of the mechanism that underlies the slow IPSP. The speculation of Libet (Libet, 1984, Libet *et al.*, 1975) as to the facilitatory or "contingent" and "consolidatory" mechanism of memory as modeled on the dopamine–cyclic nucleotide–muscarinic system of the ganglia is an important one; it is related to the classic speculation on facilitatory and reverberating processing of memory (Hebb, 1949) that also induces brain plasticity, whether chemical (macromolecular coding of memory; Hydén, 1967, 1978) or morphological (dendritic-spine response to memory; see McGeer *et al.*, 1978) in nature. It must be noted that just about every transmitter and modulator is being accused nowadays of being involved in memory phenomena (cf., e.g., Marsan and Matthies, 1982).

VI. CONCLUSIONS

The foregoing paragraphs indicate clearly why the autonomic ganglia can serve readily as a model for convenient study of the brain. The accessibility of ganglia, the dependability of their responses, and the feasibility, in their case, of analytical procedures that cannot be employed in the case of the CNS—all these factors constitute only a partial demonstration of the suitability of the ganglia as a model of the CNS; the real relevance of ganglionic studies for the CNS concerns the multiplicity of neurotransmitter and modulator ganglionic processes, on both the pre- and postsynaptic levels, their interactions, the involvement of second messengers in some of these ganglionic processes, and, finally, availability of ganglionic preparations (such as male rat pelvic ganglia) that are particularly suitable for the study of all these phenomena. In this chapter, it was attempted to adduce specific examples of the significance of the ganglia as "little brains" and as a model for the study of the CNS; actually, only a partial list of such specific examples could be given here. Yet, just as the ganglia were instrumental in the past—in the clarification of, for example, the role of dopamine in the generation of inhibitory postsynaptic potentials vs. its presynaptic function, or of the ionic mechanisms of a number of ganglionic responses to neurotransmitters and to modulators—they will serve in the future for the clarification of moment-to-moment regulation of synaptic transmission and, more speculatively, as a model for the study of higher, mental functions.

REFERENCES

Adams, P. R., Brown, D. A., and Halliwell, J. V.: Cholinergic regulations of M-current in hippocampal pyramidal cells. *J. Physiol.* (*London*) **317**:29–30P (1981).

Adams, P. R., Brown, D., Jones, S., Pennefather, P., Koch, C., Lancaster, B., Constanti, A., and Halliwell, J.: Slow cholinergic tranmission in sympathetic ganglion and hippocampal pyramidal cells, in: *Neurobiology of Acetylcholine* (N. J. Dun, ed.) Plenum Press, New York, (in press) (1986).

Aghajanian, G. K., Haigler, H. J., and Bennett, J. L.: Amine receptors in CNS. III. 5-Hydroxytryptamine in brain, in: *Handbook of Psychopharmacology*, Vol. 6 (L. L. Iversen, S. C. Iversen, and S. H. Snyder, eds.), pp. 63–96, Plenum Press, New York (1975).

Akasu, T., Hirai, K., and Koketsu, K.: Modulatory actions of ATP on membrane potentials of bullfrog sympathetic ganglion cells. *Brain Res.* **258**:313–317 (1983).

Akasu, T., Shinnick-Gallagher, P., and Gallagher, J. P.: Adenosine mediates a slow hyperpolarizing synaptic potential in autonomic ganglia. *Nature (London)* **311**:62–65 (1984).

Ashe, J. H., and Libet, B.: Modulation of slow postsynaptic potentials by dopamine, in rabbit sympathetic ganglion. *Brain Res.* **217**:93–106 (1981).

Baker, J. L., Smith, T. G., Jr., and Neale, J. H.: Multiple membrane action of enkephalin revealed using cultured spinal neurons. *Brain Res.* **154**:153–158 (1978).

Baxter, D. A., Bittner, G. D., and Brown, T. H.: Quantal analysis of long-term synaptic potentiation. *Proc. Natl. Acad. Sci. U.S.A.* (in press) (1985).

Ben-Ari, Y., Krnjevic, K., Reinhardt, W., and Ropert, N.: Intracellular observations on the disinhibitory action of acetylcholine in the hippocampus. *Neuroscience* **6**(12):2475–2485 (1981).

Bloom, F. E.: Amine receptors in CNS. I. Norepinephrine, in: *Handbook of Psychopharmacology*, Vol. 6 (L. L. Iversen, S. D. Iversen, and S. H. Snyder, eds.), pp. 1–22, Plenum Press, New York (1975).

Briggs, C. A., Brown, T. H., and McAfee, D. A.: Neurophysiology and pharmacology of long-term potentiation in the rat sympathetic ganglion. *J. Physiol.* (London) **359**:503–521 (1985).

Brown, J. H., and Masters, S. B.: Muscarinic regulation of phosphatidylinositol turnover and cyclic nucleotide metabolism in the heart. *Fed. Proc. Fed. Am. Soc. Exp. Biol.* **43**:2613–2617, (1984).

Burnstock, G.: Purinergic nerves. *Pharmacol. Rev.* **24**:509–581 (1972).

Cole, A. E., and Nicoll, R. A.: Acetylcholine mediates a slow synaptic potential in hippocampal pyramidal cells. *Science* **24**:1299–1301 (1983).

Dail, W. G., Jr.: Histochemical and fine structural studies of SIF cells in the major pelvic ganglion of the rat, in: *SIF Cells: Structure and Function of the Small Intensely Fluorescent Sympathetic Cells* (O. Eränkö, ed.), pp 8–18, Fogarty International Center, Proceedings, No. 30, U.S. Government Printing Office, Washington, D. C., (1976).

Dail, W. G., Jr., Evans, A. P., and Eason, H. R.: The major ganglion in the pelvic plexus of the male rat: A histochemical and ultrastructural study. *Cell Tissue Res.* **159**:49–62 (1975).

Dail, W. G., Moll, M. A., and Dziurzynski, R. A.: Increase in immunoactive vasoactive intestinal polypeptide (VIP) in major pelvic ganglion following interruption of the pelvic nerve. *Soc. Neurosci. Abstr.* **9**(Part I):292 (1983).

Dale, H. H.: Acetylcholine as a chemical transmitter of the effects of nerve impulses: The Wiliam Henry Welch Lecture, 1937. *J. Mt. Sinai Hosp.* **4**:401–429 (1938).

DeGroat, W. C., and Saum, W. R.: Adrenergic inhibition in mammalian parasympathetic ganglia. *Nature (London)* **231**:188–189 (1971).

DeGroat, W. C., and Saum, W. R.: Sympathetic inhibition of the urinary bladder and pelvic ganglionic transmission in the cat. *J. Physiol.* (London) **220**:297–314 (1972).

DeGroat, W. C., and Saum, W. R.: Synaptic transmission in parasympathetic ganglia in the urinary bladder of the cat. *J. Physiol.* (London) **256**:137–158 (1976).

DeGroat, W. C., Kawatani, M., Bosth, A. M., and Whitney, T.: Enkephalinergic modulation of cholinergic transmission in parasympathetic ganglia of the urinary bladder of the cat, in: *Dynamics of Cholinergic Function* (I. Hanin, ed.) Plenum Press, New York, (in press) (1986).

Dodd, J., and Horn, J. P.: Muscarinic inhibition of sympathetic C neurones in the bullfrog. *J. Physiol.* (London) **334**:271–291 (1983).

Dun, N. J., and Jiang, Z. G.: Non-cholinergic excitatory transmission in inferior mesenteric ganglia of the guinea-pig: Possible mediation by substance P. *J. Physiol.* (London) **325**:145–159 (1982).

Eccles, J. C.: Presynaptic inhibition in the spinal cord, in: *Progress in Brain Research*, Vol.

12, *Physiology of Spinal Neurons* (J.C. Eccles and J. P. Schade, eds.), pp. 65–91, Elsevier, Amsterdam (1964).

Eccles, J. C.: Historical introduction. *Fed. Proc. Fed. Am. Soc. Exp. Biol.* **28**:90–94 (1969).

Eccles, J. C., Fatt, P., and Koketsu, K.: Cholinergic and inhibitory synapses in a pathway from motor-axon collaterals to mononeurons. *J. Physiol.* (*London*) **123**:524–562 (1954).

Fisher, S. K., Van Rooijen, L. A. A., and Agranoff, B. W.: Renewed interest in the polyphosphoinositides. *Trends Biochem. Sci.* **9**:53–56 (1984).

Glavinovic, M., Ropert, N., Krnjevic, K., and Collier, B.: Hemicholinium impairs septohippocampal facilitatory action. *Neuroscience* **9**:319–330 (1983).

Glisson, S. N., Karczmar, A. G., and Barnes, L.: Effects of diisopropyl phosphofluoridate on acetylcholine, cholinesterase and catecholamines of several parts of rabbit brain. *Neuropharmacol.* **13**:623–632 (1974).

Greengard, P.: *Cyclic Nucleotides, Phosphorylated Proteins, and Neuronal Function.* Raven Press, New York (1978).

Haas, H. L., Felix, D., Celio, M. R., and Inagami, T.: Angiotensin II in the hippocampus. A histochemical and electrophysiological study. *Experientia* **36**:1394–1395 (1980).

Hadjiconstantinou, M., Potter, P. E., and Neff, N. H.: Trans-synaptic modulation via muscarinic receptors of serotonin-containing small intensely fluorescent cells of superior cervical ganglion. *J. Neurosci.* **2**:1836–1839 (1982).

Hebb, D. O.: *The Organization of Behavior.* Wiley, New York (1949).

Henon, B. K., and McAfee, D. A.: The ionic basis of adenosine receptor actions on postganglionic neurons in the rat. *J. Physiol.* (*London*) **336**:607–620 (1983).

Hoffer, B. J., Jiggins, G. R., Oliver, A. P., and Bloom, F. E.: Activation of the pathway from the locus coeruleus to rat cerebellar Purkinje neurons: Pharmacological evidence of noradrenergic central inhibition. *J. Pharmacol. Exp. Ther.* **184**:553–569 (1973).

Hökfelt, T., Kellerth, J. O., Nilsson, G., and Pernow, B.: Substance P localization in the central nervous system and in some primary sensory neurons. *Science* **192**:889–890 (1975).

Hökfelt, T., Elfvin, L.-G., Elde, R., Schultzberg, M., Goldstein, M., and Luft, R.: Occurrence of somatostatin-like immunoreactivity in some peripheral sympathetic noradrenergic neurons. *Proc. Natl. Acad. Sci. U.S.A.* **74**:3587–3591 (1977).

Hökfelt, T., Johansson, O., and Goldstein, M.: Chemical anatomy of the brain. *Science* **225**:1326–1334 (1984).

Hydén, H.: Biochemical changes accompanying learning, in: *The Neurosciences* (G. C. Quarton, T. Melnechuk, and F. O. Schmitt, eds.), pp. 765–771, Rockerfeller University Press, New York (1967).

Hydén, H.: Protein changes in neuronal membranes and synapses during learning. *Biosci. Commun.* **4**:143–150 (1978).

Javoy, F., Euvrard, C., Bockaert, J., and Glowinski, J.: Effects of GABAergic and serontoninergic drugs on the activity of cholinergic interneurons in rat striatum, in: *Neuropsychopharmacology* (P. Deniker, C. Radouco-Thomas, and A. Villeneuve, eds.), pp. 549–558, Pergamon Press, Oxford (1978).

Jiang, Z., Dun, N. J., and Karczmar, A. G.: Substance P: A putative sensory transmitter in mammalian autonomic ganglia. *Science* **217**:739–741 (1982).

Karczmar, A. G.: Central actions of acetylcholine, cholinomimetics, and related drugs, in: *Biology of Cholinergic Function* (A. M. Goldberg and I. Hanin, eds.), pp. 395–449, Raven Press, New York (1976).

Karczmar, A. G.: Brain acetylcholine and animal electrophysiology, in: *Brain Acetylcholine and Neuropsychiatric Disease* (K. L. Davis and P. A. Berger, eds.), pp. 265–310, Plenum Press, New York (1979).

Karczmar, A. G.: Multifactoral and multitransmitter basis of behaviors, and their cholinergic concomitants. *Psychopharmacol. Bull.* **17**:68–70 (1981a).

Karczmar, A. G.: Basic phenomena underlying novel use of cholinergic agents, anticholinesterases and precursors in neurological including peripheral and psychiatric disease, in: *Cholinergic Synapses* (G. Pepeu and H. Ladinsky, eds.), Adv. in Behav. Biol., Vol. 25, pp. 853–889, Plenum Press, New York (1981b).

Karczmar, A. G.: Summary and comments, in *Dynamics of Cholinergic Function* (I. Hanin, ed.), Plenum Press, New York (in press) (1986).

Karczmar, A. G., and Dun, N. J.: Cholinergic synapses, Physiological, pharmacological, and behavioral considerations, in: *Psychopharmacology: A Generation of Progress* (M. A. Lipton, A. DiMascio, and K. F. Killam, eds.), pp. 293–305, Raven Press, New York (1978).

Karczmar, A. G., Nishi, S., and Blaber, L. C.: Investigations, particularly by means of anticholinesterase agents, of the multiple peripheral and central cholinergic mechanisms and of their behavioral implications. *Acta Vitaminol. Enzymol.* **24**:131–189 (1970).

Karczmar, A. G., Nishi, S., and Bluber, L. C.: Synpatic modulations, in: *Brain and Human Behavior* (A. G. Karczmar and J. C. Eccles, eds.), pp. 63–92, Springer-Verlag, Berlin (1972).

Karczmar, A. G., Nishi, S., and Dun, N. J.: Physiology and pharmacology of ganglionic synapses as models for central transmission, in: *Advances in Pharmacology and Therapeutics* Vol. 2, *Neurotransmitters* (P. Simon, ed.), pp. 69–85, Pergamon Press, Oxford (1978a).

Karczmar, A. G., Richardson, D. L., and Kindel, G.: Neuropharmacological and related aspects of animal aggression. *Prog. Neuro-Psychopharmacol* **2**:611–632 (1978b).

Katayama, Y., and Nishi, S.: The non-cholinergic excitatory transmission in sympathetic ganglia, in: *Adv. Physiol. Sci.* Vol. 4, *Physiology and Excitable Membranes* (J. Salanki, ed.), pp. 323–327, Pergamon Press, Oxford (1980).

Kelly, J. S.: Electrophysiology of peptides in the central nervous system. *Br. Med. Bull.* **38**:283–290 (1982).

Kimura, H., McGeer, P. L., Ping, J. H., and McGeer, E. G.: The central cholinergic system studied by choline acetyltransferase immunohistochemistry in the cat. *J. Comp. Neurol.* **200**:151–201 (1981).

Kobayashi, H.: Roles of cyclic nucleotides in the synaptic transmission in sympathetic ganglia of rabbits. *Comp. Biochem. Physiol.* **72**:197–202 (1982).

Koelle, G. B.: A proposed dual neurohumoral role of acetylcholine: Its function at the pre- and post-synaptic sites. *nature (London)* **190**:208–211 (1961).

Koelle, G. B.: Cytological distributions and physiological functions of cholinesterases, in: *Handbuch der experimentellen Pharmakologie*, Ergänzungswerk, Vol. 15, *Cholinesterases and Anticholinesterase Agents* (G. B. Koelle, ed.), pp. 187–298, Springer-Verlag, Berlin (1963).

Koketsu, K.: Cholinergic synaptic potentials and the underlying ionic mechanisms. *Fed. Proc. Fed. Am. Soc. Exp. Biol.* **28**:101–112 (1969).

Koketsu, K., and Nishi, S.: Characteristics of the slow inhibitory postsynaptic potential of bullfrog sympathetic ganglion cells. *Life Sci.* **6**:1827–1836 (1967).

Koshtojantz, O., Antipina, M., and Koltsova, A.: Changes in the cholinosensitivity of the sensory motor cortex neurons during extinction of conditioned reflex (in Russian). *J. Higher Nerv. Activ.* **29**:923–929 (1979).

Koshtojantz, O. Ch., Jankel, M., Kruglikov, R. I., Schmidt, J., and Markarowa, M. J.: Über die Veränderung chemoreaktiver Eigenschaften von Neuronen im Neokortex und Hippokampus unter dem Einfluss einer elektrischen Reizung des Locus coeruleus. *Acta Biol. Med. Ger.* **39**:837–844 (1980).

Krnjevic, K.: Central cholinergic pathways. *Fed. Proc. Fed. Am. Soc. Exp. Biol.* **28**:113–120 (1969).

Krnjevic, K.: Chemical nature of synaptic transmission in vertebrates. *Physiol. Rev.* **54**:418–540 (1974).

Krnjevic, K.: Acetylcholine receptors in vertebrate CNS, in: *Handbook of Psychopharmacology*, Vol. 6 (L. L. Iverson, S. D. Iverson, and S. H. Snyder, eds.), pp. 97–125, Plenum Press, New York (1976).

Krnjevic, K.: Role of acetylcholine in the cerebral cortex, in: *Neurobiology of Acetylcholine* (N. J. Dun, ed.), Plenum Press, New York, (in press) (1986).

Krnjevic, K., and Ropert, N.: Electrophysiological and pharmacological characteristics of facilitation of hippocampal population spikes by stimulation of the medial septum. *Neuroscience* **9**:2165–2183 (1982).

Krnjevic, K., Pumain R., and Renaud, L.: The mechanism of excitation by acetylcholine in the cerebral cortex. *J. Physiol. (London)* **215**:247–268 (1971).

Libet, B.: Generation of slow inhibitory and excitatory postsynaptic potentials. *Fed. Proc. Fed. Am. Soc. Exp. Biol.* **29**:1945–1956 (1970).

Libet, B.: Long-lasting modulation of slow-excitatory-postsynaptic-potential (s-EPSP), by catecholamines, in: *Physiology of Excitable Membranes* (J. Salánki, ed.), pp. 305–309, Pergamon Press, Oxford Advances Physiological Sciences, (1981).

Libet, B., and Tosaka, T.: Dopamine as a synaptic transmitter and modulator in sympathetic ganglia: A different mode of synaptic action. *Proc. Natl. Acad. Sci. U.S.A.* **67**:667–673 (1970).

Libet, B., Kobayashi, H., and Tanaka, T.: Synaptic coupling with the production and storage of a neuronal memory trace. *Nature (London)* **258**:155–157 (1975).

Libet, B.: Heterosynaptic interaction at a sympathetic neuron as a model for induction and storage of a postsynaptic memory trace, in: *Neurobiology of Learning and Memory* (G. Lynch, J. L. McGaugh, and N. M. Weinberger eds.), pp. 405–430, The Guilford Press, New York, 1984.

Marsan, C. A., and Matthies, H. (eds.): *Neuronal Plasticity and Memory Formation*, International Brain Research Organization Monograph Series, Vol. 9, Raven Press, New York (1982).

McGeer, P. L., Eccles, J. C., and McGeer, E. G.: *Molecular Neurobiology of the Mammalian Brain*. Plenum Press, New York (1978).

McGeer, P. L., McGeer, E. G., and Peng, J. H.: Choline acetyltransferase: Purification and immunohistochemical localization. *Life Sci.* **34**:1319–2338 (1984).

McLennan, H., and York, D. H.: Cholinergic mechanisms in the caudate nucleus. *J. Physiol. (London)* **187**:163–175 (1967).

Minneman, K. P., Pittman, R. N., and Molinoff, P. B.: β-Adrenergic receptor subtypes: Properties, distribution and regulation. *Annu. Rev. Neurosci.* **4**:419–461 (1981).

Nestler, E. J., and Greengard, P.: Distribution of protein I and regulation of its state of phosphorylation in the rabbit superior cervical ganglion. *J. Neurosci.* **2**(8):1011–1023 (1982).

Nestler, E. J., Walaas, S. I., and Greengard, P.: Neuronal phosphoproteins: Physiological and clinical implications. *Science* **225**:1357–1364 (1984).

Nicoll, R. A., Schenker, C., and Leeman, S. E.: Substance P as a transmitter candidate. *Annu. Rev. Neurosci.* **3**:227–268 (1980).

Nishi, S.: Ganglionic transmission, in: *The Peripheral Nervous System* (J. I. Hubbard, ed.), pp. 225–255, Plenum Press, New York (1974).

Nishi, S., Minota, S., and Karczmar, A. G.: Primary afferent neurons: The ionic mechanism of GABA-mediated depolarization. *Neuropharmacology* **13**:215–219 (1974).

Nishi, S., Karczmar, A. G., and Dun, N. J.: Physiology and pharmacology of autonomic synapses as models of central transmission, in: *Advances in Pharmacology and Therapeutics*, Vol. 2 (P. Simon, ed.), pp. 69–85, Pergamon Press, New York (1978).

Nishizuka, Y.: Turnover of inositol phospholipids and signal transduction. *Science* **225**:1365–1370 (1984).

Nordstrom, O., Westlind, A., Hedlund, B., Unden, A., and Bartfai, T.: On the ionic mechanism of presynaptic muscarinic receptor action in rat hippocamus, in: *Cholinergic Mechanism* (G. Pepeu and H. Ladinsky, eds.), pp. 579–586, Plenum Press, New York (1981).

North, R. A., and Egan, T. M.: Electrophysiology of peptides in the peripheral nervous system. *Br. Med. Bull.* **38**(3):291–296 (1982).

Owman, C., Alm, P., and Sjoberg, N.-O.: Pelvic autonomic ganglia: Structure, transmitters, function and steroid influence, in: *Autonomic Ganglia* (L.-G. Elfvin, ed.), pp. 125–143, John Wiley, New York (1983).

Perkins, M. N., and Stone, T. W.: 4-Aminopyridine blockade of neuronal depressant responses to adenosine triphosphate. *Br. J. Pharmacol.* **70**:425–428 (1980).

Perkins, M. N., and Stone, T. W.: In vivo release of [3H]-purines by quinolinic acid and related compounds. *Br. J. Pharmacol.* **80**:263–267 (1983).

Pernow, B.: Pharmacology of substance P. *Ann. N. Y. Acad. Sci.* **104**:393–402 (1963).

Phyllis, J. W., and Kirkpatrick, J. R.: The actions of mofilin, luteinizing hormone releasing hormone, cholecystokinin, somatostatin, vasoactive intestinal peptide, and other peptides on rat cerebral cortical neurons. *Can. J. Physiol. Pharmacol.* **58**:612–623 (1980).

Puil, E., and Krnjevic, K.: Neurotransmitter actions and interactions, in: *Neuropsychopharmacology* (P. Deniker, L. Radouco-Thomas, and A. Villeneuve, eds.), pp. 527–538, Pergamon Press, Oxford (1978).

Purington, T., Fletcher, T., and Bradley, W.: Gross and light microscopic features of the pelvic plexus in the rat. *Anat. Rec.* **175**:697–706 (1973).

Ryall, R. W., and Haas, H. L.: On the physiological significance of muscarinic receptors on Renshaw cells: A hypothesis, in: *Cholinergic Mechanisms* (P. G. Waser, ed.), pp. 335–341, Raven Press, New York (1975).

Sejnowski, T. J.: Peptidergic synaptic transmission in sympathetic ganglia. *Fed. Proc. Fed. Am. Soc. Exp. Biol.* **41**:2923–2928 (1982).

Simmons, M. A.: The complexity and diversity of synaptic transmission in the prevertebral sympathetic ganglia. Prog. *Neurobiol.* **24**:43–93 (1985).

Snyder, S. H.: Neurosciences: An integrative discipline. *Science* **225**:1255–1257 (1984).

Szerb, J. C.: Characterization of presynaptic muscarinic receptors in central cholinergic neurons, in: *Cholinergic Mechanisms and Psychopharmacology* (D. J. Jenden, ed.), pp. 49–60, Plenum Press, New York (1977).

Walaas, S. I., and Greengard, P.: DARPP-32, a dopamine and adenosine 3′:5′-monophosphate-regulated phosphoprotein enriched in dopamine-innervated brain regions. I. Regional and cellular distribution in the rat brain. *J. Neurosci.* **4**:84–98 (1984).

Wood, P. L., Cheney, D. L., and Costa, E.: Interactions of neuropeptides with cholinergic septal–hippocampal pathway: Indication for a possible trans-synaptic regulation, in *Cholinergic Mechanisms* (G.-C. Pepeu and H. Ladinsky, eds.), pp. 715–722, Plenum Press, New York (1981).

York, D. H.: Amine receptors in CNS. II. Dopamine, in: *Handbook of Psychopharmacology* (L. L. Iversen, S. D. Iversen, and S. H. Snyder, eds.), pp. 23–61, Plenum Press, New York (1975).

Index

GPSR Compliance

The European Union's (EU) General Product Safety Regulation (GPSR) is a set of rules that requires consumer products to be safe and our obligations to ensure this.

If you have any concerns about our products, you can contact us on ProductSafety@springernature.com

In case Publisher is established outside the EU, the EU authorized representative is:

Springer Nature Customer Service Center GmbH
Europaplatz 3
69115 Heidelberg, Germany

Batch number: 10844669

Printed by Printforce, the Netherlands